T0184858

Artificial Intelligence

Charu C. Aggarwal

Artificial Intelligence

A Textbook

 Springer

Charu C. Aggarwal
IBM T. J. Watson Research Center
Yorktown Heights, NY, USA

A Solution Manual to this book can be downloaded from https://link.springer.com/book/10.1007/978-3-030-72357-6

ISBN 978-3-030-72359-0   ISBN 978-3-030-72359-0   (eBook)
https://doi.org/10.1007/978-3-030-72359-0

This Springer imprint is published by the registered company Springer Nature Switzerland AG
The registered company address is: Gewerbestrasse 11, 6330 Cham, Switzerland

To my wife Lata, my daughter Sayani,
and all my computer science instructors

Preface

"AI is likely to be either the best or the worst thing that happened to humanity." – Stephen Hawking

Artificial intelligence is a field that spans work from multiple communities, including classical logic programming, machine learning, and data mining. Since the founding of the field, there has a clear dichotomy between the *deductive reasoning* and the *inductive learning* forms of artificial intelligence. In the deductive reasoning view, one starts with various forms of domain knowledge (which are often stored as knowledge bases), and these forms of domain knowledge are used in order to make inferences. Such methods are often highly interpretable. The domain knowledge can be used in order to create hypotheses, which are then leveraged to make predictions. For example, in a chess game, the domain knowledge about the importance and position of pieces can be used to create a hypothesis about the quality of a position. This hypothesis can be used to predict moves by searching a tree of possible moves up to a specific number of moves. In learning methods, data-driven evidence is used to *learn* how to make predictions. For example, it is possible to generate data from chess games using self play, and then learn which moves are best for any particular (type of) position. Since the number of possible alternative move sequences in chess is too large to evaluate explicitly, chess programs often use various types of machine learning methods to relate typical patterns of pieces on the board to make predictions from carefully selected sequences. This approach is somewhat similar to how humans make chess moves. In the early years, deductive reasoning methods were more popular, although inductive learning methods have become increasingly popular in recent years. Many books in artificial intelligence tend to focus predominantly on deductive reasoning as a legacy from its dominance during the early years. This book has attempted to strike a balance between deductive reasoning and inductive learning.

The main disadvantage of inductive learning methods is that they are not interpretable, and they often require a lot of data. A key point is that humans do not require a lot of data to learn. For example, a child is often able to learn to recognize a truck with the use of a small number of examples. Although the best solutions to many problems in artificial intelligence integrate methods from both these areas, there is often little discussion of this type of integration. This textbook focuses on giving an integrated view of artificial intelligence, along with a discussion of the advantages of different views of artificial intelligence.

After presenting a broad overview in Chapter 1, the remaining portions of this book primarily belong to three categories:

1. *Methods based on deductive reasoning:* Chapters 2 through 5 discuss deductive reasoning methods. The primary focus areas include search and logic.

2. *Methods based on inductive learning:* Learning methods are discussed in Chapters 6 to 10. The topics covered include classification, neural networks, unsupervised learning, probabilistic graphical models, and reinforcement learning.

3. *Methods based on both reasoning and learning:* Chapter 11 to 13 discuss a number of methods that have aspects of both reasoning and learning. This include techniques like Bayesian networks, knowledge graphs, and neuro-symbolic artificial intelligence.

A number of topics of recent importance, such as transfer learning and lifelong learning, are also discussed in this book.

Throughout this book, a vector or a multidimensional data point is annotated with a bar, such as $\overline{X}$ or $\overline{y}$. A vector or multidimensional point may be denoted by either small letters or capital letters, as long as it has a bar. Vector dot products are denoted by centered dots, such as $\overline{X} \cdot \overline{Y}$. A matrix is denoted in capital letters without a bar, such as R. Throughout the book, the $n \times d$ matrix corresponding to the entire training data set is denoted by D, with n data points and d dimensions. The individual data points in D are therefore d-dimensional row vectors, and are often denoted by $\overline{X}_1 \ldots \overline{X}_n$. On the other hand, vectors with one component for each data point are usually n-dimensional column vectors. An example is the n-dimensional column vector $\overline{y}$ of class variables of n data points. An observed value y_i is distinguished from a predicted value $\hat{y}_i$ by a circumflex at the top of the variable.

Yorktown Heights, NY, USA

Charu C. Aggarwal

The original version of the book has been revised. A correction to this book can be found at https:// doi.org/10.1007/978-3-030-72357-6_14

Acknowledgments

I would like to thank my family for their love and support during the time spent in writing this book. A book requires a lot of time, and it requires a lot of patience on the part of family members. I would also like to thank Nagui Halim for encouraging many of my past book-writing efforts and Horst Samulowitz for his support during the writing of this book.

I would like to thank all those who have taught me various aspects of computer science, especially in the fields of algorithms and artificial intelligence. In particular, I gained a lot from Asish Mukhopadhyay, Ravindra K. Ahuja, Thomas Magnanti, and James B. Orlin. I have had the good fortune to work with several colleagues, which has helped me gain better insights into the evolution of the field of artificial intelligence. The book has benefitted from my research collaborations with several of my colleagues, and in this context, I would like to specially call out Jiawei Han, Huan Liu, Saket Sathe and Jiliang Tang. I also learned a lot from my collaborators in machine learning over the years. In particular, I would like to thank Tarek F. Abdelzaher, Jinghui Chen, Jing Gao, Quanquan Gu, Manish Gupta, Jiawei Han, Alexander Hinneburg, Thomas Huang, Nan Li, Huan Liu, Ruoming Jin, Daniel Keim, Arijit Khan, Latifur Khan, Mohammad M. Masud, Jian Pei, Magda Procopiuc, Guojun Qi, Chandan Reddy, Saket Sathe, Jaideep Srivastava, Karthik Subbian, Yizhou Sun, Jiliang Tang, Min-Hsuan Tsai, Haixun Wang, Jianyong Wang, Min Wang, Suhang Wang, Wei Wang, Joel Wolf, Xifeng Yan, Wenchao Yu, Mohammed Zaki, ChengXiang Zhai, and Peixiang Zhao.

Several people provided feedback on the book. In particular, I received feedback from several of my IBM colleagues from the field of artificial intelligence. Lata Aggarwal helped me with several figures in this book, and I would like to thank her for her support.

Contents

Author Biography

Charu C. Aggarwal is a Distinguished Research Staff Member (DRSM) at the IBM T. J. Watson Research Center in Yorktown Heights, New York. He completed his undergraduate degree in Computer Science from the Indian Institute of Technology at Kanpur in 1993 and his Ph.D. from the Massachusetts Institute of Technology in 1996.

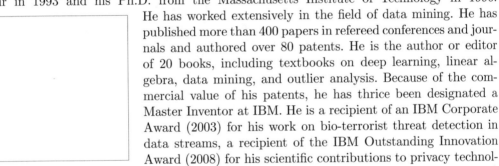 He has worked extensively in the field of data mining. He has published more than 400 papers in refereed conferences and journals and authored over 80 patents. He is the author or editor of 20 books, including textbooks on deep learning, linear algebra, data mining, and outlier analysis. Because of the commercial value of his patents, he has thrice been designated a Master Inventor at IBM. He is a recipient of an IBM Corporate Award (2003) for his work on bio-terrorist threat detection in data streams, a recipient of the IBM Outstanding Innovation Award (2008) for his scientific contributions to privacy technology, and a recipient of two IBM Outstanding Technical Achievement Awards (2009, 2015) for his work on data streams/high-dimensional data. He received the EDBT 2014 Test of Time Award for his work on condensation-based privacy-preserving data mining. He is also a recipient of the IEEE ICDM Research Contributions Award (2015) and the ACM SIGKDD Innovation Award (2019), which are the two most prestigious awards for influential research contributions in data mining. He is also a recipient of the W. Wallace McDowell Award, which is the highest technical honor given by the IEEE Computer Society across all of computer science.

He has served as the general co-chair of the IEEE Big Data Conference (2014) and as the program co-chair of the ACM CIKM Conference (2015), the IEEE ICDM Conference (2015), and the ACM KDD Conference (2016). He served as an associate editor of the IEEE Transactions on Knowledge and Data Engineering from 2004 to 2008. He is an associate editor of the IEEE Transactions on Big Data, an action editor of the Data Mining and Knowledge Discovery Journal, and an associate editor of the Knowledge and Information Systems Journal. He has served as the editor-in-chief of ACM SIGKDD Explorations. He is currently serving as the editor-in-chief of the ACM Transactions on Knowledge Discovery from Data as well as a co-editor-in-chief of ACM Books. He serves on the advisory board of the Lecture Notes on Social Networks, a publication by Springer. He has served as the vice-president of the SIAM Activity Group on Data Mining and has been a member of the SIAM industry committee. He is a fellow of the SIAM, ACM, and the IEEE, for "contributions to knowledge discovery and data mining algorithms."

Chapter 1

An Introduction to Artificial Intelligence

"Success in creating AI might be the biggest event in human history. Unfortunately, it might also be the last, unless we learn how to avoid the risks."–
Stephen Hawking

1.1 Introduction

While computers are excellent at performing computationally intensive tasks, the ability to match human intelligence and intuition with computer algorithms has always been an aspirational goal. Nevertheless, significant progress has been made on algorithms that can perform predictive tasks that would have been considered unimaginable a few decades back. The work on artificial intelligence started almost as soon as general-purpose computers came into vogue. Although there was great excitement around the field of artificial intelligence in the early years, the limited computational power of computers during those early years made their successes look modest in comparison with the inflated expectations of the time. However, as the capabilities of computer hardware increased along with data availability, successful outcomes were achieved in increasingly difficult applications. As a result, the frontier of the field has advanced rapidly, particularly in the previous decade.

In the early years, researchers saw a natural way of developing artificial intelligence in terms of machines that could reason from a knowledge base of facts. This was the *deductive reasoning approach* to artificial intelligence in which facts and logical rules were used to make conclusions. In other words, one tries to *reason* using logic (or other systematic methods like search) with a well-established base of hypotheses in order to make provably correct logical conclusions. However, it was soon realized that this approach to artificial intelligence was inadequate beyond inferring the obvious. This is because much of human intelligence is gleaned from *evidentiary* experience in day-to-day life that supports intuitive choices (but may not result in provably correct conclusions). Making non-obvious inferences often requires the sacrifice of provable correctness. Soon, approaches based on *evidentiary* learning from data were developed. In a sense, data instances provide the *evidence* needed

© Springer Nature Switzerland AG 2021
C. C. Aggarwal, *Artificial Intelligence*, https://doi.org/10.1007/978-3-030-72357-6_1

to *build* hypotheses and make predictions from them (rather than assume pre-defined hypotheses). This resulted in an *inductive learning approach* to artificial intelligence. As we will see throughout this book, these two approaches form the dominant themes of artificial intelligence. In fact, the future of artificial intelligence largely depends on being able to integrate these two points of view, just as humans depend both upon the faculties of evidentiary inferences and logic in order to make intelligent choices.

This chapter is organized as follows. The next section describes the two schools of thought in artificial intelligence. A discussion of general forms of artificial intelligence and the Turing test is provided in Section 1.3. The concept of an agent is fundamental to artificial intelligence, and it is introduced in Section 1.4. Deductive reasoning systems are discussed in Section 1.5. Inductive learning methods are discussed in Section 1.6. Methods that are based on principles from biology are discussed in Section 1.7. A summary is given in Section 1.8.

1.2   The Two Schools of Thought

From the early years of artificial intelligence, there were two primary schools of artificial intelligence, corresponding to *deductive reasoning* and *induction learning*, respectively. These differences have to do with how *hypotheses* associated with inferences are treated. The dictionary definition of a hypothesis is that of "a tentative explanation for an observation, phenomenon, or scientific problem that can be tested by further investigation." Hypotheses that are derived from existing theories (e.g., Newton's theory of gravitation) are already well established (and therefore not really tentative from a practical point of view), but they can be still considered hypotheses in the sense that they can be put to further test for confirmation. For example, Newton's theory of gravitation may not work perfectly in some restricted (relativistic) settings. In general, *all hypotheses are considered approximations of models that predict the state of the world perfectly*. The main difference between the two schools of thought lies in how these hypotheses are constructed and used. These schools of thought are defined by the following principles:

- **Deduction versus induction:** In deduction, we move from the general to the specific in a systematic way, whereas in induction we move from the specific to the general. A deduction is as follows: "All canine animals have four legs. All dogs are canines. Therefore, dogs have four legs." An induction is as follows: "I saw a couple of dogs yesterday. Both had four legs. Therefore, all dogs have four legs." Note that the deductive approach always provides mathematically accurate conclusions, whereas the inductive approach is a form of "faulty" logic based on generalizing specific experiences. However, the inductive approach is more powerful precisely because of its potential inaccuracy — combining it with statistical methods to ensure robustness can result in non-obvious inferences that are often accurate.

- **Reasoning versus learning:** The aforementioned examples of dogs are those of reasoning methods, wherein a chain of assertions are connected. Learning methods try to use statistical inference on many examples in order to make conclusions. For example, one might collect data about thousands of sea creatures and whether they lay eggs or give birth to live babies. From this data, one might try to infer that sea creatures are most likely to lay eggs (even though whales and dolphins do not). Collecting more data, such as the presence of blowholes on the creature's body can increase the accuracy of prediction. The fact that this is a (possibly erroneous) statistical inference based on relating different types of properties is fully understood up front.

As a practical matter, inductive methods are usually learning methods, whereas deductive methods are almost always reasoning methods. This is because it is not natural to use induction with reasoning, because the conclusions in reasoning methods are intended to be absolute certainties rather than statistical likelihoods. Similarly, since learning methods work with individual pieces of evidence, it is not natural to use logical deduction in combination with bits and pieces of evidence. Therefore, the two *primary* schools of thought in artificial intelligence are deductive reasoning and inductive learning (although there are indeed some methods that can be considered inductive reasoning or deductive learning). This book will focus only on the primary schools of thought.

Deductive reasoning methods often *reason* using a base of known hypotheses, which are *assumed* to be a base of incontrovertible truths (although this assumption often evolves with changes in knowledge about the state of the world). The school of deductive reasoning is heavily dependent on using systematic methods that start with this base of facts directly or indirectly and use systematic algorithms to arrive at conclusions. While deductive reasoning methods are often thought of as algebraic methods like *symbolic logic*, they sometimes use systematic algorithms like *graph search* that are not algebra-based. Deductive reasoning methods are used for inferring conclusions that are derived directly from a base of assertions. Practical applications of this type of approach in real-world settings include *expert systems* (e.g., tax software) and *knowledge-based recommender systems*. These types of systems often depend on either hard-coded methods or on methods derived from mathematical logic, such as *first-order logic*. These types of methods work best in settings, where the base of facts (e.g., tax law) is unquestioned. On the other hand, inductive learning methods typically use *learning from examples* to build hypotheses. These hypotheses are then used in order to make predictions about new examples. These methods work best where (possibly noisy) data are available (e.g., images with labels) but a clearly defined hypothesis (e.g., a hand-crafted mathematical description of the pixels that are guaranteed to create a banana) is hard to define. We formally define these two schools of thought below:

- *Deductive reasoning methods:* In deductive reasoning methods, we work from more general facts to specific facts (or even examples). This school of thought starts with a *knowledge base* of assertions and hypotheses, and then uses *logical inferences* in order to reason about unknown facts. Therefore, it starts with a set of hypotheses in the knowledge base (which are typically drawn from well-established theories or known facts), and then uses these hypotheses in order to make specific conclusions. For example, the knowledge base could contain the hypothesis that animals lacking canines are herbivores. When this is combined with the assertion that elephants do not have canines, one would come to the conclusion that elephants are herbivores. Methods like first-order logic provide a mechanism to express such statements in symbolic form in order to make inferences with the help of the rules of symbolic logic. Note that the conclusion that elephants are herbivores is a *logical certainty* based on the facts in the knowledge base (such as the fact that elephants do not have canines).

- *Inductive learning methods:* In inductive learning methods, one moves from the specific to the general. For example, one might have examples of many animals, together with information about presence or absence of different types of teeth. Herbivores will often have different characteristics of the teeth and claws, as compared to carnivores. One can then develop a *learning algorithm*, which makes a general hypothesis from these specific examples. This hypothesis is often in the form of a mathematical function of a numerical description of the examples. Often the hypothesis is defined in terms of *statistical likelihoods* relating the observed characteristics of the input data rather than in the form of logical certainties obtained from assertions (like deductive reasoning methods). The ability to learn likely predictions rather than certain predictions actually gives more power to the model, since a wider range of non-obvious conclusions become possible. After the learning algorithm has created a model, we might use this general hypothesis to again make inferences about another example that one has not seen before. This process is referred to as *generalization*. Such algorithms are often referred to as statistical learning algorithms, and the hypotheses are often defined in the form of mathematical functions of the observed input data such as the following:

$$\text{Probability(Herbivore)} = f(\text{Canine teeth length, Claw length})$$

The function $f(\cdot)$ is often constructed by using a *machine learning algorithm* or a *neural network* that starts with a generic form (i.e., a *prior assumption*) of the function and then learns its specific details in a data-driven way. By using the examples, the learning algorithm may be able to deduce the fact that carnivores have sharp claws and long canine teeth, whereas herbivores do not. Even when there are some exceptional animals that do not exhibit their class-specific characteristics, the learning algorithm will be able to capture such variations in terms of statistical likelihoods. These variations will be indirectly represented in the learned function $f(\cdot, \cdot)$. Therefore, in inductive learning, a hypothesis is (generally) a mathematical function that *approximately maps* inputs to outputs, and it uses examples to learn this function.

The conclusions in deductive methods are often[1] indirectly contained in the knowledge base of assertions as logical certainties. Therefore, different implementations of a deductive reasoning model will typically yield the same result (with some minor variations based on conventions). This is typically not the case for inductive learning models, where one creates mathematical models in terms of statistical likelihoods. The learning model and its predictions are heavily influenced by prior assumptions made about the function $f(\cdot, \cdot)$. This is natural because working from incomplete forms of evidence to hypotheses is not expected to create logically certain predictions. While the certainty of deductive reasoning inferences might seem like a strength at first glance, it is actually a constricting straitjacket because it prevents creative inferences from incomplete evidence. Most forms of advancements in human scientific endeavors arise from the creative hypotheses generated from incomplete evidence. Therefore, one may summarize this difference as follows:

> In deductive reasoning, hypotheses are stated as absolute facts that are then used to make further inferences with the use of various types of logical procedures. In inductive learning, hypotheses are not fully formulated up front (and may be stated as incompletely specified functions with undefined parameters); these hypotheses must then be completed with the use of data in the learning process. Therefore, the final hypothesis is an output of the process rather than being an input. Deductive reasoning, therefore, makes stronger assumptions about the knowledge of the world up front, because it is assumed that much of the relevant learning needed to make these assumptions has already occurred before (by a human or a machine).

---

[1]This is not the case for some deductive methods. An example is fuzzy logic.

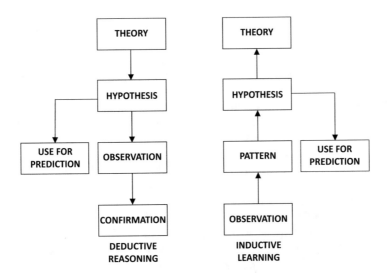

Figure 1.1: The two schools of thought in artificial intelligence

From this point of view, inductive learning can be considered a more fundamental process in artificial intelligence than deductive reasoning because it tends to find more non-obvious conclusions; in a sense, deductive reasoning often finds conclusions that are already present in the knowledge base in an indirect form. On the other hand, inductive learning may find less obvious conclusions, with the price of that process being that the learned hypotheses may turn out to be approximate rather than absolute truths. The latter is closer to how humans function by giving up precision for intuition; surprisingly, however, traditional artificial intelligence had generally placed much greater importance on deductive reasoning during the early years. The two schools of thought in artificial intelligence are illustrated in Figure 1.1. In each case, it is evident that the hypothesis is used in order to make predictions on new examples. However, the main difference lies in terms of how this hypothesis is created in the two cases. In deductive reasoning methods, the hypotheses represent a symbolic coding up of the prior knowledge about the domain at hand. In inductive learning methods, this hypothesis needs to be learned from the patterns in the underlying data. Some parts of this model (e.g., basic form of the function) are assumed up front, whereas other parts of this model (e.g., details of the function such as its parameters) are learned in a data-driven manner.

The processes in classical science use both inductive learning and deductive reasoning. For example, Newton first made real-world observations about falling objects, and then used it to create a general hypothesis about gravitation. This process is obviously inductive because it moves from specific observations to a general hypothesis. This hypothesis eventually became a *theory* through the process of repeated *confirmation*. This theory now regularly used as established *knowledge* to make predictions about the trajectory of modern space flights. This process of using a knowledge base is, of course, a form of deductive reasoning, where one uses known theories to make predictions. Inductive learning methods in artificial intelligence are similar in that they create hypotheses (mathematical functions that map inputs to outputs) from observations (examples of data instances). In the scientific world, a greater level of certainty is required, and therefore, one also has to *confirm* the general hypothesis on new examples in order to ascertain that the hypothesis is indeed correct. This confirmed hypothesis then becomes a theory only after it has been confirmed a sufficient number of times. However, in inductive learning forms of artificial intelligence, one is often happy with hypotheses that are true for a sufficiently large percentage of the cases. This type of partial confirmation is often carried out with a portion of the data, referred to as *validation data*. For example, a trained neural network that takes image pixels as input and uses a mathematical function to predict the category of the image (e.g., carrot, banana, or car) can be considered a hypothesis, although this hypothesis will typically make mistakes on many examples. Therefore, the hypotheses cannot be considered theories in the strictest sense (in the way we view them in the sciences as absolute truths). However, given the fact that the hypotheses in the sciences are also derived from observation and learning (by human scientists), they are not infallible either with the arrival of new data. For example, Newton's theories on gravitation eventually needed to be revised by relativistic physics, as new observations[2] were made. In this sense, inductive learning and deductive reasoning can be viewed as two parts of a broader scientific process that feed into one another.

The deductive school of thought primarily focuses on *knowledge engineering*, wherein concrete facts about the world are stored, in much the same way as a human expert can be considered to be repository of facts in a subject matter. This base of facts is also referred to as a *knowledge base*. By reasoning with this base of facts in conjunction with logical inferences, one can often come to non-obvious conclusions. These conclusions are usually logical certainties, based on the facts in the knowledge base.

There are many ways in which one can perform this reasoning, including the use of search or logic-based methods. Many of these logic-based methods encode the knowledge in terms of logical rules. When rules are used, the system us also referred to as an *expert system*, and the portion of the engine that uses the rules for reasoning is also referred to as the *inference engine*. Expert systems are very popular in various business-centric applications. Some examples of inductive and deductive systems in artificial intelligence are as follows:

- A medical expert system that was developed in the early years based on deductive reasoning was *MYCIN* [162]. The system uses a knowledge base of bacteria and antibiotics, as well as a set of rules indicating their relationship. Based on the responses of a physician to a set of questions, it uses the knowledge base and the rules to make recommendations for specific patients. The advantage of this system is that the recommendations of the system are highly explainable. However, the recommendations of the system are limited by whatever knowledge is provided by the human experts. Therefore, it is often hard for the system to make recommendations that cannot be derived from the knowledge base via a chain of inferences. The knowledge base is populated with the help of knowledgeable experts from the medical field. On the other hand, an inductive system would use feature representations (e.g., chemical compositions) of antibiotics and representations of bacteria in order to make inferences about new antibiotics that might work on previously unseen strains of bacteria.

- A chess program that uses the rules of the game, looks ahead on the tree of possibilities of the moves of both players, and chooses the optimal sequences of moves from the perspective of both players is a knowledge engineering system. Note that the method would require knowledge in the form of a human crafted evaluations of board positions at the terminal nodes of the tree. The most famous chess program of this type is IBM's Deep Blue [33], which was the first computer program to defeat a world champion in a match played under standard time controls. Most chess programs developed before

---

[2]The Michaelson-Morley experiments on the speed of light played a key role as observations that could not be explained by Newtonian physics.

2017 are *primarily* knowledge-based systems, though some amount of inductive learning has crept into such systems. The hand-crafted evaluations are often programmed with the help of chess grandmasters, and they can be viewed as hypotheses about the goodness of positions (although these hypotheses are obviously imperfect). On the other hand, in inductive systems like *AlphaZero* [168], one uses learning methods to determine the goodness of positions as mathematical functions of an encoding of the board position. The system learns the evaluation of board positions, via self-play and the experience of the program in winning or losing from specific positions.

- A machine translation system that encodes the rules of grammar, and uses it as a knowledge base in order to perform translation can be considered a knowledge engineering system. Expert linguists are often deeply involved in the design of such a system. On the other hand, an inductive system would use examples of sentences in two languages to build a mathematical model (e.g., a model encoded within the parameters of a neural network) that is able to generate sentences in one language from sentences in another language.

The definition of what is considered a deductive reasoning system is rather broad. Almost any computer program that accomplishes a particular goal with a well-defined and logical control flow (starting with a set of hypotheses) and coming to logically provable conclusions can be considered a deductive reasoning system. However, in practice, programs that implement day-to-day functions in business processes are often not considered forms of artificial intelligence, even though they are deductive reasoning systems. This is because tasks that were once considered hard-to-do are now considered routine tasks. For example, even though *TurboTax* can be considered a deductive reasoning system (and expert system in particular) by this definition, it is not considered a form of artificial intelligence in the modern era because of the routine nature of the function it performs. Nevertheless, it provides an excellent case study of how expert systems can be easily constructed, as long the rules are clearly defined. In this sense, the definition of what is considered artificial intelligence has slowly evolved over time. Examples of both inductive and deductive methods are shown in Table 1.1.

It is noteworthy that in the real-world, one often uses inductive learning and deductive reasoning in combination. Inductive learning often enables creativity, whereas deductive reasoning often enables routine use. Most scientific theories have come about using inductive learning from real-world observations. The discovery of well-known theories in physics, such as Newtonian mechanics and relativity, are examples of this process. Multiple laboratory experiments may be required in order to *confirm* this hypothesis, so that it becomes a theory. However, once these theories have been confirmed, they are used in a deductive manner to make inferences for useful applications. For example, the National Aeronautics and Space Administration uses Newtonian mechanics and the theory of relativity (among others) in order to make various computations and predictions about the trajectory of rockets in space travel. It is not difficult to see that one can chain together the two diagrams in Figure 1.1 to create an integrated process of inductive learning and deductive reasoning.

A point about deductive methods is that their capabilities are often more narrow. This is because all facts that can be derived with deductive methods are already present in some implicit form in the knowledge base. The inductive approach is arguably more creative, and it often allows one to come up with intuitive conclusions/predictions that are *generalized* hypotheses from facts in the data. This generalized hypothesis might not always be correct and it might be expressible only as a mathematical function rather than as an interpretable, logical statement. The lack of interpretability is often considered a weakness of inductive systems, although it is a hidden strength because many human capabilities are enabled

Table 1.1: Examples of inductive learning and deductive reasoning systems [Both in biology and artificial intelligence]

System	Inductive learning or deductive reasoning?
TurboTax	Deductive reasoning
WebMD symptom checker	Deductive reasoning
Deep Blue chess	Deductive reasoning
AlphaZero chess	Inductive learning
Flag all emails from blacklisted senders as spam	Deductive reasoning
Flag spam by comparing email content with that of previous spam/non-spam emails	Inductive learning
Using a grammar book to learn a language	Deductive reasoning
Picking up a language by conversation	Inductive learning
Combining a grammar book with conversational practice	Combining induction and deduction
Perusing the mathematical rules of algebraic manipulation	Deductive reasoning
Perusing a worked example to learn algebraic manipulation	Inductive learning
Using prior knowledge to reduce data requirements in machine learning (also called *regularization*)	Combining induction and deduction

by freeing them from the straitjacket of interpretability. For example, the choices made by a grandmaster in a game of chess can be mapped to intuition and experience but not interpretability. Inductive learning methods sometimes capture this intuition as a learned mathematical function, whereas deductive reasoning systems are unable to do so.

1.2.1   Induction and Deduction: A Historical View

In the early years, deductive reasoning methods and symbolic logic were highly favored in artificial intelligence. Several computer languages such as LISP and Prolog were explicitly developed to support symbolic and logic-based methods. A small number of the earliest efforts were inductive/learning methods, although they quickly fell out of favor in spite of their greater long-term promise. In 1943, Warren McCulloch and Walter Pitts [121] proposed an artificial model of the neuron, and they showed that any function could be computed using a network of computed neurons. Furthermore, the function computed by this network could be learned by using modifiable weights within the network.

In 1950, two undergraduate students at Harvard University, Marvin Minsky and Dean Edmonds, proposed SNARC, which was the first neural-network architecture. Ironically, Minsky eventually came to the conclusion that this area of work was a dead end, and went on to become one of the foremost opponents of neural networks. Soon, deductive reasoning and symbolic artificial intelligence became increasingly popular. In 1958, John McCarthy proposed LISP (which is an acronym for (LISt Processor)). This language was inherently designed to use symbolic expressions in order to search for solutions to problems in an approach that was aligned with the deductive school of thought. At around the same time (in 1959), Newell, Shaw, and Simon [128], proposed the *General Problem Solver*, which was designed to solve complex, general-purpose tasks. The approach could solve any problem that could be formulated as a directed network of transformations between source nodes (axioms) and sinks (desired conclusions). However, it turned out that this approach could not do much more than solving simple problems such as[3] the *Towers of Hanoi*. In the early years, there was much excitement around being able to solve toy problems (which were referred to by Marvin Minsky as *microworlds*), in the hope that these solutions would eventually generalize to larger-scale reasoning problems. This promise was never realized by deductive artificial intelligence methods because of the combinatorial explosion of solution complexity with problem complexity. For example, tic-tac-toe, chess, and Go are all board games with increasing complexity. Deductive systems work perfectly for tic-tac-toe, extremely well for chess, and rather poorly for Go. Deductive game-playing systems explicitly search the tree of possible moves, while evaluating the leaves of the tree with a hand-crafted evaluation function. This worked poorly for Go because of the large branch factor of the tree, and the limited ability to evaluate intermediate board positions in ways that can be coded up in a semantically interpretable way. The game of chess is a particularly interesting case study, because search-based (deductive) systems like *Stockfish* and *Fritz* had always outperformed learning systems until recently. The recent advancements of inductive learning occurred only when it became possible to perform large-scale learning of the system with improved computational capabilities. In fact, the first computer program to defeat a human champion was *Deep Blue* [33] from IBM, which was a deductive system. At that time, learning systems were not adept at playing games like chess, and it was assumed that too much domain knowledge was known by humans about chess — the supposition was that this domain knowledge provided deductive systems a decisive advantage over learning systems in chess. However, as computational power increased, inductive-learning systems were able to process large amounts of data via self-play and reinforcement learning. Algorithms such as *AlphaZero* [168] began to outperform traditional systems by learning intuitive characteristics of positions that could not be enunciated with human domain knowledge. This phenomenon has been repeated in many other problem domains such as machine translation, where inductive learning systems have recently outperformed deductive reasoning systems. The recent preponderance of inductive learning methods is an outcome of technological advances in which computers have become increasingly powerful and are able to process large amounts of data. Nevertheless, there are always applications where the data and computational requirements are too onerous even for modern hardware, and therefore deductive reasoning systems continue to retain their usefulness.

Many early systems used *predicate logic* (also referred to as first-order logic) for problem solving. Languages like LISP and Prolog were designed to make inferences that are common in logical reasoning, and these languages were proposed in the early years with such applications in mind. Conventional programming languages like FORTRAN are *procedural*, whereas languages like LISP and Prolog are *declarative*. In procedural languages, the control flow is the key aspect of the computer program, whereas declarative languages focus on the program logic and what to execute, rather than on control flow. Prolog is very explicitly logic-based, and it stands for PROgramming in LOGic.

However, logic is not the only approach for problem solving with deductive reasoning

---

[3]https://mathworld.wolfram.com/TowerofHanoi.html

methods. For example, many chess-playing programs use *combinatorial search* in order to infer moves. Such systems are more easily implemented with procedural programming languages. The hypothesis in this case is a (possibly imperfect and hand-crafted) evaluation function of chess positions. The key point is that deductive reasoning systems work with some pre-defined notion of truths and hypotheses. Although these truths may not be perfect in reality (such as the evaluation of a chess position), they are treated as "perfect" domain knowledge, and used to reason for subsequent inferences (such as moves). Inductive learning systems, on the other hand, start with *labeled* examples (e.g., specific chess positions and their win-loss outcome labels) and use these examples in order to create a mathematical model of the label of an unseen instance (e.g., win-loss prediction for an unseen position in chess). Often, a neural network may be used in order to map a position into an evaluation.

At the same time that advancements in deductive reasoning were being proposed in the early years, there were also several advancements in inductive learning. An important learning model was Arthur Samuel's checkers program [155] was implemented in 1959, and it was the first implementation of a learning-based algorithm for playing games. In fact, this idea was a precursor to a well-known class of modern learning algorithms, referred to as *reinforcement learning*. The first major advancement in neural networks (after Minsky's SNARC) program was Rosenblatt's perceptron [149] in 1958. This machine was based on the McCulloch-Pitts model of the neuron, and it has the functionality of being able to classify a multidimensional instance into one of two categories. This model initially created great excitement, and the New York Times, with Rosenblatt's tacit support, exaggerated descriptions of its future capability as a machine that *"will be able to walk, talk, see, write, reproduce itself and be conscious of its existence."* However, this overoptimistic assumption soon led to disappointment. In 1969, Minsky and Papert published a book on perceptrons [125], which was largely negative about the potential of being able to properly train multilayer neural networks. The book showed that a single perceptron had limited expressiveness, and no one knew how to train multiple layers of perceptrons anyway. Minsky was an influential figure in artificial intelligence, and the negative tone of his book contributed to the first winter in the field of neural networks. Unknown to the artificial intelligence community, it was indeed possible to train neural networks using ideas drawn from control theory [32]. In fact, Paul Werbos suggested one such method for training neural networks in 1974 [200]. Werbos tried hard to popularize this idea; however, the opinion against neural networks (and learned methods in general) had hardened to such an extent that the research community was not interested in these advancements. It was only much later (in 1986), that Rumelhart and Hinton's well-written paper on backpropagation [150] that headway was made on the feasibility of training neural networks. Therefore, for most of the seventies and the eighties, the deductive school of thought remained the dominant force in artificial intelligence.

By the late eighties, it was realized that deductive reasoning methods were too narrow to fulfill all of what had been promised. In general-purpose domains, they required huge amounts of knowledge in order to function well (just as inductive methods require huge amounts of data), although they could sometimes do the job well in narrowly defined domains. Another problem with building knowledge bases is that they can work only in highly interpretable settings. This property is embedded into the inherent nature of knowledge bases, as humans construct them with their semantic insights. While an interpretable knowledge base might seem to be a virtue at first glance, it is also a problem because many intelligent decisions made by humans cannot be easily enunciated in words. For example, a chess grandmaster or an expert player of the game of Go might sometimes be unable to concretely explain why they choose a particular move, beyond the fact that their experience from previous games translates into an intuitive but hard-to-enunciate understanding of favorable spatial patterns. Trying to handcraft this type of intuitive knowledge into a semantically interpretable board evaluation function is often a source of inaccuracy, as it misses the intangibles in the decision-making process.

In the early years, most leading researchers like Marvin Minsky and Patrick Henry Winston were fervent supporters of deductive forms of artificial intelligence. Minsky and many senior researchers were also virulent opponents of neural networks (and, to some extent, the broader ideas of inductive learning in general). Minsky's stance was particularly ironic, considering the fact that the SNARC machine could be considered one of the earliest neural network efforts. However, at a later stage of his career, Minsky himself admitted that the deductive reasoning-based AI strategies that had been most popular in the eighties had come to a dead end. John McCarthy, the founder of the logic-based programming language LISP, also heavily criticized expert systems in 1984 for their limited success. In 1987, the LISP programming language collapsed, which was a bad sign for what lay ahead for the deductive school of artificial intelligence. By the late eighties both the inductive and deductive schools were in trouble, although the deductive school was beginning to look more and more like a dead end. However, the period is considered a winter of artificial intelligence from a broad point of view; this was exacerbated by the fact that government funding in various countries such as the United States and Britain had been cut during this time.

Throughout the nineties, progress on inductive leaning methods like support vector machines and neural networks [41, 77] continued unabated. Although the initial performance was poor on many types of data sets because of limited data availability and computational power, computer hardware continued to improve rapidly in the two decades from 1990 to 2010. Machine learning methods like least-squares regression and support vector machines seemed to perform well with multidimensional data but not with highly structured data like images. By 2010, a neural network was winning the premier image classification competition [107] by using a neural network model that was fundamentally not different from the one proposed two decades back (at a conceptual level). The main difference was that it was now possible to meaningfully train a deeper and larger neural network with the available hardware. More recent systems [72, 73] were able to categorize images that had an accuracy better than human performance. The fact that a machine was more accurate than a human at recognizing the category of an image would have been considered unthinkable only a decade earlier. This success eventually led to an explosion in interest in inductive learning in general and neural networks in particular.

The primary strength of inductive learning systems lies in their being able to capture the inexpressible part of human cognition. For example, when a child learns a language through speaking, she does not start by memorizing the rules of grammar, but by "picking up" the language in a way that cannot be fully explained even today. Therefore, *the child grows in linguistic ability by learning through examples.* The child might occasionally receive some knowledge from her parents, such as specific concepts in grammar or vocabulary, but it is rarely the primary form of learning for native languages. A systematic process of learning grammar does reduce the number of examples one might need, but examples seem to be more essential than systematic learning. Humans are naturally inductive learners.

Although inductive learning systems have become increasingly popular, the reality is that the strengths of the two types of systems are complementary. Without some level of guidance, a *purely* inductive system might require too much data in many challenging and open-ended tasks. Similarly, it might be wasteful to use a purely learning system in a situation where domain knowledge is helpful. There have also been cases in which inductive learning has been led astray by adversarial or biased examples. As an example, Microsoft's chatbot, Tay, was quickly trained to make offensive statements by mischievous individuals, once it was released into the "wild." This example shows that inductive systems, by virtue of having the freedom to "grow" also do not have guard-rails to prevent undesirable outcomes. Indeed, the most successful and safe learning systems are likely to be achieved by a combination of inductive and deductive learning (as in the case of human learning). For example, IBM's Watson also uses a knowledge base, but it integrates the approach with inductive machine learning in order to obtain accurate performance [76].

1.3   Artificial General Intelligence

Most modern forms of artificial intelligence are application- and domain-specific, where the system is taught to perform a relatively narrow task. Examples of such tasks could be categorize an instance or play chess. The *TurboTax* application and the WebMD symptom checker can be viewed as domain-specific expert systems belonging to the deductive school of artificial intelligence. Similarly, an image categorization systems [72, 73, 107] belongs to the school of inductive learning in artificial intelligence. In many of these specialized tasks, a high level of success has been achieved. However, the original goal of artificial intelligence was to develop general-purpose intelligence; the subfield of artificial intelligence that deals with the development of general forms of intelligence is referred to as *Artificial General Intelligence (AGI)*. An example of a system aspiring to achieve this goal was Newell, Shay, and Simon's general-purpose problem solver. Unfortunately, most such systems have not been able to progress beyond simple toy settings, which are referred to as *micro-worlds* in the terminology used by the artificial intelligence community. The most recent example of a general-purpose artificial intelligence is the system *Cyc* [113], which is a large-scale project that has been in progress since the eighties. While this system does have some commercial applications in specific domains, its utility as general-purpose reasoning system is quite rudimentary (and it fails to achieve the reasoning capabilities of a small child). The lack of success of general-purpose systems leads to the natural question as to what the ultimate goal of general-purpose artificial intelligence might be. This goal was clearly enunciated in the early years in the form of a test, referred to as the *Turing test*.

1.3.1   The Turing Test

Very early on, the computer pioneer, Alan Turing, proposed a test to decide when a machine could be considered to possess the capabilities of a human being. This test is referred to as *Turing test*. The Turing test contains three terminals, which are separated from one another physically. Two terminals are occupied by humans, whereas the third is occupied by a computer. One of the humans serves as a questioner, whereas the human and the computer at the other two terminals are respondents. Since the terminals are separated from one another physically, the human questioner does not know which of the two terminals is occupied by the respondent human and which terminal is occupied by the computer that is being tested for its artificial intelligence abilities. Similarly, the human and the computer respondent do not have access to each other's responses. The questioner asks the same question from the human and the computer to test how "human-like" their responses might be. If the human questioner is unable to tell the difference between the human respondent and the computer respondent, the computer respondent is said to have passed the Turing test.

The construction of an artificial intelligence system that can pass the Turing test is sometimes seen as the ultimate aspirational goal of artificial intelligence; however, it is as yet unclear whether it is achievable or even desirable in practice. A machine smart enough to pass the Turing test might also be smart enough to knowingly fail it, be deceptive, or otherwise dangerous. To date, we do not have any system or machine that comes close to passing the Turing test. This also implies that we have no real way of judging whether a particular system has arisen to the standards of what would be reasonably considered artificial general intelligence.

1.4 The Concept of Agent

The concept of *agent* is used often in artificial intelligence, when a sequence of decisions need to be made in order o achieve a particular goal. For example, tasks such as playing a game of chess, performing a robotic task, or finding a path in a maze require a sequence of decisions from the agent. An example of an agent might be a robot, a chatbot, a chess-playing entity, or a route planner. Agents are sometimes referred to as *intelligent agents* or *intelligent autonomous agents*.

The decisions of an agent are made in the context of an *environment*, which provides the platform for the interactions of the agent. The agent interacts with the environment through perception (taking in information) and actions (making changes to the environment). This is similar to how humans interact with their environment. The primary way in which agents interact with the environment is with the notion of *states*. In the context of an artificial intelligence application, a *state* corresponds to the current configuration of the variables of the application at hand. For example, the state might correspond to the position of the robot and the exact configuration of its limbs in a robotics application, the position of the pieces on a chess board in a chess-playing system, or the spatial location of the agent in a routing application. In summary, the state tells the agent everything it needs to know in order to take effective actions, just as a human takes actions based on what their senses tell them about the environment.

At this point, it is useful to define some terminologies associated with an agent that will be used throughout the book. The two key concepts associated with the agent's interaction with the environment are those of *perception* (i.e., converting information about the environment into internal representations), and *action* (i.e., changing the state of the environment). An agent interacts with the environment through a set of *sensors* (for perception) and a set of *actuators* (for action). The inputs received by the agent from the environment are referred to as *percepts*. For humans, our sensors correspond to our senses with which we take in information from the environment, which are the percepts received through various senses such as the eyes or ears. Our actuators include our hands, legs, or any other body parts with which we accomplish a particular task. A similar analogy holds true for artificial agents. For a robot, its sensors correspond to its motion camera, auditory inputs, or even electronic signals directing it to perform particular tasks. Its actuators correspond to the various artificial limbs with which it performs its functions. The sensor in a medical diagnosis system would be the input system that allows the recording of the patient symptoms or answers to physician questions. The actuator would be a display of the possible patient symptoms. Many traditional knowledge engineering and inductive learning systems are implemented using traditional programming methods on a desktop machine. In such cases, the sensors and actuators often turn out to be the input and output interfaces, respectively.

Table 1.2: Agent examples

Agent	Sensor	Actuator	Objective	State	Environment
Cleaning Robot	Camera Joint sensor	Limbs Joints	Cleanliness evaluation	Object/joint positions	Cleaned space
Chess agent	Board input interface	Move output interface	Position evaluation	Chess position	Chess board
Self-driving car	Camera Sound sensor	Car control interface	Driving safety and goals	Speed/Position in traffic	Traffic conditions
Chatbot	Keyboard	Screen	Chat evaluation	Dialog history	Chat participants

It is noteworthy that one does not always encounter the notion of an agent in all forms of artificial intelligence. For example, the concept of agent rarely arises for *single-decision* problems in machine learning; an example is that of *classification*, where one is trying to categorize an instance into one of many classes. Such environmental settings are referred to as *episodic*, and even though the notion of agent is implicit, it is rarely used in practice. In these cases, the solution to this problem is a single-step process (e.g., categorization of a data instances), and therefore the question of system state or environment is not as important as it would be in a sequential process where an action affects subsequent states. Therefore, in such cases, the agent receives a single percept and performs a single action. The next percept-action pair is independent of the current one. In other words, the individual *episodes* are independent of one another. However, in the case of real-world interactions like robots, it makes a lot more intuitive sense to talk about agents that interact (and affect) the environment. It is only in the case of problems associated with a *sequence of decisions* that it becomes important to use the concept of agent and a state of the system. Such settings are referred to as *sequential*. An intelligent agent plays the same role that a human plays in a biological system, based on a sequence of choices that achieve a desired goal. It is noteworthy that the term "episodic" is slightly overloaded, because a finite sequence of actions and percepts are also sometimes referred to as episodic, as long as the individual sequences are independent of one another. From this perspective, episodic and sequential tasks are not completely disjoint from one another (although different books seem to use different terminologies in this respect). Tasks that do not have a finite point of termination are referred to as *continuous*. For example, a chess game contains a finite set of moves (and each game is treated as an independent episode), whereas a robot may continuously operate over a (relatively long) period of time and can therefore never be considered episodic. This distinction is particularly important in some learning-based settings (like *reinforcement learning*), because it affects the nature of the algorithms that can be used in each case. In reinforcement learning, settings like chess are referred to as episodic. Therefore, we will use the terms episodic non-sequential, episodic sequential, and continuous sequential to distinguish among these three types of tasks.

The agent typically interacts with the environment in a sequence of actions in order to accomplish a particular goal. This goal is also referred to as the *objective* of the artificial intelligence applications. In learning applications, this objective is explicitly quantified with the use of a *loss function* or *utility function* although the *performance metric* (e.g., classification accuracy), which is used to evaluate the effectiveness of the agent, may be different from the loss function used by the agent as a (typically simpler) surrogate. In some applications, a quantifiable and explicit performance metric may not be used, and the evaluation may be subjectively done by a human in order to provide feedback. The key point in most artificial intelligence applications is to select the correct choices of actions in order to achieve a particular goal or optimize a particular objective function. Examples of agents, environments, states, sensors, and actuators for different types of settings are illustrated in Table 1.2.

Each action of the agent leads to a *transition* to either the same state or a different state. A transition of the agent could also lead to a reward. In some applications, rewards are not received at each transition, but at the end of entire sequence of actions (e.g., a chess-playing agent). The goal of a reward is to direct the actions of the agent in a way that is useful for the application at hand.

1.4.1 Types of Environments

The types of environments that are encountered by agents are quite different, depending on the problem setting. For example, an environment like a self-driving car has a single agent, whereas a chess environment might have multiple agents. Another important factor is the level of *uncertainty* in the environments. Environments that are either probabilistic or not fully observable have uncertainty built into them. A chess-playing agent can fully observe the effect of its actions on the board position, therefore it is a *deterministic* and *fully-observable* environment. Note that the uncertainty arising from the action of the adversarial agent is not considered to violate determinism, since the agent is fully in control of their own actions and their effect on the environment. On the other hand, a card game is often non-deterministic when the agent may select a card from a pile, and the choice of the card is not under the control of the agent. Another issue is that of *observability* of environments. For example, a self-driving car is not a fully observable environment. This is because all parts of the road conditions may not be fully observable by the sensors of the agent. This is certainly true for human agents, where the sensory perceptions are often forward facing. Full observability is different from the *deterministic* nature of environments, although the two may sometimes occur simultaneously. In a card game, an agent's act of drawing a card from a pile has an uncertain outcome, which affects the state in a probabilistic way. Furthermore, the environment might also be partially observable if the agent is unaware of the cards that its adversary is holding. Therefore, the agent has to work with incomplete information about the true state of the game. In computer video games, there is usually some randomness to ensure that the game stays interesting for the player. Therefore, when a player makes an action, the state of the system may change in a somewhat unpredictable way. This is a non-deterministic environment. On the other hand, a two-player game of dice might be non-deterministic but fully observable, as long as the position of each player in the game is fully observable to both. It is noteworthy that multi-agent environments (like chess) might appear to be probabilistic from the perspective of each agent (because they cannot predict each other's moves). However, such environments are still treated as deterministic because each agent is in control of its actions. An environment that is not deterministic is referred to as a *Markov decision process*. Therefore, in a Markov decision processes, the state outcomes are only partially in the control of the agent though their actions. It is noteworthy that non-deterministic environments are often hard to fully distinguish from partially observable environments from the perspective of mathematical modeling, and the difference between them is mainly a semantic one.

The temporal nature of an environment can either be discrete or it can be continuous in nature. For example, in chess, the act of making a move is discrete, where each move is an action. Changes to the state occur only at these discrete time stamps. On the other hand, a self-driving car is considered a continuous-time environment in which changes to the state may occur at any moment in time. We distinguish between continuous and continuous-time environments in that the former only refers to whether or not the environment is divided into independent episodes.

Finally, environments can be either *static* or *dynamic*. In static environments, such as crossword puzzles, the state does not change, unless the agent does something. On the other hand, in the case of *dynamic environments* like self-driving cars, the environment could change without the agent having done something (e.g., changes in traffic). Some environments (like timed board games) are *semi-dynamic*, because the state of the board might not change without the agent making a move. On the other hand, the agent score changes because their chances of winning reduces, as time runs out.

1.5　Deductive Reasoning in Artificial Intelligence

Deductive reasoning methods start with a knowledge base of facts and use logic or other systematic methods in order to make inferences. There are several approaches to deductive reasoning, including search and logic-based methods. Furthermore, there are numerous applications of these types of methods, such as theorem proving or game playing. It is noteworthy that the *types* of problems that are solved using inductive and deductive methods are often quite different, although some methods could be addressed with either approach. In general, problems with clearly defined tasks as logical inferences of known facts, those requiring large amounts of known domain knowledge, or those requiring unreasonable amounts of data/computational power (for inductive learning) are often solved using deductive reasoning methods. In some cases, problems that are solved using deductive reasoning methods move into the inductive learning domain with progress in computational power and data availability. In recent years, inductive learning methods have increasingly encroached upon the types of problems that were earlier solved using deductive reasoning; this is largely a result of increased computational power that enables the use of more and more data. For example, game-playing methods such as chess were earlier addressed using (mostly) deductive reasoning methods (like *adversarial search*), but recent advancements have allowed (mostly) inductive learning methods (like *reinforcement learning*) to outperform the former.

1.5.1　Examples of Deductive Reasoning in Artificial Intelligence

There are certain types of problems that repeatedly reappear in deductive forms of artificial intelligence. These are important representatives of "typical" problems, and their solutions can often be generalized to other similar problems. Therefore, studying these problems can provide insights into solving more general problems in the deductive setting.

1.5.1.1　Constraint Satisfaction Problem

The constraint satisfaction problem is really a family of problems satisfying a similar type of structure. Broadly, the problem has to do with instantiating a set of variables to particular values, so that a pre-defined set of constraints among the variables is satisfied. The variables and the constraints may be of various types, which leads to different versions of the constraint satisfaction problem. These different versions of the constraint satisfaction problem may be more useful in different application settings. The canonical example of a constraint satisfaction problem is the Boolean satisfiability problem, which is also referred to as SAT. For example, consider the Boolean variables a and b, each of which could take on one of the two Boolean values of *True* and *False*. Now consider the following Boolean expression:

$$a \wedge \neg b$$

The operator $\neg$ stands for the negation, which flips *True* to *False* and vice versa. The operator $\wedge$ is a binary conjunction operator, which takes on the value of *True*, if both the operands on either side of the operator are *True*. Otherwise, the expression takes on the value of *False*. We want to find values for a and b, so that the entire expression evaluates to *True*. This is the *constraint* that needs to be *satisfied*. In this case, setting a to *True* and b to *False* yields a value of *True* for the entire expression. Therefore, this particular expression yields a satisfiable solution by choosing appropriate values for the underlying operands. Now, consider the following expression:

$$a \wedge \neg a$$

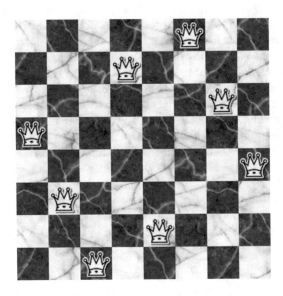

Figure 1.2: A solution to the eight-queens problem

This expression always evaluates to *False*, irrespective of what value of a we might select. This is because one of the two operands on either side of the expression is always *False*. Therefore, no solution exists for this expression, and the problem is not satisfiable. When the expressions become more complex, it becomes harder to determine whether or not a satisfiable solution exists. The SAT problem is *NP-complete*, which means that it is strongly suspected not to be polynomially solvable (although it is an open question as to whether a polynomial-time algorithm exists). In fact, the SAT problem was the first problem to be shown to be NP-complete, which is a class of problems that are strongly suspected not to be polynomially solvable (although it is an open question as to whether a polynomial-time algorithm exists). Therefore, the SAT problem has also had a fundamental contribution to developing the theory of NP-hardness. It is one of the first problems that was developed from the perspective of the theory of NP-hardness [61]. An important property of these problems is that it can be *checked* in polynomial time as to whether or not a given solution is correct. For example, in the case of the SAT problem, it is easy to check whether or not a given assignment of the variables results in an evaluation of *True* for the entire expression.

However, it is hard to determine a specific assignment that results in satisfiability without checking all possible assignments of variables.

The SAT problem belongs to a class of problems, referred to as *constraint satisfaction*. The problem of constraint satisfaction is not restricted to the use of only Boolean variables. The general forms of the constraint satisfaction problem allow the use of variables drawn from arbitrary domains of values, such as numerical or categorical data values. The choice of the type of variables depends heavily on the problem at hand. Similarly, an arbitrary constraint can be posed that is suited to the domain of values of these variables. In general, the constraint satisfaction problem is posed as values:

Definition 1.5.1 (General Constraint Satisfaction) *Given a set of variables, each of which is drawn from a particular domain, and a constraint, find an assignment of values to the variables, so that the constraint is satisfied.*

The interesting variations of the problem (which typically arise in artificial intelligence) are also NP-hard. Many puzzles in artificial intelligence, such as the *eight queens problem*, can be viewed as special cases of constraint satisfaction on a problem with numerical variables. In the *eight-queens problem*, one has to place eight queens on a chess board, so that none of the queens attacks any other queen. One can formulate the problem is that of selecting the values of eight pairs of integer numerical variables, each of which corresponds to the row and column positions of a queen. Queens are powerful pieces that can move along rows, columns, or diagonals, and, therefore, they cover a rather large number of squares. As a result, most placements of queens on the board will lead to violation of the constraint of queens not attacking one another. A solution to this problem is shown in Figure 1.2. Note that no pair of queens is aligned along a row, column, or diagonal in this case.

In order to solve the eight-queens problem, one can define variables based on the positions of the eight queens on the chessboard, and one can define constraints to ensure that none of these queens attack one another. Consider the case where (x_i, y_i) corresponds to the row and column positions of a queen, where each x_i and y_i are drawn from $\{1, 2, \ldots, 8\}$. Then, the constraints are as follows:

$$x_i \neq x_j \quad \forall i \neq j \quad \text{[Row indices are different]}$$
$$y_i \neq y_j \quad \forall i \neq j \quad \text{[Column indices are different]}$$
$$x_i - y_i \neq x_j - y_j \quad \forall i \neq j \quad \text{[Not on same diagonal]}$$
$$x_i + y_i \neq x_j + y_j \quad \forall i \neq j \quad \text{[Not on same diagonal]}$$
$$x_i, y_i, x_j, y_j \in \{1, 2, \ldots 8\}$$

A solution to this problem is, therefore, a set of mutually consistent states of the variables that satisfy these constraints. A harder variation of the problem is to place a subset of queens on the board and then find the remaining positions that need to be filled in order to satisfy all the required constraints. The constraint satisfaction problem arises in all kinds of problems in artificial intelligence, such as solving Sudoku or crossword puzzles.

Many problems in artificial intelligence suffer from the combinatorial explosion in the number of possible solutions. For example, in the case of the eight queens problem, one has to choose eight positions out of 64 positions. The number of possible solutions is given by $\binom{64}{8} = 4,426,165,368$. Therefore, there are more than four billion possible solutions to what is considered a toy problem in artificial intelligence parlance. When one moves to more practical puzzles, it becomes impossible to even enumerate all the possibilities, let alone evaluate them. For example, generalizing the eight queens problem to the *n*-queens problem leads to an NP-hard setting [62]. Therefore, a wide variety of heuristic search-based methods are used to solve such problems. These types of methods will be discussed in detail later in the book.

1.5.1.2 Solving NP-Hard Problems

The theory of NP-hardness developed in parallel with the development of algorithms in artificial intelligence, and they are naturally posed as optimization problems. NP-hard problems find optimal solutions to problems under pre-specified constraints, and the search space is of exponential size. These problems are suspected to not have a polynomial-time solution, although a formal proof of the lack of polynomial solvability has not been proposed. An *NP-complete problem* is a *decision version* of an NP-hard problem, in which one has to determine whether a valid solution exists for a constrained problem. One of these constraints could be defined in terms of the quality of the objective function. The Boolean satisfiability problem was one of the first examples of an NP-complete problem, and many constraint satisfaction problems are NP-complete as well. A generalization of the n-queens problem is NP-complete as well; in this generalization, a subset of n queens (i.e., less than n queens) have already been placed on the board in a valid configuration, and it is desirable to complete this configuration by placing additional queens in valid positions. The NP-hard version of this problem is to determine the maximum number of additional queens that can be placed on the board without violating any constraint. In general, NP-hard problems can be posed as optimization problems, whereas NP-complete problems are always posed as decision problems in which a constraint is placed on the value of the solution. Many of these problems have significant practical applications. A classical example of an NP-complete problem with wide applicability is that of the *traveling salesperson problem.*

The traveling salesperson problem is defined over a set of n cities, associated with a cost c_{ij} of traveling between city i and city j (in either direction). The traveling salesperson has to start at a given city, visit each city exactly once, and then arrive back at the starting point. The cost of this tour is the sum of the costs of the edges traversed. The goal for the traveling salesperson is to find a cycle of cities in this network, so that traversing this cycle has the least cost. This problem is widely known to be NP-hard, and it represents the *optimization version* of the problem. The *decision version* of the problem is one in which the traveling salesperson has to find whether a path exists in which the cost is at most C. The decision version of the problem might seem much easier at first glance, whereas this is not really the case. If we had a solution to the decision version of the problem, one would perform binary search over C in order to find a solution to the optimization of the problem in polynomial time. The decision version is easier only in the sense that one can easily check in polynomial time whether a given solution has cost at most C. This type of checking cannot be done for the optimization version of the problem. The decision version of the problem is therefore NP-complete. The Boolean satisfiability problem provided the impetus for the development of the theory of NP-hardness.

1.5.1.3 Game Playing

Many board games like chess and Go have an extremely high level of complexity, and the problem can be seen as one of trying to find the best choices of moves from a tree of possibilities corresponding to the board positions obtained by making successive moves. The children of each node correspond to the board positions reached using individual moves. These board positions are the states encountered by the game-playing agent. Such games

Figure 1.3: A position in the chess game between *AlphaZero* (white) and *Stockfish* (black)

are often addressed by materializing the tree of possible moves, and then selecting the best moves from the perspective of both players. Ideally, one would like to materialize the entire tree of moves down to a final outcome, although this is often not possible in practice (because of the combinatorial explosion of the tree size). Therefore, a natural approach is to materialize the tree only up to a restricted depth and then evaluate the positions at the lowest level of the tree using a heuristic function designed by a human domain expert (e.g., chess grandmaster). The heuristic function can be viewed as a form of domain knowledge (as in all deductive reasoning systems), and imperfections in the choice of the heuristic function often lead to errors in the choices made by the game-playing agent.

Deductive systems for game playing often depend heavily on human domain knowledge that is coded up in the form of position evaluations. This is often quite hard to achieve in many cases, and it turns out to be an Achilles heel for the quality of play. For example, consider the above chess position, which is extracted from a game between the inductive chess program *AlphaZero* (white) and the deductive chess program *Stockfish* (black). In this case, black has an extra pawn, but its light-squared bishop is so boxed in by its own pieces that it is hard to foresee how it can enter the game. As a result, white's position is superior in spite of its material disadvantage. Such a position can often be correctly evaluated by an experienced human grandmaster, but it is more challenging for a machine to evaluate accurately with the use of a coded evaluation function. A part of the problem is that the proper evaluation of this position requires both experience and intuition that can be perceived by a sufficiently skilled human player but hard to code up in a general and interpretable way by the same human – unfortunately, human-coded evaluations are often created using oversimplified (but interpretable) heuristics. Although the evaluation functions in chess programs are of high quality in the modern era, such positions continue to be a major challenge for deductive learners. This is one of the reasons that this particular chess game was won by the inductive learner *AlphaZero*, but could not be properly evaluated by the deductive program. This is an example of the fact that the interpretability of a deductive reasoning system also turns out to be a hidden weakness, because many real-world choices require decisions that cannot be coded up concretely in an interpretable way.

1.5.1.4 Planning

The planning problem always corresponds to the determination of a sequence of decisions that achieve a particular goal. An example of an agent in the planning problem can be a robot or it can even be an automated player in a chess game. By making a sequence of actions, an agent might complete a particular task (e.g., robot moving objects from one place to another or winning a chess game). In such a case, the states might correspond to the position and configuration of the robot at a particular time in the case of the robot example or the positions of the pieces on the chess board in the chess example. The actions of the agent might correspond to the various choices of movements available to the robot or the chess moves made by the automated player. Another example of a planning problem would be that of finding a path through a maze. Most planning problems use a reward function (or utility function) in order to control the sequence of actions of the agent. It is noteworthy that most sequential environments create planning problems in one form or another.

In the *classical planning problem*, the states represent deterministic choices along with rewards and are also fully observable. The simplest example of a classical planning problem is the desktop map application on your phone that finds the optimal route from one point to another. Like all agent-based applications, planning problems have many variations, depending on with the environment is single-agent or multi-agent, whether it is fully or partially observable, whether it is deterministic or probabilistic, and whether it is discrete or continuous. These variations are discussed in Section 1.4.1. Planning problems can either be single-agent planning problems, or they can be multi-agent planning problems. For example, an agent playing a video game such as *PacMan* is a single agent planning problem. On the other hand, a game like chess can either be played by an agent with a human opponent, or it can be played by two agents. In multi-agent planning problems, there is often (but not always) an *adversarial* element involved in the planning process.

Planning can either be domain independent or it can be domain dependent. Domain independent planning tasks can work for a variety of tasks drawn from different settings, whereas domain dependent planning is applicable only for a particular domain of tasks. An example of a domain dependent planning task is to find the route between a pair of cities. Planning problems often leverage deductive reasoning methods. For example, planning problems can be reduced to the Boolean satisfiability problem, and many of the agent-centric sequential decision problems are indirect forms of planning problems. The main distinguishing characteristic of a planning problem is that it typically requires a large number of sequential decisions to achieve a particular goal.

It is noteworthy that planning problems can be solved not only by deductive reasoning methods, but they can also be solved by inductive learning. A methodology that had become particularly popular in recent years is that of *reinforcement learning*, which is discussed in detail in Chapter 10. Reinforcement learning algorithms can often learn a long sequence of actions with the use of experience-driven training, which is a form of planning. Many reinforcement learning-based chess programs show a high level of long-term planning, which cannot be matched by tree-based (deductive reasoning) methods.

1.5.1.5 Expert Systems

Expert systems were originally intended to simulate the task of a human expert in performing a specialized task, such as making a diagnosis in a medical application. This *specialized* class of systems was proposed by Edward Feigenbaum in the sixties in response to the failure of the initial work on the general-purpose solver. An expert system contains two critical parts, corresponding to the knowledge base and the rule inference engine. A set of IF-THEN rules are used by the inference engine to reach a particular conclusion for the problem at hand. The rules are typically coded into the knowledge base with the use of *first-order logic*. Expert systems typically used a chain of inferences based on the input provided by the user query in order to reason from facts and rules.

For example, consider a patient, John, who comes to a doctor, while presenting the following facts about their situation:

John is running a temperature
John is coughing
John has colored phlegm

Now imagine a case where the expert system contains the following subset of rules:

IF coughing AND temperature THEN infection
IF colored phlegm AND infection THEN bacterial infection
IF bacterial infection THEN administer antibiotic

A doctor can then enter John's symptoms in the expert system and then use a *chain of inferences* in order to conclude that an antibiotic needs to be administered to John. There are two types of chaining, referred to as forward chaining and backward chaining in order to make these types of inferences. In practice, one would have to enter a very large number of rules and cases in order to make the system work well. Several examples of medical expert systems exist in various domains, such as MYCIN [162] in medicine, and *TurboTax* for tax law. It is important to note that expert systems are primarily suited to very specific domains, where the knowledge base can be limited to a well-defined set of facts. On the other hand, general-purpose systems such as Cyc [113] have met with only limited success. While expert systems have met with limited success in terms of general-purpose artificial intelligence, they have been quite successful and have been commercialized in a number of specific domains. However, in many of these domains, the expert systems implement relatively straightforward business logic, which does not meet the original expectation of a high level of non-obvious intelligence. An example of an expert system that is used in the modern era is *TurboTax*, which computes the taxes for an individual for a series of questions. This type of application is ideal for an expert system because tax law tends to be exact, well-defined, and the corresponding riles can be easily encoded within the knowledge base. However, in spite of being an expert system, most people do not see *TurboTax* as a form of artificial intelligence because of the routine nature of the underlying application.

1.5.2   Classical Methods for Deductive Reasoning

In this section, we discuss some of the common methods that are used for deductive reasoning in artificial intelligence. Among the various methods, search-based techniques and logic programming methods are the most popular.

1.5.2.1   Search-Based Methods

Search is one of the most common methods in artificial intelligence, because it is often used to selecting one solution out of a large number of possibilities. The domain knowledge is often encoded into the transition structure of the environment that is fed to the agent, as well as heuristic evaluation functions that are given to the agent for guiding the search. Some examples of artificial-intelligence applications that require search are as follows:

1. Finding a path in a maze from a particular source to a destination can be considered a search-based method. The entrance to the maze defines the starting state, whereas the final target of the search is designated as the goal. Note that this type of approach works without a global view of the graph structure of the network. On the other hand, finding a path from a particular source to a destination in a road network is also a search-based method, but it works with complete knowledge of the graph structure of the road network (since maps of most road networks are already available unlike the case of a maze).

2. Solving any kind of crossword puzzle can be considered an ideal candidate for search-based methods, since one must try various possibilities for filling in slot values in order to solve the puzzle. One can define a variety of ways of performing transitions between states (partial solutions to the puzzle) in order to enable the search towards a complete solution.

3. Most games like chess can be solved by searching over the tree of possibilities for moves. The search is often guided with the use of domain-specific evaluation functions, which can estimate the goodness of a particular position on the board from the perspective of each player. This type of search is also referred to as *adversarial search*, where alternate moves are made from the perspective of opponents optimizing their own objective functions.

Search-based methods are explored in classical graph theory, since the states of such a problem (e.g., chess positions on a board) can be represented as nodes and transitions among them (e.g., moves of chess) can be represented as edges of the graph. Such graphs are massive, and cannot even be fully materialized. The most common search methods in graphs are depth-first search and breadth-first search. However, depending on the amount of knowledge available to the agent, different strategies for search may be more or less effective. For example, when the agent wants to find a path in a maze, only a limited amount of information about the underlying structure of the maze is available. However, in a road network, a lot of information about the structure of the network is available. In most artificial intelligence problems, the underlying graph is so large that it cannot even be fully materialized within the storage availabilities of the computer. As a result, there is often a significant level of myopia about the effect of exploring specific nodes. In these cases, one must work on performing knowledge with limited information about the effect of making particular choices. For example, in a chess program, one can (heuristically) evaluate a board position up to a particular number of moves made by each opponent, but there is uncertainty about the possibilities beyond that set of moves. This uncertainty is exacerbated by the fact that the evaluation function for board positions is imperfect as well. For example, the board position in Figure 1.3 shows a material advantage for black, although black is positionally weaker because its light-squared bishop has been boxed in by its own pieces. Such subtle points often cannot be captured by heuristically designed hypotheses (like hand-crafted position evaluations), and it is only after evaluating a very deep tree of possibilities that this weakness becomes apparent. However, inductive methods can discover such subtle aspects of positions. All these trade-offs and various other characteristics of search are discussed in Chapters 2 and 3.

1.5.2.2　Logic Programming

Logic programming methods are also based on knowledge engineering, except that they focus on formal logical expressions, such as IF-THEN rules, in order to make deductions. There are several forms of logic programming, including *propositional logic* and *first-order logic*. First-order logic is more advanced than propositional logic, and it is more focused on representing complex relationships among objects in a particular domain. Common applications of such logical agents include automated theorem proving and the design of *expert systems*. Expert systems were among the earliest systems developed by artificial intelligence researchers, and they are used quite popularly in order to automate many tasks in the business world (e.g., automating the processes on a factory floor). Therefore, expert-systems are almost always designed in a domain-specific manner. Some of these applications are so simple that they are not even considered forms of artificial intelligence. Logic programming methods are discussed in Chapters 4 and 5. The methods used in logic programming, such as *forward chaining* and *backward chaining*, are closely related to those used in search.

As the complexity of applications increased, knowledge bases became increasingly complex, with objects and hierarchical relationships among them. In fact, many paradigms like *object-oriented programming* developed in parallel with these advancements in artificial intelligence. The combinations of objects together with their relationships are also referred to as *ontologies*, and these types of representations are captured as *knowledge graphs*. Knowledge graphs represent an informal extension of the ideas inherent in first-order logic, as they can represent relationships between objects in graphical form, and are more amenable to machine learning techniques. Knowledge graphs are discussed in Chapter 12.

1.5.3　Strengths and Limitations of Deductive Reasoning

The greatest strengths of deductive reasoning are also its greatest limitations. Deductive reasoning requires a way to code up expert knowledge in a knowledge base. Coding up such knowledge requires a human to understand an interpret this knowledge. This results in highly interpretable systems, which are obviously desirable. However, this interpretability is also an Achilles heel in the goal towards human-like behavior, because many human decisions rely on a high level of understanding that is not easily interpretable.

The most important advantage of deductive reasoning is that *it provides a path to incorporating knowledge we already know*. This provides a shortcut to learning well-known facts from scratch. There is no reason to use inductive learning in order to arrive at hypotheses that are already known. Deductive reasoning methods work best in specialized domains, where a modest amount of knowledge suffices for inferring useful conclusions, or in cases where the underlying knowledge is concrete and unquestioned. In recent years, deductive methods have often been used as an additional component of inductive learning systems in order to reduce the data requirements by supplementing it with a knowledge base of modest size. This is a natural approach from a biological perspective, because human behavior is also a combination of deductive reasoning and inductive learning.

It is noteworthy that IBM's Watson system uses a combination of knowledge bases and machine learning in order to obtain high-quality results [76]. Deductive reasoning methods are needed at various phases such as natural language parsing, and a knowledge base is used to identify important facts. At the same time, machine learning is used in order to make predictions, and score various choices. In general, a purely deductive reasoning system or a purely inductive learning system is often unable to perform satisfactorily in the full range of tasks. This is also consistent with the experiences of biological intelligence, where a combination of the (deductive) knowledge gained from concrete instruction and the (inductive) insight obtained from past experience seems to provide the best results.

1.6 Inductive Learning in Artificial Intelligence

While deductive reasoning systems try to encode domain knowledge within a knowledge base to make hypotheses, inductive learning systems try to use data in order to create their own *data-dependent* hypotheses. In inductive learning a mathematical *model* is used to define a hypothesis, the resulting model is used for prediction on examples that have not been seen before. This process is referred to as *generalization*, because one is creating a more generalized hypothesis (that applies to all examples including unseen ones) from specific sets of examples. The general idea of using examples in order to learn models for prediction is also referred to as machine learning.

A simplified set of "important" tasks for inductive learning have been identified by researchers and practitioners over the years; these tasks recur repeatedly as the various building blocks of application-centric solutions. This section will introduce these tasks together with the types of data to which they apply. To a large extent, many of the machine learning tasks are non-sequential and episodic in nature, where each action is independent of previous actions (e.g., classifying whether a part in an assembly line is defective). Therefore, one often does not encounter the notion of "agent" or "environment" when working with these types of machine learning problems. A notable exception is *reinforcement learning*, in which the environments are inherently sequential (e.g., playing a game of chess), and one always uses the concept of an agent (e.g., player agent).

There is a wide range of data types with which inductive systems can create models. In the following, we provide examples of some of the more common data types used by inductive learning applications:

- The most common type of data is *multidimensional data* in which each data point is represented by a set of numerical values, referred to as *dimensions, features*, or *attributes*. For example, a data point might correspond to an individual, and a feature might be an individual's age, salary, gender, or race. Although features might be categorical, they are often converted into numerical data types by various types of encoding schemes. For example, consider an attribute like color, which takes on one of three values from *red, blue,* and *green*. In such a case, one can create three binary attributes each of which belongs to one of the colors. Only one of these attributes takes on the value of 1, and other attributes take on the value of 0. A multidimensional data set can always be represented as a numerical matrix of values, which makes the development of inductive learning algorithms particularly simple.

- Sequence data might corresponds to text, speech, time series, and biological sequences. Each datum can be viewed as a set of features in a particular order, and the number of features in each datum might vary. For example, the lengths of two different series might vary, or the number of words in two sentences might vary. Note that neighboring features (e.g., consecutive words in a sentence or consecutive values in a time series) are closely related, and machine learning models need to account for this fact during the construction of mathematical models.

- Spatial data have features that are organized spatially. For example, images have pixels that are arranged spatially, and neighboring pixels have closely related values.

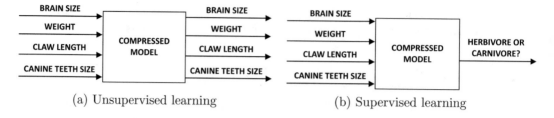

(a) Unsupervised learning   (b) Supervised learning

Figure 1.4: Unsupervised learning models common interrelationships and patterns among data attributes, whereas supervised learning models relationships of attributes to an all-important property (the supervisory label)

In an image, two adjacent pixels are highly likely to have the same value, and most of the information about the image is often embedded in a small number of pixels that lie in the regions of high variability. Machine learning models need to account for these relationships. This is a general characteristic of machine learning algorithms that work most effectively, when the specific characteristics of the data domain are used in the learning process.

Other relevant types of data include graphs or other heterogeneous data types containing different types of attributes. Reinforcement learning settings generate sequential data. Inductive learning methods for handing sequential and spatial data are discussed in Chapters 8 and 10.

Most of the examples in this section will use multidimensional data because of its simplicity and ease of use. In the following, we will introduce the notations needed to represent the vectors associated with multidimensional data. A vector $\overline{X}_i = [x_i^1, \ldots, x_i^d]$ is a set of d numerical attributes $x_i^1, \ldots, x_i^d$ of the ith data point $\overline{X}_i$. For example, if the ith data record $\overline{X}_i$ contains values associated with the ith individual, the different components $x_i^1 \ldots x_i^d$ might contain numerical properties such as age, salary, number of years of education and so on. The subscript i in $\overline{X}_i$ refers to the fact that we are talking about the ith individual, and there are a total of n individuals $\overline{X}_1 \ldots \overline{X}_n$. The value d refers to the number of dimensions. Therefore, the multidimensional data set can be expressed as an $n \times d$ matrix.

1.6.1   Types of Learning

There are two primary types of learning, which are *supervised* and *unsupervised*. First, we will describe unsupervised learning. Imagine a thought experiment in which you have never seen an animal other than a human in your life. You are then given a limited opportunity to closely examine various animals from a particular group (e.g., mammals) and create a mental model of all the different animals you have seen. Then, you are asked to give an abbreviated description of the various mammals encountered by you during this process. Since you can only retain concise memories of the various animals, you might only be able to give abbreviated descriptions of the key features of animals and may even create a compact internal model or taxonomy of the different types of animals you have seen. This is the task of unsupervised learning. Note that the *conciseness* of the description of the mammals is a key to the process, since the unusual animal-specific artifacts are typically ignored in this process. In artificial intelligence, an unsupervised learning system will (often) try to create *compressed* representations of the data (such as *clusters* or *low-dimensional representations*, which can also be used to reconstruct specific examples approximately. This process is referred to as *unsupervised*, because one is looking for the generic characteristics of an entire class of examples, rather than trying to look for characteristics that distinguish between different classes of examples.

Now consider a different situation in which each animal is tagged with a collar label indicating whether it is a herbivore or a carnivore. After the examination, you are shown a completely new animal, and are asked to predict whether the animal is a herbivore or carnivore. The animals shown to you at the time you constructed a mental model represented the "training data," which you used to learn characteristics about the animal that discriminate between the herbivores and carnivores (e.g., type of teeth and sharpness of claws). The new animal you are shown is the "test" datum. Note that in this case, your understanding of the animal's characteristics is completely guided by whether or not the animal is a herbivore. Therefore, you will pay particular attention to giving each animal a dental and claw check (unlike in the unsupervised case). This form of guidance is referred to as "supervision." In the unsupervised case, you will pay no special attention to specific features of the animal (like teeth) and treat each attribute as an equal citizen along with other attributes of the animal, as long as that attribute describes the entire group of examples well. Therefore, the labels perform the role of a supervisory teacher.

The difference between supervised and unsupervised methods is much smaller than one might think. In unsupervised models, we are trying to learn *all* characteristics of the data points and their interrelationships, whereas in supervised learning we are trying a learn how to predict a subset of important ones from the other characteristics. Both types of learning are mappings from a set of inputs to one or more outputs, and this similarity becomes particularly evident when working with neural networks (where the inputs and outputs are explicitly specified). In both cases, a compressed model of the data is constructed, but with different goals. This difference is illustrated in Figure 1.4. Note that the compressed models of Figure 1.4 is often encoded using the mathematical parameters of a machine learning model, such as a neural network. This model is also the (sometimes uninterpretable) hypothesis found by the inductive learning model. For example, the unsupervised model of Figure 1.4(a) using a neural network could be an *autoencoder*, whereas the supervised model of Figure 1.4(b) could be a *classifier* like a *linear support vector machine*. In the case of Figure 1.4(a), the compression process ensures that one can reconstruct only the key elements of the input data, while losing some details. Most human learning is of the unsupervised type. We take in data all the time through our senses and unconsciously store "important" facts and experiences in some form without regard to what we might use it for at a later date. Choosing important facts and experiences to store is a form of compression. This knowledge often turns out to be useful when trying to perform more specific, goal-oriented tasks, which correspond to supervised learning.

1.6.2 Unsupervised Learning Tasks

Unsupervised methods generally model the data using the aggregate trends among the attributes and data records. Examples of such problems include *data clustering, data distribution modeling, matrix factorization*, and *dimensionality reduction*. Some unsupervised models can also be used to generate "typical" examples of synthetic records using the model of the probabilistic data distribution. Typical applications of unsupervised methods include the following:

1. An important application is to create a summarized representation of the full data set for better understanding and application-centric analysis. An example is the problem

of data clustering, which is used to create segments of similar data points. Some forms of clustering like *Kohonen self-organizing maps* [103] are used to create visual representations of the data for better understanding.

2. A second example of a useful application is unsupervised dimensionality reduction, in which each data point is represented in a reduced number of dimensions by using a linear or nonlinear mapping. The basic idea here is to use the dependencies among the attributes to represent the data in compressed form. In the example discussed in the previous section, animals with powerful canine teeth are also likely to have long and sharp claws, whereas animals with rudimentary canine teeth are not likely to have sharp claws. These dependencies can be used to create more concise representations of the data, since some attributes can be approximately predicted with combinations of others.

3. The compressed representations created by unsupervised models can (sometimes) also be used to generate examples of synthetic data records, representing typical examples of data points from that distribution. Such classes of unsupervised models are also referred to as *generative models*.

It is noteworthy that the above applications are often enabled by similar (and sometimes the same) models.

In the following, we provide a brief description of the clustering problem as a representative of unsupervised methods. The problem of clustering is that of partitioning the different records $\overline{X}_1 \ldots \overline{X}_n$ into groups of similar records. For example, consider a situation where each $\overline{X}_i$ contains the attributes corresponding to the demographic profile of a customer. Then, an e-commerce site might use a clustering algorithm to segment the data set into groups of similar individuals. This is a useful exercise, because the similarity in demographic profile of each segment might sometimes reflect itself in the similarity of their buying patterns for various items. One can often use the segmentation created by clustering as a preprocessing step for other analytical goals. For example, on closer examination of the clusters, one might learn that particular individuals are interested in specific types of items in a grocery store, although any particular individual might have bought only a small subset of these items. This information can be used by the grocer to make recommendations for likely items to be bought by the individuals in the group. In fact, a variation of this approach is used commonly for recommendations in e-commerce applications, and it is referred to as *collaborative filtering* [5].

A popular clustering algorithm is the k-means algorithm, where we start by randomly sampling k points from the data as *cluster representatives* and assigning each point to its closest cluster representative. This results in k clusters. Then, we recompute the centroid of each cluster by averaging all the points in the cluster, and designating the centroid as the updated cluster representative. All points are again assigned to their closest cluster representative. The process is repeated iteratively, until the cluster representatives and cluster memberships of points stabilize. This results in a clustering of the data in which similar points are placed in each cluster.

It is noteworthy that clustering is a form of data compression, because each cluster can often be described in a concise way with the use of its statistical properties. For example, each point can simply be replaced with its centroid (in the k-means algorithm), and therefore one can represent the entire data set using the count of each centroid — this yields a compressed representation of the data (albeit with some loss of information). The loss of information is natural, because unsupervised models try to extract the most important patterns in the data, which can be used to reconstruct the data approximately but not exactly. Some forms of clustering, which create a probabilistic distribution of the points in each cluster, can also be used to generate synthetic (representative) points from a real data set for other applications. Another example of an unsupervised application that works particularly well for high-dimensional data is to represent each of the d-dimensional data points in $k \ll d$ dimensions by using dimensionality reduction. This is achieved with the use of a parameterized model that maps data points to their low-dimensional representations. This is a common characteristic of many unsupervised models, which build compressed models of the data with the use of a parameter vector $\overline{W}$. The basic idea is to express each point $\overline{X}_i$ approximately as a function of itself after passing through a compression stage:

$$\overline{h}_i = F_{compress}(\overline{X}_i), \quad \overline{X}_i \approx F_{decompress}(\overline{h}_i)$$
$$\overline{X}_i \approx G(\overline{X}_i) = F_{decompress}(F_{compress}(\overline{X}_i))$$

The compressed representation $\overline{h}_i$ can be viewed as a concise description containing the most important characteristics of $\overline{X}_i$. The "hidden" representation $\overline{h}_i$ is constructed inside the box of Figure 1.4(a), and it is typically not "visible" to the end user at least from the input-output perspective (although it can be extracted from the model). Its attributes are often not interpretable either. This is not always in the case. For example, in k-means clustering, the value $\overline{h}_i$ is simply the centroid of the cluster containing $\overline{X}_i$, and the decompression process simply reproduces this centroid. In other applications like dimensionality reduction $\overline{h}_i$ might be a linear or nonlinear transformation of $\overline{X}_i$, which is harder to interpret in terms of the original attributes.

The reconstruction of a data point from this hidden representation is approximate, and might sometimes drop unusual artifacts from the point. For example, representing each point by its clustering centroid in the k-means algorithm clearly loses information. In the mammalian example of the previous section, if an unusual lion without claws is input to the model of Figure 1.4(a), it might still output a lion with very similar features in other ways but with added claws (based on other typical examples seen by the model). The (non-redundant) concise description $\overline{h}_i$ does not capture rare artifacts, and it has fewer interrelationships among its attributes. Therefore, it is often a simpler matter to model the hidden data with a probability distribution and directly generating examples of $\overline{h}_i$ from this probability distribution; these examples can then be used to generate new data points as $F_{decompress}(\overline{h}_i)$. In our analogy of mammals discussed earlier, one can view this process as the generation of the description of typical (but fantasy) mammals that one has not seen before. A discussion of unsupervised learning methods is provided in Chapter 9.

1.6.3 Supervised Learning Tasks

The two most common supervised learning tasks are classification and regression modeling. In the classification task, each data point $\overline{X}_i$ is associated with a *target variable* y_i, which is also referred to as the *class label*. The class label is the indicator variable that denotes the identity of the group that the point $\overline{X}_i$ belongs to. In the simplest case of binary classification, the target variable y_i is drawn from $\{-1, +1\}$, whereas the target variable is drawn from one of k unordered categorical values (e.g., blue, green, or red). The value k represents the number of groups in the data just like clustering. Unlike the case of clustering, the groups are identified in a supervised way with the use of specific criteria (e.g., herbivore or carnivore animals). Note that a clustering algorithm may not always naturally cluster the data into these specific types of groups. The values of $y_1 \dots y_n$ are provided for n training data points. This data is then used to build a model of how the features relate to the group identity (e.g., carnivores have specific types of teeth). In many cases, the group identity may be a complicated function of the features, which may not be easily explainable, but may be expressed as a somewhat complicated mathematical function of the numerical features in $\overline{X}_i$. The group identities are also referred to as *classes* or *categories*. In the case of multiway classification, it is important to note that the class labels are not ordered with respect to one another.

A concrete example of the classification problem is the setting in which the input features in $\overline{X}_i$ correspond to the pixel values of an image, and the labels in y_i correspond to the categories of the image (e.g., categorical identifier of banana, car, or carrot). Since individual pixels are rather primitive features, it is often hard to express the category y_i as an explainable function of the values of the individual pixels. Nevertheless, inductive methods are usually able to learn this function in a data-driven manner, whereas no deductive reasoning methods are known that can categorize images accurately from pixels.

The *regression modeling problem* is closely related to classification. The main difference is that the value of y_i is numerical in the regression modeling problem. For example, one might try to predict an individual's credit scores from their past history of credit card balances and payments. In both classification and regression, the goal is to *learn* the target variable y_i (whether it is categorical or numerical) as a function of the ith data point $\overline{X}_i$:

$$y_i \approx f(\overline{X}_i)$$

It is possible for the function $f(\cdot)$ to take on a highly complicated and difficult-to-interpret form, especially in the case of neural network and deep learning models. The function $f(\overline{X}_i)$ is often parameterized with a weight vector $\overline{W}$. Consider the following example of binary classification into the labels $\{-1, +1\}$:

$$y_i \approx f_{\overline{W}}(\overline{X}_i) = \text{sign}\{\overline{W} \cdot \overline{X}_i\} \tag{1.1}$$

Note that we have added a subscript to the function to indicate its parametrization. How does one compute $\overline{W}$? The key idea is to penalize any kind of mismatching between the *observed value* y_i and the predicted value $f(\overline{X}_i)$ with the use of a carefully constructed *loss function*. Therefore, machine learning models often reduce to the following optimization problem:

$$\text{Minimize}_{\overline{W}} \sum_i \text{Mismatching between } y_i \text{ and } f_{\overline{W}}(\overline{X}_i)$$

Once the weight vector $\overline{W}$ has been computed by solving the optimization model, it is used to predict the value of the class variable y_i for instances in which the class variable is not known. In the case of classification, the loss function is often applied on a continuous relaxation of $f_{\overline{W}}(\overline{X}_i) = \text{sign}\{\overline{W} \cdot \overline{X}_i\}$ in order to enable the use of differential calculus for optimization of a continuous function. An example of such a loss function for binary classification is the least-squares loss function, which is also referred to as Widrow-Hoff loss:

$$\text{Minimize}_{\overline{W}} \sum_i (y_i - \overline{W} \cdot \overline{X}_i)^2$$

Note that the discrete sign function has been dropped in front of $\overline{W} \cdot \overline{X}_i$, which creates a continuous objective function. On computing the optimal value of $\overline{W}$ by solving this optimization problem, Equation 1.1 can be used to make predictions for new test instances that the algorithm has not seen before. Among supervised learning algorithms, neural networks have become particularly popular in recent years because of their superb performance in some domains such as images and text. Neural networks use a network of computational nodes with weights associated with edges. Learning occurs by changing the weights in the neural network, in response to errors in prediction. This process of changing weights has a direct analog in biology, because learning occurs in living organisms by modifying the strengths of the synapses connecting biological neurons.

Reinforcement Learning

Whereas unsupervised learning and supervised learning define rather simplified and narrow tasks, reinforcement learning algorithms try to enable end-to-end learning for specific tasks. The core idea is to provide feedback to the learner via a *trial-and-error process*, such as in teaching a robot how to walk. Note that a reinforcement learning algorithm will often used a supervised learning algorithm (like a neural network) as a subroutine, and will continuously train it, as it receives positive or negative feedback about its moves. Reinforcement learning has the strongest parallels to biological learning. A mouse can be trained to learn a path through a maze by giving it appropriate rewards (e.g., food) on reaching the exit point of the maze. Over many tries, a mouse will gradually learn which paths are correct, and which are not. Each time the mouse receives a reward, the synaptic strengths inside its biological neural network are updated. The cumulative effect of the updates results in learning. The process used in reinforcement learning is very similar.

AlphaZero is a reinforcement learning based chess program[4], that learns the best moves by repeatedly playing against itself and updating a learning model that predicts moves using the current position of the chessboard as input. The learning model uses a neural network, and the weights of the neural network are updated depending on the outcomes of games. It learns more like humans than programs that rely on a tree of moves (i.e., programs that use knowledge engineering). As a result, its playing style is far more enterprising and human-like than conventional chess engines. The main challenge associated with reinforcement learning is that it requires *extremely* large amounts of data. This is possible to generate in closed systems (like game playing), where one can generate unlimited amounts of data via self-play. Similar successes have been reported in video games or e-robots, where one can generate unlimited amounts of data via simulation. However, the performances in case of limited data availability (like actual robots) are far more modest. There are practical issues in exposing a system to a trial-and-error process in which failures can often damage the system. This has been the main challenge associated with the use of reinforcement learning in real-world settings, in spite of its outstanding performances in scenarios that can be simulated to generate large amounts of data. A discussion of reinforcement learning is provided in Chapter 10.

1.7 Biological Evolution in Artificial Intelligence

All of biological intelligence is owed to the Darwinian process of evolution among living organisms. In biological evolution, organisms compete for survival, as a result of which fitter animals tend to mate more often. Over time, organisms adapt to their environment in order to have better chances of survival. Humans stand at the pinnacle of the evolutionary hierarchy primarily because of their superior intelligence, which provides them the best chances of survival. Therefore, evolution can be considered the master algorithm in biology.

---

[4]It can also play Go and Shogi.

Note that biological evolution is a large-scale form of reinforcement learning, where a large number of organisms evolve via their interactions with one another as well as the environment. All reinforcement learning algorithms are inductive methods, which learn from examples/experiences. Therefore, we are ourselves the products of inductive learning.

The success of biological evolution leads to the following natural question. If evolution has done wonders for biology, why hasn't it done so for artificial intelligence? It is certainly not for lack of trying. Evolutionary paradigms exist both for optimization and for artificial intelligence. The paradigm in optimization most related to biological evolution is that of *genetic algorithms* [85], whereas the corresponding paradigm in artificial intelligence is that of *genetic programming* [106]. The main problem with these methods is that they have (so far) not been able to provide encouraging results beyond some restricted problem domains. In this context, genetic algorithms have been far more useful, as compared to genetic programming methods.

In genetic algorithms, the solutions to an optimization problem are encoded as strings. For example, consider a problem in d dimensions in which we want to find a set of k centroids (central points) of clusters so that the average distance of each point to its closest cluster centroid is as small as possible. A solution to this problem may be encoded as a string of length $k * d$ containing the k centroids (each of dimensionality d). This string is referred to as a *chromosome*, and the objective function of this string is referred to as a *fitness function*. In a clustering problem, the fitness would be the average distance of each point to its closest centroid. It is sometimes possible to combine two solutions (i.e., two sets of centroids) in order to create two new solutions that share the important characteristics of their parents. These are the offsprings of the parents, and the process is referred to as *crossover*. The parents are selected for recombination in a biased (Darwinian) way, by selecting the fitter individuals, so that the children solutions are also likely to be more fit. The selection is often done by using biased random sampling, where the fitness function is used to incorporate the bias into the selection process. This process is analogous to that of natural selection in biology, and it is also referred to as *selection* in genetic algorithms. In addition, a process called *mutation* introduces greater variations among the solutions. For example, one might add some noise to the centroids in the solution set in order to encourage diversity. Like biological evolution, one always works with a *population* of solutions, and repeats the process of selection, crossover, and mutation, until all individuals in the population tend to become more similar. The result is a heuristically optimal set of solutions to the problem at hand. A discussion of genetic algorithms is provided in Chapter 2.

Genetic programming is a branch of genetic algorithms, in which the solutions to the optimization problem are computer programs themselves, and the fitness function is the ability of a computer program to perform a specific task. By using this approach, it is expected that the computer program will evolve with time, so as to be able to perform the task more effectively. The paradigm of genetic programming is very similar to that of biological evolution.

Practical successes with the genetic programming paradigm have been somewhat limited because of computational challenges. Biological evolution has had access to billions of years of a genetic algorithm *in parallel* over an innumerable number of small and large animals. For example, even the number of ants is over ten-thousand trillion, and the number of insects runs into the quintillions. Each of these organisms can be viewed as a tiny "program" that interacts with other programs, recombines, mutates, and reproduces itself based on the process of natural selection. Furthermore, the entire evolutionary system has had access to the "computational" power of the sun for billions of years. Although the process is quite inefficient, its scale is mind boggling. Clearly, we have no technology today that matches the scale and computational power of biological evolution. As a result, there are many critics of genetic programming, who suggest that the field tries to be over-ambitious in working from "first principles" without a proper understanding of the limitations under which it functions. Whether this criticism is well founded remains to be seen, especially as more powerful computational paradigms like *quantum computing* emerge from the horizon. Genetic programs do have limited successes in narrowly defined domains. This is, again, true of most fields in artificial intelligence, where general-purpose intelligence (of the type passing a Turing test) remains elusive. In this sense, the future potential of genetic programming (with increased computational power) remains largely unknown.

1.8 Summary

Artificial intelligence has two primary schools of thought, depending on the types of methods that are used to solve particular problems. These two schools correspond to the deductive reasoning and the inductive learning schools of thought. Deductive reasoning methods like logic programming were favored in the early years, but they reduced in popularity over time, as they failed to deliver in more challenging settings. In later years, inductive learning methods (such as machine learning and deep learning) became more popular. These classes of methods learn from experience and evidence, and they are much closer to how humans learn. Among the inductive learning methods, reinforcement learning and genetic programming remain the most ambitious forms of artificial intelligence, as they try to simulate general-purpose artificial intelligence. These methods are also closest to the biological paradigm. However, these methods have shown promise only in restricted domains because of their massive data and computational requirements.

1.9 Further Reading

An excellent and detailed resource on artificial intelligence is the Russell and Norvig book [153]. The book is generally more focused on traditional (deductive) forms of artificial intelligence, although one of the sections of the book is also devoted to learning. For inductive methods, there are several books focused on machine learning [6, 7, 20, 21, 67, 71], as well as on the mathematics of machine learning [8, 45, 176]. Books on programming languages like LISP [133, 163] and Prolog [37] are also available.

1.10 Exercises

1. Classify each of the following as examples of either inductive learning or deductive reasoning: (a) Using a combinatorial algorithm to find the shortest path on a map from one point to another; (b) Using past experience at driving in the city to construct the shortest path from one point to another; (c) Enumerating the *entire* tree of possibilities in tic-tac-toe and defining moves based on this tree; (d) Making a move in tic-tac-toe based on the outcomes of moves made in the past in similar positions; (e) Classifying a machined slab as defective or non-defective by comparing its measurements with those of other defective and non-defective slabs; (f) Classifying a machined slab as defective or non-defective by comparing its measurements with a set of ideal measurements.

2. It is known from the theory of gravitation that the distance y in meters fallen by an object released from a tower with zero initial velocity is approximately given by $y =$

$4.9x^2$, where x is the time in seconds. Suppose that Mr. Ignoramus, the scientist, knows nothing about the theory of gravitation, but suspects that y is a polynomial function of x of degree at most three. He models the function as $y = a_0 + a_1x + a_2x^2 + a_3x^3$. He then estimates the values of a_0, a_1, a_2, and a_3 by repeatedly dropping a ball from a tower and measuring y at different values of x using a video of the dropping a ball. He then uses this model to estimate the value of y in meters when x is 10^6 seconds (because the tower is not tall enough to yield this value of y).

(a) Is this process deductive reasoning or inductive learning?

(b) Will Mr. Ignoramus obtain the theoretical model $y = 4.9x^2$ as a result of his experiments exactly? If there are any variations from the theoretical model, explain what might be the reason.

(c) Suppose that Mr. Ignoramus performs all the experiments and the estimation of y at $x = 10^6$ independently on two separate days. Will he obtain the same estimate on the two days?

3. Discuss whether each of the following agent settings below are episodic non-sequential, episodic sequential, or continuous sequential: (a) Classifying each of a set of images into one of three categories; (b) Playing a chess game; (c) Driving a car for a long time; (d) Playing a video game of PacMan; (e) Predicting the first-year sales volume of a book by an electronic retailer by using the number of pre-orders; (f) Predicting whether a machine slab is defective based on its dimensions.

4. Consider a data set of primary school students containing their date of birth, age in days, and the grade of the student. Discuss a very simple procedure for compressing the data into a single dimension, and the most likely errors resulting from such a compression.

5. Are the following examples of static/dynamic and discrete/continuous-time environments for the following agents: (a) A robot walking in a hallway; (b) An game of tic-tac-toe; (c) A chatbot agent talking to a person through a keyboard interface. The answers to the above might depend on the assumptions you make. Therefore, please state the assumptions you make (if needed).

6. The solution to the eight queens problem is not unique. Provide an alternative solution to the eight queens problem beyond the one that is given in the text of the chapter.

7. Suppose that it is identified by an astute data scientist that spam emails often have the words *"Free Money"* embedded in them. Subsequently, the data scientist implements a system that identifies spam emails by removing all emails containing both these words. Discuss why this process involves both inductive learning and deductive reasoning.

Chapter 2

Searching State Spaces

"Between truth and the search for it, I choose the second." – Bernard Berenson

2.1 Introduction

The goal of an agent in artificial intelligence is (often) to reach a particular class of states (e.g., a winning state in chess), or to reach a state with high "desirability" based on specific criteria defined by the application at hand. In other words, the agent needs to *search* through the space in order to reach a particular goal or to maximize its reward. In some cases, the path chosen by the agent through the search space has an impact on the earned reward.

In most real-world settings, the state space is very large, which makes the search process very challenging. This is because the number of possible states that can be reached by a small number of actions is very large, and an agent can often reach irrelevant parts of the space. While search problems are naturally suited to settings in which sequential actions are performed on states (e.g., moves in chess), their use is not exclusive to these cases. Even in settings in which the order of decisions does not matter, one can use search in order to explore the space of possible solutions by creating an artificial sequence to the process of making key decisions. For example, in the case of the constraint satisfaction problem, the variables of the problem can be set sequentially, as one searches through the space of possible variable assignments. There are two variations of search-centric settings, which are as follows:

- Given a particular starting state, reach a particular end state as a goal. The solution of many mazes and puzzles (including the constraint satisfaction problem) can be modeled with this scenario. The starting state corresponds to the initial configuration of the puzzle, and the goal state(s) correspond to any of the desired configurations. In some cases, one might also want to minimize the cost of the path required to reach the end state from the start state. This is often the case in routing applications. Therefore, costs are associated with the *paths* used to reach from one state to another, and some states are considered goal states. Note that the specific approach chosen to plan the

© Springer Nature Switzerland AG 2021
C. C. Aggarwal, *Artificial Intelligence*, https://doi.org/10.1007/978-3-030-72357-6_2

path from the start state to the end state might depend on the background knowledge of the agent and the details of the environment at hand. For example, an agent who has a map of a maze will use a different search-based algorithm than one that does not have this information.

- Given a particular starting state, and a loss function associated with each state, reach an end state with the least loss. Therefore, costs are associated with *states*. This type of setting usually requires a different type of algorithm than the one in which costs are associated with paths. Typical examples of algorithms used in this case include hill climbing, simulated annealing, tabu search, and genetic algorithms.

The first of these scenarios is goal-oriented search, although losses might also be occasionally built into the search in terms of the cost of traversals. The second of these scenarios tries to optimize state-centric losses, and it is the focus of methods such as simulated annealing and genetic algorithms. Some versions of search are also closely related to *planning*, wherein a specific sequence of actions achieves a specific goal. In general, planning corresponds to settings in which the sequence of actions required to achieve a particular goal is relatively long. While the planning problem is sometimes treated as a separate field of artificial intelligence, its differentiation from other settings in artificial intelligence is only in terms of the increased complexity caused by the need to find the correct sequence of actions. For example, a chess player may look for a long sequence of moves that is likely to guarantee victory.

The methods discussed in this chapter fall in the class of deductive reasoning methods. However, many search and planning scenarios can also be addressed with the use of *reinforcement learning algorithms*, which is an inductive learning method. A key point is that the search-based algorithms in this chapter use domain knowledge about the state space in order to make decisions about paths that are likely to be rewarding. However, reinforcement learning algorithms use the prior experience of the agent during its traversal of the state space in order to make decisions about the actions to be used in order to earn rewards. The use of prior experience in order to make decisions is a data-driven method, which can be viewed as a form of inductive learning. Deductive reasoning methods are preferable when it is expensive to collect data about the prior experiences of the agent. Reinforcement learning algorithms are discussed in the next chapter and in Chapter 10.

This chapter focuses on the single-agent setting, in which all actions are made by the same agent. This setting does not cover two-agent or multi-agent settings like chess. Multi-agent settings are more complex because the objective functions to the optimized might be different from the perspective of different agents. These situations arise often in all types of game-playing applications. Because of fundamental differences in the methods used for single-agent and multi-agent settings, multi-agent settings are discussed in the next chapter.

This chapter is organized as follows. The remainder of this section discusses how the state space may be conceptually modeled as a graph. Goal-directed search algorithms are discussed in Section 2.2. The use of path costs and state-specific objective function values to improve goal-directed algorithms is discussed in Section 2.3. Local search with state-specific loss values is discussed in Section 2.4. This includes the methods of hill climbing and simulated annealing. Genetic algorithms for search and optimization are discussed in Section 2.5. The constraint satisfaction problem is discussed in Section 2.6. A summary is given in Section 2.7.

2.1.1 State Space as a Graph

The modeling of the state space as a graph enables the development of algorithms that leverage the graph structure of the space for search. In other words, state-space search is transformed to graph search. The advantage of this approach is that the problem of graph search is a well studied problem in graph theory, mathematics, and computer science. Therefore, modeling the state space search problem as one of graph search enables the use of various types of off-the-shelf algorithms, once the transformation of the problem to a graph-centric setting has been achieved. Search methods in graph theory have been studied for more than fifty years, and a wide variety of methods are available to address various settings.

The state space for search can be treated as a *directed graph*, in which each state can be treated as a node of the graph. In directed graphs, the edges between nodes have a direction, which is denoted by an arrow. Two nodes i and j are connected by a directed edge (i, j), if an agent can move from state i to state j with a single action. The direction of the edge in the modeled graph mirrors the direction of the transition of the agent from state i to state j. In other words, the *tail* of the edge[1] is at node i and the *head* of the edge is at state j. Note that in some problems, the definition of a node and the structure of the graph is heavily defined by how one defines the actions of the agent (which is a part of the design of the strategy for search). We will give an example of this flexibility in this section.

In order to provide an understanding of this concept, consider the eight-queens problem as a case study. In this problem, the goal is to place eight queens on an 8×8 chessboard, so that none of the queens attack one another. We advise the reader to refer to the eight-queens problem discussed in the previous chapter before perusing further (see Figure 1.2 of Chapter 1). Note that any placement of the k queens for $k \leq 8$ on the board can be treated as a state. However, only states in which none of the queens attacks the other and for which $k = 8$ are considered goal states. The agent continually tries to put additional queens on the board, and therefore an edge exists from state i to state j, if and only if state j has an additional queen placed on the board with respect to state i (and is otherwise identical to state i). Examples of two nodes connected by an edge are shown in Figure 2.1. While the graph might technically contain nodes that violate the attack constraint, only edges between two nodes that do not violate the "attack constraint" will be traversed, as the agent continues to try to build on a solution that does not violate validity so far. Such an edge traversal is also referred to as a *transition*, because it defines how the action of an agent changes the state. Since the attack constraint is never violated in any traversal, it is also possible to consider a graph in which only nodes not violating the attack constraint are included; therefore, the valid states correspond to those positions in which no pair of queens attack each other. The node at the starting point of the transition is referred to as the tail of the edge, and the node at the ending point of transition is referred to as the head of the edge. The node at the tail of the edge is referred to as the *predecessor* of the node at the head, and the node at the head of the edge is referred to as the *successor* of the node at the tail. Since the node at the head of an edge always has one more queen than a node at the tail of the edge, such a graph cannot contain directed cycles in which a sequence of edges forms a closed loop; in other words, the state-space graph is a *directed acyclic graph* in this particular case. It is noteworthy that the state-space graph of most application-centric settings is not acyclic in nature.

---

[1]In graph theory, the head of an edge corresponds to the arrow-head, whereas the tail of the edge corresponds to the other end. The edge points from the tail to the head, which corresponds to the direction of the transition.

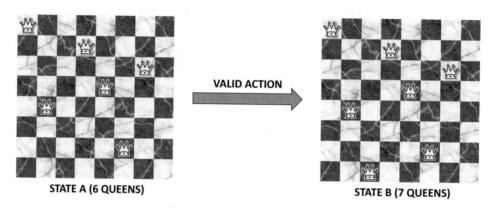

Figure 2.1: A valid transition caused by an agent action of placing a queen on the chessboard

The graph traversal process starts with $k = 0$ queens, and successively increments k while placing more queens on the board. The *initial state* might therefore correspond to a solution in which the board is empty, and therefore the value of k is 0. The goal for the agent is to keep traversing the state space, until a valid solution is reached. When the value, k, of the number of queens is much smaller than 8, it is easy to find a placement in which the queens do not attack one another. However, as k increases, it becomes successively harder to find a square on which the queen may be placed without violating the attack constraint. In some cases, the agent might reach a state, where it is no longer possible to place a queen on the board without violating further constraints. For example, the solution on the right-hand side of Figure 2.1 has seven queens, but it is impossible to place an eighth queen on any square without it being attacked by one of the seven other queens. Therefore, this state is a dead end, where the agent cannot proceed further on an edge in order to find a valid solution. For a graph representation containing only valid states, such a node would not have an outgoing edge. Therefore, in such cases, *backtracking* becomes necessary, where an earlier position (with fewer queens) is expanded in a different way. The goal of search algorithm is to explore the state space in a systematic manner in order to find a valid state containing eight queens. The abstraction of state spaces in terms of graphs enables this process to be treated with a certain level of simplicity, because the design of search algorithms on graphs has been widely explored in classical mathematics and computer science.

As discussed earlier, the design of the state-space graph may depend on the strategy used by the search, and therefore the definition of a state or a transition may vary as well (depending on the exploration strategy at hand). Even in the case of the eight-queens problem, it is possible to design a completely different strategy for search, and with a correspondingly different graph structure representing the different way in which the states and transitions are defined. For example, it is possible to conceive a different graph structure on the state space in which any solution containing eight queens is a state, whether the queens attack one another or not. Two states are connected by an *undirected* edge, if one can transition between these two states by moving a single queen from one square to another and vice versa. Instead of an undirected edge, one can also use two directed edges in both directions, since the transition occurs in both directions. The goal is to keep moving queens until a valid solution is reached. Therefore, the states are defined by all possible positions containing eight queens, whether they attack one another or not, and the key point is to move from one state to another until a position with minimal number of attacks is reached. A key point is that the agent strategy is different here, both in terms of the choice of the initial state, as well as the actions used, and therefore the structure of the underlying graph is different as well. In fact, these cases are best solved by defining heuristic loss functions that penalize the number of mutual attacks in that state, and then trying to move from one state to another in a hill-climbing or simulated annealing approach. Such methods are discussed in Section 2.4.

Finally, it is important to note that most artificial intelligence problems are such that the state space is huge. For example, the number of ways in which $k = 8$ queens can be placed on a board with 64 squares is $\binom{64}{8} \approx 4.43 \times 10^8$ (including solutions in which the queens might attack one another). Which states are included in the state space graph depends on the algorithm at hand. Some algorithms might use only states in which queens do not attack one another, whereas other algorithms may find a final solution only by going through states that violate the constraints of the problem (such as the queens attacking one another). Therefore, the efficiency of the process depends heavily on the specific strategy used for constructing the state-space graph. The eight-queens problem is considered a toy problem in artificial intelligence. A very complex state space, such as the valid positions in a game of chess, has a larger number of states than the number of atoms in the universe. In such cases, it is impossible to hold the entire graph structure on any storage device, and one must devise clever solutions that visit only a small part of the state space. Therefore, agents usually have to work with partial (local) knowledge of the graph during exploration. This creates many challenges and trade-offs associated with the types of search algorithms that can be used.

2.2 Uninformed Search Algorithms

In this section, we will discuss algorithms for goal-oriented search where agents move from state to state in a search process, but they do not have any information about the heuristic goodness of intermediate states. As a result of the lack of information about the desirability of intermediate states, these algorithms are referred to as *uninformed*. Uninformed search algorithms perform rather poorly in practice, because they tend to meander around aimlessly in a large search space. Nevertheless, these algorithms are important because they define the base on top of which more sophisticated algorithms are built.

As discussed in the previous section, the state-space graphs are implicitly defined by the choice of the states, as well as the choices of the actions in each state. For goal-directed algorithms like the eight-queens problem, there is typically an initial state s and one or more conditions $\mathcal{G}$, which define whether a state corresponds to the goal. These values are inputs to the search algorithm. In many cases, the choice of state space is not unique; the same problem can be defined by multiple state-space graphs, depending on the nature of the initial state (e.g., initial position in eight-queens problem), the definition of a state, and the specific choice of actions used to reach the goal (e.g., placing queens sequentially or shuffling pairs of queens on the board). The actions for moving from state to state are often constructed in such a way, so as to enable an efficient search process. In other words, one wants to create a search process that maximizes the chances of states reaching the goal condition through an ordered traversal process. However, once the graph and the transition strategy has been chosen, the specific design of the search process also has a bearing on the efficiency of the approach. Furthermore, the graph is often dynamically created as it is explored, because it is often impossible to pre-store the large state-space graph up front. In most practical applications involving large state spaces, the graphs are often locally expanded by selecting a current state from a set of potential states to be visited, and then

Algorithm *GenericSearch*(Initial State: s, Goal Condition: $\mathcal{G}$)
begin
  LIST= { s };
  repeat
    Select current node i from LIST based on pre-defined strategy;
    Delete node i from LIST;
    Add node i to the hash table VISIT;
    for all nodes $j \in A(i)$ directly connected to i via transition **do**
    begin
      if (j is not in VISIT) add j to LIST;
      $pred(j) = i$;
    end
  until LIST is empty or current node i satisfies $\mathcal{G}$;
  if current node satisfies $\mathcal{G}$ **return** success
        else return failure;
  { The predecessor array can be used to trace back path from i to s }
end

Figure 2.2: The generic search algorithm over a large state space

expanding the nodes adjacent to the current state with a transition. The set of potential states to be visited are stored in a list, which is denoted by the variable LIST, and the states in it are referred to as *active states*. The set of active nodes define a *frontier* (i.e., boundary) of the search process, as it currently stands. The specific way in which nodes from LIST are selected has a significant effect on the computational efficiency and memory requirements of the search process.

In the following, we provide a detailed description of a generic version of the search algorithm (cf. Figure 2.2). The search algorithm starts with the initial state, and adds it to LIST. Subsequently, it selects this node from LIST, deletes this node from LIST, and adds its neighbors to LIST. This node is then marked as "visited." All the visited states are added to a hash table, referred to as VISIT. It is important to keep track of these nodes in order to avoid exploration of redundant states that have already been explored earlier. This can happen often when there are multiple paths to the same node, or when the state-space graph contains cycles. Such redundant paths are avoided by not adding those specific nodes to LIST during search expansion, if those nodes have already been visited by the algorithm. In each iteration, the algorithm selects a node from LIST and designates it as the *current node*. If the current node (state) already satisfies all the goals required by the agent, then termination has been reached, and this node is reported (along with the path to the current node if needed). The algorithm checks whether the current node selected from LIST is a goal state. In the very first step, the selected node is unlikely to be a goal state, because it is the source node s.

If the current node does not satisfy the criteria for a goal state, all possible states that can be reached from the currently selected node (by a single transition), and which have also not been visited so far, are added to LIST. Nodes that have already been visited will be available in the hash table VISIT. At the same time, the current node is deleted from LIST. It is possible for the frontier (i.e., LIST) to reduce in size during expansion of a node, when no unvisited node is added to LIST, but that node is itself deleted from LIST. This situation is referred to as *backtracking* because the next node to be visited is always a neighbor of a state that was visited earlier (and not the state that was just deleted). After examination of a node and its deletion from LIST, another node needs to be selected from LIST for expansion. The choice of which node is selected defines the search strategy, which will be discussed later. This process of successive expansion of nodes from LIST is continued, until the goal is reached, or the LIST becomes empty. At this point, the algorithm terminates. If the LIST becomes empty, then it implies that a state satisfying the goal criteria cannot be reached from the initial state. In some cases, it might imply that either the approach for defining the state-space graph or the choice of initial state is a suboptimal one. For example, choosing a starting state for the eight-queens problem with seven queens already placed in carelessly selected positions will often lead to a dead end. Alternatively, it might also imply that no valid solution exists for satisfying the goals of the agent. An overview of the generic search algorithm is shown in Figure 2.2. It is noteworthy that the variable LIST, which contains the active nodes, represents the *frontier* of the state space, which is currently being explored. The word "frontier" corresponds to the fact that this LIST of nodes contains all unvisited nodes *directly* reachable (by a single edge) from all nodes that have been visited so far. Therefore, the nodes contain the boundary between visited and unvisited nodes (on the unvisited side). When a node is selected from LIST during the search process, one is essentially selecting a node from the boundary of the currently explored graph in order to expand it further. The set VISIT contains the nodes in the explored portion of the graph, whereas the set LIST contains the frontier nodes adjacent to nodes in VISIT that have been observed as candidates for exploration (by virtue of direct adjacency to explored nodes) but have not yet been fully explored.

As discussed earlier, the choice of how nodes are selected from LIST is important. This choice defines the size of LIST and the shape of the frontier in the graph being explored. It also has an effect on both computational efficiency and memory requirements. Some examples of the ways in which nodes can be selected for expansion are as follows:

- If the elements of LIST are sorted in order of addition of nodes, and the *latest* node added to LIST is selected for further exploration and expansion, the search strategy is referred to as *depth-first search*. Depth-first search explores the nodes in Last-In-First-Out (LIFO) order. Therefore, as current nodes are explored, one continues to explore then deeply until either a goal node or a dead end is reached. If a dead end is reached, it becomes necessary to backtrack. Back-tracking occurs automatically in the context of selecting the next node on LIST, which is a node added earlier to the list than the most recently explored node. It is possible for backtracking to occur quite frequently in depth-first search (which tends to keep a lid on the size of LIST as back-tracked nodes are deleted from LIST without any neighbors being added). Depth-first search tends to extend into long paths of visited nodes, which results in a frontier that is *deep* rather than *broad*. The length of the path from the starting state to the current state is referred to as the depth of the current state. Note that in depth-first search, LIST will only contain nodes that are directly adjacent connected to the current path being explored (i.e., the path from the source node to the currently explored nodes). As a result, the size of LIST is rather limited, since the maximum path length in most real-world graphs is quite limited.

- If the elements of LIST are sorted in order of addition of nodes, and the *earliest* node added to LIST is selected for further exploration and expansion, the search strategy is referred to as *breadth-first search*. Breadth-first search explores the nodes in First-In-First-Out (FIFO) order. Breadth-first search tries to visit nodes in the same order as they are added to LIST, which leads to a large backlog of nodes waiting to be processed in LIST. Breadth-first search generally results in a broad frontier of nodes to be explored. Another property of breadth-first search is that it tends to find paths to nodes that have the fewest number of edges; this is because this strategy will always visit all nodes at distance d from the starting state before visiting nodes at distance $(d + 1)$. Furthermore, when exploring at depth d, the frontier list will contain all unexplored nodes at depth d and the nodes directly connected to explored nodes at depth d. It is noteworthy that the number of nodes at depth d can increase exponentially with d for large state spaces. As a result, breadth-first search tends to have high memory requirements in terms of the size of the frontier list.

- Distance-oriented expansions generalize breadth-first search in that edges are associated with a cost, and one chooses to expand the node for which the shortest path found (so far) from the source node to the currently selected node is as small as possible. Choosing the lowest cost node for further expansion is referred to as *uniform cost search*. In this case, LIST is implemented as a sorted priority queue.

- For goal-oriented expansions that are associated with a cost, the nodes are often associated with a heuristic cost value assigned by the agent, and this cost is a heuristic or domain-specific estimate of the distance to the *goal*. Sorting the LIST in this order leads to *best-first search*. Note that this approach is different from uniform-cost search, because distance *estimates* from the *goal* are used in best-first search, whereas distances from the *source* are used in uniform-cost search. Using heuristic distance estimates from the goal to guide search naturally leads to quicker discovery of goal nodes (as compared to other nodes), especially when the goal node is located farther away from the source node as compared to most other nodes in the graph. Since best-first search uses heuristic estimates of the distance to the goal node, it is considered an informed search algorithm. Therefore, the framework of uninformed search algorithms also supports informed search with minor modifications.

Several methods such as A^*-search combine the ideas of uniform cost search and best-first search. Some of these methods will be discussed later in the section on informed search algorithms.

These different strategies are associated with different trade-offs. For example, the LIST size expands rapidly in case of breadth-first search, and the memory requirements are therefore rather high. This is primarily caused by the fact that the number of nodes at depth d can be as large as b^d for a state space in which the agent has b actions[2] available at each state. The value of b is referred to as the branch factor. Breadth-first search often finds short paths from the source to the goal, whereas depth-first search might sometimes find long paths to the destination (because it visits all nodes reachable from a particular node before backtracking from that node). Depth-first search tends to reach dead-end nodes periodically at earlier stages of the search (as compared to other search strategies). In such cases, depth-first search backtracks to earlier nodes for further search. This type of backtracking keeps the size of LIST in check, because backtracking reduces the size of the frontier list by 1. Since breadth-first search typically expands LIST more rapidly at early stages of the search, the size of the frontier list is much larger. Examples of frontier lists in the case of breadth-first and depth-first search are shown in Figures 2.3(a) and (b), respectively. The visited nodes and frontier nodes (in LIST) are denoted by 'V' and 'F', respectively, in each case. The nodes that have not yet been visited and are also not currently in the frontier list are left blank. All of these nodes, which are reachable from the initial state, will eventually be added to the frontier list and visited (for finite state spaces).

---

[2]In practice, the number of actions available to the agent may vary with the state at hand.

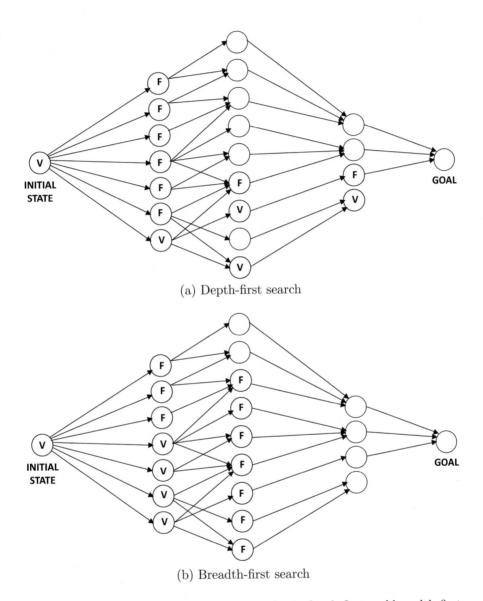

(a) Depth-first search

(b) Breadth-first search

Figure 2.3: The frontier nodes after visiting five nodes in depth-first and breadth-first search

In search-based algorithms, the agent has to keep track of a few other pieces of information that are needed for reconstructing the path that the agent took in order to reach from the source to the destination (while ignoring the intermediate dead ends the agent might have reached). In order to achieve this goal, the agent always keeps track of the *predecessor* of each node. This predecessor is maintained in the array $pred(\cdot)$, which is shown in Figure 2.2. The predecessor of a node is defined as the node at the tail of the edge that was used to add that node to the LIST in the first place. Since each node is added at most once to LIST, it follows that a single path will exist from the source node to each visited node. In other words, the set of edges defined by the predecessors create a directed tree rooted at the source. Once the goal state has been reached, the agent traces the predecessors backwards from the goal state to the initial (source) state. This provides the agent with the knowledge needed to reconstruct the valid path from the initial state to the goal state. For example, in the case of the eight-queens problem, the agent will be able to reconstruct the order and positions at which the queens were added to the board in order to create the final configuration. While this path is not particularly important in the case of the eight-queens problem (since other sequences of placing queens work equally well), it turns out to be more important in some other tasks (such as that of finding a path through a maze). In particular, such reconstructions are important in settings in which the sequence of actions used is critical (unlike the case of the eight-queens problem).

Time and Space Complexity

The running time of this type of search algorithm is bounded above by the number of edges in the graph. For directed graphs, each edge is scanned exactly once, when the state corresponding to the tail of the node is expanded. Therefore, the search process is linear in the size of graph. Although this might seem efficient at first glance, the reality is that it can be quite expensive in practice. It is noteworthy that the main challenge associated with search algorithms is the large state space that they try to explore. Without a proper design of the search, it is possible for the algorithm to explore for a long time without finding a meaningful solution. Therefore, the time complexity largely depends on how quickly one finds the goal state — it is often possible to reduce the time required for finding the goal state only by using *informed* search strategies in which there is some additional information about the quality of intermediate solutions found by the approach. In general, it is often hard to predict which approach (e.g., depth-first search or breadth-first search) will reach the goal more quickly.

Another important issue is that of space complexity. The space complexity is defined by the size of LIST, and it is far more sensitive to the specific approach used for search, as compared to the time complexity. For example, in depth-first search, the frontier list contains only the adjacent nodes to those on the current path being followed. Some graph structures are such that the maximum path length in the graph is quite limited. For example, in the eight-queens problem, the maximum path length is eight when each action (edge in state-space graph) is that of placing a queen on the board. For graphs with limited maximum path length p and limited number of actions a per state, the maximum size of the LIST is $p \cdot a$. On the other hand, the size of LIST can often become uncontrollably large in the case of breadth-first search.

Offline Versus Online Search

Many search problems are formulated in offline settings, wherein the agent first performs the search over the state space in order to find the goal state, and then simply reports the path found in the solution from the start state to the goal state. In other words, the initial search can be considered an offline process, and only the solution obtained from the search is reported in real time. The intermediate steps performed in order to find this final sequence of actions are not considered important. This is natural for settings such as that of trying to find a path from one city to another with a map. This process can be performed before actually embarking on the journey. After the path has been found, it is used in real time, when the path from one city to another is physically traversed. Another example is the eight-queens problem in which only the final solution (of how the queens are placed) is important, and the intermediate steps used to determine a solution are not quite as important. In fact, for a particular solution to the eight-queens problem, one can place the queens on the board in any order to reach the same solution (as long as the squares are chosen correctly).

However, in many settings of artificial intelligence, the agent is required to perform the search in real time, while having limited visibility about the long-term effect of choosing a particular state for exploration. This situation is considered a more challenging setting. In these cases, the agent might reach a dead end after exploring a state, in which case the agent needs to backtrack to an earlier (predecessor) state and then move forward to another candidate node in LIST for further expansion and exploration. For example, if one has to solve the problem of finding paths in a maze from an initial point to a final destination (without access to the map of the maze), one might often reach dead ends, and then backtrack to an earlier fork in the path where a different choice is made. Therefore, the backtracking is often done in real time (as opposed to doing it with a better global view of the state space graph in offline algorithms). Online algorithms present special challenges for informed search algorithms, because of partial knowledge about the distance of different states from the goal state.

Online settings arise often in applications like robotics, where the choices are made by the robot in real time. Making an incorrect choice at a particular stage might result in the robot reaching a dead end. In some cases, an incorrect decision might result in a damage to the robot (which is modeled by using negative rewards associated with those states). In general, online settings are more complex than offline settings because of the unforeseen effects of specific actions, and the need to adjust the actions to these types of events. As a result, real-time settings are better addressed with inductive learning, where past experience can help in shaping future adjustments. Such settings are discussed in Chapter 10.

2.2.1   Case Study: Eight-Puzzle Problem

The eight-puzzle problem is defined using 3×3 tiles, and a single empty space that allows adjacent tiles to slide into. Each tile is annotated by a single number from 1 through 8. Tiles can be exchanged with an adjacent empty space (sharing a side) by sliding the tile onto the empty space. This type of sliding represents a transition from one state to the other, which creates a different arrangement of the numbered tiles. By repeated slides of the tiles, the positioning of the numbered tiles in the puzzle can be changed significantly. The goal of the eight-puzzle problem is to rearrange the tiles via repeated transitions, so that the tiles are arranged in the order 1 through 8, when visiting the tiles in each row from left to right, and

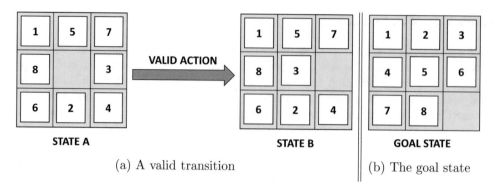

(a) A valid transition   (b) The goal state

Figure 2.4: The Eight-Puzzle Problem

rows from top to bottom. The number of states is bounded[3] above by $9! = 362880$, which creates a graph of relatively modest size. Furthermore, the degree of each node is at most 4, since each blank tile can be moved to one of at most four positions. This particular state space corresponds to an *undirected graph* in which each of a pair of adjacent states is always reachable from one another. Alternatively, one can create a directed graph with symmetric transitions in either direction. An example of a transition is shown in Figure 2.4(a), and the final state (i.e., the goal state) is shown in Figure 2.4(b).

For this problem, the starting state can be a scrambled arrangement of the puzzle tiles, at which one is handed the puzzle. Whenever a tile is moved to an adjacent empty slot, it results in a transition. One can use any of the search methods discussed in the previous section in order to find a solution to the eight-puzzle problem. Even though the graph is of modest size, it is still challenging to use an uninformed search algorithm, since there is only one goal state. In such a case, it is possible for the search to meander around aimlessly for a long time without reaching the goal state. Therefore, it becomes important to include some kind of objective function associated with a state, so that the objective function improves as one moves closer to the goal state. In other words, the objective function performs the role of guiding the search process, so that fruitless paths are avoided more often. One can view this process as that of including domain-specific information with the definition of the problem so as to improve the likelihood of success in a search. After all, search methods belong to the class of knowledge-based methods, and therefore the use of domain-specific information is common. Such strategies are referred to as *informed search strategies*. This will be the topic of discussion in Section 2.3.

2.2.2   Case Study: Online Maze Search

In this problem, we have a maze along which the agent can traverse, such as the one shown in Figure 2.5(a). A state is defined by the position of the agent. In this sense, it may be considered a continuous state space, although it is possible to discretize the state space by choosing important intermediate points in the maze as states. These intermediate points are the ones at which a change in direction of movement of the agent *might* occur. In particular, these points could be either be points at which a turn is available, or those at which a dead

---

[3]Not all positions of the tiles can reach the goal state. Therefore, when the puzzle is physically manufactured, the tiles are always located in their correct position. This way of manufacturing the puzzle automatically restricts the "randomized" arrangement from which one starts solving the puzzle.

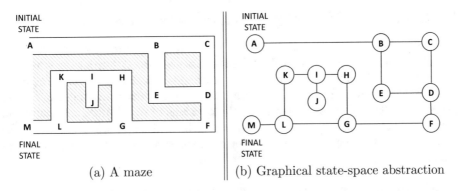

(a) A maze

(b) Graphical state-space abstraction

Figure 2.5: A maze and its state-space abstraction

end has been reached. These states are marked in Figure 2.5(a). At turning points, the agent has an option to choose one or more paths, and this corresponds to the action performed by the agent (which results in a transition to a new state). A maze can be naturally modeled as a graph by treating turning points or dead ends as nodes, and the direct connections between them as edges. The corresponding graph on the state-space structure is shown in Figure 2.5(b).

A natural assumption in maze-based problems (which is different from real-world routing problems) is that the agent does not have a map of the full maze at hand. Furthermore, we consider the (more difficult) case, where the agent must search for a solution in real time, while exploring the maze. Therefore, the agent only knows the choices of actions (e.g., choosing from a fork in the path) at the current state, and has limited visibility about what will occur further down the maze as a consequence of selecting a particular action. In some cases, the agent might reach a dead end. The only assumption that is being made is that the agent has enough visibility to see when adjacent states (i.e., adjacent forks or dead ends) have already been visited. Therefore, the agent works with partial knowledge of the state space. However, this setting can be handled by uninformed search algorithms, which do not use the global structure of the state space in making choices.

For mazes of modest size, such problems are relatively simple for the agent to explore, and a straightforward depth-first search is guaranteed to reach the final (goal) state, as long as a path exists from the initial state to the ending state. Because of the limited visibility, the agent will be forced to perform repeated explorations and perform backtracking over various paths until the final state is reached. Note that such problems are referred to as *online* problems when the agent's actions occur in real time. During backtracking, the agent needs to have sufficient memory about its previous movements through the maze. The main point is that when a node on LIST is selected for expansion, it might not be directly connected to the current node; in such cases, the agent might have to backtrack from its current node i to reach the node on LIST that is next selected for exploration. This is achieved by following the predecessor nodes both from the current node and LIST in order to find their common ancestor. Therefore, the predecessor information will need to be maintained in order to enable effective exploration. Some variations of the problem also associate costs with edges, in which case the agent might have to continue to explore the maze, until the shortest path is found.

2.2.3 Improving Efficiency with Bidirectional Search

One issue in goal-directed search is that it can take a very long time when the state space is huge. In bidirectional search, one explores in the forward direction from the starting state, and in the backward direction from the goal state. The bidirectional search is continued until both the forward and backwards directions of movement reach the same node. The key insight in bidirectional search is that both directions of search need to stumble on *only one common node*, as they expand their visited lists. This makes it more likely that far fewer nodes need to be expanded in order for the two searches to eventually stumble on at least one common node.

> **Algorithm** *BiSearch*(Initial State: s, Goal State: g)
> **begin**
>   FLIST= { s }; BLIST= { g };
>   **repeat**
>   Select current node i_f from FLIST based on pre-defined strategy;
>   Select current node i_b from BLIST based on pre-defined strategy;
>   Delete nodes i_f and i_b respectively from FLIST and BLIST;
>   Add nodes i_f and i_b to the hash tables FVISIT and BVISIT, respectively;
>   **for** each node $j \in A(i_f)$ not in FVISIT reachable from i_f **do**
>   Add j to FLIST; $pred(j) = i_f$;
>   **for** each node $j \in B(i_b)$ not in BVISIT **do**
>   Add j to BLIST; $succ(j) = i_b$;
>   **until** FLIST or BLIST is empty or (FLIST $\cap$ BLIST)$\neq$ {};
>   **if** (either FLIST or BLIST is empty) **return** failure;
>   **else return** success;
>   { Reconstruct source-goal path by selecting any node in FLIST $\cap$ BLIST
>   and tracing predecessor and successor lists from that node; }
> **end**

Figure 2.6: Bidirectional Search Algorithm

In bidirectional search, one maintains two lists, FLIST and BLIST, corresponding to forward and backward search, respectively. The forward search is similar to the search algorithm discussed in Figure 2.2. However, one also performs backward search starting from the goal state. For directed graphs, this means that edges are traversed in the reverse direction. In each iteration, a single node of FLIST is expanded in the forward direction and a single node of BLIST is expanded in the backwards direction. Similarly, hash tables of nodes that are visited in forward and backward search are maintained and denoted by FVISIT and BVISIT, respectively. The set $A(i)$ denotes the forward adjacency list corresponding to nodes outgoing from i. The set $B(i)$ denotes the reverse adjacency list of nodes incoming into i. In each iteration, one node from each of FLIST and BLIST are selected and expanded in the backwards and forwards directions, respectively. These nodes are added to FVISIT and BVISIT, respectively. The moment that FVISIT and BVISIT have one common node, the search is terminated because one can create a path from the starting state to the goal state through this common node. This bidirectional search algorithm is illustrated in Figure 2.6. The predecessor and successor data structures need to be maintained separately; the predecessor data structure is maintained by the forward search and the successor data structure is maintained by backwards search. These two data structures are used to reconstruct the path from the source to the goal state. The idea is to select a node that is common between FLIST and BLIST and then use the predecessor and successor lists to trace paths to the source and the goal states, respectively. The algorithm of Figure 2.6 is designed to work with a single goal state. It is also possible to modify the algorithm to work with a pre-defined set of goal states. This is achieved by adding all the goal states to BLIST at the beginning of the algorithm.

Even though the bidirectional search algorithm is more efficient than the unidirectional search algorithm in terms of the number of nodes explored, it does require the computation of an intersection between FLIST and BLIST, which can be expensive. In other words, each step of the algorithm is more expensive. Furthermore, the approach requires the identification of specific goal states for initiating the backwards direction of the search, which can be problematic when the number of possible goal states is large, and is defined in an indirect way (such as via algorithmic computation of a particular constraint on the state). In many problems, an exhaustive list of goal states may not even be known.

2.3 Informed Search: Best-First Search

For many goal-directed algorithms, uninformed search algorithms perform rather poorly. The reason is that the search space is extremely large, and the agent meanders around aimlessly in this large space, while trying to find a needle in a haystack. The uninformed search approach is also sensitive to the specific way in which the state-space graph is modeled and the actions for the agent have been designed in order to transition from one state to another in this space. There are multiple ways of performing this modeling for the same problem. For example, in the case of the eight-queens problem, it is possible to restrict the actions performed at a particular state (such as placing queens only on successive columns in successive steps), so as to reduce the size of the search space. This choice is rather ad hoc, and other ways of placing queens may be more expensive. Therefore, a large part of the efficiency may depend on ad hoc heuristics and the initial modeling used to define state spaces and corresponding actions in many problems. For larger problems, such heuristic choices may exhibit only limited improvements and the overall strategy of uninformed search may be too expensive to be considered useful. In informed search, an additional *domain-specific function* $l(i)$ is added to each node i, which plays a key role in guiding the direction of the search.

In many informed search algorithms, the path lengths from the starting state to the goal state also plays an important role. In many such problems, each path between a pair of nodes is associated with a cost, and the goal of the informed search process is to find a path of the smallest cost from the source to the goal node. The cost of a path is assumed to the sum of the costs on the individual edges, although more general path costs are possible. In such cases, the following two assumptions are made:

1. The agent is able to compute the best cost from the source node to any current node in the frontier list using only nodes that have been visited so far. For any particular node i in LIST, this value of this cost is denoted by $c(i)$. For now, we assume that the cost $c(i)$ is the lowest sum of the values of the costs d_{jk} associated with each individual edge (j, k) on the path from the starting node to node i. Note that the value of $c(i)$ is initialized to ∞ for all non-starting nodes, and 0 for the starting node. The value of $c(i)$ continually updated, as more nodes are visited by the algorithm. Therefore, $c(i)$ is only a *current* estimate of the best-cost path from the starting node to node i, because the corresponding path is constrained to pass only through nodes that have been visited so far. This is achieved by adding a single line to the portion

of the pseudocode in Figure 2.2 that scans the adjacent nodes of current node i as follows:

```
for all nodes j ∈ A(i) directly connected to i via transition do
begin
    if (j is not in VISIT) add j to LIST;
    pred(j) = i;
    c(j) = min{c(j), c(i) + d_{ij}}  [New addition of single line]
end
```

This current estimate $c(i)$ is obviously a pessimistic bound on the best cost of the path from the source to that state, because more nodes continue to be explored over the course of algorithm; the best cost may improve in cases where better paths emerge from the starting node to node i through the discovery of alternate routes via newly explored nodes.

2. The agent is able to compute an *estimate* on the best cost from any state in LIST to the goal state. This is an all-important heuristic (and domain-specific) quantity that plays an important role in guiding the agent's choices. For any particular node i, this value is denoted by $l(i)$. In general, the estimate could either be an overestimate or an underestimate on the best cost. However, for some algorithms such as A^*-search, it is important for $l(i)$ to be an underestimate (i.e., optimistic bound) in order for the algorithm to work correctly.

Since the costs d_{ij} are inherently nonnegative quantities, a natural assumption is that $c(i)$ and $l(i)$ are nonnegative for all states. This also means that the value of $l(i)$ is 0 at the goal state, when it is a lower bound or an estimate on the distance to the goal state. This is because we do not allow negative values of the lower bound $l(i)$, and the true cost to get from the goal state to itself is already known to be 0.

The quantity $l(i)$ *must* be used in any informed search algorithm in order to ensure that the algorithm does not meander around the state space graph discovering states that are irrelevant to the state at hand. Note that the quantity $c(i)$ provides no insight about the proper direction of search, at least from the perspective of finding goal states. It is important to note that the value of $l(i)$ is a heuristic one, and it is a way to incorporate domain knowledge about the problem at hand. It is normal for the value of $l(i)$ to be an underestimate on the distance to the goal state (because of particular types of desirable properties of underestimates as discussed later). The specific way in which the underestimate is computed depends on the problem at hand. For example, in the case of the routing problem, one might choose $l(i)$ to be a straight line distance from current state i to the goal state. In the case of the eight-puzzle problem, the value of $l(i)$ might simply represent an estimate of how far the state i is from the goal state by aggregating the "distances" of each tile from their ideal positions in the goal state. This can also be an underestimate, if the "distance" represents the minimum number of times that the tile needs to be moved to reach the goal state. *The goal of using the domain-specific function $l(i)$ is to minimize the number of expanded nodes in order to reach the goal state.*

The path cost $c(i)$ and the domain-specific cost $l(i)$ can be used in order to create an estimate $e(i)$ on the best cost from the source state to the goal state among all paths that pass through state i:

$$e(i) = F[c(i), l(i)]$$

The value of $e(i)$ is an increasing function $F[\cdot, \cdot]$ of $c(i)$ and $l(i)$. As we will see later, the A^*-algorithm uses the sum of $c(i)$ and $l(i)$. The two factors $c(i)$ and $l(i)$ respectively reduce the path cost (of the final path from source and goal nodes) and the exploration cost (of the number of fruitless nodes that need to be expanded on the way to the goal node). Weighting either of the two factors to a greater degree will find a different trade-off between path cost and node-exploration cost. Best-first search can be implemented with minor modifications to the uninformed search algorithm of Figure 2.2; the main refinement is that nodes from LIST are selected using $e(i) = F[c(i), l(i)]$. Therefore, the node selection strategy in this modified version of Figure 2.2 is as follows:

Choose the node i with the smallest value of $e(i) = F[c(i), l(i)]$ for further expansion in the algorithm of Figure 2.2.

In the event that $l(i)$ is not used in the function $F[\cdot, \cdot]$, the approach reduces to an uninformed search method. For example, consider a simple linear cost function that adds up the costs on the edges. For such linear cost functions, choosing the node from LIST with the smallest value of $c(i)$ in Figure 2.2 for further expansion amounts to a simple algorithm known as the *Dijkstra algorithm*, which is guaranteed to find a shortest path from the source to each node; however, it will find goal nodes later than many other nodes if they are located at large distances from the source nodes. This algorithm is also referred to as *uniform cost search*. Uniform cost search amounts to setting $l(i) = 0$ and selecting the node with the smallest value of $e(i) = c(i)$ from LIST in each iteration of Figure 2.2. However, uniform cost search is not seen as an informed search algorithm because it fails to use any domain knowledge of how far one is from the goal state while selecting nodes. As a result, uniform cost search is essentially blind to the proper direction of search while selecting the next node to expand, and it only uses the best known cost of traversals to currently visited nodes (and a single additional edge cost) in selecting subsequent nodes. However, it does not use any estimation of how far a candidate might be from the goal state in the selection process. This characteristic makes the uniform cost search algorithms require the processing of too many nodes before termination at a goal state (although it does have the merit of returning a shortest path).

When the function $F[c(i), l(i)]$ does use $l(i)$ in order to perform node selection, the goal state is reached quickly relative to other nodes, and the algorithm is referred to as *best-first search*. There are many alternatives available on the choice of the function $F[\cdot, \cdot]$. We explore two common variations of best-first search, which are greedy best-first search and A^*-search, respectively.

Comments on Termination Criteria

In some cases, informed search requires changes to the termination criteria. In Figure 2.2, the algorithm terminates when the first goal state is found. For the large part, informed search continues to use the same termination criterion. However, it is critical to note that the algorithm *might not* terminate with the shortest-cost path when the function $F[\cdot, \cdot]$ is not independent of $l(i)$. In order to ensure that the shortest-cost path is found, the values of $l(i)$ and $c(i)$ need to satisfy particular criteria. The most important criterion is that $l(i)$ and $c(i)$ must be nonnegative; additional criteria and their effects are discussed in a later section.

2.3.1 Greedy Best-First Search

In some applications, such as the eight-puzzle problem, one only cares about getting to the goal state (while exploring as few intermediate states as possible), and (unlike routing problems) the specific cost of the path between the starting state and the goal state is only a secondary concern. As a result, it does not make sense to use $c(i)$ in order to decide which node to expand first, because $c(i)$ represents the cost of the path required to reach state i from the starting state. In such a case, it makes sense to use only the value $l(i)$ in order to select which node to expand from LIST, because $l(i)$ represents an estimate on the cost of the best path to the goal state. Making this choice increases the likelihood that one will move closer to a goal state by selecting node i for further expansion. Therefore, the rule for greedy best-first search is as follows:

> Choose the node i with the smallest value of $l(i)$ for further expansion in the algorithm of Figure 2.2. In other words, we set $F[c(i), l(i)] = l(i)$.

The function $l(i)$ is typically hand-crafted with domain knowledge about the problem at hand, and it represents an estimate of the lowest cost path to the goal. For example, in the case of the eight-puzzle problem, one might set $l(i)$ to be the sum of the number of rows and the number of columns by which each tile is separated from its ideal position. Therefore, actions that move the tiles as close as possible to their ideal position (in the aggregate) are preferred by this approach. The use of greedy best-first search is not guaranteed to reach the goal state quickly. This is because best-first search uses only the effect of a single move from the currently explored position in order to examine the effect on the domain-specific function $l(i)$. In such cases, it is often possible for the search process to reach a dead end (and to have to backtrack to other nodes in LIST). However, choosing the nodes for expansion based on this criterion reduces the number of times that such a backtracking may need to be performed, because the general bias of the search is towards goal states. Therefore, best-first search typically reduces the time for exploration in comparison with an uninformed search strategy.

2.3.2 A^*-Search

The A^*-search algorithm is designed for cases where the cost of the path from the starting state to the goal state needs to be optimized, while also keeping the number of irrelevantly explored nodes in check while reaching the goal state. A key property of the A^*-algorithm is that even though it uses both $l(i)$ and $c(i)$ in the node selection process, the cost of the first path reaching the goal state is always the shortest one. The A^*-algorithm uses the value $e(i) = c(i) + l(i)$ for further expansion. We write this rule in a concrete way below:

> Choose the node i with the smallest value of $e(i) = c(i) + l(i)$ for further expansion in the algorithm of Figure 2.2. In other words, we set $F[c(i), l(i)] = c(i) + l(i)$.

Note that the value of $l(i)$ is always zero, when the node i is a goal node, and therefore the value of $e(i) = c(i) + l(i)$ is always the same as that of $c(i)$ for goal nodes. The A^*-algorithm can be shown to terminate at an optimal value of $c(i)$ when it reaches a goal node i under some conditions on $l(i)$, which are discussed later in this section.

A^*-search combines the principles underlying uniform cost search and greedy best-first search, but does not lose the ability to find shortest paths (like uniform cost search). Uniform cost search uses only $c(i)$, and greedy best-first search uses only $l(i)$. As a result, the former performs poorly in terms of the number of nodes explored, whereas the latter performs poorly in terms of path costs (when they are present). Therefore, A^*-search combines $c(i)$ and $l(i)$ in order to select the relevant node from LIST. It is also noteworthy that the appropriate choice of the search method heavily depends on the application at hand; it is possible for uniform cost search or greedy best-first search to be a more appropriate choice than the (more sophisticated) A^*-search for certain types of applications. For example, greedy best-first search is a more appealing choice for applications where path costs are not quite as relevant, and it is desirable to discover goal nodes as quickly as possible. A^*-search is capable of addressing generalized costs (beyond additive costs over edges), which satisfy particular rules of *cost algebra* [51]. This may be desirable in some applications.

The A^*-algorithm can be shown to terminate at an optimal path cost to the goal node under some reasonable assumptions on the nature of $l(i)$. As in the case of the class of all informed search algorithms, the values of $c(i)$ and $l(i)$ are always nonnegative, and the value of $l(i)$ is zero for goal states. Furthermore, two additional assumptions are made in this case:

- *Optimistic assumption:* The value $l(i)$ is a lower bound on the best cost of a path from the node i to the goal state, and is therefore an optimistic representation of the cost required to reach the goal state. This criterion is also referred to as *admissibility*.

- *Generalized triangle inequality:* The estimate $l(i)$ can be written as the pairwise estimated cost $L(s, i)$, where s is the starting state. The generalized triangle inequality is defined in terms of pairwise costs between any set of three states i, j, and k as follows:

$$L(i, k) \leq L(i, j) + L(j, k)$$

This condition is also referred to as *consistency* or *monotonicity*.

A consistent choice of $l(\cdot)$ is *always* admissible, whereas an admissible choice of $l(\cdot)$ is *often* consistent. We make the following assertions:

- If the cost-to-goal estimate $l(i)$ is admissible, then the algorithm of Figure 2.2 always reaches an optimal solution (with the lowest cost) to the goal state.

- If the cost estimate $c(i)$ is consistent, it can be shown that the algorithm of Figure 2.2 will select nodes from LIST in increasing order of $e(i)$. It is worthwhile to keep in mind that the value of $l(i)$ is always zero when i is a goal node, and therefore the nodes visited before the goal state will always have lower values of $c(i)$.

In the case of consistent costs, the algorithm selects nodes in increasing order of $c(i)$, and this property is also satisfied by the Dijkstra algorithm with nonnegative edge costs. In fact, the A^*-algorithm can be shown to be a generalized version of the Dijkstra algorithm.

An important point is that A^* search never expands any node for which the optimal cost from the source to that node is greater than the cost to the goal. This property is important in ensuring that the shortest path has already been found when the algorithm reaches the goal state for the first time (and there are no alternative paths of smaller cost through unexplored nodes). If $l(i)$ is hand-crafted in an insightful way, the A^*-algorithm will expand a relatively number of nodes in order to find a shortest path to the goal. In spite of its heuristic efficiency, A^*-search can show heuristic performance in the worst case, which is exponential in the depth of the search. Therefore, it is possible for the algorithm to be expensive in pathological cases. However, it often does quite well in practical settings, which is why this algorithm is so popular in classical artificial intelligence.

2.4 Local Search with State-Specific Loss Functions

The A^*-search algorithm is best suited to applications in which the cost of the route from the starting state to the goal is the primary objective in the search process. However, in many cases, one only needs to find the final state, without worrying (too much) about the path length — often, the final state has all the information encoded that we needed in the first place. Among the algorithms in the previous section, only greedy best-first search is agnostic to the path length required to reach a particular state. In some other cases, one may not have a crisp definition of a goal state, and one only wants to find a state with the least *loss* (i.e., objective function), which needs to be minimized. This can occur in settings, where the state space is too large to discover a specific goal state, a state least loss is good enough for the application at hand.

What are scenarios in which the final state matters rather than the path required to reach it? Consider the case of the eight-queens problem. In this case, one only cares about the specific positioning of the queens in the final state, rather than the number of specific steps required to reach that state. In the case of the eight-puzzle problem, although one would not prefer a solution with a large number of redundant transitions from the starting state, one might not distinguish too much between two solutions with slightly different lengths. This is the reason that loss values are often associated with states in these cases. Such loss values are analogous to the value of $l(i)$ used in the previous section, although there is no notion of costs and distances in the framework considered in this section.

Although it is possible to handle scenarios like the eight-queens or eight-puzzle problem with greedy best-first search, a richer class of problems can be solved with the use of local-search algorithms. In problems like the eight-queens or the eight-puzzle setting, there are one or more desired goal states, and heuristic loss values can be defined to measure closeness to these states. The loss functions are defined in such a way that the goal states have zero loss. In other applications, one does not have a notion of a goal state at all, but one attempts to simply find the state with the least possible loss. In the optimization version of NP-hard problems, each solution can be viewed as a state with an objective function that one is trying to optimize; however, one would not have a notion of a goal state. For example, in the case of the traveling salesperson problem one would be trying to find a cycle with the minimum cost. *This is extremely common for loss-centric settings, where one is trying to optimize an objective function rather than reach a goal.* In general, optimization is a more general problem that of reaching a goal, because most goal-centric problems can be converted to optimization-centric problems with the use of appropriately defined loss functions. The optimization-centric view is more general than the goal-centric view, because one can often model goal-centric problems in the optimization-centric setting by using the distance of a state from the goal as the relevant objective function to be minimized (like the heuristic value $l(i)$ used in greedy best-first search). The main difference in the settings considered in this section is that the objective function is a part of the problem definition in the optimization-centric setting, whereas it needs to be heuristically defined in the goal-centric setting. Associated with each state i, we have a loss value $L(i)$ which can be considered a heuristic loss over its value at the goal state. For example, in the case of the eight-puzzle problem, one might use the sum of the distances of the tiles from their ideal positions in order to define $L(i)$.

The optimal value of the loss $L(i)$ may or may not be known, depending on the application at hand. Some examples are as follows:

1. Consider the eight-puzzle problem in which the tiles need to be placed in their proper positions. A state is defined by any valid placement of the tiles. A transition is caused by movement of any of the tiles. The loss $L(i)$ of a state i is simply the sum of the distances of each tile from its ideal position. Therefore, the optimal loss value is 0 in this setting, when the tiles are placed in their correct positions; since the distance

of each tile from its ideal position is 0, the aggregated loss value is 0 as well. The optimal value of the loss is known in this case, because a specific goal state needs to be reached.

2. Consider the eight-queens problem discussed earlier in this chapter in which no pair of queens can attack one another. In this case, any placement of eight queens on the chessboard is a valid state. The loss of a state is defined as the number of pairs of queens that attack one another, based on that particular placement of queens on the chessboard. Since no pair of queens attack one another in the ideal position, the optimal loss value is 0. The optimal value of the loss is known in this case, because a specific goal state needs to be reached.

3. Consider a traveling salesperson problem in which we wish to find a tour of a set of cities of the least cost. In such a case, the cost of a solution is the cost of the path and the state-space graph contains edges between solutions obtained by interchanging a pair of cities. It is noteworthy that the optimal cost of a solution is not known, unless one exhaustively enumerates all possible tours (which may be intractable in practice).

In cases where the cost of the path to reach the final state is only of secondary importance (or is not defined at all), it makes sense to define state-specific loss values, so that the loss value of a state depends on how far it is from the goal state. These algorithms are referred to as *local*, because they only depend on the local characteristics of the states during the search, such as the state-specific losses in a particular locality. On the other hand, the search algorithms of the previous section are very focused on path-specific information in the search process, which requires a global view of the structure of the graph. Because there is no longer a need to maintain a global view of the structure of the graph, it becomes possible to use a larger family of algorithms in these cases.

The general principle in all of these local-search algorithms is very similar; one searches for neighboring states of a given state in order to improve the state loss with successive transitions. There are several algorithms that are designed for optimizing state-specific losses with the use of this type of local transition approach. The most popular ones among them are hill climbing, Tabu search, simulated annealing, and genetic algorithms. The first three of these methods are very similar, because they work with an individual solution, which is improved over time. On the other hand, genetic algorithms work with the biological paradigm, where a *population* of solutions is improved by allowing them to interact with one another. Nevertheless, all these methods can be considered different forms of local search. This section will introduce hill climbing, Tabu search, and simulated annealing, whereas the next section will introduce genetic algorithms.

2.4.1 Hill Climbing

Although the term hill-climbing naturally refers to maximization-centric objective functions, it is now universally applied to both maximization- and minimization-centric objective functions. After all, one can convert any minimization-centric objective function to a maximization-centric objective function by negating the objective function value. Therefore, we will continue to work with the consistent convention of minimization-centric objective functions, as used in other parts of this chapter. As a result, hill-climbing will really represent descent rather than ascent.

Hill climbing derives its name from the analogy of climbing a hill when searching for the top of the hill; the top of the hill signifies an optimal solution to an optimization problem (in the context of maximization-centric objective functions). Each step of this "climb" is an action made by the agent of moving to an adjacent solution in the state space graph. However, this action must always improve the objective function value. In a maximization problem, the objective function must increase because of a transition, whereas in a minimization problem, the loss must decrease. In other words, one always moves to an adjacent state so as to "improve" the objective function value, whether it is a maximization or a minimization problem. Since one always improves the objective function value with each agent action, one is guaranteed not to revisit a state. The step of improving the objective function value is repeated until all actions from the current state worsen the objective function value. Note that the algorithm is guaranteed to terminate in any problem with a finite number of states, as no state is ever repeated during hill climbing. The overall algorithm for hill-climbing is shown in Figure 2.7.

Algorithm *HillClimb*(Initial State: s, Loss function: $L(\cdot)$)
begin
  CURRENT= $\{s\}$;
  repeat
    Scan adjacent states to CURRENT until all states are scanned or a
      state NEXT is found with lower loss than CURRENT;
    if a state NEXT was found with lower loss than that of CURRENT
        set CURRENT=NEXT;
  until no improvement in previous iteration;
  return CURRENT;
end

Figure 2.7: The hill-climbing algorithm

As shown in Figure 2.7, the algorithm always maintains a current node, and it tries to improve the objective function continuously by moving to an adjacent state with a lower value of the objective function. For this purpose, the adjacent states are scanned, until a state is found with lower objective function value. This is the basic form of hill climbing. The simplest approach is to scan the adjacent states until the *first* state is found with a lower objective function value. In *steepest* hill climbing, one uses the adjacent state with the *best* improvement in objective function value. The specific implementation of hill climbing in Figure 2.7 uses the first state (rather than the state with best improvement) for the interchange, as long as it has a better objective function value. Using the first state often provides greater efficiency, since a small fraction of the states need to be scanned in order to improve the objective function value (although the number of steps are often greater). Such an approach can also be randomized by varying on the order of scanning states (if multiple executions of hill climbing are performed in order to select the best one). There are many variations of the basic hill climbing algorithm, depending on the precise details of how a state is chosen to be interchanged with the current state. However, in all cases, an adjacent state with better objective function is found for performing the transition. For a finite state-space graph, hill-climbing is always guaranteed to terminate. However, as discussed in Section 2.4.1.1, the termination state is not guaranteed to be optimal.

The definition of an adjacent state is based on the presence of an edge between two states in the state-space graph. Some examples of loss functions in various problem formulations are as follows:

- *Eight-queens problem:* The previous sections discussed a search-based method that places queens one by one on the board, while satisfying all constraints. Instead of

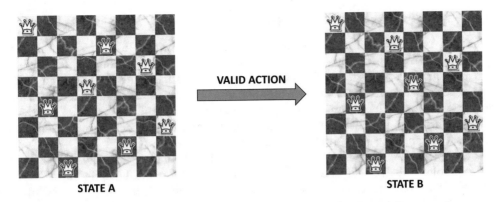

Figure 2.8: A valid transition in which the column positions of the queens on rows 2 and 4 are interchanged. The interchange reduces the number of diagonal attacks from two to one

placing queens one by one on the board, hill climbing defines a valid state as one with all eight queens on the board, including those that violate diagonal constraints. However, all row-wise and column-wise constraints are respected. The loss function of a state is the count of the number of pairs of queens that attack one another along a diagonal. Each action by the agent (i.e., transition) corresponds to the exchange of a pair of rows on the chess board. By "exchanging" rows i and j, we refer to the fact that the columnwise positions of the queens in row i and j are interchanged during a transition. In hill climbing, the goal is to move from one state to another until no constraint is violated. For example, Figure 2.8 shows a valid transition in which rows 2 and 4 are interchanged, thereby reducing the number of diagonal attacks. Although it might seem at first glance that this approach is a quick way to get to an optimum solution, the reality is that one can get stuck in *local optima*, where no move improves the loss function. This issue will be discussed in the next section.

- *Eight-puzzle problem:* In this case, each state is a position of the tiles, and each transition corresponds to the movement of a tile to an adjacent empty slot. The loss function is the sum of the "distances" of each tile from its ideal position. The distance of each tile from its ideal position is the sum of its row-wise and column-wise distances from its ideal position — this value also represents the minimum number of times the tile needs to be moved in order to get to its ideal position.

- *Traveling salesperson problem:* In the traveling salesperson problem, each state corresponds to a valid solution to the problem, which is an ordered sequence of cities (defining the tour of the salesperson). A transition to an adjacent state correspond to an interchange of any pair of cities. The loss function of a state corresponds to the cost of the tour representing that state.

It is possible to define states and transitions in a variety of ways, and this choice crucially affects the efficiency of the heuristic. The main problem with hill climbing is that it often tends to get stuck in local optima. When the hill-climbing approach is stuck in a local optimum, no transition from the current state can improve the loss function, although a global loss value has not yet been found.

2.4.1.1   The Problem of Local Optima

The crucial problem with hill climbing is that of local optima. In many cases, it is not possible to improve on the loss of a state by a single transition, although it is possible to do so with multiple transitions. For example, the state B in Figure 2.8 is not an optimal position because two queens attack one another. However, this state is also a local optimum. The reason is that no interchange between any pair of rows of state B can improve[4] the loss function. Therefore, hill climbing cannot find the global optimum in this case, even though one does exist multiple steps away from the current solution.

Local optima are especially common in NP-hard problems, such as the traveling salesperson problem. In fact, in almost all attempts to perform hill climbing with NP-hard problems, it is common to get stuck in local optima. After all, it is hard to find optimal solutions to such problems, and it would be too simplistic to expect a simple algorithm like hill climbing to find an optimal solution. One can view local optima in a similar way to how one views dead ends in search — the main difference is that search provides a way to perform backtracking with the use of a frontier list, and therefore one can try to find another path to the goal state. On the other hand, a local optimum is a final solution in hill climbing, where there is no option to perform backtracking. For example, if one were to examine the decision version of the eight queens problem in which we want to find a state with zero loss (i.e., no queens attacking one another), a local optimum provides a dead end containing a position where two queens attack one another (cf. Figure 2.8), and one cannot reach the desired goal with a single transition. In this sense, pure hill climbing may find a poorer solution than a backtracking search-based method if an infinite amount of time were available for the search process (although hill climbing arrives at a local optimum rather quickly). In such cases, the use of greedy best-first search makes a lot of sense, when compared to pure hill climbing as an alternative. One can view greedy best-first search as an enhancement of steepest hill climbing, in which the frontier list provides alternative routes of exploration when one reaches a local optimum. The inability to perform backtracking after making an incorrect choice is one of the important weaknesses of hill climbing, which causes it to get stuck in local optima. On the other hand, hill climbing can often find solutions of reasonably good quality in a modest amount of time. For a state-space graph with n nodes and maximum degree d, a hill climbing algorithm can be shown to terminate in at most $O(nd)$ evaluations of the loss function. At any particular state, the loss of at most $O(d)$ states needs to be evaluated. Furthermore, a maximum of $O(n)$ such evaluations need to be performed, since no state can be revisited. As a result, the overall time amounts to a maximum of $O(nd)$ evaluations. In practice, the number of evaluations is much smaller, since the longest path length $m \ll n$ in a graph is orders of magnitude smaller than the number of states n. The hill-climbing algorithm never revisits a state and would therefore need to traverse a number of states defined by the longest path in the graph. In other words, the true time complexity of the hill-climbing algorithm is $O(md)$ rather than $O(nd)$.

In spite of the efficiency of hill climbing, the ability to perform backtracking provides methods like greedy best-first search an advantage over hill climbing. Even simple problems like the eight-puzzle problem can frequently get stuck in local optima with hill-climbing methods. It turns out that it is indeed possible to create an algorithmic alternative to backtracking in the context of the hill-climbing family of solutions. A deterministic approach to achieve this goal is referred to as *Tabu search*, and a probabilistic approach is referred to as *simulated annealing*. In Tabu search, one is allowed to get out of local optima by choos-

---

[4]The reader can try the 13 possible interchanges that only involve one of the two violating queens in state B. None of these interchanges results in an improvement of the loss function.

Algorithm *TabuSearch*(Initial State: s, Loss function: $L(\cdot)$)
begin
  CURRENT= $\{s\}$;
  VISIT= $\{\}$; { VISIT is the Tabu [taboo] list }
  repeat
    Add CURRENT to VISIT;
    if all neighbors of CURRENT are in VISIT **then** set RANDOM= ϕ;
      else set RANDOM to a random neighbor of CURRENT that is not in VISIT;
    Scan adjacent states to CURRENT until all states are scanned or a
      state NEXT $\notin$ VISIT is found with lower loss than CURRENT;
    if a state NEXT$\notin$ VISIT was found with lower loss than that of CURRENT
        then set CURRENT=NEXT **else if** RANDOM$\neq \phi$ set CURRENT=RANDOM;
  until RANDOM= ϕ;
  return CURRENT;
end

Figure 2.9: Tabu Search

ing poorer solutions, but solutions visited earlier are not allowed to be visited again. The simulated annealing approach draws analogies from physics, and allows the loss function to worsen with some probability. This probability reduces over time, as the algorithm progress. Although most of these local search methods do not find a global optimum in real-world problems, they do improve the quality of the solution sufficiently that they become more useful in the context of practical applications.

2.4.2 Tabu Search

The Tabu search algorithm is quite similar to hill climbing, except that it allows solutions to worsen when they reach local optima, or are "almost" local optima. A state is considered to be an "almost" local optimum, if the only states that improve the objective function were visited earlier. States visited earlier might have better objective functions, since Tabu search does not monotonically improve the loss function (as in hill climbing). These states are the "taboo" states, which is how this method derives its name. In such cases, Tabu search selects a random state other than one of the taboo states, even though selecting such a state might worsen the objective function value.

The set of taboo states is contained in the hash table VISIT, and it grows over time, as more states are visited. The hash table VISIT can be considered similar to the variable VISIT in other generic search algorithms like that in Figure 2.2. In most practical applications of Tabu search, this list continually grows with time, and will therefore need to be pruned when the list size becomes unmanageable. We will first discuss a simplified version of the algorithm, and then we will discuss the modifications used in standard Tabu search to handle these types of practical challenges.

The simplified version of the Tabu search algorithm is illustrated in Figure 2.9. We have used a very similar pseudocode structure as the hill-climbing algorithm of Figure 2.7 in order to make the similarities and differences between these methods very clear. Unlike the hill-climbing algorithm of Figure 2.7, a hash-table VISIT is used to keep track of previously visited states. Every time a state is visited, it is added to VISIT, which ensures[5] that this state will not be visited again. At any given node, the algorithm checks whether any

---

[5]This is true only when nodes are not deleted from VISIT. In practice, nodes are almost always deleted from VISIT.

unvisited node can improve the objective function value. If such a node exists, then the algorithm moves to that node. Otherwise, the algorithm checks whether the current node has at least one unvisited neighbor. In such a case, the algorithm selects a random unvisited node for the next move. If no unvisited node exists in the neighborhood of the current node, the algorithm terminates since the random node is automatically set to ϕ. The current node is returned by the algorithm. It is possible for this node to be a local optimum, but it is almost always much better than the result of a hill-climbing algorithm.

The algorithm of Figure 2.9 is a grossly simplified version of Tabu search. It is not optimized for either computational or storage efficiency. In practice, there are two modifications that are commonly used in order to achieve better performance:

- Instead of selecting any neighbor that can improve the loss function, the best possible unvisited neighbor is commonly used, as defined by the loss function. This will lead to quicker improvements in each step, but the implementation of each step is slower because of the need to evaluate all unvisited neighbors.

- Instead of allowing the hashtable VISIT to grow in an unchecked manner, a bound on the number of nodes in hashtable is maintained both for computational and space efficiency. When this bound is reached, the node that was added to the hashtable the earliest is deleted. The node that was visited the earliest is deleted, because it is more likely to be suboptimal in terms of its loss value (because the loss value generally improves over time). This also makes it less likely to be visited again. Note that this type of approach requires some bookkeeping on the order in which nodes are visited and added to VISIT.

- The use of a hash table VISIT that allows deletions is akin to the use of a short-term memory. Tabu search also allows various ways of keeping intermediate and long-term memories. This is achieved with the use of rule-based methods. We refer the reader to [65] for a discussion of such techniques.

Tabu search was proposed by Fred Glover in 1986. In general, there are many variations of Tabu search, depending on the specific choice of hill climbing or "taboo" methodology. Tabu search is not guaranteed to reach a global optimum, and pathological examples can be constructed that get trapped in local optima. This can occur when all neighbors of a locally optimal node have already been visited, and therefore RANDOM is set to ϕ. A tutorial on Tabu search may be found in [65].

2.4.3 Simulated Annealing

Simulated annealing [101] can be viewed as a stochastic version of hill climbing, in which the moves from one state to another are probabilistic in nature. This probabilistic approach allows a non-zero probability of allowing the loss value of the current state to worsen; paradoxically, this apparently suboptimal approach of (occasionally) worsening loss values in the short term allows simulated annealing to avoid local optima. After all, there is no way to move out of a local optimum, if one is only allowed improve the loss value with a single transition — this is also the main reason that hill climbing gets stuck in local optima. Simulated annealing shares many conceptual similarities with Tabu search, in which poor moves are allowed, albeit in a probabilistic way.

Simulated annealing derives its name from the concept of annealing in metallurgy, wherein a metal is heated to a high temperature, and then allowed to cool down in order to reach a low-energy crystalline state. The heating to higher temperature is similar to

Algorithm *Simulated-Annealing* (Initial State: s, Loss function: $L(\cdot)$)
begin
  CURRENT= $\{s\}$; BEST= $\{s\}$;
  Set initial value of $T = T_0$ in the same magnitude as the
    typical differences between adjacent losses;
  $t = 1$;
  repeat
    Set NEXT to randomly chosen adjacent state referred to as CURRENT;
    $\Delta L = L(\text{NEXT}) - L(\text{CURRENT})$;
    Set CURRENT=NEXT with probability $\min\{1, \exp(-\Delta L/T)\}$;
    if $L(\text{CURRENT}) < L(\text{BEST})$ set BEST=CURRENT;
    $t \Leftarrow t + 1$; $T \Leftarrow T_0/\log(t + 1)$;
  until no improvement in BEST for N iterations;
  return BEST;
end

Figure 2.10: The simulated annealing algorithm

allowing the solutions to move in the wrong direction with some probability. This probability is high in the initial phases in order to enable the solution to move out of local ruts, but it reduces with passage of time as the solution generally moves to "better" regions of the state space. This basic principle in simulated annealing in making suboptimal choices in order to avoid local optima is used in all types of optimization, including continuous optimization methods[6] like gradient descent.

In order to understand the basic principle behind simulated annealing, think of a marble on a surface full of potholes. A marble will naturally settle in the pothole it is closest to. This pothole might not necessarily be the deepest pothole in the space of solutions, and therefore there is an incentive to sometimes allow the marble to come out of these types of potholes. If the marble is moving sufficiently fast initially, its initial velocity (akin to temperature) will allow it to come out of shallow potholes. But it might not be able to come out of deeper potholes. Over time, the marble will slow down because of friction, and will often settle in a pothole whose depth depends both on the initial velocity and the topography of the surface. In most cases, the marble will not settle in the shallowest potholes, although it might not settle in the deepest potholes either. This is achieved in simulated annealing by using a *temperature parameter* T that regulates the probability of moving in the wrong direction from a given state. As in the case of the marble, simulated annealing is not guaranteed to find a global optimum.

Simulated annealing uses a similar algorithmic framework as hill climbing of moving to adjacent states, although the solution is not guaranteed to improve from the current state when moving to an adjacent state. Let i be the current state, and j be a candidate state obtained by using a random action from the current state. Then, the loss difference is $\Delta L = L(j) - L(i)$. Note that if the value of ΔL is negative, the adjacent state provides an improvement in the loss value. Therefore, the probability of a move from state i to j at temperature T is defined as follows:

$$P(i, j, T) = \min\{1, \exp(-\Delta L/T)\}$$

When the move from i to j reduces the loss value, the probability of a move is 1. On the other

---

[6]In gradient descent, methods like the bold driver algorithm and the momentum method allow the temporary worsening of solutions. We refer the reader to [8] for a discussion of these methods.

Algorithm *GeneticAlgorithm*(Initial Population: $\mathcal{P}$, Loss function: $L(\cdot)$)
begin
  $t = 1$;
  repeat
    $\mathcal{P} = Select(\mathcal{P})$; { Select procedure uses loss function }
    $\mathcal{P} = Crossover(\mathcal{P})$; { A subset of solutions may remain unchanged }
    $\mathcal{P} = Mutate(\mathcal{P})$; { Uses mutation rate as a parameter }
    $t \Leftarrow t + 1$;
  until convergence condition is met;
  return $\mathcal{P}$;
end

Figure 2.11: The basic form of the genetic algorithms

hand, when the move from i to j increases the loss value, the probability of a move is less than 1. This probability is a decreasing function of the loss increase ΔL and an increasing function of the temperature T. For very large values of the temperature, the probability of all moves is nearly 1, and therefore any random move is accepted. With reduced temperatures, moves that worsen the loss have intermediate probability values, depending on how much the loss is worsened. Therefore, slight increases in loss are relatively frequent, although large increases in loss are rare. When the temperature approaches 0, increases in loss values are discouraged to the point that the approach becomes similar to hill climbing. The value of T is initially chosen to be of the same order of magnitude as typical values of ΔL. This specific approach for updating states is also referred to as the *Metropolis algorithm*, although many variations of this paradigm are available.

The overall approach of simulated annealing works as follows. The algorithm starts by setting to T to T_0. The value of T_0 is initialized by randomly sampling pairs of adjacent states and computing the difference in loss. The value of T_0 can be initially set to a small constant factor larger than this average difference in loss. In each iteration a random state that is adjacent to the current state is sampled. If this state improves the objective function value, then this change is accepted. Otherwise, a worsening rate is accepted based on the aforementioned annealing probability. Since transitions can worsen the quality of the state, the algorithm always keeps track of the best solution found so far. In each iteration, the temperature is lowered slightly by dividing the initial temperature by $\log(t + 1)$, where t is the index of the iteration. There is a significant body of literature on different ways in which the temperature can be lowered with increasing iteration index. The specific approach used for lowering the temperature is referred to as the *cooling schedule* of simulated annealing.

The simulated annealing algorithm is illustrated in Figure 2.10. It is also noteworthy that this particular implementation of the annealing algorithm is one of the many popular variations of the approach. Many alternative annealing schedules exist that provide different trade-offs, between being greedy in terms of preferring better solutions, and in avoiding local optima. In most cases, the trade-off is regulated with the use of a properly chosen cooling schedule. The stochastic approach of simulating annealing can be viewed as a stochastic alternative to the backtracking methods often used in search in order to get out of undesirable states.

2.5 Genetic Algorithms

The simulated annealing approach of the previous section borrows a paradigm from physics and metallurgy in order to perform search. It is common to borrow paradigms from the sciences in order to design algorithms in optimization and artificial intelligence. Genetic algorithms borrow their paradigm from the process of biological evolution [66, 85]. Biological evolution can be considered nature's ultimate optimization experiment, wherein living organisms evolve and become fitter over time through the difficult process that was referred to by Darwin as the *"survival of the fittest."* In genetic algorithms, each solution to an optimization problem can be viewed as an individual, and the fitness of the "individual" is equal to the objective function value of the corresponding solution. Although it is natural for objective functions in genetic algorithms to be defined as maximization problems (in line with the concept of "fitness"), it is equivalently possible to say that solutions with lower loss values are more fit. In this section, we will assume that the loss of state i is denoted by $L(i)$, and its inverse defines the fitness of the solution.

Both hill climbing and simulated annealing work with a single solution that is improved over the course of the algorithm. However, genetic algorithms work with a *population* of solutions, whose fitness is improved over time (as in biological evolution). This population of solutions is denoted by $\mathcal{P}$. The solutions (i.e., states) are encoded as strings, which are the analogs of chromosomes in biology. For example, in the traveling salesperson problem, a representative string could be a sequence of indices of the cities, which defines the tour of the traveling salesperson. The "fitness" of the string may be defined by the cost of the tour corresponding to that sequence of cities. As in biological evolution, the three processes of *selection, crossover,* and *mutation* are defined in order to improve the fitness of the population of solutions. These processes are defined as follows:

- *Selection:* In selection, the members of the population $\mathcal{P}$ are re-sampled in a biased way, so that the new population is more likely to contain a larger number of copies of solutions with better (lower) loss value. The process of selection is literally akin to concept of ensuring the survival of the fittest, because poor solutions tend to have fewer representatives after selection. There are several ways in which selection can be performed, some of which directly use the objective function value, and others use only the rank of the objective function value. For example, if $L(i)$ is the nonnegative loss function, one could create a roulette wheel for which the probability of state i is $1/(L(i) + a)$, where $a > 0$ is an algorithm-specific parameter. The roulette is spun $|\mathcal{P}|$ times, and all the members of the population that win in a roulette spin are then included. In rank-based selection, the roulette wheel is biased by the rank of the solution, wherein solutions with higher rank in order of fitness are preferred.

- *Crossover:* In crossover, pairs of solutions are selected, and their characteristics are recombined in order to create new solutions. Typically, the string representations of the two solutions are used in order to create a recombined solution. However, it is important to perform this process carefully in order to ensure that infeasible solutions are not created. For example, if one simply randomly samples from the pairs of aligned positions in two strings for the traveling salesperson problem, the sampled string will most likely contain repetitions of cities. Instead, one might treat the rank of the city as its position in the string; then, one can sample the rank of the city from each of the two solutions, and sort the cities using the sampled rank. It is noteworthy that the process of creating children solutions from parents is a heuristic one, and it provides domain experts the opportunity to encode their understanding of the problem into the crossover process. Two children solutions can be created from each pair of parent solutions, and these two children replace their parents in the population. The crossover process can cause the genetic algorithm to fail if it is not properly implemented. Therefore, it is important to use domain-specific knowledge in order to combine the parent solutions in a meaningful way. The crossover process often uses only a percentage of the population, so that a subset simply passes from one generation to the next.

- *Mutation:* The process of mutation is not very different from how neighbors are chosen in simulated annealing. As in simulated annealing, a neighboring state of the solution is chosen at random. However, the mutated solution always replaces its parent, whether its solution fitness is better or not. The responsibility of weeding out bad mutations is left to the selection process, as in biological evolution. The fraction of solutions that is mutated in each generation is controlled by the mutation rate. In some variations, the mutation rate is initially kept high, but it reduces with time. This type of approach is reminiscent of simulated annealing.

Genetic algorithms proceed by a cyclic process of selection, crossover, and mutation. Each such cycle is referred to as a *generation*. Because of the process of selection, which creates multiple copies of fitter solutions, the individuals in the population tend to become more similar over time. De Jong defined convergence of the genetic algorithm when 95% of the positions in the population became identical [46]; in other words, for any randomly sampled pair of strings in the population, their corresponding positions match 95% of the time. However, other termination criteria are available, such as the use of a fixed number of generations, or the loss of the solutions reaching particular convergence properties (wherein the average loss value of the population does not improve significantly over time). The basic pseudocode for a genetic algorithm is illustrated in Figure 2.11.

It is noteworthy that there is some disagreement in the research community as to how well crossover really works. One problem is that crossover often results in infeasible or suboptimal solutions, unless it is carefully controlled in a domain-specific way. On the other hand, it is much easier to control mutations, because it is usually not difficult to define neighboring states in most problem domains. Indeed, even if crossover is dropped, genetic algorithms tend to work quite well. This is not particularly surprising, given the similarity of the resulting mutation-based approach to simulated annealing. After all, mutations can be seen a form of stochastic neighborhood search, when combined with the bias of the selection mechanism. However, with a careful design of the crossover process, improvements can often be obtained over pure mutation-based algorithms. The main point is that crossover must be able to exchange useful portions of the solution to create children solutions. Therefore, some methods use optimized crossover mechanisms in order to recombine important portions of the solutions and create a solution with an even better objective function value [9]. In this sense, it is important to note that blind variations of genetic algorithms that arbitrarily code solutions as strings in order to implement off-the-shelf genetic algorithms do not always work well.

2.6   The Constraint Satisfaction Problem

The constraint satisfaction problem is one of the fundamental problems in artificial intelligence, wherein a configuration of a system must be found, which satisfies a pre-defined set of variables. Many puzzles in artificial intelligence can be modeled as constraint satisfaction problems. Mathematically, the constrained satisfaction problem is modeled over a set of variables, each of which takes on values from a particular domain. A set of constraints is defined over these variables, and the goal of the constraint satisfaction problem is to find an assignment of values to the variables, so that each constraint is satisfied. The canonical form of the constraint satisfaction problem is that of Boolean satisfiability in which all variables are drawn from the Boolean value of *True* or *False*. The constraint satisfaction problem is a rather general formulation, and most NP-complete problems can be represented as constraint satisfaction problems. This is not particularly surprising because most NP-complete problems can be formulated by defining constraints over integer variables and then finding a feasible solution to these constraints — this general area of mathematics and optimization is referred to as *integer programming*.

Formally, a constraint satisfaction problem is denoted over a set of variables $x_1 \ldots x_d$. The variable x_i is drawn from the domain $\mathcal{D}_i$, which is typically a set of discrete values. In addition, we have a set of constraints $\mathcal{C}_1 \ldots \mathcal{C}_n$. The goal of the constraint satisfaction problem is to find assignments for the variables $x_1 \ldots x_d$, so that all constraints are satisfied. Many of the problems discussed earlier in this chapter, such as the eight-queens problem and the eight-puzzle problem can be formulate as constraint satisfaction problems over integer variables.

2.6.1 Traveling Salesperson Problem as Constraint Satisfaction

The traveling salesperson problem discussed earlier in this chapter can be expressed as constraint satisfaction. In this case, we have a set of cities denoted by $\{1, \ldots, n\}$, and there are $(n(n-1))$ variables $z_{ij} \in \{0, 1\}$ corresponding to whether or not the traveling salesperson travels from city i to city j. Therefore, the domain of each variable is drawn from $\{0, 1\}$, which corresponds to a binary integer program. The cost of traveling from city i to city j is denoted by d_{ij}. The goal is to find a tour of the cities of cost at most C. Then, the constraint satisfaction problem needs to find a solution satisfying the following constraints:

$$\sum_{j=1}^{n} z_{ij} = 1 \quad \text{[Exit each city exactly once]}$$

$$\sum_{i=1}^{n} z_{ij} = 1 \quad \text{[Enter each city exactly once]}$$

$$\sum_{i=1}^{n} \sum_{j=1}^{n} z_{ij} d_{ij} \leq C \quad \text{[Cost of tour is at most C]}$$

This problem is a *binary* integer program, because each variable is drawn from $\{0, 1\}$. A feasible solution to this integer program is a valid tour of cost at most C.

2.6.2 Graph Coloring as Constraint Satisfaction

The graph coloring problem works with a graph $G = (V, E)$ with n nodes in V, and undirected edges in E that connect nodes in V. Each node in V needs to be assigned with at one of q colors, so that the colors of nodes i and j at the opposite ends of edges in E are different. Then, it is NP-complete to determine whether a coloring exists with at most q colors.

Next, we define the variables and the constraints corresponding to the graph coloring problem. First, we define $n \times q$ variables z_{ij}, so that i varies from 1 through n, and j varies

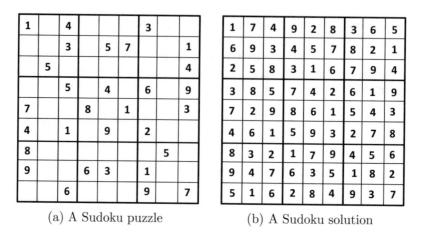

(a) A Sudoku puzzle   (b) A Sudoku solution

Figure 2.12: A Sudoku puzzle and its solution

from 1 through q. The domain of each variable z_{ij} is $\{0, 1\}$. The value of z_{ij} is 1 if node i is assigned color j. Then, the constraint satisfaction problem may be defined as follows:

$$\sum_{k=1}^{q} z_{ik} = 1 \quad \text{[Color each node exactly once]}$$

$$z_{ik} + z_{jk} \leq 1 \quad \forall (i,j) \in E, \ \forall k \in \{1 \ldots q\} \quad \text{[No two adjacent nodes have the same color]}$$

The graph coloring problem is closely related to the *map coloring problem*, wherein we wish to color the regions (e.g., provinces or states) in a map, so that each pair of adjacent regions has different colors. The map coloring problem has practical applications in cartography, because it helps visually distinguish the different regions in the map. If you examine a colored map of the United States, you will notice that all pairs of adjacent states have different colors. One can convert a map coloring problem to a graph coloring problem by representing each region with a node, and each neighborhood relationship between regions with an edge between the corresponding nodes. The map coloring problem is simpler than the graph coloring problem, because the underlying graph is planar. A coloring with $q = 4$ colors always exists (and it can be found in polynomial time). This result is referred to as the *four color theorem*.

2.6.3   Sudoku as Constraint Satisfaction

Sudoku is a well-known puzzle, which creates an incompletely specified grid of numbers in an 9×9 array. Each number in the grid should be a value from 1 to 9. A subset of the numbers in the grid is specified, whereas other numbers are missing. The goal of the problem is to fill in the missing numbers, so that the following constraints are satisfied:

- Each row contains a single occurrence of each of the numbers 1 through 9.

- Each column contains a single occurrence of each of the numbers 1 through 9.

- The 9×9 grid is divided into 3×3 grids, each of which is of size 3×3. Each of these 9 grids must contain a single occurrence of each of the numbers 1 through 9.

It is possible for an arbitrary initial assignment of numbers to not have a solution at all. However, a *well-posed* problem of Sudoku will always have a grid of values that can be filled in. An example of a well-posed Sudoku puzzle and its solution are provided in Figures 2.12(a) and (b), respectively. The Sudoku puzzle can be modeled as a constraint satisfaction problem by defining z_{ijk} to be 1 if the (i, j)th entry takes on the integer value of k. The value of k ranges from 1 through 9. Otherwise, the value of z_{ijk} is 0. Then, the Sudoku problem can be defined as a constraint satisfaction problem as follows:

$$\sum_{k=1}^{9} z_{ijk} = 1 \quad \forall i, j \quad \text{[Each square takes on one value]}$$

$$\sum_{i=1}^{9} z_{ijk} = 1 \quad \forall j, k \quad \text{[Each column/digit pair occurs once]}$$

$$\sum_{j=1}^{9} z_{ijk} = 1 \quad \forall i, k \quad \text{[Each row/digit pair occurs once]}$$

$$\sum_{i=1}^{3} \sum_{j=1}^{3} z_{i+r,j+s,k} = 1 \ \forall r, s \in \{0, 3, 6\}, \forall k \quad \text{[Each digit occurs once in a } 3 \times 3 \text{ subgrid]}$$

One can also define the Sudoku problem for larger grids, and it becomes increasingly difficult to solve with increasing grid size. For example, solving the Sudoku problem with subgrid size $n \times n$ and grid size $n^2 \times n^2$ is known to be NP-complete, when the integers used to fill the Sudoku grid are drawn from $\{1, \ldots, n\}$. This is because the number of ways in which the grid can be filled increases exponentially with increasing size of the grid. There is no known systematic way of filling the grid, so that a solution can be found in polynomial time.

2.6.4   Search Algorithms for Constraint Satisfaction

In the following, we will introduce a number of search algorithms for constraint satisfaction. We will use the notations discussed at the beginning of this section, in which we have d variables $x_1 \ldots x_d$ and n constraints $\mathcal{C}_1 \ldots \mathcal{C}_n$. A variety of search algorithms can be used for constraint satisfaction, depending on how one chooses to define the states.

A natural way to perform search is by successive instantiation of variables. In this case, each state corresponds to an instantiation of each variable x_i to a value from $\mathcal{D}_i \cup \{*\}$, where '*' indicates a "don't care." A value of "don't care" indicates that the corresponding variable has not yet been instantiated. States are defined only for cases where the instantiated variables satisfy all the constraints that apply to them. Note that some of the constraints may not need to be applied, if they contain variables corresponding to "don't care" values. The starting state sets each variable to the "don't care" value. In the eight-queens problem, this initial state corresponds to the case in which no queens have been placed on the board.

The resulting state-space graph is a directed acyclic graph, which connects only pairs of nodes with an additional instantiated variable. It is noteworthy that one can reach the same state after instantiating the variables in any order. This is because multiple paths exist to a particular state from the starting state. For example, in the eight-queens problem, one can place the eight queens in any random order to reach the same state. However, the choice of which variable is instantiated first has an effect on the efficiency of the search. In general, if one chooses a search path that is not fruitful, it is helpful to find out about it as early as possible. Therefore, it makes sense to select the variable with the smallest possible branch factor (i.e., valid values) first in order to instantiate. This ensures that one is less likely to make an incorrect decision, since one has fewer choices to make, and one would be more likely to make a correct choice on from a probabilistic point of view. For example, if one places the queens on the board one column after another, it would make sense to select the column with the smallest number of remaining legal positions first in order to instantiate the position of the next queen. However, while selecting the value to instantiate for this variable, it makes sense to select the value that yields the largest possible branch factor in the next step. This ensures that later steps have more options available, and it reduces the chances of being forced to backtrack from a dead-end state. This type of application of search algorithms to constraint satisfaction is also referred to as *backtracking search*. As we will see later, these types of search algorithm also play a key role in techniques associated with propositional and first-order logic. Therefore, some of these methods will be revisited in Chapters 4 and 5.

2.6.5 Leveraging State-Specific Loss Values

It is also possible to perform search for solutions to the constraint satisfaction problem with the use of state-specific loss values. In this case, one works with a possibly inconsistent but complete assignment of values to variables, and gradually reduces the inconsistency over time with the use of a loss function. In the case of the eight-queens problem, one might start with all eight queens on the board, but in some inconsistent state. The loss may be defined in terms of the number of pairs of queens that violate a constraint. Subsequently, each transition allows the movement of a queen, and the goal is to make transitions in order to reduce the loss. There are multiple ways in which one can define states and transitions, and the specific effectiveness of a particular method depends heavily on this choice. An example of a different way of defining the states and performing the transitions in the eight-queens problem is discussed in Section 2.4.1.

Once the states and transitions have been defined, there are several ways in which one can search the state space. One possibility is to use best-first search with the use of the loss function. In pure hill climbing, one will make transitions only to reduce the objective function value, but this will often lead to local optima. Therefore, one can use other methods, such as simulated annealing or genetic algorithms in order to search for better solutions. The choice of the loss function is highly application-specific, and it depends on the domain at hand. In general, the constraint satisfaction problem provides the template for most of the difficult combinatorial problems, which are NP-complete, and have often been tackled with the use of search methods.

2.7   Summary

Search algorithms were among the earliest methods used in artificial intelligence, since most problems in the domain can be viewed in terms of graph-structured state spaces. The simplest forms of search are uninformed search, such as depth-first and breadth-first search. These algorithms, however, usually take a long time to find a goal state because they tend to meander around aimlessly in large search spaces. Therefore, the search process can be improved with the use of state-specific loss values, which provide the search algorithm with useful hints about good directions of search. Many algorithms, such as best-first search, hill climbing, Tabu search, simulated annealing, and genetic algorithms, can be viewed as variations of this broad approach. These methods are also closely related to one another. Numerous applications of search algorithms are discussed in this chapter, such as the eight-queens problem, solving mazes, and constraint satisfaction. This chapter discusses the case of single-agent search. The next chapter discusses multi-agent search, which is relevant for applications like game playing.

2.8 Further Reading

Basic algorithms for search can be found in algorithm textbooks [40] as well as in network-flow algorithm books [12]. Numerous algorithms for search can also be found in the classical artificial intelligence book by Russell and Norvig [153]. A tutorial on Tabu search methods may be found in [65]. An excellent summary of simulated annealing may be found in the overview paper by Bertsimas and Tsitsiklis [19]. The classical book for genetic algorithms is Goldberg's book [66]. The extension of genetic algorithms to genetic programming may be found in the book by Koza [106].

2.9 Exercises

1. The algorithm in the text proposes a non-recursive algorithm for search. Propose a recursive algorithm for depth-first search.

2. Propose a recursive algorithm for a modification of depth-first search that imposes a limit d on the depth of the search. In other words, the algorithm always backtracks when its current path from the starting state has d edges. How would you achieve this by modifying the non-recursive algorithm in Figure 2.2?

3. Modify best-first search in order to find the best *alternative* path from the start state to the goal state using at most d edges.

4. Provide the order of traversal of nodes when performing both breadth-first and depth-first search in Figure 2.5(b). Perform tie-breaking in lexicographic order.

5. Express the eight-puzzle problem as a constraint satisfaction problem.

6. Suppose that the Boolean satisfiability problem is solved using genetic algorithms, in which the values of the variables are encoded as a string and the fitness is equal to the Boolean value of the resulting expression. Discuss why a straightforward use of genetic algorithms is unlikely to work very well.

7. Suppose that you modify best-first search so as to restrict the number of nodes added to LIST in each iteration (cf. Figure 2.2). Discuss how restricting the number of nodes added to LIST in each iteration along with a best-first criterion provides a continuum between hill climbing and best-first search.

8. Create a variation of best-first search in which the size of LIST is restricted by adding only the top-r successor nodes (caused by a transition from the current node) to LIST. Discuss the trade-offs associated with such a modification.

9. Consider a graph a containing n nodes in which $s < n$ nodes have been explored in the forward direction and backwards direction by bidirectional search (Figure 2.6). The

graph contains a single starting state and a single goal state. What is the probability that forward search contain the single goal state? What is the probability that the forward and backwards search have at least one node in common, assuming that forward and backward search each reach a random subset of s nodes. Use this analysis to discuss why bidirectional search is likely to be more efficient.

10. Write the pseudocode of Tabu search in which the algorithm always moves to the best possible state from each position. You can use the pseudocode of the book as a starting point.

Chapter 3

Multiagent Search

"Friendships born on the field of athletic strife are the real gold of competition. Awards become corroded, friends gather no dust."– Jesse Owens

3.1   Introduction

The previous chapter focuses on search involving a single agent. However, many real settings are associated with environments involving multiple agents. In multiagent environments, there are multiple agents interacting with the environment and affecting the state of the environment, which in turn affects the actions of all agents. Therefore, an agent cannot take actions without accounting for the effects of the actions of other agents. Since the actions of other agents cannot always be exactly predicted, a multi-agent environment works with partial knowledge; however, some information is available about the goals and utility functions of the other agents. Therefore, this type of setting is considered to be related to but distinct from uncertain or probabilistic environments.

The different agents might be mutually cooperative, independent, or they may be competitive (depending on the application at hand). Some examples of multi-agent environments are as follows:

- *Competitive agents:* Games like chess represent competitive environments in which the agents are in competition with one another. The actions of the agents cause transitions in the state of the environment (which correspond to the position of the pieces on the chessboard). Such environments are also referred to as *adversarial*. Adversarial agents do not communicate with one another beyond affecting the environment in an adversarial way.

- *Independent agents:* In a car-driving environment, one might have multiple car-driving agents, each of which is optimizing its own objective function of safely driving on the road and reaching its specific destination. However, the agents are not adversarial; in fact, the agents are partially cooperative because neither agent would like to have an

© Springer Nature Switzerland AG 2021
C. C. Aggarwal, *Artificial Intelligence*, https://doi.org/10.1007/978-3-030-72357-6_3

accident with the other. However, the cooperation is only implicit because each agent is optimizing its own objective function, and the cooperation is a side effect of their mutual benefit.

- *Cooperative agents:* In this case, the agents are communicating with one another, and they are optimizing a global objective function that defines a shared goal. For example, it is possible to envisage a game-playing environment in which a large number of agents perform a collaborative task and keep each other informed about their progress. In the event that the game is defined by two teams playing against one another, the setting has aspects of both competition and cooperation.

This chapter will propose methods that can work in these different types of environments. Some methods are able to work in all three types of environments, whereas other methods are designed only for adversarial environments. We will discuss both generic methods (which can work in all types of environments) as well as specific methods that are geared towards adversarial environments. The special focus on adversarial environments is caused by its importance in key artificial intelligence applications like game playing.

In many settings, the agents might alternate in their actions, although this is not always necessary. For example, in the case of games like chess, the agents usually alternate to make moves. However, in some environments, the agents do not alternate in their move choices. Since much of the focus of this chapter is on adversarial environments, it will be assumed that the moves of the agents alternate. Furthermore, we will specially focus on environments containing two agents, although many of the ideas can be generalized to environments containing multiple agents.

Multiagent environments necessitate different types of search settings as compared to single-agent environments. One issue is that a given agent has no control over the choices made by another agent, and therefore it might have to *plan for all possible contingencies* arising from the other agent's action. In other words, it is no longer sufficient to find a single path from the starting state to the goal in order to determine whether it is possible to achieve success with a particular action — rather, one must consider all possible choices that other agents might make while evaluating the long-term effect of a particular action. For example, while evaluating whether a particular move in chess will yield a win, one must consider all possible moves that the opponent might make (and not just a single one). Ideally, the agent must assume that the adversary will make the best possible move available to it. It is noteworthy that this setting is very similar to what happens in probabilistic environments where an action made by an agent has consequences that are stochastic in nature. Therefore, one might need to consider all possible stochastic outcomes while evaluating the effect of a particular action.

Multiagent environments are inherently online because one cannot make a decision about future actions of an agent without considering *all possible* choices made by the other agents (i.e., all possible contingencies). Since it is usually too expensive to enumerate all possibilities (i.e., all contingencies), one simply has to look ahead for a few transitions and then perform a heuristic evaluation in order to determine which action is the most appropriate one in the current state. Therefore, multiagent algorithms are mostly useful for only making choices about the next action in a particular state; subsequent actions require one to observe the actions of other agents before reevaluating how to make further choices. For example, a chess-playing agent can only make a decision about the next move, and there is considerable uncertainty about its next move without knowing what the adversary might play. It is only after the adversary makes the next move that the chess-playing agent may reevaluate the specific choice of move that it might make. This type of search is referred to as *contingency-centric search*.

One useful way of performing contingency-centric search is to use AND-OR trees. Such trees have OR nodes corresponding to the choices made by the agent making a particular decision (where the agent has free choice and can select any one), and they have AND nodes corresponding to the actions of the other agents (i.e., contingencies where the agent has no choice). For the case of competitive agents, these trees can be converted into utility-centric variants, which are referred to as minimax trees. Before discussing such types of adversarial search, we will address the more general issue of contingency planning, which occurs in situations beyond adversarial environments. Most of this chapter will, however, focus on competitive environments and game playing, which are adversarial environments. The focus on game-playing environments is because of their special importance as microcosms of artificial intelligence settings, where many real-world situations can be considered in a manner similar to games. While games are obviously far simpler (and more crisply defined) than real-life situations, they provide a testing ground for the basic principles that can be used for more complex and general scenarios.

This chapter is organized as follows. The next section introduces AND-OR trees for addressing contingencies, which occur quite frequently in the context of multiagent environments. The use of AND-OR search represents a scenario of uninformed search, which can take a long time in most realistic environments. Therefore, the use of informed search with state utilities is discussed in Section 3.3. This approach is then optimized with alpha-beta pruning in Section 3.4. Monte Carlo tree search is introduced in Section 3.5. A summary is given in Section 3.6.

3.2 Uninformed Search: AND-OR Search Trees

Single-agent environments correspond to search settings in which finding *any* path from the starting state to the goal state can be thought of as a success. In other words, one can choose *any* action at a given state that eventually leads to a goal node. Therefore, *all* nodes can be thought of as OR nodes, and finding a single path from the starting node to the goal is sufficient. This is not the case in multi-agent environments, where one cannot control the actions of other agents. One consequence of the multiagent setting is that finding a single path through the state-space graph is no longer sufficient to ensure that choosing a particular course of action will yield success. For example, if a particular move in chess leads to a win only because of the opponent making suboptimal moves, whereas another move leads to a guaranteed win, the first of the two moves may not be preferable. As a result, one has to account for the different choices (i.e., paths) that result from this uncertainty, while choosing a course of action in a particular state. In other words, one must work with a tree of choices for the paths from any particular state to the goal; the adversarial nodes are AND nodes, where all possible paths must be followed to account for the worst-case scenario where the opponent makes the choices that are most suitable to her. Therefore, unlike single-agent environments, a multi-agent environment contains both AND nodes and OR nodes.

In this section, we will primarily consider the case of the two-agent environment. In the two-agent environments, each node of the tree corresponds to the choices made by a particular agent, and alternate levels of the tree correspond to the two different agents. The decisions made by the agents are represented by a tree of choices (such as a tree of moves in chess). The placement of AND and OR nodes depends on the choice of the agent from the perspective of which a particular action is being performed. We refer to the two agents as the *primary agent* and the *secondary agent*, respectively. The root of the tree corresponds to the choices made by the primary agent, and it is always an OR node. The entire tree is being constructed from the perspective of the primary agent. At the levels of the tree, where the secondary agent is due to make a transition, the primary agent cannot predict what action the secondary agent might take, and therefore has to account for all possibilities of actions taken by the secondary agent; such a node is, therefore, an AND node since the tree is constructed from the perspective of the primary agent. The use of the AND node ensures that one explores *all* possibilities for the actions chosen by the secondary agent. This leads to AND-OR search, in which the OR nodes correspond to the primary agent (from whose perspective the problem is being solved), and the AND nodes correspond to the choices made by the secondary agent(s). For simplicity, we first consider a two-agent problem in which the goal is defined from the perspective of (primary) agent A, whereas (secondary) agent B might make transitions that agent A cannot control. Furthermore, agent A and agent B make alternate transitions. Although this model might seem simplistic, it is the model for large classes of two-person-game settings. can capture more powerful settings, such as multi-agent settings, by relatively minor modifications (cf. Section 3.2.1). In the AND-OR tree, alternate actions are made by agent A and agent B using nodes at alternate levels. Therefore, directed edges exist only between the nodes corresponding to the transitions of the two agents. We assume that the state-space graph is structured as a tree, which is a simplification of the assumptions made in the previous chapter. One consequence of this simplification is that the same state may be repeated many times in the tree, whereas each state is generally represented only once in the state-space graphs of the previous chapter. For example, the same position in chess may be reached by using different sequences of moves, and this is especially evident during the opening portion of the game, where alternate move sequences to reach the same opening position are referred to as *transpositions*. However, each of these states is a different node in the tree, and therefore a given state may be represented multiple times in the tree. It is also possible to collapse this tree into a state-space graph in which each state is represented exactly once (see Exercise 9), and doing so can lead to a more efficient representation, which is typically not a tree. The simplification of using a tree structure assures that one does not have to keep track of previously visited nodes with a separate data-structure. In practice, keeping such a data structure can lead to considerable computational savings, although it leads to increased algorithmic complexity and loss of simplicity in exposition.

The starting state s is the root of the tree corresponding to the state-space graph being examined. The initial state for the problem is denoted by s, in which agent A takes the first action. The goal state is assumed to be reached when the condition $\mathcal{G}$ is assumed to be *True* from the perspective of agent A. The overall procedure is shown in Figure 3.1, and it is a modification of the procedure used in Figure 2.2 of Chapter 2. This algorithm is structured as a recursive algorithm and its output is the Boolean value of *True* or *False*, depending on whether or not any goal node (defined by constraint set $\mathcal{G}$) is reachable from the starting node s. Note that returning a Boolean value is sufficient in order to determine which course of action to take at the current node, because one can always print out the branch that was followed at the top level of the recursion in order to determine what action agent A should make. While it is possible to identify the relevant branch at the top level of the tree, it is no longer possible to report the full path of actions from the starting state to the goal node because of the inability to predict the actions of the secondary agent B. After all, all AND nodes need to be explored, even though OR nodes allow the flexibility of following only a single path. In general, one can only predict the next action from the current node, and the approach is typically used in online environments (as the secondary agent takes actions and subsequent states become known).

Algorithm *AO-Search*(Initial State: s, Goal Condition: $\mathcal{G}$)
begin
  $i = s$;
  if i satisfies $\mathcal{G}$ **then return** *True*;
      else initialize resultA= *False*;
  for all nodes $j \in A(i)$ reachable via agent A action from i **do**
  begin
   resultB= *True*;
   if (j satisfies $\mathcal{G}$) **then return** *True*;
   for all nodes $k \in A(j)$ reachable via agent B action from j **do**
   begin
    resultB= (resultB AND (*AO-Search*(k, $\mathcal{G}$)));
    if (resultB = *False*) **then** break from inner loop;
   end;
   resultA = (resultA OR resultB);
   if (resultA= *True*) **then return** *True*;
  end;
  return *False*;
end

Figure 3.1: The AND-OR search algorithm

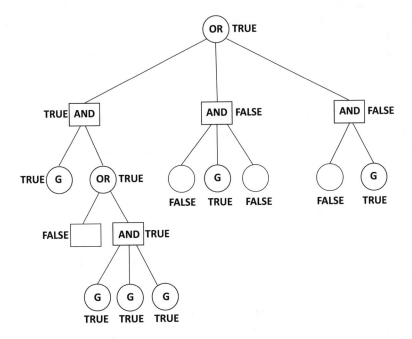

Figure 3.2: The AND-OR tree for multiagent search

A single recursive call of the AND-OR search algorithm processes two levels of the tree, the first of which corresponds to the actions of agent A and the second corresponds to the actions of agent B. Therefore, two levels are processed in each recursive call. The algorithm starts at node s and checks whether the starting node satisfies the goal conditions in $\mathcal{G}$. If this is indeed the case, the algorithm returns the Boolean value of *True*. Otherwise, the algorithm initiates a recursive call for each node two levels below the current node. This is achieved by first retrieving all states $j \in A(i)$ that are accessible via a single action from current state i, and then calling the procedure from each child of j. Note that each such child of i is also one in which the primary agent is due to make an action. The set S of grandchildren of node i is defined as follows:

$$S = \cup_{j \in A(i)} A(j)$$

The recursive call is initiated for each such node $k \in S$, unless the goal condition $\mathcal{G}$ is satisfied by a recursive call from one of these nodes. The recursive call returns either *True* or *False*, depending on whether or not the goal is reachable from that node using the AND-OR structure of recursive search. The results of all the recursive calls at the nodes two levels below are then consolidated into a single Boolean value of *True* if for even one of the actions of agent A from the current node, all actions for the agent B at the child node return a value of *True*. In other words, the OR condition is applied at the current node, and the AND condition is applied at the nodes one level below. Therefore, the two Boolean operators are applied at alternate levels of the tree. The recursive description of Figure 3.1 only returns a single Boolean value, and it can also be used to decide the action at the top level of the tree. In particular, if the Boolean value of *True* is returned by the algorithm, the top level of any OR branch that yields the value of *True* can be reported as the action choice. Note that subsequent choices of actions of either agent at deeper levels of the tree are not reported; this is because the choices made by the secondary agent are not predictable and can sometimes be suboptimal.

An example of an AND-OR tree is shown in Figure 3.2. The goal nodes are marked by 'G.' The goal node is always a node that returns the value of *True*. It is possible for a goal node to occur at the action of either agent A or agent B, although goal nodes might correspond to only one of the two agents in some game-centric settings. For example, a win (i.e., occurrence of goal node) in the game of tic-tac-toe occurs only after the move by the primary agent. It is evident that many more branches and paths of the tree need to be traversed because of the AND condition (as compared to single-agent settings in which only a single path needs to be traversed). Note that a single-agent setting can be viewed as a search tree with only OR nodes, which is the main reason that only a single path needs to be traversed.

For large state spaces, the need to traverse multiple paths in AND-OR trees can be a problem, because one would need to evaluate a large number of paths in order to determine whether the goal condition can be reached. The number of possible paths is directly proportional to the number of nodes in the tree. Since the state space can be massive in real-world settings, this can be impractical in many cases.

3.2.1 Handling More than Two Agents

The description of the AND-OR search algorithm in the previous section involves only two agents. What happens in settings where there are more than two agents? A multi-agent environment might seem to be quite onerous at first glance, especially if the secondary agents can make transitions in arbitrary order, which are not necessarily alternately interleaved with those of the primary agent. For example, when one has three agents, A, B, and C, the transition by agent B could occur before agent C and vice versa. In such cases, the transitions made by multiple secondary agents can be modeled by using the union of all possible actions by other agents and treating it as the action made by a single dummy (secondary) agent; similarly, a missing step by the primary agent between consecutive steps by the secondary agents can be modeled with a dummy action that transitions from a state to itself. The resulting problem can then be reduced to the two-agent problem in which alternate steps are taken by the two agents. There are, of course, limitations to this approach; the limitations are caused by the fact that combining multiple actions into a single action can blow up the number of possible actions at a particular state. This will result in a tree with very high degree, whose size increases rapidly with increasing depth. As a result, one can practically address this type of scenario only when the underlying trees are shallow.

3.2.2 Handling Non-deterministic Environments

Interestingly, AND-OR search is also useful in non-deterministic environments, where the transitions resulting from a single agent are probabilistic in nature. In probabilistic settings, an agent cannot fully control the state resulting from choosing a particular outcome. This is similar in principle to the multi-agent setting in which the primary agent cannot fully control the state they will find themselves in as a result of choices made by other agents. In other words, each action made by an agent has a different distribution of outcomes in terms of the final state in which the agent lands after making a transition. This type of randomness can be captured by allowing each action a of the agent in state i to move from state i to a dummy action-specific state (i, a). This dummy state is an AND node. From this node, one moves to the different states corresponding to the various possibilities allowed by action a in state i. The probability of moving to a particular state from a dummy state is governed by a probability distribution specific to state (i, a). This setting is exactly similar to the two-agent setting discussed in the previous section. Therefore, multiagent environments are quite similar to non-deterministic settings in many ways (from the perspective of a single agent who cannot predict the other agent); however, we do not formally consider an environment to be probabilistic simply by virtue of being a multi-agent environment. Similarly, a probabilistic environment is not formally considered a multiagent environment. Nevertheless, the algorithms in the two cases turn out to be quite similar from the perspective of a single agent (who cannot completely control the transitions at all levels of the tree).

3.3 Informed Search Trees with State-Specific Loss Functions

The AND-OR trees discussed in the previous section represent a case of uninformed search in which large parts of the search space may need to be explored. Unfortunately, this type of setting is not very suitable for large state spaces, in which the size of the tree is very large as well. For example, consider the two-agent game of chess. There are more than 100 billion possible states after each player has made four moves. The number of states for a 40-move game of chess is more than a number of atoms in the observable universe. Clearly, one cannot hope to search the entire tree of moves with the use of AND-OR search. The main problem is that the depth of the tree can be quite large in the worst case, as a result of which it is impossible to explore the tree down to *leaf nodes* which have termination outcomes and crisply defined Boolean values. Therefore, it is useful to have a method by which one can control the depth (and number of states) of the tree that need to be explored. In other words, there needs to be aa way to explore the upper portions of the tree effectively and make decisions about transitions without having access to the eventual outcomes at leaf nodes.

In such cases, it is only possible to perform search over a subset of the tree with the help of state-specific loss functions. State-specific loss functions represent the heuristic "goodness" of each node from the perspective of each agent, and therefore free the agent from the intractable task of having to know the termination outcomes at leaf nodes. By using state-specific loss functions, one is able to evaluate intermediate states in which a lower numerical value is indicative of a more desirable state. This evaluation can be used for various types of pruning, such as for reducing the depth of the explored tree or for pruning specific branches of the tree. Note that the utility of a particular state from the perspective of each agent is typically different; in the case of adversarial game-playing settings, the utilities of the two players might be completely antithetical to each other. A state that is good for one agent is poor for the other, and vice versa. In fact, in the case of adversarial settings like chess, the evaluation of a particular position from the perspective of one agent is exactly the negative of the evaluation from the perspective of the adversary.

In the following, we will work with a two-agent environment, denoted by agent A and agent B respectively. Each state is associated with a loss function. The loss of state i for agent A is denoted by $L(i, A)$, whereas the loss of state i for agent B is denoted by $L(i, B)$. Smaller loss values are considered desirable from the perspective of each agent. *An important assumption in this case is that both players are aware of each other's evaluation functions.* This knowledge can be used by an agent to explore the tree from the perspective of the other agent. The cases in which agents are unaware of each other's evaluation functions is generally not feasible to address with the use of reasoning methods; in those cases, it becomes important to use learning methods in which agents learn optimal choices from past experiences.

Loss functions associated with states are often constructed using heuristic evaluations of a state with the aid of domain knowledge. In order to illustrate this point, we will use the game of chess as an example. In such a case, a state corresponds to a board position, and an agent corresponds to a chess player. In such a case, an evaluation of the board position corresponds to an evaluation of the goodness of a position from the perspective of the player who is about to make a move. These types of goodness values are often constructed on the basis of algorithms by human experts, who use their knowledge of chess in order to give a numerical score to the position from the corresponding player's perspective (since evaluations are specific to each player in an adversarial game). This evaluation might aggregate numerical scores for material, positional factors, king safety, and so on. The evaluation function is almost always constructed and coded into the system with the help of a human domain expert; the use of domain knowledge is, after all, the hallmark of all deductive reasoning methods. One problem is that such evaluations are hard to encode in an interpretable way with explicit material and positional factors (without losing some level of accuracy). Human chess players often make moves on the basis of intuition and insight that is hard to encode in an interpretable way. This general principle is true of most real-world settings, where domain-specific proxies can encode relevant factors to the problem at hand in an incomplete way. In other words, such evaluations are inherently imperfect by design and are therefore bound to be a source of inaccuracy in the decisions made by the underlying system. Recent years have also seen the proliferation of inductive learning methods for board evaluation in the form of reinforcement learning. This topic will be discussed later in this chapter and also in a dedicated chapter on reinforcement learning (see Chapter 10).

For some adversarial settings like chess, we might have $L(i, A) = -L(i, B)$. In such cases, these types of search reduce to *min-max search* in which the objective of search at alternative levels is either minimized or maximized by defining a single loss. However, we do not make this assumption in this section, and work with the general case in which the objectives of the two agents might be independent of one another. Therefore, no relationship is assumed between $L(i, A)$ and $L(i, B)$. The agent taking the action at the root of the tree is referred to as the primary agent A, and the agent taking action just below the root is the secondary agent B. In game-playing situations, it is possible for the system to use one of the two agents (say, agent B) to take decisions about actions, and the other agent (say, agent A) to play an anticipatory role in predicting the optimum actions the human might make. However, in other situations, it is possible for each agent to be full automated (and part of the same system). Furthermore, the objectives of the two agents can be partially or wholly independent. For appropriate choices of the losses, it is even possible for the agents to cooperate with one another partially or completely.

The use of loss functions associated with states helps in creating an informed exploration process in which the number of states visited is greatly reduced with pruning. A key point is that such informed search algorithms also use the domain-specific loss functions in order to explore the tree of possibilities up to a particular depth d (as an input parameter) in order to reduce the size of the tree being explored. The depth d corresponds to the number of actions on the path of maximum length from the root to the leaf. When d is odd, the last transition at the lowest level of the tree corresponds to the primary agent A. For example, using $d = 1$ corresponds to making the best possible move in chess after using a heuristic evaluation with the primary agent making a single move. When d is even, the last transition is made by the secondary agent B. The computational advantages of doing so are significant, because most of the nodes in search trees are at the lower levels of the tree.

By restricting the depth of exploration, one is exposed to the inaccuracies caused by imperfections in the evaluation function. The goal of search is only to sharpen the evaluation of a possibly imperfect loss function by using lookaheads down the tree of moves. For example, in a game of chess, one could simply apply the evaluation function after performing each possible legal move and simply selecting the best evaluation from the agent's perspective. However, this type of move would be suboptimal because it fails to account for the effects of subsequent moves, which are hard to account for with an evaluation that is inherently suboptimal. After all, it is hard to evaluate the goodness of a position in chess, if a long sequence of piece exchanges follow from the current board position. By applying a *depth-sharpened evaluation*, one can explore the entire tree of moves up to a particular depth, and then report the best move at the top level of the tree (where each move is optimal from the perspective of the player being considered at a particular depth). This evaluation is of much better quality because of the use of lookaheads. Therefore, the choice made at the root of the tree is also of much better quality (because of deep lookaheads) than a move made only with a difficult-to-design loss function (but no lookahead).

The overall algorithm for informed search with two agents, denoted by agent A and agent B, is shown in Figure 3.3. The input to the algorithm is the initial state s and the maximum depth d of exploration of the tree. Note that each action of an agent contributes 1 to the depth. Therefore, a single action by each agent corresponds to a depth of 2. The losses from the perspective of agent A and agent B in state i are $L(i, A)$ and $L(i, B)$, respectively. In adversarial environments, these losses are the negations of one another, but

Algorithm *SearchAgentA* (Initial State: s, Depth: d)
begin
    $i = s$;
    if $((d = 0)$ or $(s$ is termination leaf$))$ **then return** s;
    $min_a = \infty$;
    for $j \in A(i)$ **do**
    begin
        $OptState_b(j) = SearchAgentB(j, d - 1)$;
        if $(L(OptState_b(j), A) < min_a)$ **then** $min_a = L(OptState_b(j), A)$; $beststate_a = OptState_b(j)$;
    end;
    return $beststate_a$;
end

Algorithm *SearchAgentB* (Initial State: s, Depth: d)
begin
    $i = s$;
    if $((d = 0)$ or $(s$ is termination leaf$))$ **then return** s;
    $min_b = \infty$;
    for $j \in A(i)$ **do**
    begin
        $OptState_a(j) = SearchAgentA(j, d - 1)$;
        if $(L(OptState_a(j), B) < min_b)$ **then** $min_b = L(OptState_a(j), B)$; $beststate_b = OptState_a(j)$;
    end;
    return $beststate_b$;
end

Figure 3.3: The multi-search algorithm in a two-agent environment

this might not be the case in general settings. The notation $A(i)$ denotes the set of the states directly reachable from state i via a single action of the agent at that state. There, $A(i)$ represents the adjacency list of node i in the tree of moves. The algorithm is structured as a *mutually* recursive algorithm in which each node for the search algorithm for agent A calls the search algorithm for agent B and vice-versa. The search call for agent A is denoted by *SearchAgentA*, and the search call for agent B is denoted by *SearchAgentB*. The main difference between the two procedures is that different losses $L(i, A)$ and $L(i, B)$ are used to make key choices in the two algorithms, and these choices govern the behaviors of the algorithms at hand in terms of the preferred actions. The two algorithms are denoted by the mutually recursive subroutine calls *SearchAgentA* and *SearchAgentB*, respectively. The former algorithm returns the best possible state obtained after exploring d moves down the tree from the perspective of agent A, whereas the second algorithm returns the best possible state from the perspective of agent B after exploring d moves down the tree. In other words, the two pseudocodes are almost identical in structure, except that they minimize over different loss functions. Each call returns the best possible state d transitions down the tree from the perspective the agent concerned. It is also possible to rewrite each pseudocode to return the value of the optimal node evaluation along with the optimal state d levels down the tree (which is how it is normally done in practice). This is achieved by keeping track of the optimal state value of each node during the current sequence of calls, and it is helpful in avoiding repeated evaluation of the same node during backtracking. *Unlike AND-OR trees, a single, optimal path can indeed be found d levels down the tree by using the loss-function.* Therefore, aside from reducing the depth of the tree being considered, the approach reduces the amount of bookkeeping required by the algorithm as compared to an AND-OR algorithm (which requires exploration of multiple paths and corresponding bookkeeping).

The recursive algorithm works as follows. If the input depth d is 0 or the current node i is a termination leaf, then the current state is returned. On the other hand, when each algorithm is called for a non-terminal nodes i, the corresponding algorithm for a particular agent recursively calls the algorithm for the other agent from each node $j \in A(i)$ with depth parameter fixed to $(d-1)$. The minimum of the evaluations of the state is returned. Therefore, a state from $d-1$ levels down the tree will be returned by each of these $|A(i)|$ recursive calls. The final state is selected out of these $|A(i)|$ possible states by selecting the state with the least loss among these states. When agent A calls $SearchAgentB$ for each state $j \in A(i)$, the optimal state from the perspective of agent B is stored in $OptState_b(j)$, and the best of these states from over the different values of j from the perspective of agent A is returned. Returning the best multi-agent state d levels down the tree does not necessarily yield the best action at the current node directly. However, one can separately keep track of which branch was followed to reach the best state d levels down the tree.

Although this algorithm can be used for adversarial environments by selecting the loss function for one agent to be the negative of the loss function for the other agent, it can also be used in other type of multiagent environments as long as the loss functions of the two agents are properly defined. In such cases, it is possible to have a certain level of cooperation between the two agents, if the corresponding loss functions are sufficiently aligned towards a particular goal.

3.3.1   Heuristic Variations

The algorithm of Figure 3.3 is the most basic version of the approach, and it is not optimized for performance. One issue is that many of the explored branches of the tree are often redundant. For example, in a game of chess some obviously suboptimal moves can be ruled out very quickly by humans only by exploring the tree of possibilities to a very shallow depth. One possibility is to explore a branch only if its evaluation is better than a particular threshold from the perspective of the agent for whom the evaluation is being performed. By using this approach, large parts of the search space are heuristically pruned early on. This can be achieved by using an additional loss threshold parameter l. When the loss function at the current node is greater than this quality threshold, it is unlikely that this particular branch will yield a good evaluation (although there can always be surprises deeper down the tree). Therefore, in the interest of practical efficiency, the algorithm simply returns the state corresponding to this node directly (by applying the loss function at this state) without exploring down further. This type of change is extremely minor, and it tends to prunes out branches that are unlikely to be germane to the evaluation process. There are many ways of further sharpening the pruning by using a shallow depth of exploration for pruning evaluations (such as 2 or 3) and a deep depth for the primary evaluation (as in the previous section).

Another natural optimization is that many states will be reached via alternative paths in the tree. For example, in a game of chess, different orders of moves[1] might result in the same position. It is helpful to maintain a hash table of previously visited positions, and use the evaluations of such positions when required. This can sometimes be tricky, as the earlier evaluation of a position at depth d_1 might not be as accurate as the evaluation of

---

[1] For those chess players who are familiar with formal move notation, the following two sequences $(1.e4, Nc6, 2.NF3)$, and $(1.Nf3, Nc6, 2.e4)$ lead to the same position.

Algorithm *MaxPlayer*(Initial State: s, Depth: d)
begin
   $i = s$;
   if $((d = 0)$ or $(s$ is termination leaf$))$ **then return** s;
   $max_a = -\infty$;
   for $j \in A(i)$ **do**
   begin
      $OptState_b(j) = MinPlayer(j, d - 1)$;
      if $(U(OptState_b(j)) > max_a)$ **then** $max_a = U(OptState_b(j))$; $beststate_a = OptState_b(j)$;
   end;
   return $beststate_a$;
end

Algorithm *MinPlayer*(Initial State: s, Depth: d)
begin
   $i = s$;
   if $((d = 0)$ or $(s$ is termination leaf$))$ **then return** s;
   $min_b = \infty$;
   for $j \in A(i)$ **do**
   begin
      $OptState_a(j) = MaxPlayer(j, d - 1)$;
      if $(U(OptState_a(j)) < min_b)$ **then** $min_b = U(OptState_a(j))$; $beststate_b = OptState_a(j)$;
   end;
   return $beststate_b$;
end

Figure 3.4: Minimax search algorithm in a two-agent adversarial environment

a position at depth $d_2 > d_1$. Therefore, the depth of evaluation of the position also needs to be maintained in the hash table. In many cases, the same position may be obtained in different branches of the tree (by simple transpositions of moves), in which case the stored evaluation in the hash table can save a lot of time. This type of position is referred to as a *transposition table*. While it is not practical to maintain all previously seen positions, it is possible to cache a reasonable subset of them, if they are deemed to be relevant in the future. For example, it does not make sense to cache a position in chess, when a subset of the pieces have already been exchanged, or if the current position has already moved sufficiently far away from a cached position. In games like chess, some positions are unreachable from others. For example, if a pawn is moved from its starting state, the corresponding state cannot be reached again.

3.3.2   Adaptation to Adversarial Environments

The informed search approach is naturally suited to adversarial two-player environments like chess. The main difference of adversarial search with the informed search algorithm is that the loss functions of the two agents are related to one another in the former case; the loss function $L(i, A)$ and $L(i, B)$ are related to one another by negation:

$$L(i, A) = -L(i, B)$$

The resulting trees are referred to as *minimax trees*, since they minimize and maximize at alternate levels. Such trees form the basis of most traditional chess programs over the past two decades, such as *Stockfish* and *DeepBlue*, although newer algorithms such as *AlphaZero* tend to use ideas from reinforcement learning. The former is a deductive method that uses domain knowledge (e.g., evaluation function), whereas the latter is an inductive learning method that learns from data and experience. Many methods combine ideas from both schools of thought.

One convention that is commonly used in chess-playing programs is to work with utility functions rather than loss functions, where the objective needs to be maximized rather than minimized. Therefore, we restate the algorithm of Figure 3.3 in minimax form, where a maximization of the utility function $U(\cdot)$ is performed in even layers (starting at the level of the base recursion) and a minimization of the same function $U(\cdot)$ is performed in odd layers (starting one layer below the base recursion). Therefore, the utility function is maximized with respect to the primary agent at the root of the tree. We conform with this different convention (as opposed to a minimization-centric loss function) in this section in order to make it consistent with the most common use of this approach in chess-playing programs.

The resulting algorithm is shown in Figure 3.4. As in the case of Figure 3.3, the input to the algorithm is the initial state s and the depth d of the search. It is assumed that agent A occurs as the maximizing agent at the base of the recursion, and agent B occurs as the minimizing agent one level below the base of the recursion. This algorithm is almost exactly identical to the pseudocode shown in Figure 3.3 as a mutually recursive algorithm, except that the two agents do not have different loss functions; the *same* utility function $U(\cdot)$ is maximized by agent A by the *MaxPlayer* call, and it is minimized by the agent B by the *MinPlayer* call. All other variables are similar in Figures 3.3 and 3.4. The control flow is also similar, except that maximization happens in one pseudocode and minimization happens in the other pseudocode. The subroutine *Maxplayer* describes the approach from the maximization player's perspective, whereas the subroutine *Minplayer* describes the approach from the minimization player's perspective. It is noteworthy that minimax trees are often implemented so that the best move is not always returned, especially if the next best moves have very similar evaluation values. After all, the domain-specific utility function is only a heuristic, and allowing slightly "worse" moves results in greater diversity and unpredictability of the game-playing program. This ensures that the opponent cannot play by simply learning from previous programs in order to identify the weaknesses in the program.

An example of a minimax tree is shown in Figure 3.5. The nodes at alternate levels of the tree are denoted by either circles or rectangles, depending on whether the node performs minimization or whether it performs minimization. The circular nodes perform maximization (player A), whereas the rectangular nodes perform minimization (player B). A leaf node might either be one at which the game terminates (e.g., a checkmate or a draw-by-repetition in chess), or it might be one at which the node is evaluated by the evaluation function (because the maximum depth of search has been reached). Each node in Figure 3.5 is annotated with a numerical value corresponding to the utility function. The numerical values at circular nodes are computed using the maximum of the values of their children, whereas the values at rectangular nodes are computed using the minimum of the values of their children. The utilities are computed using the domain-specific function at the leaf nodes, which might either be terminal nodes (e.g., checkmate in chess), or they might be nodes present at the maximum depth at which the minimax tree is computed. The computation of the domain-specific evaluation function, therefore, needs to be efficiently computable, as it regulates the computational complexity of the approach. The tree of Figure 3.5 is not perfectly balanced, because the terminal state might be reached early in

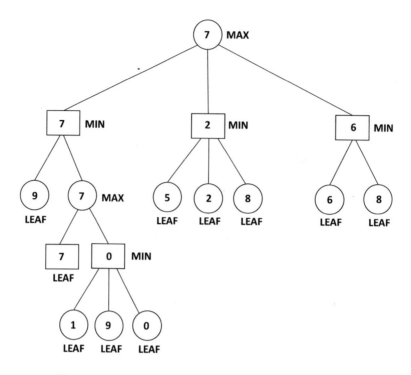

Figure 3.5: The minimax tree for adversarial search

many cases. For example, it is possible to arrive at a checkmate in chess after less than 10 moves or more than 100 moves. Therefore, such trees may be highly unbalanced in real settings. If the recursive algorithm of Figure 3.4 is used, the nodes will be visited in depth-first order in order to compute the values at various nodes. It is also possible to organize the algorithm of Figure 3.4 in order to use other strategies for exploring the nodes (such as breadth-first search). However, depth-first search is desirable because it is space efficient, and it enables a particular type of pruning strategy, which will be discussed later. For problems like chess, breadth-first search is too inefficient to be used at all.

3.3.3 Prestoring Subtrees

In many settings, certain portions of the tree are repeated over and over again. For example, in chess, the upper portion of the tree is fairly standard in terms of the portions of the tree that are known to be nearly optimal. There are a fairly large number of paths in the opening portion of the tree that result in nearly equal positions for both opponents. Each of these sequences is often named in chess by a distinctive name such as *Sicilian Defense*, *Queen's Indian*, and so on. This set of sequences is collectively referred to as *opening theory*, and they can be strung together in a tree structure that is a very sparsely selected subset of the (opening portion of the) minimax tree. Human grandmasters spend a significant amount of time studying opening theory, and also discovering novelties (i.e., small variations of known theory) that can provide them an advantage in an actual game because of the complexity of analyzing the long-term ramifications of a particular opening. It is generally hard for search trees (or other deductive methods) to discover such novelties easily because the effect of choosing a particular opening persist long into the game (beyond the depth to which a search typically works). At the same time, it is also hard for minimax trees to discover known theory (and the best opening moves) with the use of the search methods, as a minimax tree can make no distinction between known theory and a novelty. A key point is that human grandmasters have collectively spent hundreds of years in understanding the long-term effects of different openings, which is hard for search methods to fully match. As a result, this portion of the tree (referred to as the *opening book*) is pre-stored up front and used to make quick moves early on by selecting one of the branches in the opening book. Doing so significantly improves the strength of the resulting game-playing system. This is a classical example of an informal knowledge base, where the collective wisdom of hundreds of years of human play is pre-stored up front as a hypothesis for the system to use for better play. Indeed, minimax trees with opening books perform significantly better than those that do not use opening books.

At the lower levels of the tree, the board contains a smaller number of pieces for which outcomes are known with perfect play. For example, a queen versus rook end game (i.e., four pieces on the board including the kings) always results in a win for the player with the queen. However, it is not always a simple matter to play such endings perfectly, and even human grandmasters have been known to make mistakes[2] in the process, and reach suboptimal outcomes. When working with computers, the lower portions of such trees are tractable enough that the trees can be expanded to the leaves. In many cases, expanding the trees might require a few days of computational effort, which cannot be achieved during real-time game play. Therefore, the optimal moves for all positions with at most five pieces have been explicitly worked out computationally and have been pre-stored in *tablebases*. These tablebases are typically implemented as massive hash tables. Simply speaking, a hash table can be used to map from the position to the optimal move. A significant number of six-piece endgames have been pre-stored as well. These optimal moves have been worked out computationally (which can be viewed as a form of inductive learning). However, since they are pre-stored as tablebases (which are informal forms of knowledge bases), they are considered deductive methods as well, where the hypothesis (tablebase) is provided as an input to the system (no matter how it was actually derived). In some of these cases, an optimal endgame sequence might be longer than 100 moves, which is difficult even for the best grandmasters to figure out over the board during play. Therefore, these additional knowledge bases greatly improve the power of minimax trees.

3.3.4 Challenges in Designing Evaluation Functions

It is not always a simple matter to design domain-specific evaluation functions for game-centric or other settings, and some states are inherently resistant to evaluation without further expansion to lower levels. In order to understand the challenges associated with evaluation function design, we provide an example based on the inherent instability of trying to predict the value of a position in chess; some positions require a sequence of piece exchanges, and an evaluation of such positions in the middle of a dynamic exchange can lead to an evaluated value that reflects the true balance of power between the players rather poorly. A key point is that it is often hard to encode the complex dynamics between pieces with the use of a particular function. This is because each exchange leads to a wild swing in any (material-centric) domain-specific evaluation, as the pieces are rapidly removed from the board. Such volatile positions are not "quiet" and are therefore referred to as *non-quiescent*

---

[2]A great example of this situation corresponds to the round 5 playoffs during the 2001 FIDE World Chess Championship in Moscow. The grandmaster Peter Svidler reached a queen versus rook end game against Boris Gelfand, who refused to resign and effectively challenged his opponent to prove that he knew how to play the end game perfectly. Peter Svidler made mistakes in the process, and the game resulted in a draw.

(a) A non-quiescent position   (b) A quiescent position after exchanges
    (white to move)   (black to move)

Figure 3.6: Black seems to have significantly more material in position (a). After a series of exchanges, the position is shown in (b), which is a draw with optimal play from both sides

in chess parlance. An example of a non-quiescent position is shown in Figure 3.6(a). In the position of Figure 3.6(a), it seems that black has significant material advantage, and also has the white king in check. However, after a series of forced exchanges, the position becomes quiet (as shown in Figure 3.6(b)), and it is theoretically equal between white and black (as a draw with optimal play from both sides). Therefore, a common approach to evaluating board positions is by first finding the "most likely" quiescent position after a short sequence of moves. Finding quiescent positions in chess is an important area of research in its own right, and it remains an imperfect science in spite of significant improvements over the years. This problem is not specific to chess and it arises in all types of game-playing settings. The algorithms used to arrive at quiescence are not perfect, and can often be fooled by a particular board position. This often has a devastating effect on the overall evaluation. A poor evaluation also results in mistakes during play, especially when a shallow minimax tree is used.

Beyond the more obvious issue of non-quiescent positions, challenges arise because of subtle aspects of evaluations that cannot be easily hand-crafted. For example, a human chess grandmaster evaluates a board position in terms of complex combinations of patterns and intuition gained from past experience. It is hard for evaluation functions to encode this type of knowledge. This is a problem with any deductive reasoning method that relies on interpretable hypotheses (e.g., numerical value of a pawn versus the numerical value of a bishop) to hand-craft knowledge into the system. Therefore, beyond overcoming these challenges, it is important to know which positions are harder to evaluate and than others so that those specific portions of the tree can be explored more deeply. Indeed, this is already done in most game-playing programs with various heuristics, which tends to make the trees somewhat unbalanced. Recent experiences seem to suggest that one can *learn* robust evaluations in a data-driven manner, rather than hand-crafting them. This is the task of *reinforcement learning* methods like *Monte Carlo search trees*, which will be discussed in a later section. However, these methods fall into the school of thought belonging to inductive learning, rather than deductive reasoning. This will be topic of discussion in Section 3.5.

3.3.5 Weaknesses of Minimax Trees

Minimax trees are typically too deep to be exhaustively evaluated down to the final outcome, which is why heuristic evaluations are used at leaf nodes. By using minimax search, one is effectively sharpening the evaluation function by looking ahead a few levels down the tree. For deep enough evaluations, a weak evaluation function can become extremely strong. For example, in the case of a chess-playing program, consider the use of the trivial evaluation function of simply setting the evaluation to either +1 or −1 when the game is in a checkmate position (depending on which player is winning), and 0, otherwise.

Chess is a game of finite length, and an upper-bound on the maximum length of a chess game is known to be 17,697 half-moves. In such a case, a minimax tree of depth 17,698 (counting moves by each player as an additional depth) will cause a chess-playing program with the aforementioned trivial evaluation to strengthen to perfect evaluations at the root node (via minimax computations). This is because a sequence of at most 17,698 moves will either lead to a checkmate or will lead to a draw based on the termination rules of chess. However, such an approach cannot be implemented in any computationally practical setting, as the number of nodes in such a tree will be larger than the number of atoms in the universe. The main problem is that it is impossible to use a very large depth of evaluation because of the explosion in the number of nodes that need to be evaluated. After all, the number of nodes increases by a factor of more than 10 for each additional level of the tree. Therefore, one needs to evaluate intermediate positions with heuristic evaluation functions. The depth of this evaluation is referred to as the *horizon* or *half-ply*[3] in the context of chess-playing programs. As a practical matter, the horizon can often be quite constrained because of the combinatorial explosion of the number of possible positions. In certain positions as the starting point of the search, a game of chess can have more than 100 billion possible outcome positions to evaluate after four moves from each player. In practice, most chess programs need to evaluate the positions much deeper in order to obtain moves of higher quality. With the trivial evaluation function discussed above, any minimax tree of computationally feasible depth will make random moves, as all leaf nodes will be usually non-terminal; therefore, the algorithm will make random moves in most positions. It is important to develop better heuristics for evaluating chess positions in order to reduce the depth of the tree at which high-quality moves are made. For example, if one had an oracle evaluation function to exactly predict the value of a position as a win, loss, or draw with optimal play, a tree of depth 1 would suffice for perfect play. The reality is that while evaluation functions have become increasingly sophisticated, there are severe limitations on how well they can be evaluated with domain-specific heuristics. Therefore, the minimax evaluation with increased depth is essential for high-quality chess play. Furthermore, executing very sophisticated evaluation functions can itself be computationally expensive, which can ultimately result in lower depth of evaluation in a time-constrained setting. In such cases, it is possible to lose the computational gains obtained from using sophisticated evaluation functions (because of reduced depth) to the additional cost of evaluating each function. Therefore, it often becomes a sensitive balancing act in designing a sophisticated evaluation function that can also be evaluated efficiently. The trade-off between spending time for better evaluation functions versus exploring the minimax tree more deeply continues to be a subject of considerable interest among practitioners and researchers in computer chess.

When the depth of the evaluation is limited, the weaknesses of imperfect evaluation functions become more apparent. Over the years, the depth of evaluation has improved in

---

[3]A ply corresponds to one move by each player. A half-ply corresponds to a single move by one of the two players.

game-playing systems (because of sophisticated hardware), and the evaluation function at the leaf nodes has also improved as programmers have learned novel ways of encoding domain knowledge into chess-playing software. The combination of the two adds significantly to the playing strength of the program. As a specific example, *Deep Blue* used specialized hardware together with carefully designed evaluation functions to defeat the world champion, Garry Kasparov, in a match of six games in 1997. The hardware was highly specialized, and the chips were specifically designed for chess-specific evaluation. *Deep Blue* had the capability of evaluating 200 million positions per second using specialized computer chips, which was very impressive for the computer hardware available at the time. This was the first time that such a powerful machine was combined with a sophisticated evaluation function in order to create a machine playing chess at the highest level. However, Kasparov did win one game and draw three games over the course of the six-game match. In the modern era, it would be unlikely[4] for a world champion to hope to draw even one game out of six games with an off-the-shelf chess software like *Stockfish* running on a commodity laptop or mobile phone. Simply speaking, artificial intelligence has far leapfrogged humans, as far as chess is concerned. A large part of this success stems from the ability to implement increasingly sophisticated evaluation functions and search mechanisms in an efficient way. Improvements in the state-of-the-art of computer hardware have also helped significantly in being able to construct and explore minimax trees in an efficient way. There have also been many advancements in designing strategies for pruning fruitless branches of minimax trees in order to reduce the computational burden of exploration.

Chess has been a success story for minimax trees, because of the large amount of interpretable domain knowledge available in order to perform utility function evaluations of high quality from given positions on the board. In spite of this fact, many weaknesses in modern chess programs do exist; these weaknesses become particularly apparent during computer-to-computer play resulting in chess positions where the evaluations require a greater level of intuitive pattern recognition. An imperfect heuristic evaluation can turn out to be an even greater problem in board positions where the tree has a large degree, and therefore one cannot create a very deep tree. Therefore, it is important to develop a general strategy to prune branches that are not very promising from each player's perspective in order to be able to evaluate the positions more deeply. One naïve way of doing this is to prune branches for which the immediate evaluation is extremely poor from the perspective of the agent that is making the next move. However, doing so can sometimes inadvertently prune relevant subtrees because more promising moves may be hidden lower down the tree. On the other hand, certain types of pruning are guaranteed to not lose relevant subtrees. This approach uses some special properties of minimax trees, and it is referred to as alpha-beta pruning.

3.4   Alpha-Beta Pruning

The main goal of alpha-beta pruning is to rule out irrelevant branches of the tree, so that the search process becomes more efficient. We present alpha-beta pruning with the use of a utility function $U(\cdot)$ which maximizes and minimizes at alternate levels of the tree. Therefore, all notations in this section are the same as those used in Section 3.3.2. The modified algorithm is illustrated in Figure 3.7. The alpha-beta method is very similar to

---

[4]The odds can be computed using the Elo rating system for both chess players and computers. As of the writing of this book, the Elo rating of the best chess program exceeded that of the current world champion by about 530 points. This difference translates to ten wins by the computer for every *draw* by the world champion.

Algorithm *AlphaBetaMaxPlayer*(Initial State: s, Depth: d, α, β)
begin
   $i = s$;
   if $((d = 0)$ or $(s$ is termination leaf$))$ **then return** s;
   for $j \in A(i)$ **do**
   begin
      $OptState_b(j) = AlphaBetaMinPlayer(j, d - 1, \alpha, \beta)$;
      if $(U(OptState_b(j)) > \alpha)$ **then** $\alpha = U(OptState_b(j))$; $beststate_a = OptState_b(j)$;
      if $(\alpha > \beta)$ **then return** $beststate_a$; **{ Alpha-Beta Pruning }**
   end;
   return $beststate_a$;
end

Algorithm *AlphaBetaMinPlayer*(Initial State: s, Depth: d, α, β)
begin
   $i = s$;
   if $((d = 0)$ or $(s$ is termination leaf$))$ **then return** s;
   for $j \in A(i)$ **do**
   begin
      $OptState_a(j) = AlphaBetaMaxPlayer(j, d - 1, \alpha, \beta)$;
      if $(U(OptState_a(j)) < \beta)$ **then** $\beta = U(OptState_a(j))$; $beststate_b = OptState_a(j)$;
      if $(\alpha > \beta)$ **then return** $beststate_b$; **{ Alpha-Beta Pruning }**
   end;
   return $beststate_b$;
end

Figure 3.7: The alpha-beta search algorithm

the minimax approach discussed in the previous section, except that it adds a *single line of code* that prunes out irrelevant branches, when it is understood that the current branch cannot find a better action at a node present at a higher level of the tree from the perspective of the corresponding player.

In order to understand alpha-beta pruning, we will provide an example of how some branches can be pruned in particular situations. Consider the example shown in Figure 3.5 in which maximization is performed at the top level of the tree, and minimization is performed at the next level. After processing the entirety of the first subtree hanging from the root of the tree, a value of 7 is returned to the root (which is the current maximization value from the perspective of agent A at the root of the tree). When the second subtree at the root is being processed, the minimization agent B at the next level encounters the first leaf, which has a value of 5. While processing further leaves can lead to an even lower value from the perspective of the minimization agent B, it knows that the maximization agent A already has a value of 7 available from the first subtree (and would never choose an action with a lower utility). Therefore, processing the entire subtree further is fruitless from the perspective of agent B, and further branches (corresponding to the leaf nodes containing values 2 and 8) can be pruned. The relevant portion of the tree from Figure 3.5 is shown in Figure 3.8.

Next, we describe the alpha-beta search algorithm more formally. The reason that the approach is historically referred to as alpha-beta search is because the best evaluations from the perspective of the two adversarial agents are often denoted by α and β, respectively. The overall algorithm is shown in Figure 3.7, and it is very similar to the algorithm in

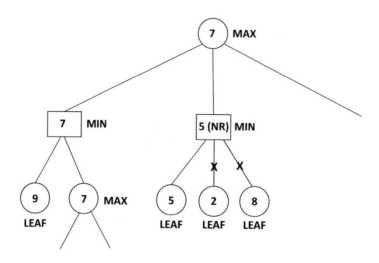

Figure 3.8: An example of how fruitless branches can be pruned

Figure 3.4 The parameters α and β serve the same role as the variables max_a and min_b in Figure 3.4, except that they are no longer initialized from scratch in each call of the mutually recursive pseudocodes. Rather α is a *running estimate* of best evaluation from the perspective of agent A over the course of the entire algorithm, and β is a running estimate of the best evaluation from the perspective of agent B over the course of the algorithm. Therefore, these parameters are passed down the recursion, as the parameters are updated. As a result, one works with tighter estimates of the optimal states from the perspective of each agent, given that more information is passed down the mutually recursive calls. The value of α is set to $-\infty$ at the base call of the algorithm, whereas the value of β is set to ∞. Throughout the course of the algorithm, we will always have $\alpha \leq \beta$ at a node or else the children of that node no longer need to be explored. The best way to understand α and β is that they are both pessimistic bounds on the evaluations from the perspectives of the maximizing and minimizing players, respectively. In an optimal minimax setting the optimal move that agent A makes is the same as the optimal move that agent B expects A to make when the same utility function is used by both. Therefore, at the end of the process, one must have $\alpha = \beta$ at the root of the tree. However, for pessimistic bounds one must always have $\alpha \leq \beta$. Therefore, whenever we have $\beta < \alpha$ at a node of the tree, the action sequence leading to such a node will never be selected by at least one of the two players and can be pruned. The overall algorithm in Figure 3.7 is very similar to the approach used in Figure 3.4, except for the additional parameters α and β, which provide the pessimistic estimates for the two players. Furthermore, a single line of code in each of the pseudocodes for the two agents performs the alpha-beta pruning, which is not present in the pseudocodes for the basic minimax algorithm. One consequence of carrying forward α and β as a recursive parameter is that the algorithm is able to use bounds established in distant regions of the tree for pruning.

An example of alpha-beta pruning is shown in Figure 3.9, and the evaluation of each node is also shown. The nodes that are not relevant for computation of the optimal state are marked as '(NR)' and the evaluations within these nodes are not exact. The acronym '(NR)' stands for "(Not Relevant)". The reason for the non-relevance of this node is that one or more of the descendent subtrees rooted at this node have already been explored, and

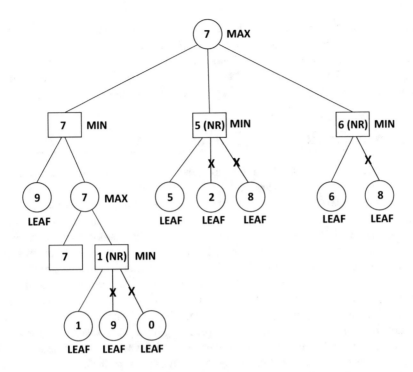

Figure 3.9: The minimax tree for adversarial search

the optimal state rooted at that node is deemed to be unacceptable to the opponent (based on other options available to the opponent higher up the tree). Therefore, other subtrees that descend from this tree are not explored further (as the entire subtree rooted at that node is deemed to be irrelevant). Therefore, the value within the node marked by '(NR)' is based on explored branches, and is a pessimistic bound. Correspondingly, many of the branches rooted at these nodes are marked by 'X', and they are not explored further. Even though only a small number of nodes are pruned in Figure 3.9, it is possible to prune entire subtrees using this approach in many cases.

In all of the examples we have shown so far, the pruning is done based on bounds developed in alternate levels of the tree. However, since α and β are carried down to lower levels of the tree, it is also possible for pruning to occur at a lower level of the tree because of bounds that were developed higher up in the tree.

3.4.1 Importance of Branch Evaluation Order

An important point is that the effectiveness of the pruning depends heavily on the order in which the branches of the minimax tree are explored. It makes a lot more sense to explore promising branches of the tree first from each agent's perspective, because it leads to sharper evaluations earlier on in the process. The notion of "promising" branches is evaluated in a myopic fashion by computing the utility of a state immediately after taking a single action from the perspective of the agent taking that action. The state with the best utility from the perspective of the agent making the transition is explored first. This is achieved by applying the domain-specific utility function (e.g., evaluation of chess position) after making one transition (e.g., move in chess) and then picking the best one among all these possibilities. More sophisticated methods for branch ordering might use multiple transitions for a shallow lookahead in evaluation. By exploring branches in this order, the agent is more likely to obtain a more favorable evaluation at deeper levels of the tree early on the process of depth-first search. This leads to a tighter pessimistic bound for each agent earlier in the process, and therefore a larger fraction of redundant branches is pruned.

3.5   Monte Carlo Tree Search: The Inductive View

Although variations of Monte Carlo tree search can be used for any decision problem, it has historically been used only for adversarial settings. Therefore, the discussion in this section is based on the adversarial setting. The main problem with the minimax approach of the previous section is that it is based on domain-specific evaluation functions, which might sometimes not be available in more complex games. The weaknesses of domain-specific evaluation functions are particularly notable in games where the evaluation of a position is not easily interpretable and depends on a large level of human intuition. This is particularly evident in games like Go, which depend on a high level of spatial pattern recognition. Even in games like chess, where domain-specific evaluation functions are available, they tend to have significant weaknesses. As a result, the tree of possibilities often needs to be evaluated to significant depth (with some level of heuristic pruning) in order to ensure that good solutions are not missed. This can be very expensive in chess, and turns out to be intractable in games like Go. Even in chess, computers tend to be weak in certain types of positions that require a high level of intuition to understand. Indeed, humans were able to consistently defeat minimax trees in chess during the early years by using these weaknesses of minimax trees. The difference in style of play by minimax trees from humans is very apparent in terms of lacking creativity in game play. Creativity is a hallmark of learning systems that can generalize past experiences to new situations in a novel way.

Monte Carlo tree search uses the inductive perspective of *learning* from past explorations down the tree structure in order to learn promising paths over time. Furthermore, Monte Carlo tree search does not require an explicit evaluation function, which is consistent with the fact that it is a learning approach rather than a knowledge-based approach. This type of statistical approach requires fewer evaluations from an empirical perspective although it is not guaranteed to provide an optimal solution from the minimax perspective. Note that the minimax tree is designed to provide an optimal solution from each opponent's perspective, if *one were to construct a tree down to termination nodes.* However, the meaningfulness of such an "optimal" solution is questionable when one has to stop at a particular depth of the tree in order to evaluate a questionable utility function. It might sometimes make more sense to explore fewer branches all the way down to goal states, and then empirically select the most promising branches.

Monte Carlo tree search is based on the principles of reinforcement learning, which are discussed in detail in Chapter 10. We will first discuss a basic version of Monte Carlo tree search, as it was proposed during the early years. This version is referred to as the *expected outcome model,* and it captures the important principles of the approach in spite of some obvious limitations; modern versions of Monte Carlo tree search are based on these principles, and they use various modifications to address these limitations. In a later section, we will discuss how modern versions of Monte Carlo tree search have improved over this model.

Monte Carlo tree search derives its inspiration from the fact that making the best move in a given position will often result in improved winning chances over multiple paths of the tree. Therefore, the basic approach expands sampled paths *to the very end* until the game terminates. By sampling the rollout to the very end, one is able to avoid the weaknesses associated with the domain-specific evaluation function. As we will see later, the methodology of sampling uses some knowledge of past experiences in order to narrow down the number of branches that need to be explored.

At the root of the tree, one considers all possible states reachable by a single transition and then selects a particular node with the use of Monte Carlo sampling. At this newly selected node, all possible actions are sampled from the perspective of the opponent and one of these nodes is randomly selected for further expansion. Therefore, this process is continually repeated until one reaches a state that satisfies the goal condition. At this point, the sampling process terminates. Note that the repeated process of node sampling will result in a single path, along with all the immediate children of the nodes on the path. The sampled path is, therefore, much deeper than any of the paths considered by a relatively balanced minimax model. This process of repeated sampling up to termination of the game tree is referred to as a *rollout*. The process of sampling paths is repeated over the course of the algorithm in order to create multiple rollouts. The multiple rollouts provide the empirical data needed to make choices at the top level of the tree. After a sufficient number of paths have been sampled, the algorithm terminates.

After the algorithm terminates, the win-loss-draw statistics are *backpropagated* along each branch right up to root. For each branch, the number of times that it is played is stored, along with the number of wins, draws, and losses. For a given branch b, which is traversed N_b times, let the number of wins and losses be W_b and L_b, respectively. Then, one possible heuristic evaluation E_b for the branch b is as follows:

$$E_b = \frac{W_b - L_b}{N_b}$$

This evaluation favors branches that have a favorable win-to-loss ratio. The value will always range between -1 and $+1$, with larger values being reached by branches that have higher win to loss ratios. At the end of the Monte Carlo simulation, the evaluation of all branches at the root of the tree is used in order to recommend a move. In particular, the branch with the largest value of E_b. This approach is essentially the most primitive form of Monte Carlo tree search, although significant modifications are required in order to make it work well. In particular, we will see that the branches sampled later on in the process are not independent from the ones that were sampled earlier on. This process of continuously learning from past experience is referred to as reinforcement learning.

It is noteworthy that Monte Carlo tree search successively builds upon the tree that it has already expanded. Therefore, it might contain multiple branches corresponding to multiple rollouts. The resulting tree is therefore bushy and unbalanced. The number of leaf nodes (with a final outcome) in the resulting Monte Carlo tree is therefore bounded above by the number of rollouts (as some rollouts might be occasionally[5] replayed). Examples of Monte Carlo trees after successive rollouts are shown in Figure 3.10.

The success of Monte Carlo tree search heavily depends on the fact that the different evaluations from a particular branch of the tree are correlated with one another. Therefore, the statistical win-loss ratio from a particular branch of the tree often leads to accurate evaluations at the top level, even though a particular rollout might not be remotely optimal from the perspective of either opponent. This means that one is often able to estimate the

---

[5]For many complex games with long sequences of moves, it is highly unlikely to replay the same rollout. Repeats are more likely in simple games like tic-tac-toe.

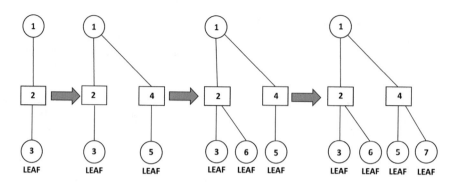

Figure 3.10: Results of Monte Carlo rollouts in the expected outcome model

value of a particular move by statistical sampling. This is different from a minimax tree, which tries to find optimal paths from the perspective of each opponent. It has been shown (under certain assumptions) that the prediction of the basic version of Monte Carlo tree search converges to that of a minimax tree *constructed all the way down to the termination nodes*. However, the number of samples required is too large for the Monte Carlo approach to even remotely approach convergence, and minimax trees cannot be (practically) constructed down to termination nodes anyway. In practice, the effects of an empirical approach are quite different from the domain-specific strategy of minimax trees.

One advantage of inductive learning methods like Monte Carlo tree search is that they do not suffer from the hardcoded weaknesses of a domain-specific evaluation function, as in the case of a minimax tree. Monte Carlo tree search is particularly suitable for games with very large branching factors (like Go) and also in the case of games that do not have well-developed theory in order to perform domain-specific evaluations. Compared to chess, Go tends to require a greater level of intuition from human players, and it is often hard to perform explicit calculation of the quality of a specific position. In such cases, simulating the game up to the very end seems like a reasonable solution, given that it is not realistically possible either to construct the tree up to a large depth (as required by minimax) or to reasonably evaluate an intermediate node in an accurate way.

While Monte Carlo tree search theoretically evaluates positions to the very end, it can lead to infinite (or very long) loops in certain types of games. Therefore, a practical limit is often imposed on the length of the tree path in order to avoid impractical situations. Such nodes can be ignored in the evaluation process.

As discussed above, convergence to a minimax tree requires a very large number of rollouts. To address this problem, many variants of Monte Carlo tree search do not perform the rollouts completely randomly. Rather, a certain level of bias may be used in choosing branches wither by using a domain-specific evaluation function, or based on the success of previous rollouts. The goal of incorporating this type of bias is to speed up convergence of the evaluation of branches; the trade-off is that the final solution may be suboptimal, and a better solution may be obtained with unbiased sampling in cases where it is possible to perform a larger number of simulations. Therefore, a trade-off between *exploration* and *exploitation* is natural. The next section discusses some of these enhancements, which defines how such trees are used in practice.

3.5.1 Enhancements to the Expected Outcome Model

The previous section describes the expected outcome model, which was the progenitor of Monte Carlo tree search, but was not actually referred to by that name. There are several problems with the expected outcome model, the most important of which is the fact that it takes too long to converge to the results obtained from a minimax tree. As a result, this basic version of the model performs rather poorly. Modern variants of Monte Carlo tree search rely on the experience of past explorations in order to grow the tree in promising directions. These promising directions are learned by identifying better win to loss ratios in the past. The precise degree of biasing is a key trade-off in the process of construction of the Monte Carlo tree, and it is based on the idea on reinforcement learning, where one *reinforces* past experiences during the learning process. Modern versions of Monte Carlo tree search improve the power of the approach by making several modifications to the original version of the expected outcome model.

The vanilla version of Monte Carlo tree search (i.e., the expected outcome model) explores branches completely randomly. The main challenge in doing so is that only a few paths might be promising, whereas most branches might be disastrously suboptimal. The drawback of this situation is that purely random search might have some of the properties of searching for a needle in a haystack. In reality, some branches are likely to be more favorable than others based on the win-loss ratio of earlier traversals down that branch. Therefore, a level of determinism in exploration is incorporated at the upper levels of the tree. However, by exploiting only the win-loss ratio of previous traversals (after performing a small number of them), one would repeatedly favor specific branches (which might be suboptimal) and fail to explore new branches that might eventually lead to better choices. Therefore, a trade-off between exploration and exploitation is used, which is based on ideas drawn from the principle of *multi-armed bandits*. In multi-armed bandits, a gambler repeatedly tries to discover the best of two slot machines to play on by trying them repeatedly (assuming that the expected payoffs from the two machines are not the same). Playing the slot machines alternately to learn their win-loss ratio is known to be wasteful, especially if one of the machines yields repeated rewards early on and the other yields no rewards. Therefore, the multi-armed bandit approach works by regulating the trade-off between exploration (by stochastically trying different slot machines) and exploitation (by stochastically favoring the one that has performed better so far).

The multiarmed bandits method uses a variety of strategies to learn the best slot machine over time, one of which is the *upper-bounding method*. This method is described in detail in Chapter 10, and it explores branches based on their "optimistic" potential. In other words, the gambler compares the most optimistic statistical outcomes from each slot machine, and selects the slot machine with the highest reward in the optimistic sense. Note that the gambler would be likely to favor machines that are played less often, since there is greater variability in their predicted statistical performance — greater variability is always favored by optimists. At the same time, machines with superior historical outcomes will also be favored, because past performance does add to the optimistic estimate of performance. In the context of tree search, this method is also referred to as the *UCT algorithm*, and it stands for *upper-bounding algorithm applied to trees*. The basic idea is that branches are evaluated as a sum of a quantity based on the earlier win-loss experience from that branch (and a bonus based on how frequently that branch has been explored earlier). Infrequently explored branches receive a higher bonus in order to encourage greater exploration. For example, a possible evaluation of node i (corresponding to a particular branch) is as follows:

$$u_i = \underbrace{\frac{w_i}{n_i}}_{\text{Exploit}} + \underbrace{c\frac{\sqrt{N_i}}{n_i}}_{\text{Explore}} \tag{3.1}$$

Here, N_i is the number of times the parent of node i is played, w_i is the number of wins starting from node i out of the n_i times that it is played. The quantity u_i is referred to as the "upper bound" on the evaluation quality. The reason for referring to it as an upper bound, is that we are adding an exploration bonus to the win record of a particular branch, and this exploration bonus favors infrequently visited branches. The key point here is that the portion w_i/n_i is a direct reward for good performance, whereas the second part of the expression is proportional to $1/n_i$, which favors infrequently explored branches. In other words, the first part of the expression favors exploitation, whereas the second part of the expression favors exploration. The parameter c is a balancing parameter. It is also important to note that the value of u_i is evaluated from the perspective of the player making the move, and the value of w_i/n_i is also computed from the perspective of that player (which is different at alternate levels of the tree).

The presence of $\sqrt{N_i}$ in the numerator (in contrast to the value of n_i in the denominator) ensures that the exploration component reduces in relative value, if both n_i and N_i grow at the same rate. In general, the denominator always needs to grow faster than the numerator for this to occur. In the extreme case, when the value of n_i is 0 and N_i is positive, the value of $\sqrt{N_i}/n_i$ is set to the value of ∞ (which is the right-hand side limit of the expression as the (positive value) n_i goes to 0^+. Therefore, a branch that has never been selected is always prioritized for exploration over those that have been selected at least once.

Another possible example of the evaluation of u_i is as follows:

$$u_i = \frac{w_i}{n_i} + c\frac{\sqrt{\ln(N_i)}}{\sqrt{n_i}}$$

This expression also satisfies the property that the numerator of the exploration component grows slower with increasing N_i than does the denominator with increasing n_i.

A tree starts off with only the root node, and increases in size over multiple iterations. Branches with the largest upper bound are *deterministically* constructed until a new node in the tree is found that was never explored earlier. This new node is created and is referred to as a *leaf node* from the perspective of the current tree. It is here that the probabilistic evaluation of the leaf node starts (which is backpropagated to each branch in the tree statistics). Note that the overall tree constructed will still have a random component to it, since the evaluations at leaf nodes are performed using Monte Carlo simulations. The value of the upper bound in the next iteration does depend on the results of these simulations, which will affect the structure of the tree. Although Monte Carlo rollouts are used for evaluation of leaf nodes, the path taken by the rollouts is not added to the Monte Carlo tree.

The above description of the tree construction process suggests that a distinction is made between the deterministic approach to each iteration of the tree construction (using the above exploration-exploitation trade-off) and the evaluation at leaf nodes using Monte Carlo rollouts. In other words, like minimax trees, the leaf nodes might be intermediate positions in the specific adversarial setting (e.g., chess game) rather than terminal positions (which correspond to wins, losses or draws). However, there are several important difference from minimax trees in terms of how the structure of these trees is arrived at and how Monte Carlo rollouts are used from these leaf nodes in order to evaluate the position down to terminal nodes. The use of Monte Carlo rollouts for evaluations ensures that inductive learning is prioritized over domain knowledge. One typically performs multiple Monte Carlo rollouts from the same leaf node for greater robustness in evaluation. This type of approach for evaluating leaf nodes uses the general principle that such rollouts can often lead to robust evaluations in the expected sense, even though the rollouts will obviously make suboptimal moves.

There are also significant differences in terms of the nature of the evaluation with respect to the expected outcome model. Unlike the case of the expected outcome model, one does not use uniform probabilities over all valid moves in a given position in order to perform the Monte Carlo rollouts. Rather, the moves are typically predicted using a machine learning algorithm that is trained to predict the probability of the best move, given the current state. The machine learning algorithm can be trained on a database of chess positions together with the evaluations of various moves as eventually resulting in a win, loss, or draw. The goodness evaluation of the machine learning algorithm is used to sample a move at each step of the Monte Carlo rollout.

Machine learning algorithms can be used not only to bias the Monte Carlo rollouts (which are performed at lower levels and do not add to the Monte Carlo tree), but they can also be used to influence the best branch to explore during the upper levels of the deterministic tree construction process. For example, the method in [168] uses a heuristic upper bound quantification of each branch, which is enhanced with a machine learning method. Specifically, a machine learning method predicts the probability of each action in a particular state (chess position). This probability is learned using a training database of the performance in similar positions in the past. Higher probabilities indicate more desirable branches. The exploration bonus for each state is multiplied with this probability p_i in order to yield a modified value of u_i as follows:

$$u_i = \frac{w_i}{n_i} + c \cdot p_i \frac{\sqrt{N_i}}{n_i}$$

This relationship is the same as Equation 3.1, except that the value of p_i is used in order to weight the exploration component. The value of p_i is computed using a machine learning algorithm. As a result, desirable branches tend to be selected more often while creating the Monte Carlo tree. This general principle is referred to as that of adding *progressive bias* [36].

Many of these enhancements were used in *AlphaZero* [168], which is a general-purpose reinforcement learning method for Go, chess, and shogi. The resulting program (and other programs based on similar principles) have been shown to outperform chess programs that are based purely on minimax search with domain-specific heuristics. An important point about inductive methods is that they are often able to implicitly learn subtle ways of evaluating states *from experience*, which goes beyond the natural horizon-centric limitations of domain-specific methods. By using Monte Carlo rollouts, they explore fewer paths, but the paths are deeper and are explored in an expected sense. In order to understand why such an approach works better, it is important to note that a poor move in chess will often lead to losses over many possible subsequent continuations of the go, and one does not really need the optimal minimax path to evaluate the position. In other words, the poor nature of the move will show up in the win-loss ratios of the Monte Carlo rollouts.

Past experience with the use of such methods in chess has been that the style of play is extremely creative. In particular, *AlphaZero* was also able to recognize subtle positional

(a) Minimax recommends moving threatened knight
(white to move)

(b) *AlphaZero* moves rook
(black to move)

Figure 3.11: A deep knight sacrifice by *AlphaZero* (white) in its game against *Stockfish* (black)

factors during play. As a specific example, a chess position from the *AlphaZero-Stockfish* game [168] is shown in Figure 3.11, where the Monte Carlo program *AlphaZero* is playing white and the minimax program *Stockfish* is playing black. This position is an opening variation of the Queen's Indian defense, and white's knight is under threat. All minimax programs at the time recommended moving the knight back to safety. Instead, *AlphaZero* moved its rook (see Figure 3.11(b)), which allowed the capture of the knight. However, a strategic effect of this choice was that it caused black's position to stay underdeveloped over the longer term with many of its pieces in their starting positions. This fact was eventually exploited by *AlphaZero* to win the game over the longer term. These types of strategic choices are often harder for minimax programs to make, because of the natural horizon effects associated with minimax computation. On the other hand, inductive methods learn form experience as they explore the Monte Carlo tree deeper than a typical minimax program, and also learn important patterns on the chess board from past games. These patterns are learned by combining the rollouts with deep learning methods. Although *AlphaZero* might not always have seen the same position in the past, it might have encountered similar situations (in terms of board patterns), and favorable outcomes gave it the knowledge that underdeveloped positions in the opponent were sufficiently beneficial for it in the longer term. This information was indirectly encoded in the parameters of its neural network. This is exactly how humans learn from past experience, where the outcomes from past games change the patterns stored in the neural network of our brain. Some of these methods will be discussed in Chapter 10.

  AlphaZero was able to discover all of the opening theory of chess on its own, unlike minimax programs that require the known opening tree of moves (based on past human play) to be hard-coded as an "opening book." This is a natural difference between a deductive reasoning method (which often leverages domain knowledge from human experience) and an inductive learning method (which is based on data-driven learning from machine experience). In some cases, *AlphaZero* discovered opening novelties or resurrected much-neglected lines of play in opening theory. This led to changes in the approach used by human players.

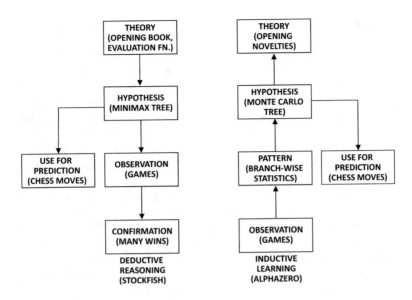

Figure 3.12: Revisiting Figure 1.1: The two schools of thought in artificial intelligence as applied to chess

The reason is that repeated Monte Carlo rollouts from specific opening positions encounter win-loss ratios that are very informative to the learning process. This is also how humans learn from playing many chess games. A similar observation was made for backgammon, where a reinforcement learning program by Tesauro [186] led to changes in the style of play by human players. The ability to go beyond human domain knowledge is usually achieved by inductive learning methods that are unfettered from the imperfections in the domain knowledge of deductive methods.

It is also noteworthy that even though Monte Carlo methods were tried well before the *AlphaZero* method by chess programs such as *KnightCap* [17], they were never successful in outperforming chess programs based on the minimax method. It was only in 2016 that *AlphaZero* was able to achieve a victory over one of the best minimax programs at the time (*Stockfish*). An important point is that inductive learning methods require a *lot of data and computational power*. While one can generate an unlimited amount of data in the chess domain by computer self-play, limitations in computational power can stand in the way of training such programs effectively. Because of this limitation, inductive methods were not trained or optimized (before 2016) at the scale required to outperform domain-specific minimax trees (which were already quire powerful in their own right). After all, the human world champion, Garry Kasparov, had already been defeated by a domain-specific minimax program (*Deep Blue*) as far back as 1997. *AlphaZero* also increased the power of Monte Carlo tree search methods by using deep learning methods to predict the best moves in each position. Currently, the best human players are typically unable to defeat either minimax chess programs or inductive Monte Carlo methods, even after handicapping the chess program with two removed pawns at the starting position. Among the two classes of methods, it is only recently that Monte Carlo methods have overtaken minimax trees.

3.5.2 Deductive Versus Inductive: Minimax and Monte Carlo Trees

Minimax trees create a hypothesis with the use of a domain-specific evaluation function (and an opening book for the initial moves) and then predict using this hypothesis. This is a deductive reasoning method because it incorporates the hypothesis of the expert chess player into the evaluation function and the opening book. While this hypothesis can be fallible, its quality can be confirmed by playing the chess program multiple times and registering many wins. While it is sometimes thought that deductive reasoning methods only deal with absolute truths, these truths hold in the context of the knowledge initially fed into the system. If the knowledge base is imperfect (such as a bad opening book or evaluation function), it will also show up in the quality of the chess playing. The model created with a minimax tree is the hypothesis, which is used to predict moves at each position. The mapping of different elements of a minimax chess program to the different steps in deductive reasoning are shown on the left-hand side of Figure 3.12.

On the other hand, Monte Carlo tree search is an inductive learning method. Monte Carlo tree search makes predictions by generating *examples* (rollouts) and then makes a prediction based on the experience of this algorithm with these examples. Modern variations of Monte Carlo tree search further combine tree search with reinforcement learning and deep learning in order to improve predictions. Deep learning is used in order to learn the evaluation function instead of using domain knowledge. For example, *AlphaZero* uses Monte Carlo tree search for rollouts, while also using deep learning to evaluate the quality of the moves and guide the search. The Monte Carlo tree together with the deep learning models form the hypothesis used by the program in order to move moves. Note that these hypotheses are created using the *statistical patterns* of empirical behavior in games. *AlphaZero* was given zero opening knowledge of chess theory, and *it discovered chess theory on its own* in a few hours of training. In many cases, it discovered exciting opening novelties that had not been known to human grandmasters. Clearly, such capabilities are outside the ambit of deductive reasoning methods, which have a ceiling to their performance based on what is already known (or considered "correct").

In general, Monte Carlo trees tend to explore a few promising branches deeper based on evaluations from previous experience, whereas minimax trees explore all unpruned branches in a roughly similar way. The human approach to chess is similar to the former, wherein humans evaluate a small number of promising directions of play rather than exhaustively considering all possibilities. The result is that the style of chess from Monte Carlo tree search is more similar to humans than that from minimax trees. The programs resulting from Monte Carlo trees can often take more risks in game playing, if past experience has shown that such risks are warranted over the longer term. On the other hand, minimax trees tend to discourage any risks beyond the horizon of tree exploration, especially since the evaluations at leaf levels are imperfect.

The traditional versions of chess-playing programs, such as *DeepBlue* and *Stockfish* were based on the minimax approach. Historically, it was always believed that chess-playing programs (which are dependent on a high-level of domain-specific knowledge found by earlier experience of humans) would always do better than an inductive approach (which starts from scratch with no domain knowledge). After all, evaluation functions use all sorts of chess-specific heuristics that depend on the past experience of humans with various types of chess positions. This trend indeed seemed to be true until the end of 2016, when *AlphaZero* outperformed[6] *Stockfish* quite significantly. This reversal was caused by increased compu-

---

[6]The original match was conducted on the basis of game conditions that were heavily criticized, since *Stockfish* was not provided an opening book. However, subsequent programs based on the same principles

tational power, which almost always improves inductive learning methods far more than deductive reasoning methods. Indeed, many of the exciting results in artificial intelligence in the previous decade have started to show that the classical school of thought in artificial intelligence (i.e., deductive reasoning) had serious limitations that were often overlooked in the early years (at the expense of inductive learning methods).

3.5.3 Application to Non-deterministic and Partially Observable Games

The Monte Carlo tree search method is naturally suited to non-deterministic settings such as card games or backgammon. Minimax trees are not well suited to non-deterministic settings because of the inability to predict the opponent's moves while building the tree. On the other hand, Monte Carlo tree search is naturally suited to handling such settings, since the desirability of moves is always evaluated in an *expected* sense. The randomness in the game can be naturally combined with the randomness in move sampling in order to learn the expected outcomes from each choice of move. On the other hand, a minimax tree would require one to follow each outcome of an uncertain state resulting from an agent action in order to perform the evaluation; this is because the same agent action (e.g., a dice throw) might result in different outcomes over different samples. In other words, there is no single action that yields a guaranteed outcome for a particular choice of action, and it is impossible to create a minimax tree whose branches can be deterministically selected by the two opponents.

A similar observation holds true for card games, where pulling a card out of a pack might lead to different outcomes, depending on the random order of the cards in the pack. Similarly, playing a particular card might have different outcomes depending on the unknown state of the cards in the opponent's hand. This is a partially observable setting. As a result, a minimax tree can never provide a guaranteed outcome of any particular choice of playing cards, because of the uncertainty in the state of the game, as observed by the agent. Because of this inability to fully control outcomes using the voluntary choices of the two opponents, there is simply no way of picking the best branch at a given state, unless one works with expected outcomes. This type of setting is naturally addressed with Monte Carlo methods, where repeated replay yields a score at each branch and the best scoring branch is always selected. After all, the Monte Carlo method provides an *empirically optimal choice* rather than a choice with *optimality guarantees*, which works well in uncertain settings. Furthermore, one does not gain much from the guaranteed optimality of minimax trees, because anomalous imperfections in the domain-specific evaluation function often show up in particular leaf nodes, and can drastically affect the move choice. This is sometimes less likely when one is using a statistical approach of picking a branch that leads to more frequent wins.

3.6 Summary

Multiagent search shares a number of similarities with non-deterministic environments with single agents, and similar techniques can be used in both cases. For example, a single-agent setting in a non-deterministic environment can be treated in a similar way to a

---

as *AlphaZero*, such as *Leela Chess Zero*, indeed outperformed all earlier minimax programs by winning the computer chess championships in 2019.

two-agent deterministic environment, where the other "agent" is used to handle the non-determinism. In such cases, AND-OR trees can be used, where the AND corresponds to the non-deterministic choices. However, AND-OR trees are not practical because of the combinatorial explosion associated with goal-oriented search; in such cases, informed search can be leveraged with the help of utility or loss functions. An important setting associated with multiagent search is the adversarial environment, which occurs often in game-playing environments. The resulting trees are referred to as minimax trees, because they maximize and minimize the utility function at alternate levels. Minimax trees do not work very well in environments with large branching factors or in non-deterministic environments. In such cases, Monte Carlo tree search is much more effective. Monte Carlo tree search is an inductive learning method, and it scales better to settings with very large branching factors.

3.7 Further Reading

The minimax method has its roots in game theory, and it is also closely related to duality theory in machine learning. In particular, John von Neumann's minimax theorem lies at the heart of duality theory in machine learning [8]. The use of alpha-beta pruning for minimax trees first appeared in Arthur Samuel's checkers program [155], and this program was also the first known use of bootstrapping methods in reinforcement learning. Several variants of Monte Carlo tree search were used in the early years for improving minimax search, and the first known use was by Abramson in his PhD thesis and subsequently published in 1990 [1]. This model was the *expected outcome model*, which was not particularly powerful, and looks nothing like the Monte Carlo tree search that is used today. In particular, Abramson's model uses randomized rollouts right from the root of the tree with random moves, whereas the modern versions of upper bounding methods first expand the tree using the upper bounding strategy, and then use biased rollouts starting from the leaves. The first modern version of the technique appeared in the method by Colulom [42]. The incorporation of the upper-bounding idea in Monte Carlo tree search was proposed by Kocsis and Szepesvari [102]. A survey of Monte Carlo search methods may be found in Browne *et al.* [31].

3.8 Exercises

1. Consider an AND-OR tree with five levels (including the root), where the root is an OR node, and all goal nodes are contained at the leaf level. The AND-OR tree has three branches at each node and is perfectly balanced.

   (a) What is the total number of leaf nodes in the tree?

   (b) What is the minimum number of leaf nodes that would need to be reached in order to terminate the algorithm with a success of accomplishing the agent's goal?

   (c) What is the minimum number of leaf nodes that would need to be reached in order to terminate the algorithm with a failure?

2. Explain how you would define the nodes of an AND-OR tree in order to handle non-deterministic environments.

3. Consider the two-agent setting in which each agent has her own loss function. Specifically, the loss function in state i of agent A is $L(i, A)$ and that of agent B is $L(i, B)$.

Discuss why it is possible for the agents to both cooperate and compete with the appropriate choice of loss function.

4. Implement a minimax tree without alpha-beta pruning for tic-tac-toe. Now add alpha-beta pruning to the program.

5. It is well known that playing with the white pieces has the "first-movers advantage" in chess, and very strong players of equal strength are more likely to win playing with white than when playing with black. However, it is not known whether the advantage of playing with the white pieces yields a *guaranteed* win with optimal play (as no computer or human can play optimally).

   (a) If playing with the white pieces results in a guaranteed winning advantage as well as a "frequently-experienced advantage," explain what this means in terms of the fundamental structure of minimax trees and Monte Carlo search trees in chess.

   (b) If playing with the white pieces does not result in a guaranteed winning advantage but it does result in a "frequently-experienced advantage," explain what this means in terms of the fundamental structure of minimax trees and Monte Carlo search trees in chess.

6. Discuss game situations in which minimax trees might outplay Monte Carlo trees in a game of chess for the same amount of computational power.

7. Implement the primitive version of the Monte Carlo tree search (which is the expected outcome model) to play a game of tic-tac-toe. Use a random move in each position for the Monte Carlo rollouts.

8. Consider a tic-tac-toe game in which the opening player puts in an 'X' at one of the corners and the opponent places a 'O' at the opposite corner. Construct the skeleton of a minimax tree (illustrating only the optimal moves for each player starting at this position) to show that the first player always wins.

9. Consider an AND-OR tree in which alternate levels represent the moves of different players in chess. Discuss how one can compress this tree into a state-space graph, which is not necessarily structured as a tree, but has fewer nodes.

10. The previous exercise discusses the conversion of AND-OR search trees to general graphs. Discuss how you can apply AND-OR search to a general state-space graph, which is not structured as a tree. You may assume that each state includes information about which agent is making the transition. [Hint: The main difference from the AND-OR search algorithm on trees is the need to keep track of states that have been visited. You may find the framework of single-agent search useful in providing guidelines.]

Chapter 4

Propositional Logic

"A mind all logic is like a knife all blade. It makes the hand bleed that uses it."–
Rabindranath Tagore

4.1 Introduction

The previous chapters have shown several examples of how domain knowledge can be used in order to solve search-oriented problems. For example, domain-specific utility and cost functions can be used to play games like chess by searching for high-quality moves. The key point in search-oriented settings is that the domain knowledge is captured in the transition graph, starting/goal states, and in the utility functions associated with the nodes of the transition graph. In general, however, the nature of domain knowledge could be more complex, and a simple transition graph may not be able to capture the complexity required in such settings. Furthermore, the nature of the queries and problem settings in many domains may be complex, which cannot be handled with this type of simplistic modeling.

In order to move beyond narrow domains, it is required to create a more general framework for storing domain knowledge. This is achieved in artificial intelligence with the use of *knowledge bases*, which contain the background knowledge on top of which various artificial intelligence applications are constructed. Stated in informal terms, a knowledge base is a repository of known facts about a problem domain. These facts are then used in order to make inferences that are relevant to a specific user-defined problem or query. Logic provides a natural way to represent knowledge bases with the use of *logical sentences*, which collectively create a *knowledge representation language* for knowledge bases. Examples of knowledge representation languages include *propositional logic* and *first-order logic*. First-order logic is the preferred language for knowledge bases, but it is built on top of the principles of propositional logic. Although propositional logic is not powerful enough to support most real-world knowledge bases, it is provides the foundation for first-order logic.

Logical languages use *sentences* in the form of propositional expressions in order to express knowledge about the state of the world. A sentence is an assertion about the world in a mathematically defined knowledge representation language (which is propositional logic in the case of this chapter). Examples of sentences could be as follows:

> When it thunders, there is also lightening.
> There is no lightening today.
> There is no thunder today.

In this case, the third sentence can be logically inferred from the first two sentences. Note that these sentences are written in natural language, which makes it difficult to *formally* derive the third sentence from the first two sentences. An artificial intelligence system that reasons about the world needs to have this capability. A knowledge base has to express these sentences in a mathematical form in order to make logical inferences. The sentences in the knowledge base represent the known facts relevant to the problem at hand, and the logical relationships among different sentences may be leveraged to infer useful facts. This process lies at the formal mathematical process of logical inferencing, which uses symbolic representations of sentences.

The sentences in the knowledge base may or may not be derivatives of one another, although it is assumed that they are not inconsistent with one another. Any inconsistency in the knowledge base must be detected and removed up front before using it for an artificial intelligence application. Some sentences, which are considered fundamental truths, are treated as *axioms*, and such sentences are not derivatives of other sentences because they are considered fundamental truths. All sentences in the knowledge base are either axioms or they can be derived from axioms. Once the knowledge base has been constructed, it can be used to make non-obvious inferences by using a sequence of laws from propositional logic. In many cases, the agent is given a knowledge base of general facts, and then is queried based on a more specific facts. In the above example, the knowledge base might contain only the first sentence, and then the agent might be provided the second sentence as an additional (more specific) fact before being queried about other derivative facts. For example, a knowledge base may contain general logical assertions about disease symptoms and treatments for various diseases. Then, a specific fact about a particular patient's symptoms can be used to make inferences about their possible treatments. Therefore, knowledge bases are often used by the agent in order to respond to various types of queries that are generated in ad hoc fashion.

Propositional logic is, however, quite limited in its expressivity. Most logic-based artificial intelligence systems use first-order logic rather than propositional logic. In spite of its limited expressiveness, it is important to understand propositional logic and its applications as a prelude to discussions on first-order logic. This chapter will introduce the fundamental mathematical machinery underlying propositional logic, which is also inherited by first-order logic. This will set the stage for a more detailed discussion on first-order logic in the next chapter.

This chapter is organized as follows. The next chapter introduces the basics of propositional logic. The laws of propositional logic are introduced in Section 4.3. The use of propositional logic in expert systems is illustrated in Section 4.4. The basics of theorem proving with the use of the laws of propositional logic are discussed in Section 4.5. Methods for automated theorem proving are introduced in Section 4.6. The method of proof by contradiction is discussed in Section 4.7. The use of special types of representations of propositional expressions to make inferencing procedures more efficient is discussed in Section 4.8. A summary is given in Section 4.9.

4.2 Propositional Logic: The Basics

A knowledge base contains logical expressions, which are informally referred to as *sentences*. A sentence or *statement* is an alternative term used by computer scientists in lieu of the term *logical expression*. The use of the word sentence arises from the fact that artificial intelligence researchers often like to view logical expressions as formal types of assertions in knowledge bases in lieu of natural language sentences. While it would be ideal to create a knowledge base with natural language sentences (and indeed there are some knowledge bases that attempt to do so), such an approach increases the complexity of processing and parsing the natural language sentences. Logical expressions are crisply defined, unambiguous, and are amenable to computational parsing, which is not always the case in natural language sentences. In addition, well-defined procedures exist for inferring logical sentences from one another using the laws of propositional logic. Therefore, such sentences can be used to make logical inferences using propositional algebra, which is very well defined. On the other hand, checking whether two natural language sentences follow from one another is rather difficult, and it continues to be a field of research in artificial intelligence.

Sentences in propositional logic are formally represented as logical expressions formed out of a set of Boolean variables, each of which takes on the value of either *True* or *False*. The values *True* and *False* are referred to as *atomic operands*, and they represent the two fundamental constants in mathematical logic (just as real values represent the fundamental constants in arithmetic). All propositional expressions eventually evaluate to one of these two values. These two values are *negations* of one another, and they represent the only two possible values in propositional logic that arise from the evaluation of propositional expressions. The *negation operator* is denoted by $\neg$, and the relationship between the two atomic operands can be expressed as follows:

$$True = \neg False, \qquad False = \neg True$$

The logical negation operator flips the truth value of the expression following it. The negation operator is also referred to as the NOT operator. Strictly speaking, the use of the "=" symbol is not meaningful in propositional logic, but is instead replaced by the equivalence operator, denoted by "$\equiv$." Therefore, the above equivalence can be represented more accurately as follows:

$$True \equiv \neg False, \qquad False \equiv \neg True$$

Propositional variables are denoted by letters such as a, and each of them may take on the value of *True* or *False*. Each variable is considered an *atomic proposition*, and it is possible to build more complex expressions from these atomic propositions with the use of logical *operators*. A generalization of the notion of an atomic proposition is that of a *literal*, which includes negations of atomic propositions. In other words, the negation $\neg a$ of an atomic proposition a is also considered a literal, whereas $\neg a$ is not considered an atomic proposition. The value of $\neg a$ is *True* if and only a is *False*, and vice versa.

Propositional logic constructs propositional expressions from atomic propositions, which might correspond to mathematical representations of complex statements connecting these simpler propositions. These mathematical representations can be used to combine the truth values of multiple sentences to derive new sentences, and the truth values of these complex sentences may sometimes be non-obvious at first glance. The machinery of propositional logic is designed to derive these truths.

In order to incorporate greater expressivity in propositional logic, two other operators are used in propositional logic, which are the "logical and" and the "logical or" operators.

These operators are denoted by $\land$ and $\lor$, respectively. The value of $a \lor b$ is *True* if and only if at least one of the two variables takes on the value of *True*. Similarly, the value of $a \land b$ is *True* if and only if both variables take on the value of *True*. The expression $a \land b$ is referred to as the *conjunction* of a and b, whereas the expression $a \lor b$ is referred to as the *disjunction* of a and b. For example, consider the following pair of sentences:

> Either it is not raining, or John is not at work.
> It is raining.

From these two sentences, one can infer that John is not at work. These sentences can be expressed in terms of the propositional variables a and b. Let a denote the fact that it is raining, and b denote the fact that John is at work. Then the first of the above sentences can be expressed as $\neg a \lor \neg b$, and the second of the above sentences is the atomic proposition a. Once we have this mathematical representation of the above sentences, it is task of propositional logic to infer that John is not at work (i.e., $\neg b$) using the symbolic manipulation rules of propositional logic. These rules will be discussed over the course of this chapter. In order for artificial intelligence to build a reasoning system, it is also needed to be able to make such inferences in a formal way. The logic-based methodology provides a formal way to express these statements in propositional algebra, and make further inferences using the laws of propositional logic. The design of propositional logic requires the introduction of propositional operators and a grammar that can appropriately express these types of logical sentences and make sequences of inferences from them.

The above examples are based on the use of a single operator in a sentence. When atomic propositions or literals are combined with such operators, they are referred to as *propositional expressions*. In practice, a logical sentence contains multiple operators. In order to meaningfully represent sentences with multiple operators, one first needs to define the rules of what constitutes a valid sentence in propositional logic. Expressions in propositional logic have a well-defined syntax, just like all algebraic expressions do in numerical algebra. Propositional logic contains well-formed formulas using the following recursive construction rules:

- Any atomic proposition is well formed.

- The negation of an atomic proposition is well formed.

- If a and b are well formed formulas, then $a \lor b$ and $a \land b$ are well formed.

- If a is well formed then (a) is well formed. The use of $(\cdot)$ allows one to evaluate specific parts of the expression earlier, as in conventional (numerical) algebra. This is critical in being able to define the precedence in evaluation of portions of the expression. It is noteworthy that the operators $\land$, $\lor$, and $\neg$ have certain natural precedences, which are similar to the precedences[1] of operators used in numerical algebra. These precedences are described in a later paragraph of this section. The use of parentheses is useful when these precedences need to be overruled (as in numerical algebra).

One can construct complex propositional expressions (or well-formed formulas) from the atomic propositions. These expressions are also referred to as propositional expressions or propositional formulas. For example, consider the Boolean variables a and b. Now consider the following propositional expression:

$$a \land \neg b$$

---

[1] Numerical arithmetic uses the PEMDAS system, which creates an ordering of parenthesis, exponents, multiplication, division, addition, and substraction.

This expression contains two operators, and it has a value of *True* when a is *True* and b is *False*. Otherwise, the value of the expression is *False*. Note that the $\neg$ operator is executed first, which is defined by the laws of precedence in propositional logic (discussed later in this section). These types of well-formed formulas play a key role in propositional logic, because they are used to formally represent sentences in knowledge bases.

Although the three fundamental operators, $\wedge$, $\vee$, and $\neg$, are sufficient to represent all logical expressions, additional operators are used in order to provide better semantic interpretability to complex propositional expressions. A very important operator that occurs in rule-based systems of propositional logic is the "implies" operator:

$$a \Rightarrow b$$

Implications are also referred to as *rules,* and are extremely common in expert systems. The expression on the left-hand side is referred to as the *antecedent* of the rule, and the expression on the right-hand side is referred to as the *consequent* of the rule. This expression means that if a is *True* then b must also be *True* for the entire expression to be true. On the other hand, if a is *False*, then b can take on any value, and the entire expression still turns out to be true. It is easy to see that this expression is equivalent to the following:

$$\neg a \vee b$$

It is also possible to include the implication operator in well-formed expressions, such as the following:

$$(a \vee b) \Rightarrow (\neg c \wedge d)$$

The implication operator is one-sided, because a value of *False* for the antecedent guarantees a value of *True* for the full expression, irrespective of the value of the consequent. Therefore, the antecedent and consequent are not necessarily equivalent to one another, when the full expression takes on the value of *True*. In order to show the equivalence of two propositional expressions, we need a different operator, which is denoted by the symbol "$\equiv$", which was introduced at the beginning of this section. In other words, one can denote this equivalence as follows:

$$a \equiv b$$

It is common to use the equivalence operator for propositional expressions that reduce to one another, such as the following:

$$(a \Rightarrow b) \equiv (\neg a \vee b)$$

Note that the implication operator "$\Rightarrow$" is unidirectional corresponding to the logical statement "if," whereas the equivalence operator is really a bidirectional implication, which stands for "if and only if." This operator can also be denoted by $\Leftrightarrow$ as follows:

$$a \Leftrightarrow b$$

This expression is true if and only if a and b take on the same Boolean value. The relationship between equivalence and unidirectional implication is as follows:

$$(a \Leftrightarrow b) \equiv (a \equiv b)$$
$$\equiv [(a \Rightarrow b) \wedge (b \Rightarrow a)]$$

Since the operators $\Leftrightarrow$ and $\equiv$ are the same, we will consistently use the latter throughout this book. The equivalence $(a \equiv b)$ between a and b can also be expressed in terms of the three fundamental operators of propositional logic as follows:

$$(\neg a \vee b) \wedge (\neg b \vee a)$$

In complex propositional expressions, it is important to understand the notion of *operator precedence* in order to know which operator is executed first. For example, in the expression $a \wedge \vee c$, should the $\wedge$ be executed first, or should the $\vee$ be executed first. The two different orders of precedence do not lead to the same result. This is a similar question to one that occurs in numerical algebra, where the multiplication operator is executed first in the expression $2 + 3 * 4$. A similar set of precedences can be defined in propositional logic as follows:

- The negation operator, $\neg$, has the highest precedence.

- The second-highest precedence is that of the $\wedge$ operator.

- The third-highest precedence is that of the $\vee$ operator.

- The $\Rightarrow$ and $\equiv$ operators have the lowest precedence.

Note that $\wedge$ is the logical analog of multiplication in numerical algebra, whereas $\vee$ is the logical analog of addition in numerical algebra. In numerical algebra, multiplication has higher precedence than addition. Similarly, in propositional expressions, the $\wedge$ operator has higher precedence than the $\vee$ operator In fact, some expositions of propositional logic use '+' in lieu of $\vee$ and '·' in lieu of $\wedge$. In other words, the expression $a \wedge b \vee c$ would be written as $a \cdot b + c$. As a result, the expression $a \wedge b \vee c$ is equivalent to $(a \wedge b) \vee c$ rather than $a \wedge (b \vee c)$. In order to show that the two expressions $(a \wedge b) \vee c$ and $a \wedge (b \vee c)$ are not equivalent, one can set a to *False* and c to *True*. In such a case, the two expressions evaluate to different values as follows:

$$[(a \wedge b) \vee c] \equiv [(False \wedge b) \vee True] \equiv True$$
$$[a \wedge (b \vee c)] \equiv [False \wedge (b \vee True)] \equiv False$$

Therefore, the two expressions are not equivalent. In cases where there is ambiguity or one wants to force a precedence that is different from the natural precedence, one can use parentheses to force it. In other words, while $a \wedge b \vee c$ automatically evaluates to $(a \wedge b) \vee c$, one can use the expression $a \wedge (b \vee c)$ to force the $\vee$ operator to be evaluated first. Again, this is a similar approach to what is done in numerical algebra.

Several other operators, such as NAND, NOR, and XOR, are also used in propositional logic, and these expressions can be expressed in terms of the operators $\wedge$, $\vee$, and $\neg$. For example, the XOR function over a and b, denoted by $a \oplus b$ takes on the value of *True* if and only if exactly one of a and b takes on the value of *True*. One can express the XOR function in terms of the three basic operators $\wedge$, $\vee$, and $\neg$ as follows:

$$a \oplus b \equiv (a \wedge \neg b) \vee (\neg a \wedge b)$$

The NAND/NOR operators are the negations of the AND/OR operators, and are denoted by $\overline{\wedge}$ and $\underline{\vee}$, respectively. Therefore, the NAND and the NOR operators are defined as follows:

$$[a \overline{\wedge} b] \equiv [\neg (a \wedge b)]$$
$$[a \underline{\vee} b] \equiv [\neg (a \vee b)]$$

Table 4.1: Truth table for $\neg a$

a	$\neg a$
True	False
False	True

Table 4.2: Truth tables for $a \wedge b$ and $a \vee b$

a	b	$a \wedge b$
True	True	True
True	False	False
False	True	False
False	False	False

(a) $a \wedge b$

a	b	$a \vee b$
True	True	True
True	False	True
False	True	True
False	False	False

(b) $a \vee b$

The explicit use of these operators in propositional logic is rare. In general, propositional expressions are expressed with the use of $\wedge$, $\vee$, $\neg$, and $\Rightarrow$ operators for simplicity. Although any propositional expression can be expressed in terms of $\neg$ and one of the other two operators ($\vee$ or $\wedge$), the readability of a propositional expression is considered important. Therefore, the $\Rightarrow$ operator is used extensively in propositional logic because of its natural ability to represent semantically interpretable rules in the knowledge base. Similarly, the equivalence operator is frequently used to convert between different types of propositional expressions.

4.2.1 Truth Tables

A *truth table* is an alternative way of writing a logical expression. A truth table that is defined in terms of k atomic variables has 2^k rows. Each row corresponds to a possible instantiation of the Boolean variables in the table. The $(k + 1)$ columns of table correspond to the values of the k input variables and the final result. For example, consider the simplest case of the $\neg$ operator with $k = 1$, which is shown in Table 4.1. In this case, the table contains $(k+1) = 2$ columns and $2^1 = 2$ rows, corresponding to the different input and output values.

Consider the slightly more complex case of the expression $a \wedge b$, where a and b are atomic expressions. Then, the truth table for this expression is shown in Table 4.2(a). The table contains $(2 + 1) = 3$ columns and $2^2 = 4$ rows. It is easy to verify that the output for each row is the same as the corresponding Boolean value of the $\vee$ expression. The truth table for $a \vee b$ is shown in Table 4.2(b). The table has 3 columns and 2^2 rows.

An expression is *satisfiable* if at least one row of the truth table evaluates to the value of *True*. On the other hand, an expression is a *tautology* if all rows of the table evaluate to the value of *True*. Determining whether an expression is satisfiable or a tautology are important problems in propositional logic; the above exposition implies that one can use truth tables in order to determine whether expressions are satisfiable or are tautologies. Unfortunately, this approach requires exponential time in terms of the number of variables k, because each combination of values of the binary variables needs to be checked; unfortunately, the number of rows in the table is 2^k. Therefore, a natural approach is to simplify a propositional expression first in order to determine whether it simplifies to a tautology or to determine whether it is satisfiable. This chapter will introduce some key algorithms for these types of simplifications.

So far, we have shown how to convert propositional expressions into truth tables. It is also possible to perform the reverse task of converting a truth table into a propositional expression. The natural way to express a Boolean table, in which n rows have a value of *True* in the final column, as a propositional expression of the following form:

$$R_1 \vee R_2 \vee \ldots \vee R_n$$

Each R_i is a propositional expression (rather than atomic expression), and it contains a logical AND of the atomic propositional variables in the k columns of the table or their negations. For example, consider the case where the variables a_2 and a_3 are set to *False* in the ith row and all other variables are *True*. In such a case, the logical expression for the ith row is as follows:

$$R_i \equiv (a_1 \wedge \neg a_2 \wedge \neg a_3 \wedge a_4 \ldots \wedge a_{k-1} \wedge a_k)$$

In other words, each R_i is a *conjunction* of literals, where each literal is an atomic proposition or its negation, depending on whether that symbol takes on the value of *True* or *False* in the truth table. Such an expression is said to be in *disjunctive normal form*, which will be discussed later in this chapter.

In order to understand how to convert a truth table into a propositional expression, consider the case of Table 4.2(a), where there is only one row with the value of *True*, and the corresponding symbols a and b take on the value of *True* in this row. Let the rows of the Tables 4.2(a) and (b) in corresponding order be denoted by R_1, R_2, R_3, and R_4, respectively. Therefore, the corresponding logical expression for Table 4.2(a) is the single term R_1, which is $a \wedge b$. On the other hand, in the case of Table 4.2(b), there are three rows that take on the value of *True*. The corresponding expression is a disjunct of three expressions R_1, R_2, and R_3, which results in the following expression:

$$R_1 \vee R_2 \vee R_3$$
$$\equiv (a \wedge b) \vee (a \wedge \neg b) \vee (\neg a \wedge b)$$

At first glance, this result would seem to be different from the expression $a \vee b$, for which the truth table of Table 4.2(b) is constructed. However, it turns out that the two expressions are equivalent, because one can *algebraically* show that $R_1 \vee R_2 \vee R_3$ is equivalent to $a \vee b$. This type of equivalence can be shown by using the laws of propositional logic, which is the topic of the next section.

4.3   Laws of Propositional Logic

The laws of propositional logic are used to convert expressions into simpler form using algebraic manipulation. In principle, this is similar to the way expressions are shown to be equivalent in numerical algebra. The laws of propositional algebra are also similar to those used in numerical algebra, although there are some key areas of differences. In the classical field of mathematical logic, these rules are used by mathematicians to prove that logical expressions are equivalent. A natural question arises as to whether this can also be done in an automated manner. This area in artificial intelligence is referred to as *automated theorem proving*. The algebraic simplification and manipulation of propositional expressions can be useful in a variety of applications in artificial intelligence, such as in the case of expert systems.

We first define a number of elementary and fundamental axioms in propositional logic, which are referred to as the *idempotence*, *identity*, and *annihilation* relations:

$$a \lor a \equiv a \qquad \text{[Idempotence of } \lor \text{]}$$
$$a \land a \equiv a \qquad \text{[Idempotence of } \land \text{]}$$
$$a \lor \textit{False} \equiv a \qquad \text{[Identity for } \lor \text{]}$$
$$a \land \textit{True} \equiv a \qquad \text{[Identity for } \land \text{]}$$
$$a \lor \textit{True} \equiv \textit{True} \qquad \text{[Annihilator for } \lor \text{]}$$
$$a \land \textit{False} \equiv \textit{False} \qquad \text{[Annihilator for } \land \text{]}$$

An important axiom associated with negation operator is that of *double negation*. The use of double negation leads to the same expression:

$$\neg(\neg a) \equiv a \qquad \text{[Double negation law]}$$

The correctness of this law follows from the fact that flipping the value of an atomic operand twice leads to the same value.

Since the truth value of a is always different from that of $\neg a$, applying the $\lor$ and $\land$ operators between a and $\neg a$ always leads to one of the two atomic truth values:

$$a \lor \neg a \equiv \textit{True} \qquad \text{[Complementarity law for } \lor \text{]}$$
$$a \land \neg a \equiv \textit{False} \qquad \text{[Complementarity law for } \land \text{]}$$

These laws are referred to as the *complementarity laws* in propositional logic.

The $\land$ and $\lor$ operators are commutative and associative. The commutativity laws are as follows:

$$(a \lor b) \equiv (b \lor a)$$
$$(a \land b) \equiv (b \land a)$$

Note that these laws are similar to the commutativity laws used in addition and multiplication for numerical algebra. Similarly, the associativity laws of propositional logic are similar to those of addition and multiplication in numerical algebra:

$$((a \lor b) \lor c) \equiv (a \lor (b \lor c))$$
$$((a \land b) \land c) \equiv (a \land (b \land c))$$

It is possible to apply the AND operator to multiple propositional variables without the use of parentheses without loss of ambiguity. This is because the grouping is irrelevant as a result of associativity:

$$(a \land b \land c) \equiv [(a \land b) \land c]$$
$$\equiv [a \land (b \land c)]$$

A similar result also holds true for the OR operator, because it is associative.

$$(a \vee b \vee c) \equiv [(a \vee b) \vee c]$$
$$\equiv [a \vee (b \vee c)]$$

It is evident that many of these rules are similar to the rules that are common in numerical algebra. As in the case of precedence rules, the $\vee$ operator behaves like the addition operator in numerical algebra, whereas the $\wedge$ operator behaves like the multiplication operator in numerical algebra. However, the laws of propositional logic are also different from those in numerical algebra in some critical ways. The first difference is that the distributive law applies in a symmetric way between the $\wedge$ and $\vee$ operators:

$$(a \wedge (b \vee c)) \equiv ((a \wedge b) \vee (a \wedge c))$$
$$(a \vee (b \wedge c)) \equiv ((a \vee b) \wedge (a \vee c))$$

The first of these laws is very similar to the distributive law in numerical algebra, although the latter does not have any analogous law in numerical algebra. This is because $a + bc$ is not equivalent to $(a + b)(a + c)$ in numerical algebra.

Finally, the De Morgan laws are the "negation push through" laws, which allow us to push a negation through AND/OR operators. The De Morgan laws are as follows:

$$\neg(a \vee b) \equiv \neg a \wedge \neg b$$
$$\neg(a \wedge b) \equiv \neg a \vee \neg b$$

The De Morgan laws are quite different from the laws of numerical algebra, and there is no precise analog in numerical algebra for these cases. One consequence of the De Morgan laws is that it shows that one does not need *both* $\vee$ and $\wedge$ operators to express any propositional statement; either one will do in conjunction with the negation operator. However, since the readability of propositional expressions is essential, all three operators are used extensively in propositional logic.

4.3.1   Useful Properties of Implication and Equivalence

Much of the propositional logic in artificial intelligence deals with rules, and therefore the properties of the implication operator are particularly important. Some of the aforementioned laws can be used to derive other laws in propositional logic involving the implication operator. These laws are used frequently enough that they are often treated as fundamental laws of propositional logic (even though they can be derived from other laws). Note that the implication and equivalence operators can be fully expressed in terms of the fundamental propositional operators AND, OR, and NOT, and therefore the laws of these (more fundamental) operators can be used to derive the laws of the implication and equivalence operators. For example, an important law is that of *contraposition*:

$$(a \Rightarrow b) \equiv (\neg b \Rightarrow \neg a)$$

This law can be shown using the definition of implication with respect to the fundamental operators $\vee$ and $\neg$, and then showing that each side is equivalent to $\neg a \vee b$. The left-hand side is equivalent to $\neg a \vee b$ by definition. Showing that the right-hand side is equivalent to the $\neg a \vee b$ requires the use of the double negation law to show equivalence to $b \vee \neg a$, and then using the commutativity of the $\vee$ operator. The contraposition law is important enough that it is treated as a fundamental axiom, even though it can be derived from other laws of propositional logic. In other words, it is technically a theorem (derived from other axiomatic laws) rather than an axiomatic law in its own right. One can see that the contraposition law makes semantic sense by considering the following pair of statements:

> If it rained today, I used my umbrella today.
> If I did not use my umbrella today, it did not rain today.

Here, the atomic propositions a and b are "it rained today" and "I used my umbrella today," respectively. Then, the first statement is equivalent to $a \Rightarrow b$ and the second statement is equivalent to $\neg b \Rightarrow \neg a$.

Another important property of implication is that of *transitivity*. By the transitivity law, the two rules $a \Rightarrow b$ and $b \Rightarrow c$ can be used to infer $a \Rightarrow c$. This can be shown by using case-wise analysis over the two different Boolean values of b. If b is *True*, then c must be *True* as well so that the rule $b \Rightarrow c$ is equivalent to *True*. Therefore, $a \Rightarrow c$ must be *True* as well. On the other hand, if b is *False*, then a must be *False* as well by contraposition of $a \Rightarrow b$. Therefore, $(a \Rightarrow c) \equiv \neg a \vee c$ must be *True* as well. Case-wise analysis is a useful tool for showing many results in propositional logic.

There are also a number of laws associated with equivalence, which can be derived as byproducts of other laws in propositional logic. All expressions are equivalent to themselves, which is the *reflexive* property.

$$a \equiv a$$

The equivalence relation is also commutative and transitive:

$$(a \equiv b) \equiv (b \equiv a)$$
$$[(a \equiv b) \wedge (b \equiv c)] \Rightarrow (a \equiv c)$$

The equivalence relation applies to the negations of the individual arguments:

$$(a \equiv b) \equiv (\neg a \equiv \neg b)$$

4.3.2 Tautologies and Satisfiability

Certain types of propositional expressions always take on the value of *True*, which are referred to as *tautologies*, and an example is as follows:

$$(a \wedge \neg b) \vee (a \wedge b) \vee (\neg a)$$

No matter how we set the values of a and b, the expression can be shown to have a value of *True*. Such tautologies can always be shown using the rules of propositional logic. For example, the first portion $(a \wedge \neg b) \vee (a \wedge b)$ of the above expression can be shown to be equivalent to $a \wedge (\neg b \vee b)$ using the distributive law. Therefore, one obtains the following expression:

$$[a \wedge (\neg b \vee b)] \wedge \neg a \equiv [a \wedge True] \vee \neg a$$
$$\equiv a \vee \neg a$$
$$\equiv True$$

The negation of a tautology (such as the entire expression above) always has a value of *False*, and it is therefore not *satisfiable*. Satisfiability refers to the fact that one can find at least one assignment of values to variables, so that the entire expression turns out to be *True*. This problem of determining whether an expression is satisfiable is known to be NP-complete, and therefore no polynomial time algorithm is known. On the other hand, a simple exponential-time algorithm does exist for testing satisfiability. Such an algorithm tries all 2^k possible assignments of values to variables to check if the expression takes on the value of *True* for at least one assignment. A table of all 2^k possible values of variables and the expression is a tautology if all outcomes yield the value of *True*. As discussed earlier, the table of all possible outcomes is referred to as a truth table, and it is an equivalent way of representing a propositional expression. In the case of a tautology, all entries in the final column of the truth table will be *True*. In the case of an unsatisfiable expression, all entries in the final column of the truth table will be *False*. A satisfiable expression that is not a tautology will contain both *True* and *False* values in the final column of the truth table. In practice, the use of a truth table to check whether an expression is a tautology or whether it is satisfiable is impractical for expressions containing many variables. The reason is that the approach requires 2^k time for an expression with k variables. Many logic-based systems use a large number of variables, as a result of which this running time becomes impractical. The problem of determining whether an expression is a tautology is NP-complete as well (see Exercise 14).

Particular types of tautologies can be used to justify well-known proof techniques used in mathematics. For example, when one wants to prove a particular result in mathematics, one often assumes the negation of that result and then proves a contradiction. Stated formally, if one assumes a and wants to show b, one really wants to show $a \Rightarrow b$. However, by assuming the negation $\neg b$, one can often show $\neg a$, which results in a contradiction to the assumption that a is *True*. One is effectively using the following tautology:

$$(a \Rightarrow b) \equiv (\neg b \Rightarrow \neg a)$$

This is, in fact, one of the key methods used in artificial intelligence in order to derive inferences in knowledge bases.

4.3.3   Clauses and Canonical Forms

The problem of propositional satisfiability is an important one in the proof systems associated with knowledge bases. The determination of whether an expression is satisfiable is often facilitated by converting them to particular types of canonical forms. In order to understand canonical forms, we first need to introduce some terms and expressions. For example, consider the case where a, b, and c are (not necessarily atomic) expressions. Then, each of a, b, and c in the expression $a \lor b \lor c$ is referred to as a *disjunct*. Similarly, each of a, b, and c in the expression $a \land b \land c$ is referred to as a *conjunct*.

Knowledge bases typically contain particular types of well-formed formulas, which is also referred to as *clausal form*. A *clause* is defined as a disjunction of literals, and it is an alternative representation of a rule. Knowledge bases are always expressed in terms of rules in order to enable inferencing procedures. These rules can be expressed as clauses as long as the antecedent of the rule only contains the $\land$ operator. Consider a rule of the following form:

$$b_1 \land b_2 \land \ldots \land b_k \Rightarrow a$$

If the antecedent of the rule is *True*, the consequent must also be *True* in order for this propositional expression (i.e., rule) to take on the value of *True*. This is because this rule can be expressed as $a \vee \neg b_1 \vee \neg b_2 \vee \ldots \vee \neg b_k$, which is a disjunction of literals. Similarly, a disjunction of literals can be converted to a rule by putting any one of them in the consequent, and the conjunction of remaining (after negation) in the antecedent. Since knowledge bases contain rules that are relevant for making inferences, it is natural to build them around clauses that can be converted into rules.

An interesting property of logical expressions is that all expressions can be expressed as a disjunction of conjuncts. This is a *canonical form* of logical expressions, and it is referred to as *disjunctive normal form (DNF)*. An example of an expression in disjunctive normal form is as follows:

$$(a \wedge \neg b) \vee (c \wedge d) \vee e \vee (b \wedge c \wedge \neg d)$$

Note that each of the formulas a, b, $\neg b$, c, d, $\neg d$, and e, must be literals in order for the above expression to be considered to be in disjunctive normal form. Furthermore, each of the disjuncts must be a conjunction of literals. It requires polynomial time to check Boolean satisfiability on a propositional formula in disjunctive normal form (in terms of the length of the DNF formula). This is because the expression is unsatisfiable if and only if each of the disjuncts is unsatisfiable; this is possible when each disjunct contains the conjunction of a literal and its negation. All these conditions can be checked very efficiently in terms of the length of the DNF expression. However, the polynomial complexity of satisfiability checking in terms of the *length* of the expression does not mean that the problem is polynomially solvable in terms of the length of *arbitrary* propositional formulas; this is because converting a propositional formula to disjunctive normal form might require exponential time. Furthermore, the DNF formula might have length that is exponentially related to the number of terms in the original version of the propositional expression. For example, the expression $\wedge_{i=1}^{n}(a_i \vee b_i)$ has length $O(2^n)$ when it is converted to disjunctive normal form. This result can be shown by induction, wherein we assume that the result holds for $n = r$, and then use this assumption to show that the result also holds for $n = r+1$. This also explains why the Boolean satisfiability problem is NP-hard, while it requires polynomial time to check the satisfiability of an expression in DNF form. Interestingly, finding whether an expression is satisfiable is equivalent to finding whether the negation of the expression is a tautology. In general, checking whether an expression is a tautology is also an NP-hard problem (like Boolean satisfiability). Therefore, the satisfiability and tautology problems are equally difficult.

Another important canonical form of an expression is the *conjunctive normal form (CNF)*, in which the expression is written as a conjunct of disjuncts. An example of an expression in conjunctive normal form is as follows:

$$(a \vee \neg b) \wedge (c \vee d) \wedge e \wedge (b \vee c \vee \neg d)$$

Each of the expressions a, b, $\neg b$, c, d, $\neg d$, and e, must be literals in order for the formula to be considered to be in conjunctive normal form. Checking whether an expression is a tautology is relatively simple, when the expression is stated in conjunctive normal form (in terms of the length of the CNF expression). This is because each of the conjuncts must contain literals of the form $a \vee \neg a$ in order for each conjunct to evaluate to a tautology. It is noteworthy that converting an expression to CNF form might create a new expression that is much longer than the original expression. Therefore, even though checking whether a CNF expression is a tautology is simple, the process of converting an expression to conjunctive normal form is what causes the increase in complexity.

4.4 Propositional Logic as a Precursor to Expert Systems

In general, expert systems cannot be constructed efficiently with propositional logic. The reason is that expert systems require a way to reason over objects. For example, some truths may be relevant to all objects, whereas others may be relevant to a specific object such as John. Therefore, one needs to a way to distinguish between generic propositional statements and those that are specific to particular objects; furthermore, one needs to be able to connect and reason between universal statements over all objects and specific statements about particular objects. However, propositional logic does not have a notion of objects over which reasoning can be performed. A more advanced form of logic, referred to as *first-order logic*, is able to perform this type of reasoning with *quantifiers* such as $\forall$ and $\exists$. These quantifiers provide mechanisms of reasoning over general and specific groups of objects. However, since propositional logic forms the basis of first-order logic, it is helpful to create a toy example of an expert system with propositional logic in order to obtain a feel for how such systems might work in general (although the process is somewhat cumbersome because of the simplistic nature of propositional logic). In the next chapter, we will show how this very example can be implemented more cleanly with first-order logic.

A knowledge base typically contains multiple sentences (logical statements), and one often wishes to use these statements in order to infer specific conclusions. For example, consider a situation where one is trying to create an expert system in order to make patient diagnoses. We revisit one of the examples discussed in Chapter 1, where a patient John comes in with a specific set of symptoms. The patient John presents the following facts about their situation:

John is running a temperature
John is coughing
John has colored phlegm

Now imagine a case where the expert system contains the following subset of rules:

IF coughing AND temperature THEN infection
IF colored phlegm AND infection THEN bacterial infection
IF bacterial infection THEN administer antibiotic

Note that these statements apply to all individuals, and therefore they can be applied to John as well. Therefore, we can create propositional variables from these statements that are specific to John. Note that this type of conversion is implicitly done is first-order logic, although we have done this explicitly in this case:

IF John is coughing AND John has temperature THEN John has infection
IF John has colored phlegm AND John has infection THEN John has bacterial infection
IF John has bacterial infection THEN administer antibiotic to John

Then, one can define the propositional variable c for John coughing, t for John running a temperature, f for John having an infection, p for John having colored phlegm, b for John having bacterial infection, and a for administering an antibiotic to John. The rules in the knowledge base can be coded up in the form of propositional rules as follows:

$$c \wedge t \Rightarrow f$$
$$p \wedge f \Rightarrow b$$
$$b \Rightarrow a$$

This set of rules only provides a snapshot of the knowledge base, because many other rules might be defined based on other medical conditions in the knowledge base. The conditions for John can be summarized as follows:

$$t \wedge c \wedge p$$

The goal of the inference engine in expert systems is to use the rules in the knowledge base in combination with the laws of propositional logic in order to make inferences. Therefore, the inference engine will start with $t \wedge c \wedge p$ in order to make a series of inferences that eventually leads to a logical variable corresponding to administering a medicine (e.g., administering an antibiotic).

We make three important observations about the knowledge engineering process that are common to such systems. The first is the fact that the knowledge base contains *general knowledge* about diseases and their specific instances. This part of the knowledge base is stable, and it changes relatively slowly over time in most cases. The second is that additional, specific information is often available in the context of a query, which was not originally present in the knowledge base. In this particular case, we have additional knowledge corresponding to John's symptoms, and the query is also related to possible treatments for John. The final part is the inferencing procedure in response to specific queries (e.g., treatments for John). Strictly speaking, distinguishing between general knowledge (about diseases and symptoms) and specific knowledge (about a particular patient) is done using *first-order logic*, so that the instantiation to a specific object is done automatically as a part of the algebra (see Chapter 5). This is because first-order logic allows the binding of propositions to particular subjects such as John, whereas we are forced to hard-code each rule in propositional logic with every possible person (such as John) in order to make the expert system work. However, in this case, we will make the simplified assumption that every proposition in the knowledge base is specific to John, and using it for a different patient requires us to appropriately modify the knowledge base. This assumption allows us to show the basic principle of inferencing, although it is not a practical approach for building expert systems.

The process of inferencing keeps adding statements to the knowledge base. Since each of the assertions c and t hold for John and we have the rule $c \wedge t \Rightarrow f$ in the knowledge base, one can also infer f for John. Furthermore, it is known that John has colored phlegm, it follows that p is true. In other words, $p \wedge f$ can be inferred, which implies that b is true. Using the rule $b \Rightarrow a$, one obtains a. In other words, the expert system recommends the administration of an antibiotic to John. In particular types of settings, one might want to show that a particular propositional expression takes on the value of *True*, and the procedure is referred to as *entailment*. In this particular case, the goal clauses correspond to conjunctions of Boolean variables that recommend specific actions. Note that reaching these types of conclusions (goal clauses) becomes a non-trivial matter when the expert system contains hundreds of thousands of rules. What we have shown is the procedure of entailment, where one statement logically follows from another. In the next section, we will formalize these ideas of using propositional logic in order to reach specific propositional expressions (which is analogous to the idea of goal states in search) by using a sequence of transformations. As we will show later in this chapter, the systematic procedure used for reaching propositional expressions is similar to algorithms used in search. Therefore, logic-based systems are fundamentally not very different from those used in the search; they provide a more formal and mathematical mechanism for achieving the same goal.

4.5   Equivalence of Expressions in Propositional Logic

In theorem proving, we try to show that one propositional expression is equivalent to another by using the laws of propositional logic. As we have already discussed, this equivalence can be shown by using truth tables. Unfortunately, the use of truth tables is computationally expensive, because an expression with k variables requires one to try 2^k possible assignments corresponding to the different rows in the truth table. Real-world propositional expressions in knowledge bases can contain large numbers of variables. Note that the problem of determining whether two expressions A and B are equivalent is the same as determining whether the expression $A \equiv B$ is a tautology. As we have discussed earlier, the problem of determining whether an arbitrary propositional expression is a tautology is NP-hard. Therefore, the problem of determining whether two propositional expressions are equivalent is NP-hard in general. However, in practice, a number of computational tricks exist for reducing the complexity of the process.

Although it is possible to prove the equivalence of propositional expressions by constructing their truth tables (and showing that the outcomes in all rows of the truth tables are equivalent), this is a rather expensive way to do so. If a propositional expression contains 20 variables, it will result in the creation of a truth table with $2^{20} = 1048576$ entries for each of the two expressions. Evaluating 1048576 pairs of expressions and comparing them is expensive. Therefore, a more efficient approach is to use the laws of propositional logic in the previous section in order to transform one expression into another. Furthermore, these laws need to be applied in a systematic way with a computer algorithm. In order to illustrate the use of propositional logic to show equivalence, we will use a few examples.

Example 4.5.1 (Equivalence of Expressions) *Show that the expression* $(a \vee c) \wedge (a \vee d) \wedge (b \vee c) \wedge (b \vee d)$ *is equivalent to the expression* $[(a \wedge b) \vee (c \wedge d)]$.

In order to show the equivalence of the above expressions, we will use a sequence of transformations using the laws of propositional logic. This sequence of steps is shown below with the appropriate explanation:

$$([a \vee c) \wedge (a \vee d) \wedge (b \vee c) \wedge (b \vee d)] \equiv [[(a \vee c) \wedge (a \vee d)] \wedge [(b \vee c) \wedge (b \vee d)]] \quad [\text{Associativity}]$$
$$\equiv [(a \vee (c \wedge d)) \wedge (b \vee (c \wedge d))] \quad [\text{Distributivity}]$$
$$\equiv [(a \wedge b) \vee (c \wedge d)] \quad [\text{Distributivity}]$$

Note that this transformation requires only three steps, whereas building a truth table would have required us to substitute $2^4 = 16$ combinations of possible values from $\{True, False\}$ for the four variables a, b, c, and d. Clearly, building a truth table would be the more cumbersome approach in order to show equivalence of the two expressions. By showing that the two expressions are equivalent, we have effectively shown that the following statement is a tautology:

$$[(a \vee c) \wedge (a \vee d) \wedge (b \vee c) \wedge (b \vee d)] \equiv [(a \wedge b) \vee (c \wedge d)]$$

Showing that two expressions are equivalent is the same as showing that the composite expression obtained by putting an "$\equiv$" operator between the two expressions is a tautology. Similarly, showing that an expression is a tautology is the same as showing that the expression is equivalent to the atomic expression *True*. Consider the following example:

Example 4.5.2 (Tautology) *Show that the following statement is a tautology:*

$$[(a \Rightarrow b) \wedge (b \Rightarrow c)] \Rightarrow (a \Rightarrow c)$$

Showing that a statement is a tautology boils down to reducing the statement to the atomic value of *True*. The above statement can be simplified in a variety of ways, and a case-wise analysis will be used in order to show the result:

Case I: The Boolean variable b takes on the value of *True*. In this case, the simplification of the expression is as follows:

$$\begin{aligned}
\{[(a \Rightarrow b) \wedge (b \Rightarrow c)] \Rightarrow (a \Rightarrow c)\} &\equiv \{[(\neg a \vee True) \wedge (\neg True \vee c)] \Rightarrow (a \Rightarrow c)\} \\
&\equiv [True \wedge c] \Rightarrow (a \Rightarrow c)\} \\
&\equiv c \Rightarrow (a \Rightarrow c)\} \\
&\equiv \neg c \vee \neg a \vee c \\
&\equiv (\neg c \vee c) \vee \neg a \\
&\equiv True \vee \neg a \\
&\equiv True
\end{aligned}$$

Case II: The Boolean variable b takes on the value of *False*. In this case, the simplification of the expression is as follows:

$$\begin{aligned}
\{[(a \Rightarrow b) \wedge (b \Rightarrow c)] \Rightarrow (a \Rightarrow c)\} &\equiv \{[(\neg a \vee False) \wedge (\neg False \vee c)] \Rightarrow (a \Rightarrow c)\} \\
&\equiv [\neg a \wedge True] \Rightarrow (a \Rightarrow c)\} \\
&\equiv \neg a \Rightarrow (a \Rightarrow c)\} \\
&\equiv a \vee \neg a \vee c \\
&\equiv (a \vee \neg a) \vee c \\
&\equiv True \vee c \\
&\equiv True
\end{aligned}$$

Therefore, one obtains a result of *True* in both cases. It is noteworthy that a truth table is the most extreme instance of case-wise analysis, where one tries *all* possible combinations of variables to obtain a value of *True* in each case. While the use of a truth table is impractical in most cases, restricted forms of case-wise analysis are common in propositional logic proofs. In most cases, it is possible to construct the proof simply without case-wise analysis, and only using the basic axioms of propositional logic. For example, the above expression can be simplified without casewise analysis:

$$\begin{aligned}
\{[(a \Rightarrow b) \wedge (b \Rightarrow c)] \Rightarrow (a \Rightarrow c)\} &\equiv \{[(\neg a \vee b) \wedge (\neg b \vee c)] \Rightarrow (\neg a \vee c)\} \\
&\equiv \neg[(\neg a \vee b) \wedge (\neg b \vee c)] \vee (\neg a \vee c) \\
&\equiv (a \wedge \neg b) \vee [(b \wedge \neg c) \vee (\neg a \vee c)]
\end{aligned}$$

Using the distributive property of AND over OR, we obtain the following:

$$\begin{aligned}
(a \wedge \neg b) \vee [(b \vee \neg a \vee c) \wedge (\neg c \vee \neg a \vee c)] &\equiv (a \wedge \neg b) \vee [(b \vee \neg a \vee c) \wedge True] \\
&\equiv \neg(\neg a \vee b) \vee (\neg a \vee b) \vee c \\
&\equiv True \vee c \equiv True
\end{aligned}$$

One issue is that the aforementioned proofs are ad hoc in nature, which is possible for a human, but not a simple mater for an automated system. When performing theorem proving with AI systems, one needs a systematic way of showing these types of results. This will be the topic of discussion in the subsequent sections.

4.6   The Basics of Proofs in Knowledge Bases

As discussed earlier, expert systems require ways of inferring conclusions from the rules in the knowledge base. The previous section also provides examples of how the laws of propositional logic can be used to make inferences. In this section, we discuss how one might automate this process in a somewhat more systematic way. In knowledge bases, the general problem may be stated as follows:

> If KB represents the conjunction of all the statements in the knowledge base and q represents a query, then we wish to show that $KB \Rightarrow q$.

In our earlier example of a medical expert system (cf. Section 4.4), KB represents the conjunction of all general statements about symptoms, their implied diseases and treatments. The query q is a sentence that John's symptoms imply a particular candidate treatment, such as the use of antibiotics. Technically, it is also possible to add John's symptoms to the knowledge base, and in that case, the query sentence simply corresponds to the candidate treatment for John.

A natural question arises as to show one can formalize the problem of automated theorem proving in terms of states and actions. This creates a similar framework to search, wherein one searches for particular states from a starting state. We somehow need to define the notion of a state that depends on the knowledge base, and then define the set of actions that allows the agent to move from one state to another. Therefore, the notions of state and action are defined as follows:

- *State:* The current knowledge base together with all its sentences/rules is referred to as a state. The state changes as more statements are added to the knowledge base as a result of the inferencing procedure. The initial state corresponds to the initial knowledge base with the initial set of sentences in it. For example, in a medical expert system, the initial knowledge base might contain all the rules relating symptoms to diseases as well as the statements corresponding to the candidate treatments for various diseases. The goal state corresponds to any knowledge base that contains sentences corresponding to suggested treatments for a patient with particular symptoms. Note that these sentences can only be added to the knowledge base with the use of actions that add statements to the knowledge base. This process is described below.

- *Action:* Actions correspond to using rules whose antecedents match one more statements in the knowledge base, thereby inferring the consequent. This consequent can then be added to the knowledge base. For example, when a symptom of a patient matches the antecedent of a rule in the knowledge base, the disease or treatment in its consequent is added to the knowledge base (as it applies to the patient at hand). Adding a consequent to the knowledge base automatically changes the state, since the knowledge base has now been augmented with additional assertions. These additional assertions might trigger further actions, if they match the antecedents of existing rules. These types of actions lie at the heart of a type of inferencing in expert systems, known as *forward chaining*. However, there are other alternatives such as proof by contradiction or backward chaining, which will also be discussed in this chapter.

The above example of adding a consequent to the knowledge base, when the antecedent is *True* is referred to as *Modus Ponens*. There are a number of other common techniques used to perform inferencing; the most common ones include *and-elimination* and *or-elimination*. We will discuss each of these techniques below.

We first discuss the rationale behind the technique of *Modus Ponens*. Formally, if $a \Rightarrow b$ is a rule, and a is known to take on the value of *True*, then one can infer that the statement b is *True* as well and add it to the knowledge base. *Modus Ponens* follows from the following tautology:

$$((a \Rightarrow b) \wedge a) \Rightarrow b$$

In order to show this tautology, one can use the following proof methodology:

$$[((a \Rightarrow b) \wedge a) \Rightarrow b] \equiv [\neg((a \Rightarrow b) \wedge a) \vee b]$$
$$\equiv [\neg(a \Rightarrow b) \vee \neg a \vee b]$$
$$\equiv [\underbrace{\neg(\neg a \vee b)}_{\neg c} \vee \underbrace{(\neg a \vee b)}_{c})$$
$$\equiv [\neg c \vee c]$$
$$\equiv True$$

In the above proof, we repeatedly substitute $\Rightarrow$ with $\vee$, and consolidate $(\neg a \vee b)$ into a single variable c for clarity. The final result of the statement turns out to be *True*, which proves that the statement is a tautology. As a general proof methodology, tautologies are sometimes useful for augmenting knowledge bases, when they appear in the form $x \Rightarrow y$. This is because the presence of x in the knowledge base can be used to augment the knowledge base with statement y. *Modus Ponens* is, therefore, an example of a tautology that can be repeatedly used for deductive inferencing and for changing the state of the knowledge base by adding statements to it.

Another useful technique that is used for inferencing in knowledge bases is *and-elimination*. The basic idea is that given $a \wedge b$, and a, one can infer the statement b. The principle of and-elimination is a consequence of the following tautology:

$$[(a \wedge b) \wedge a] \Rightarrow b$$

This tautology can be inferred as follows:

$$([(a \wedge b) \wedge a] \Rightarrow b) \equiv (\neg[(a \wedge b) \wedge a] \vee b)$$
$$\equiv (\neg(a \wedge b) \vee \neg a] \vee b)$$
$$\equiv ([(\neg a \vee \neg b) \vee \neg a] \vee b)$$
$$\equiv [(\neg a \vee \neg a) \vee (b \vee \neg b)]$$
$$\equiv \neg a \vee True$$
$$\equiv True$$

The principle of *or-elimination* states that given $a \vee b$ and $\neg a$, one can infer b. Or-elimination corresponds to the following tautology:

$$[(a \vee b) \wedge \neg a] \Rightarrow b$$

The or-elimination tautology is equivalent to *Modus Ponens*, because one can replace $a \vee b$ with $\neg a \Rightarrow b$ in the above tautology:

$$[(\neg a \Rightarrow b) \wedge \neg a] \Rightarrow b$$

It is easy to see that this tautology is the same as *Modus Ponens*, except that $\neg a$ is used instead of a in the statement. Therefore, or-elimination is largely redundant with respect to *Modus Ponens*, although it continues to be used quite frequently in automated systems.

One can combine *Modus Ponens* with and-elimination in order to add statements to the knowledge base. Any of the propositional laws discussed in the previous section can also be used. One can, therefore, model the problem of automated theorem proving as a search over all possible states that can inferred from the statements in the knowledge base with the use of the various laws of propositional logic. This process is continued, until one arrives at the goal statement. This procedure is guaranteed to be *sound*, in the sense that it always derives valid inferences. One challenge with the procedure is that an infinite number of possible statements can often be added, which leads to an unrestrained expansion of the knowledge base and the corresponding number of states. For large knowledge bases, this can sometimes create significant challenges in terms of reaching the goal state — however, a systematic approach can be used to reach the goal in finite time. As we will show later, there are multiple ways to achieve this goal. The first method that we will discuss is that of *proof by contradiction*.

4.7   The Method of Proof by Contradiction

You might have often encountered proofs in high school mathematics, where in order to prove a particular statement, one often assumes its negation to be true. One can then arrive at a contradiction by deriving a statement that we already know to be false. This broad approach is based on a principle in propositional logic, which is referred to as *proof by contradiction*. By assuming the negation of the statement to be proven as a pre-condition, the goal is to show that some expression and its negation have a value of *True* in the knowledge base. Since both a statement and its negation cannot be true at the same time, it means that the negation of the conclusion must have been false in the first place (contrary to the original assumption). Therefore, one is able to prove that the conclusion must have been true.

As an example, consider the case of a simple knowledge base in which the propositional expressions $a \Rightarrow b$ and a are present, and we wish to show that b is *True* as well. The additional expression b is the query in this case, and it might be a complex propositional expression in its own right. Note that we can show this result immediately using *Modus Ponens*, although our goal here is to illustrate proof by contradiction. In proof by contradiction, one can simply assume that the statement $\neg b$ (which is the negation of the statement to be proven) takes on the value of *True* for *any setting of the propositional variables* in which the knowledge base statements are *True*. Therefore, the aim is to show that a consistent configuration of the knowledge base and the negation of the query clause (additional assumption) cannot be found. In other words, *no matter what the setting of propositional variables might be*, either some statement in the knowledge base is *False*, or the negation of query clause is *False* (or both).

In the language of propositional logic, the formal way of doing this is to show that the conjunction of the statements in the knowledge base and the additional (negated) assumption must always evaluate to *False* irrespective of the settings of propositional variables. In the simple example of the knowledge base and query discussed above, we need to show that applying the AND operator to the three expressions a, $(a \Rightarrow b)$, and $\neg b$ yields a value of *False*, no matter what the settings of the propositional variables might be. In other words, we obtain the following propositional expression, and we must show that this expression is

False irrespective of the settings of the propositional variables:

$$a \wedge (a \Rightarrow b) \wedge \neg b$$

One can replace the $\Rightarrow$ operator in the propositional expression with the $\vee$ operator to obtain the following:

$$a \wedge (\neg a \vee b) \wedge \neg b$$

Applying the distributive law of $\wedge$ over $\vee$ in the propositional expression and re-grouping, one obtains the following:

$$[\underbrace{(a \wedge \neg a)}_{False} \wedge \neg b] \vee [a \wedge \underbrace{(b \wedge \neg b)}_{False}]$$

The conjunction of a propositional variable and *False* is *False* according to the annihilator law. Therefore, we obtain the following propositional expression:

$$(False \vee False)$$

Note that the expression *False* $\vee$ *False* evaluates to *False* (according to the idempotence law). However, our original assumption was that there is some setting of propositional variables in which a, $(a \Rightarrow b)$, and $\neg b$ all turn out to be *True*. Therefore, we arrive at a contradiction. This contradiction must have come from the additional assumption, since the knowledge base is always assumed to be self-consistent (i.e., it has some setting of variables so that every statement in the knowledge base can be *True*). Therefore, the expression $\neg b$ cannot be *True* under any of these settings of variables (in which the statements of the knowledge base are *True*, and b must be *True* in these settings. This completes the proof.

One advantage of proof of contradiction is that it can be reduced to a case of the propositional satisfiability problem. In the previous example, one would need to show that the following propositional expression is not satisfiable:

$$a \wedge (a \Rightarrow b) \wedge \neg b$$

In general, if KB represents the conjunction of all the statements in the knowledge base, and we wish to prove the statement q (which is the goal clause), then one must show that $KB \wedge \neg q$ is not satisfiable.

The above proof is somewhat ad hoc in nature, which works well for a human theorem prover but not quite as well for an automated system. What we need is a systematic approach to proof by contradiction, which can be effectively coded with a computer algorithm. This is achieved by expressing $KB \wedge \neg q$ in conjunctive normal form, and showing that the expression always evaluates to the value of *False*. The reason that the conjunctive normal form of the propositional expression is useful is that it allows the use of a systematic procedure called *resolution* for proving unsatisfiability. Before discussing resolution, we will show how one can express the statements in a knowledge base in conjunctive normal form.

Converting a Knowledge Base to Conjunctive Normal Form

The statements in a knowledge base might be contain complicated propositional expressions. Therefore, one needs a sequence of well-defined steps that can be used in order to convert the knowledge base into conjunctive normal form. We assume that all sentences in the knowledge base are expressed in terms of the operators $\equiv$, $\Rightarrow$, $\wedge$, $\neg$, and $\vee$. The following sequence of steps is used, starting with the propositional expression $KB \wedge \neg q$:

1. Replace each $\equiv$ in $KB \wedge \neg q$ with an expression containing only the operator "$\Rightarrow$". For example, the statement $a \equiv b$ can be replaced with the following expression:

$$(a \Rightarrow b) \wedge (b \Rightarrow a)$$

Note that a and b might themselves be propositional expressions rather than literals.

2. Replace each "$\Rightarrow$" operator in $KB \wedge \neg q$ with $\vee$. For example, if we have the expression $a \Rightarrow b$, it can be replaced with $\neg a \vee b$. Note that each a and b could itself be a propositional expression rather than a literal.

3. When a propositional expression is expressed in conjunctive normal form, the negation operator must appear only in front of literals. For example, a subexpression such as $\neg(a \vee b)$ would never appear in $KB \wedge \neg q$, because the negation operator appears in front of a complex expression rather than a literal. However, in the original set of statements in the $KB \wedge \neg q$ (which is not necessarily in conjunctive normal form), this condition might not hold. This problem can be addressed by using De Morgan's laws. In this case, we push the negations into the subexpressions, so that *all negations apply only to literals.*

4. After pushing through the negations to the level of the literals, one might have a nested expression containing both the $\wedge$ and $\vee$ operators. At this point, we use the distributivity law, in which the operator $\vee$ is distributed over $\wedge$. For example, the expression $a \vee (b \wedge c)$ is replaced by the following:

$$(a \vee b) \wedge (a \vee c)$$

By performing this distribution, one was able to convert the expression to conjunctive normal form. In general, one might need to perform repeated distribution of $\vee$ over $\wedge$, so that one eventually obtains an expression in which there are no longer any $\vee$ operators that can be distributed over the $\wedge$ operator. It can be shown that the resulting expression is always in conjunctive normal form. An example of an expression that would require two successive distributions is $(a \vee b \wedge c \wedge d)$.

Once $KB \wedge \neg q$ has been converted to conjunctive normal form, a *resolution procedure* is used to decide whether the expression is satisfiable.

We will provide an example of the process of converting an expression into conjunctive normal form. Consider the following propositional expression, which is not originally in conjunctive normal form (but we wish to convert it to this form):

$$[(d \wedge e) \Rightarrow f] \equiv [g \vee \neg h]$$

First, we replace the "$\equiv$" operator with the conjunction of two rules corresponding to the following:

$$\{[(d \wedge e) \Rightarrow f] \Rightarrow [g \vee \neg h]\} \wedge \{[g \vee \neg h] \Rightarrow [(d \wedge e) \Rightarrow f]\}$$

One can sequentially replace each "$\Rightarrow$" with "$\vee$" along with De Morgan's laws to obtain the following:

$$\equiv \{[(\neg d \vee \neg e) \vee f] \Rightarrow [g \vee \neg h]\} \wedge \{[\neg g \wedge h] \vee [(\neg d \vee \neg e) \vee f]\}$$
$$\equiv \{[(d \wedge e) \wedge \neg f] \vee [g \vee \neg h]\} \wedge \{[\neg g \wedge h] \vee [(\neg d \vee \neg e) \vee f]\}$$

At this point, the negations are placed directly in front of the literals. However, the expression is still not in conjunctive normal form, and one has to repeatedly use the distributive property in order to express it as a conjunction of clauses. One can then repeatedly use the distributive property to obtain the following expression:

$$[d \vee g \vee \neg h] \wedge [e \vee g \vee \neg h] \wedge [\neg f \vee g \vee \neg h] \wedge [\neg g \vee \neg d \vee \neg e \vee f] \wedge [h \vee \neg d \vee \neg e \vee f]$$

It is noteworthy that the conversion to CNF has caused a significant expansion of the size of the output. This is a common situation, where the final length might be exponentially related to the length of the initial input (in terms of the number of literals). The increased length of the CNF output in terms of the input is not particularly surprising; if the length of the CNF expression were not significantly longer than the input, it would provide an avenue to solving an NP-hard problem in polynomial time. In other words, the NP-hardness of the problem of identifying tautologies is hidden in the size of the CNF formula that is created by the approach.

Resolution Procedure

The resolution procedure is designed to use the conjunctive normal form to infer a contradiction; it works by repeatedly looking at pairs of conjunctions that differ only in terms of the negation in front of a single literal. An example of such a pair of conjuncts would be $(f \vee \neg g \vee h)$ and $(f \vee \neg g \vee \neg h)$. These two conjuncts differ only in terms of the final literal, which is h in one case and $\neg h$ in another case. From these two statements, one can infer $(f \vee \neg g)$. The reason is that one can use the distributive property to infer the following:

$$[\underbrace{(f \vee \neg g}_{d} \vee h) \wedge \underbrace{(f \vee \neg g}_{d} \vee \neg h)] \equiv [(f \vee \neg g) \vee (h \wedge \neg h)]$$

$$\equiv [(f \vee \neg g) \vee \mathit{False}]$$

$$\equiv (f \vee \neg g)$$

Therefore, this step leads to the elimination of a literal from a pair of conjuncts in $KB \wedge \neg q$. Repeating this step to reduce the number and size of conjuncts lies at the core of the resolution procedure. By continually repeating this procedure, one of two situations will occur at some point. The first situation is that no further literal can be eliminated. In such a case, it can be formally shown that $KB \wedge \neg q$ is satisfiable, and therefore KB does not entail q. On the other hand, it is possible for the repeated reduction in the sizes of the individual conjuncts to lead to a situation where an atomic proposition and its negation are present in the augmented knowledge base. This situation leads to a contradiction, and one can conclude that $KB \wedge \neg q$ is not satisfiable. Since $KB \wedge \neg q$ is not satisfiable, it means that its negation $\neg[KB \wedge \neg q]$ always takes on the value of *True*. In other words, the statement $\neg KB \vee b$ takes on the value of *True*, which in turn implies that $KB \Rightarrow q$ is a tautology. This means that KB entails q. Therefore, the overall algorithm for resolution may be described using the following steps:

1. If no pair of clauses exist that differ in the negation of a single literal in the CNF form of $KB \wedge \neg q$, then report that KB does not necessarily imply q and terminate.

2. If a pair of clauses exists in $KB \wedge \neg q$ that are atomic propositions and are negations of one another, then report that KB implies q and terminate.

3. Select a pair of clauses differing in the negation of a single literal and apply the resolution procedure to create a new, shorter clause. Add the new clause to KB and return to the first step.

It is noteworthy that the resolution procedure can also be used to construct a proof by tracing how one obtains the atomic propositions that are negations of one another. An important point is that eliminating the variable b from $(b \vee a) \wedge (\neg b \vee c)$ is simply an application of transitivity of the two rules $\neg a \Rightarrow b$ and $b \Rightarrow c$ to create the new rule $\neg a \Rightarrow c$. Therefore, one can write the proof in terms of repeated application of transitivity. Furthermore, resolving $b \vee a$ with $\neg b$ to yield a is simply a form of *Modus Ponens*.

It is also noteworthy that the expression for the query sentence q will always be used at some point of the entailment procedure when one is trying to show a contradiction for $KB \wedge \neg q$. If this is not the case, it would imply that a contradiction exists in the knowledge base itself, and *any* propositional expression can be entailed. A general assumption of these methods is that we work with knowledge bases that are self-consistent. For example, no reasonable knowledge base would contain two statements like the following:

> If it rains, it pours. $(r \Rightarrow p)$
> It rains and it does not pour. $(r \wedge \neg p)$

With such a knowledge base containing contradictions, a resolution procedure will always entail any other statement, irrespective of its content.

In order to understand how the resolution procedure works, we will provide an example. Consider the situation where the knowledge base $KB \wedge \neg q$ is represented in the following form:

$$(c \vee d) \wedge (\neg d \vee e \vee b) \wedge (d \vee e \vee b) \wedge (\neg d \vee \neg e \vee b) \wedge (d \vee \neg e \vee b) \wedge \neg b$$

One can pair clauses that differ in the negation in front of a literal as follows:

$$(c \vee d) \wedge \underbrace{(\neg d \vee e \vee b) \wedge (d \vee e \vee b)}_{\text{Pair}} \wedge \underbrace{(\neg d \vee \neg e \vee b) \wedge (d \vee \neg e \vee b)}_{\text{Pair}} \wedge \neg b$$

On simplifying by collapsing pairs of clauses, one obtains the following:

$$(c \vee d) \wedge \underbrace{(e \vee b) \wedge (\neg e \vee b)}_{\text{Pair}} \wedge \neg b$$

On simplifying further, one obtains the following:

$$(c \vee d) \wedge \underbrace{b \wedge \neg b}_{\text{Contradiction}}$$

Note that the pair $b \wedge \neg b$ evaluates to *False* (or the *empty* clause), which is a contradiction. This implies that KB must entail q, as including $\neg q$ in the knowledge base leads to a contradiction. The implicit assumption is that the original knowledge base KB is satisfiable, and it is only the inclusion of $\neg q$ that causes the contradiction. Therefore, one must be careful while building the initial knowledge base to make sure that it is satisfiable.

An important property of the resolution procedure is that it is *complete*. What completeness means is that if KB entails q, the resolution of $KB \wedge \neg q$ will always arrive at a contradiction irrespective of the order in which the operations are implemented. The procedure is also known to be *sound*. Soundness refers to the fact that if KB does not entail q, then the procedure will terminate at some point without being able to add further clauses. The fact that the resolution procedure is both complete and sound means that the procedure is *decidable*. The notion of decidability means that one can make a clear judgement one way or another whether KB entails q (given sufficient computational time). As we will see in Chapter 5, other advanced forms of logic, such as first-order logic are not fully decidable.

4.8 Efficient Entailment with Definite Clauses

Each conjunct in the CNF form of a knowledge base is a clause that can be expressed as a disjunction of literals. Therefore, knowledge bases are naturally represented in terms of clauses for the purposes of inferencing. A special type of clause, referred to as a *definite clause*, is commonly used in knowledge bases; the use of definite clauses is leveraged to implement a special case of the CNF form. A definite clause is defined in the following way:

1. A *body* is defined as either an atomic proposition, or the conjunction of $k > 1$ atomic propositions $b_1 \dots b_k$. Therefore, an example of a body is $b_1 \wedge b_2 \wedge \dots \wedge b_k$.

2. A definite clause is either an atomic proposition a, at a rule of the form $b_1 \wedge b_2 \wedge \dots \wedge b_k \Rightarrow a$. In the latter case, the atomic proposition is referred to as the *head* of the rule.

Note that if we express the definite clause as a disjunction of literals, it can be expressed as $a \vee \neg b_1 \vee \neg b_2 \vee \dots \vee \neg b_k$. A noteworthy point is that this disjunction contains *exactly one* positive literal. Another type of clause, referred to as the *Horn clause*, is a slight generalization of this idea. When expressed as a disjunction of literals, a Horn clause has *at most* a single positive atomic proposition, and the remaining disjuncts are negations of atomic propositions. Therefore, a clause containing only negative literals, such as $\neg b_1 \vee \neg b_2 \vee \dots \vee \neg b_k$ is a Horn clause, but it is not a positive definite clause. However, all positive definite clauses are Horn clauses. Horn clauses containing only negative literals are considered goal clauses, because the negation of a set of positive assertions $b_1 \wedge b_2 \wedge \dots \wedge b_k$ leads to the disjunction of the negations of these literals (which is the expression $\neg b_1 \vee \neg b_2 \vee \dots \vee \neg b_k$). It is noteworthy that resolving two Horn clauses leads to another Horn clause. Furthermore, the result of the resolution of a Horn clause and goal clause (with no positive literals) will always be a (simplified) goal clause. Therefore, Horn clauses are closed under resolution. This type of recursive property makes Horn clauses particularly useful from the perspective of the resolution procedure. As a result, Horn clauses lead to a systematic procedure for efficient inferencing. For example, resolving goal clauses with positive definite clauses leads to quick reduction in the size of the goal clauses. As a result, one can often reach the null clause rather quickly.

Several other procedures, such as forward chaining and backward chaining, work particularly well with positive definite clauses. This will be the topic of discussion in the next few sections.

4.8.1 Forward Chaining

A useful resolution algorithm that works with positive definite clauses is the *forward chaining algorithm*. The forward chaining algorithm starts with the known facts in the knowledge base. These known facts are always represented as positive literals in the database, which are contained in LIST. We assume that an atomic proposition q corresponds to the goal clause for simplicity (although the idea can be extended to more complex propositional expressions by introducing a new atomic proposition that is equivalent to the goal). The

Algorithm *ForwardChain*(Knowledge Base: KB, Goal proposition: q)
begin
   LIST= Positive facts (atomic propositions) in KB;
   Initialize *unmatched*[r] for each rule r in KB;
   { Set *unmatched*[r] to number of atomic propositions in antecedent of r; }
   repeat
     Select proposition p from LIST based on pre-defined strategy;
     If p matches q **return**(*True*);
     { Success in inferring q from KB }
     $\mathcal{R}$ = All rules in KB containing p in antecedent;
     for each rule $r \in \mathcal{R}$ **do**
     begin
       Reduce *unmatched*[r] by 1;
       if *unmatched*[r] = 0 add consequent of r to LIST;
     end
     Delete node p from LIST;
   until LIST is empty;
   return(*False*);
   { Failure in inferring q from KB }
end

Figure 4.1: The forward chaining algorithm

basic idea is to start with the known facts in LIST and repeatedly use *Modus Ponens* to derive other facts in the consequents of rules. These facts are then added to LIST as well. The positive-definite Horn clauses are used to create the rules. Since positive-definite Horn clauses always contain exactly a single positive literal, it follows that the rules can be represented by a conjunction of positive literals in the antecedent and a single positive literal in the consequent. At the initialization of the algorithm, only the known facts (atomic propositions) are added to LIST. Since rules imply from positive literals to other positive literals, this property of LIST always containing only positive literals (because of modification by *Modus Ponens*) is always maintained by the algorithm.

A key point is to systematically match the conjunction of the propositions in the rule antecedents with the atomic propositions in a dynamically changing LIST. Note that a rule for which all propositions are not matched currently might eventually be fired later using *Modus Ponens* as new propositions enter LIST. For each rule r in the knowledge base, one maintains the value *unmatched*[r], which is the number of atomic propositions in the antecedent of r that are so far unmatched in the database. This value is initialized to the number of atomic propositions in the antecedent of the corresponding rule. For example, the value of *unmatched*[r] for the rule $b_1 \wedge b_2 \wedge \ldots \wedge b_k \Rightarrow a$ is k at the time of initialization.

Subsequently, we select any atomic proposition p from LIST, and check if it matches q. If p indeed matches q, then one can terminate the algorithm after reporting that the knowledge base entails q. Otherwise, we select all rules $\mathcal{R}$ in the knowledge base KB for which the antecedent contains p. The value of *unmatched*[r] for each rule $r \in \mathcal{R}$ is decreased by 1. If the value of *unmatched*[r] for any rule $r \in \mathcal{R}$ reaches 0, it means that all propositions in its antecedent are true facts, and therefore, the consequent of r can also be added to LIST. At this point, the atomic proposition p is removed from LIST. In the event that LIST becomes empty because of this deletion, one can infer that q is not entailed by the knowledge base, and terminate after reporting this fact. On the other hand, if LIST is not empty, one selects

Algorithm *BackwardChain*(Knowledge Base: KB, Goal proposition: q, List: MUST-PROVE)
begin
  MUST-PROVE= MUST-PROVE $\cup\{q\}$;
  { MUST-PROVE contains all facts that need to be proved }
  if q appears as a positive fact (atomic proposition) in KB **return**(*True*);
  $\mathcal{R}$ = All rules in KB containing q as consequent;
  if $\mathcal{R}$ is empty **return**(*False*);
  for each $r_i \in \mathcal{R}$ **do**
  begin
    Let $p_1 \ldots p_k$ be literals in antecedent of r_i not in MUST-PROVE;
    if $\wedge_{j=1}^{k} BackwardChain(KB, p_j, \text{MUST-PROVE}) \equiv True$ **then return**(*True*);
  end
  return(*False*);
  { Failure in inferring q from KB }
end

Figure 4.2: The backward chaining algorithm

the next proposition available in it and continues the same procedure as described above. The forward chaining algorithm is illustrated in Figure 4.1. Note that the basic structure of the forward chaining algorithm shares some similarity to search algorithms, where a list maintains the nodes to be explored further. It is noteworthy that the algorithm of Figure 4.1 can be viewed as a form of breadth-first search. It is assumed that the goal clause is stated as an atomic proposition q.

Forward chaining is, therefore, a repeated application of *Modus Ponens*, and, therefore, it is a valid procedure. Furthermore, the procedure is closely related to the resolution procedure by contradiction (and collapsing literals). While the literal-collapsing approach uses both transitivity and *Modus Ponens*, the forward chaining algorithm uses only the latter. This is possible because it works with Horn clauses. If an assertion is entailed by the knowledge base, this procedure will always be able to reach this conclusion in a finite amount of time. Therefore, forward chaining is a *complete* algorithm. Furthermore, if a statement is not entailed by the knowledge base, then LIST will eventually become empty in a finite amount of time. The reason is that the number of unmatched propositions in the various rules keep reducing over the course of the algorithm, and a lower bound on the number of unmatched propositions is 0. Forward chaining is sound and complete for a database expressed as Horn clauses. One important point is that even though this algorithm is efficient in terms of the size of the knowledge base, the knowledge base itself can become quite large when it is constrained to be expressed in terms of Horn clauses.

4.8.2 Backward Chaining

Backward chaining is a procedure that works backward from the desired conclusion to the known facts. Consider the situation whether one wants to test if $KB \Rightarrow q$. In such a case, the algorithm first checks the knowledge base to determine if q is *True*. If this is the case, then the algorithm terminates. Otherwise, the antecedents of at least one of the rules that contain q in the consequent must take on the value of *True*. Note that this is inherently an OR operation over the available rules. However, for a given rule all its atomic propositions in the antecedent must be true, which is an AND operation over the available atomic propositions in the antecedent. This is because the rules in the database always contain the conjuncts of atomic propositions in the antecedent. We again check the knowledge base to determine if all conjunct literals of at least one rule are satisfied (i.e., present as positive facts in the knowledge base). If so, we terminate with a success. Otherwise, we repeat the process recursively by searching for rules that contain the literals (with unknown truth values) in the consequent. Therefore, we need to prove *all* the conjuncts of *at least one* rule. In other words, the algorithm boils down to AND-OR search, which is discussed in Section 3.2 of Chapter 3. Like the AND-OR search algorithm, backward chaining is inherently a depth-first search procedure. This is different from forward-chaining method, which is a breadth-first search procedure. Therefore, backward chaining is presented more naturally as a recursive algorithm.

For example, consider the case where the knowledge base contains the rules $a \wedge c \Rightarrow q$, $d \Rightarrow a$, $e \Rightarrow c$, as well as the positive facts d and e. Starting with the goal q, backward chaining will discover the literals a and c, which both need to be satisfied. Then, the algorithm will call the procedures using a and c each as the new goals, which respectively yields d and e as the new goals. Since the knowledge base contains d and e as positive facts, the procedure will terminate as a success.

The overall procedure for backward chaining is illustrated in Figure 4.2. Unlike the forward chaining procedure, the backward chaining algorithm has been presented as a recursive procedure (although the forward chaining algorithm can also be presented[2] as a recursive procedure). In addition to the same input parameters as the forward chaining procedure (the knowledge base KB and goal q), the backward chaining algorithm has a list called MUST-PROVE, which is the set of propositions that are already among the requirements to the proven, based on earlier depths of the recursion. This list is required in order to properly handle situations where there is circularity, such as the rules $c \Rightarrow d$ and $d \Rightarrow c$. Without properly keeping track of such cases, it is possible for the backward chaining procedure to be stuck in a circular loop of recursions forever. The list MUST-PROVE is initialized to the empty set. After initiating the call, the goal q is added to MUST-PROVE, so that future depths of the recursion are aware that this goal was already encountered earlier on. In the event that q occurs as a positive fact (atomic proposition) in the knowledge base, the algorithm returns the value of *True*. Then, the algorithm finds all rules for which q is the consequent. For each such rule r_i, it finds all the literals $p_1 \ldots p_k$ in the antecedent of r_i, which are not present in MUST-PROVE. For each such literal, it calls the backward chaining algorithm recursively k times, where the jth recursive call uses p_j as the new goal. If all calls return the value *True*, then the backward chaining procedure returns the value of *True* as well. This process is repeated for each of the rules r_i, for which q occurs as a consequent. After processing all these rules, if the procedure has not yet returned the value of *True*, then the backward chaining procedure terminates with a returned value of *False*.

4.8.3 Comparing Forward and Backward Chaining

Forward and backward chaining are two search approaches that achieve the same result. However, there are significant differences in terms of the efficiency and interpretability of these two approaches. First, forward chaining blindly tries to reach all possible goals in the forward direction. As a result, forward chaining often generates many intermediate inferences that have nothing to do with the goal at hand. This is particularly noticeable when the knowledge base is large, and contains many disconnected facts. The tendency of forward chaining to explore unrelated facts to the goal q makes the approach rather slow. On

---

[2]All recursive algorithms can be systematically converted to non-recursive variants by using a standardized methodology [40].

the other hand, backward chaining starts with a narrow goal, and uses depth-first search in the backwards direction in order to explore different ways in which this specific goal can be proven using the rules in the knowledge base. As a result, the new goals that are generated in the backward direction are limited, and are very relevant to the goal at hand. When the knowledge base contains many disconnected facts, the approach will often not explore those portions of the knowledge base at all.

The intermediate results in forward chaining also tends to be less interpretable because of the indiscriminate way in which large numbers of inferences are made. In such cases, it becomes difficult to separate the relevant inferences from the irrelevant ones. This is primarily because the forward chaining inferences are often not specifically directed towards goal clauses. On the other hand, the focused approach of backward chaining tends to generate relevant paths, which are highly interpretable in terms of their relationships to the goal. As a result, backward chaining methods are often used in theorem proving systems.

4.9 Summary

This chapter introduces propositional logic, which is the most basic form of logic in mathematics. Although propositional logic is too simple to be of direct use in expert systems, it provides a foundation on top of which first-order logic is constructed; the latter is the preferred language of expert systems and other logic-based approaches in artificial intelligence. The chapter introduces the basics of propositional logic, together with its various laws. These laws have been used in conventional mathematics to show various types of proofs. However, in order to apply these methods to artificial intelligence, more systematic procedures are needed.

A number of key procedures of propositional logic are introduced in this chapter, which help in automated theorem proving. A key concept is entailment, where one attempts to prove specific facts using a set of initial conditions and rules in the knowledge base. Entailment is enabled by the laws of propositional logic, and some special proof techniques such as *Modus Ponens*. A key idea in entailment is to use proof by contradiction, where the negation of the goal expression is added to the knowledge base in order to show a contradiction. The underlying computational method is referred to as resolution. Furthermore, knowledge bases that are expressed in terms of Horn clauses can be leveraged in combination with forward and backward chaining procedures for efficient inference. These broad ideas of propositional logic are carried over to first-order logic, which is discussed in the next chapter.

4.10 Further Reading

The basics of logic may be found in [88]. An early discussion on the role of logic in artificial intelligence may be found in [131]. An excellent chapter on propositional logic may be found in Aho and Ullman's classical book [11] on foundations of computer science. In addition, Russell and Norvig's book [153] provides an overview from the artificial intelligence perspective.

4.11   Exercises

1. Show that if $(\neg a \vee (b \wedge c))$ and $\neg b \vee d$ each take on the value of *True*, then $a \Rightarrow d$ must take on the value of *True* as well.

2. Use truth tables to show that the expressions $a \Rightarrow b$ and $\neg b \Rightarrow \neg a$ are equivalent.

3. Construct the truth table for each of the following propositional expressions. Also point out if any of these expressions is a tautology:

   (a) $(a \wedge b) \Rightarrow a$
   (b) $(a \vee b) \Rightarrow a$
   (c) $a \Rightarrow (a \wedge b)$
   (d) $a \Rightarrow (a \vee b)$

4. For each of the expressions that is shown to be a tautology using the truth table method in Exercise 3, show using propositional laws that it is a tautology.

5. Construct the truth table for $(a \vee b \vee c)$. Now use the truth table to create a propositional expression with seven terms. Use the laws of propositional logic to simplify this expression to $(a \vee b \vee c)$.

6. Show that the following propositional expression is a tautology:

$$[a \Rightarrow b_1 \wedge b_2 \wedge \ldots \wedge b_k] \equiv [\wedge_{i=1}^{k}(a \Rightarrow b_i)]$$

7. Show that the following expression is not satisfiable:

$$\neg[(a \wedge b) \Rightarrow a] \wedge b$$

Construct a truth table to show that the expression is not satisfiable.

8. Convert the following expression to conjunctive normal form:

$$(a \equiv b) \vee \neg[c \wedge d]$$

9. Show that the following expression is a tautology:

$$[\wedge_{i=1}^{k} a_i] \Rightarrow [\vee_{i=1}^{k} a_i]$$

10. Show the equivalence of the following expressions:

$$[[\vee_{i=1}^{k} a_i] \Rightarrow b] \equiv [\wedge_{i=1}^{k}[a_i \Rightarrow b]]$$

11. Consider a knowledge base containing the following rules and positive facts:

$$a \wedge c \wedge d \Rightarrow q$$
$$e \Rightarrow a$$
$$e \Rightarrow c$$
$$f \Rightarrow e$$
$$d$$
$$f$$

Simulate a backward chaining procedure on this toy knowledge base to show that it entails the goal q.

12. For the knowledge base and goal clause of Exercise 11, simulate a forward chaining procedure to show that the knowledge base entails the goal q.

13. Create truth tables for the XOR, NAND, and NOR logical operators.

14. Suppose that you had an algorithm to determine in polynomial time whether an expression is tautology. Use this algorithm to propose a polynomial-time algorithm whether an expression is satisfiable. Given that satisfiability is NP-complete, what can you infer about the computational complexity of the tautology problem?

15. Consider the following two statements:

> If Alice likes daisies, she also likes roses. Alice does not like daisies.

Do the above sentences entail the following?

> Alice does not like roses.

Chapter 5

First-Order Logic

"People who lean on logic and philosophy and rational exposition end by starving the best part of the mind." – William Butler Yeats

5.1 Introduction

First-order logic is a generalization of propositional logic, which is also referred to as *predicate logic*. Predicate logic is a more powerful extension of propositional logic, which can perform more complex reasoning tasks in artificial intelligence that are not possible with the use of only propositional logic. The main reason that propositional logic does not work well for reasoning tasks in artificial intelligence is that many statements in a knowledge base are true across an entire set of objects, which then need to be applied to specific cases. For example, consider the following pair of statements:

All mammals give birth to live babies. A cat is a mammal.

From the above two statements, one can infer that a cat gives birth to live babies. While it is possible to infer the fact that cats give birth to live babies, it would require us to define a rule that relates a cat to a mammal. This is a problem when one has a large domain of objects and one has to define rules for each and every mammal in the domain; doing so can blow up the size of the knowledge base in terms of the number of rules.

This type of situation occurs frequently in expert systems, where some statements are true for entire domains of objects, whereas others are true only for specific members of that domain. For example, a medical diagnosis system might make statements that are true across all patients, and then one might want to apply these statements to a specific patient like John. In other words, we want a way to refer to specific types of objects, make general statements about them, and then use these general statements about specific objects. First-order logic provides the flexibility to move between statements about domains of objects to specific objects and vice versa.

C. C. Aggarwal, *Artificial Intelligence*, https://doi.org/10.1007/978-3-030-72357-6_5

A *predicate* is a function that takes as argument one or more *objects* or *entities*, and returns a Boolean value. For example, the propositional variable a is returned by applying the predicate $F(\cdot, \cdot, \cdot)$ on the objects x, y, and z:

$$a \equiv F(x, y, z)$$

The output of this predicate is a truth value contained in a, which is either *True* or *False*. The number of arguments of a predicate is referred to as its *arity*. Note that the variables x, y, and z are not logical variables, but they represent the objects to which these statements apply. It is only after predicates are applied to these objects that one obtains a logical expression. For example, consider the case where the object x is director Steven Spielberg, the object y is the movie *Saving Private Ryan*, and the object z is the *Best Director Award*. The predicate function $F(x, y, z)$ is *True* when the director Steven Spielberg wins the best director award for *Saving Private Ryan*. In other words, we have the following:

$$F(\textit{Steven Spielberg}, \textit{Saving Private Ryan}, \textit{Best Director Award}) \equiv \textit{True}$$

Allowing these types of predicates creates a more expressive form of logic, which can be the basis of designing more complex deductive reasoning systems in artificial intelligence (like expert systems). This is because, if we have a different movie directed by Steven Spielberg or a different triplet of movie, director, and award, we can use the *same* predicate, except that we can change the binding of the object variables:

$$F(\textit{Steven Spielberg}, \textit{The Terminal}, \textit{Best Director Award}) \equiv \textit{True}$$

$$F(\textit{John Williams}, \textit{The Terminal}, \textit{BMI Music Award}) \equiv \textit{True}$$

By using this approach, we are able to tie together different truth statements, based on the objects that they apply to.

One can even make statements about entire populations of objects; an example is the case where $B(x)$ corresponds to the fact that mammal x gives birth to live babies. By allowing this statement to be true for all mammals, one can make the statement that all mammals give birth to live babies by simply defining the domain of mammals (and without having a separate propositional statement for each mammal in the domain of mammals). This type of generic statement is not possible in the case of propositional logic, where a set of similar statements about a domain of objects needs to be decomposed into separate assertions about individual objects. This tends to make the process of inference somewhat unwieldy and computationally expensive. Furthermore, in such cases, it becomes impossible to discern any relationships among similar assertions about different objects with the use of propositional logic. Reasoning systems are almost always used in the context of statements about objects, and the ability to tie them together with such predicates is critical in being able to make complex inferences about objects. For example, the above set of statements implies that the movie *The Terminal* has won at least two major awards; this would be impossible to infer with the use of only propositional logic in an efficient way. However, in order to introduce the notion of counting the number of movies, additional rules would need to be added to the knowledge base that define the notion of counting. First-order logic is better able to represent the semantics of natural language because of its ability to work with the notions of objects and relations.

First-order logic has an overall algebraic structure that is similar to propositional logic; however, an important difference is that propositional logic only deals with the truth of assertions, whereas first-order logic deals with the truth of assertions in the context of their relevance to *objects* and assertions indicating there relations and interactions with one another. At the same time, it is also possible to have assertions that are not related to specific objects (like propositional logic). First-order expressions can include propositional variables, such as the following:

If it rains tomorrow, Tommy will eat his carrots.

The propositional variable for raining tomorrow is r, and the predicate expression for Tommy eating carrots is $E(Tommy, Carrots)$. Then, the above statement can be expressed using both propositional and first-order variables as follows:

$$r \Rightarrow E(Tommy, Carrots)$$

Note that the above sentence includes both propositional and predicate expressions, and the broader syntax is similar to that of propositional logic. This makes first-order logic a richer formalism to express and derive truth statements about the relationships among objects than propositional logic. In fact, the latter is a special case of the former. As we will see later in this chapter, many inferencing procedures of first-order logic are generalizations of those used in propositional logic (with suitable changes made to handle predicates).

It is also possible to enhance first-order logic with different classes of objects and hierarchical relationships among them. Such hierarchical relationships are referred to as *ontologies*. When first-order logic is used together with such set-theoretic concepts, it is referred to as *second-order logic*. As we will see in Chapter 12, knowledge graphs encode some of these aspects of second-order logic, although knowledge graphs cannot themselves be considered a formal part of the symbolic paradigm. In fact, knowledge graphs provide a convenient route to integrating deductive reasoning and inductive learning methods. A further generalization of first-order logic is *temporal logic*, which allows us to derive truth statements about the temporal relationships among objects that happen to be events. The greater complexity and richness of first-order logic (compared to propositional logic) enables the capability of constructing knowledge bases and their associated proof systems. This chapter will discuss these issues in detail.

This chapter is organized as follows. The next section introduces the basics of first-order logic. The process of populating a knowledge base is discussed in Section 5.3. An example of a toy expert system with the use of first-order logic is discussed in Section 5.4. Systematic procedures for inferencing in first-order logic are discussed in Section 5.5. A summary is given in Section 5.6.

5.2 The Basics of First-Order Logic

In this section, we will discuss the basic formalisms of first-order logic, and the different types of operations that can be performed on the underlying objects. An important observation is that by using predicates in place of propositional variables to perform logical operations, we are able to express relationships between objects in a more detailed way. For example, consider the assertion stating that *Steven Spielberg* received the *Best Director Award* for *Saving Private Ryan*. One can also view the triplet as a relationship between the entities *Steven Spielberg*, *Saving Private Ryan*, and *Best Director Award* as follows:

⟨*Steven Spielberg, Saving Private Ryan, Best Director Award*⟩

One can create a set of triplets that show the relationships between person, movie, and award categories as follows:

$$\{\langle Steven\ Spielberg, Saving\ Private\ Ryan, Best\ Director\ Award\rangle,$$
$$\langle Steven\ Spielberg, The\ Terminal, Best\ Director\ Award\rangle,$$
$$\langle John\ Williams, The\ Terminal, BMI\ Music\ Award\rangle, \ldots, \}$$

In other words, it is now possible to populate the knowledge base with the raw data about the entities that have won awards for various movies, and with the use of a single predicate. Note that one can also achieve this in propositional logic with the use of a single statement about each triplet. However, in propositional logic, each of these statements is an independent assertion, and it becomes harder to relate the similarity in structure between the different assertions in an interpretable way. In first-order logic, these assertions are naturally related because they are expressed in terms of object-wise instantiations of the same predicate; this is particularly helpful when one is working with proofs corresponding to the relationships between various objects. Furthermore, one can make generic assertions in terms of object variables like x that can be bound to any value from a particular domain of interest, and this issue will be discussed slightly later. The domain over which the various objects are defined is also referred to as the *domain of discourse*. The domain of discourse might correspond to people, places, or specific types of things.

Consider the case where we want to assert that any entity who wins an award is automatically invited to the Oscars, and the predicate $O(x)$ indicates the fact that the entity x has been invited to the Oscars. In such a case, we can add the following assertion to the knowledge base, *which is true for all values of x, y, and z*:

$$F(x, y, z) \Rightarrow O(x)$$

In practice, these types of general statements about entire domains of objects are preceded with the mathematical notation $(\forall x, y, z)$ in order to show that it holds for all entities in the domain at hand. This mathematical notation is referred to as a quantifier, which will be discussed in the next section. If one can show that $F(x, y, z)$ is true by instantiating x to the value of *Steven Spielberg*, one can automatically show that Steven Spielberg was invited to the Oscars. In other words, one can infer that the assertion $O(Steven\ Spielberg)$. Note that one would need to have a separate statement for each triplet of person, movie, and award in the case of propositional logic in order to infer the fact that Steven Spielberg was invited to the Oscars. This can be computationally challenging when one is dealing with a large domain containing millions of objects, and therefore a large number of assertions would need to be scanned and compared with one another to make the same inference. In some settings, the domain of discourse can even be of infinite size, which makes it impossible to express a rule for each object in the domain (using a finite knowledge base). An important advantage of first-order logic is that one is able to combine the information in multiple logical statements (which could be *either general or specific statements about objects*) in order to make *specific inferences about objects*. This is particularly important in applications like expert systems, where one must start with general statements about the state of the world, and use these statements to make more specific inferences about specific cases or individuals. This is not possible with the use of propositional logic, unless one chooses the cumbersome option of creating a rule like $F(x, y, z) \Rightarrow O(x)$ for *every possible instantiation* of x, y, and z. Note that the number of statements increases exponentially with the number of arguments in a predicate, as one must instantiate each variable independently to every member from its domain of discourse.

5.2.1 The Use of Quantifiers

Consider the example of the rule discussed in the previous section:

$$F(x, y, z) \Rightarrow O(x)$$

Note that the assertion is about all objects in the domain of discourse. In some cases, assertions may be relevant to "at least one" object in the domain of discourse. Therefore, we need a way to formally denote whether the assertion is about all objects, or only about some objects. This is achieved with the notion of *quantifier*, which does not have a corresponding counterpart in propositional logic. The two key quantifiers in first-order logic are ($\forall$) and ($\exists$). The former is referred to as a *universal quantifier*, whereas the latter is referred to as an *existential quantifier*. A universal quantifier can make an assertion about all objects in the knowledge base, whereas an existential quantifier makes statements about some (at least one) object. Our earlier example of the rule, $F(x, y, z) \Rightarrow O(x)$, was implicitly intended for all values of x, y, and z in the domain of discourse (though the quantifier was not explicitly shown), and therefore a more formal and correct way of expressing this statement with a universal quantifier is as follows:

$$\forall x, y, z \; [F(x, y, z) \Rightarrow O(x)]$$

Similarly, there might be a particular value of one argument for which an expression might become true, irrespective of other arguments. For example, consider the case where $E(x, y)$ refers to the fact that person x eats food y. Furthermore, $N(x)$ is a predicate indicating that person x is non-vegetarian. Consider the following statement:

For any person, if that person eats Beef then that person is non-vegetarian.

This statement can be formally expressed with the use of the universal quantifier as follows:

$$\forall x \; [E(x, Beef) \Rightarrow N(x)]$$

Note that the $\forall x$ quantifier applies to each occurrence of x within square brackets, which is referred to as the *scope* of the quantifier. Any occurrence of x within this scope of the quantifier makes the variable *bound* to the quantifier. Any other occurrence of the variable x outside the scope is said to be *free*. The free variable will typically be bound by a different quantifier defined in an outer nest of the logical expression, and therefore the resulting statement will often be a subexpression of a larger sentence bound by multiple quantifiers. Formulas without free variables are referred to as *closed formulas* or *sentential forms*. Knowledge bases built upon first-order logic typically contain sentential forms of statements, and the occurrences of free variables are rare (if any) in knowledge bases (although it is a useful concept from an analytical point of view during proofs). Operations such as universal and existential quantification provide important expressive power in order to endow greater flexibility to first-order logic in relation to propositional logic. However, first-order logic is built on top of the fundamental algebraic formalisms used in propositional logic. All the rules and laws of propositional logic also apply to first-order logic by using the predicates in lieu of propositional variables. Just as expressions of propositional logic are built using propositional variables, the expressions of first-order logic are constructed from predicates using the operators and formalism of propositional logic. However, first-order logic can also support propositional variables, which makes it a strict superset of propositional logic. An atomic proposition of propositional logic is simply be treated as an atomic formula in first-order logic with zero arguments.

The operators of first-order logic that are common to propositional logic have the same order of precedence as in propositional logic. The key set of operators that are fundamentally different from propositional logic are the quantifiers; these operators have the highest precedence among all operators. The operator precedence in first-order logic is as follows:

- The quantification operators have the highest precedence.

- The negation operator, $\neg$, has the next highest precedence.

- The third-highest precedence is that of the $\wedge$ operator.

- The fourth-highest precedence is that of the $\vee$ operator.

- The $\Rightarrow$ and $\equiv$ operators have the lowest precedence.

Leaving aside the quantifier precedences, it is noteworthy that all other precedences are similar to those used in propositional logic. In order to emphasize the importance of quantifier precedence, we note that the following pair of first-order statements is not the same:

$$\forall x \; [E(x, Beef) \Rightarrow N(x)]$$
$$\forall x \; E(x, Beef) \Rightarrow N(x)$$

In the first statement, the quantifier applies to the entire expression because of the use of parentheses. In the second statement, the quantifier only applies to $E(x, Beef)$, because the quantifier has a higher precedence than the $\Rightarrow$ operator; therefore, the second occurrence of the variable x in the predicate $N(x)$ occurring in the second statement is free.

The aforementioned example uses the same notation x for a free and a bound variable. Although the expression is syntactically correct, it can often cause confusion because of the difficulty in distinguishing between the two different uses of the variable x. As a consequence of the confusion caused by using the same notation for a free and a bound variable, it is common to use a different notation for each variable, whether it is free or bound. For clarity, a different notation will be used to distinguish each occurrence of a variable that is specific to a particular quantifier. A clearer way of writing the previous example of free and a bound variable is as follows:

$$\forall x \; E(x, Beef) \Rightarrow N(y)$$

In this case, it is clear that the variable x is a bound variable, whereas the variable y is free. In a sense, using different notations for variables in different scopes makes the expression more readable. Furthermore, the inclusion of a parenthesis does not change the meaning of the assertion in this case:

$$\forall x \; [E(x, Beef) \Rightarrow N(y)]$$

This way of using distinct notations for each variable with a different binding is referred to as *standardization*, and it is extremely common (and recommended) in first-order logic. Most algorithmic procedures in first-order logic use standardization as a preprocessing step to improve readability and avoid confusion.

The second form of the quantifier, referred to as the *existential quantifier* states that at least one value of a variable exists for which a statement is *True*. For example, the statement that there is at least one person on the planet that eats beef can be expressed as follows:

$$\exists x \; E(x, Beef)$$

It is possible to combine the universal and existential quantifiers in a single statement, such as the following:

$$\exists y \forall x \; E(x, y)$$

It is noteworthy that the order of the operators is important. For example, the operator ($\exists$) has precedence over ($\forall$) in the above assertion. Consider the following two first-order expressions:

$$\exists y \forall x \; E(x, y)$$
$$\forall x \exists y \; E(x, y)$$

The assertion $(\exists)y(\forall)xE(x, y)$ means that there is at least one particularly amazing food that everyone on the planet eats (which is highly doubtful). On the other hand, the statement $(\forall)x(\exists)yE(x, y)$ means that each person on the planet eats at least one food (although that food could be different over the various people). This is a much weaker statement, and is obviously true in general. One can clarify the difference between the two statements by adding brackets between the two quantifiers as follows:

$$\exists y [\forall x \; E(x, y)] \quad \text{[Everybody eats at least one special thing]}$$
$$\forall x [\exists y E(x, y)] \quad \text{[Each person eats something]}$$

In practice, this type of explicit bracketing is rarely used, since the semantic meaning of a sentence is clear from the order of the operators.

Now consider our earlier statement that all people who eat beef should be considered non-vegetarian:

$$\forall x \; [E(x, Beef) \Rightarrow N(x)]$$

We wish to generalize this statement to the assertion that if there is at least one meat (e.g., beef or chicken) eaten by person x, then the person x is non-vegetarian. Clearly, we need some way to defining the subset of foods that can be considered meats. Therefore, we introduce the additional predicate $M(y)$ indicating that the food y is a meat in order to filter out the meats within a first-order expression. Then the definition of a non-vegetarian person can be rephrased with the use of the existential quantifier as follows:

$$\forall x \exists y \; [E(x, y) \wedge M(y) \Rightarrow N(x)]$$

In this case, the $\forall$ quantifier has precedence, and therefore the statement translates to the following:

> For each person, if they eat at least one meat (possibly different for each person), then the person in question is non-vegetarian.

It is important to get the order of operators right in order to interpret the statement correctly. Reversing the order of quantifiers would imply a society in which each person would have to eat the same meat (say, chicken) in order for the person to be considered non-vegetarian. Furthermore, more than one such special meat could exist, since an existential quantifier allows the occurrence of more than one instance of an object satisfying an assertion.

In some cases, the use of an existential quantifier over the $\Rightarrow$ operator might result in a semantic interpretation that is not intended. For example, consider the case where we want to say that there is some non-vegetarian person who eats Beef. Then, it might seem that a natural way of expressing this assertion is as follows:

$$\exists x \ [N(x) \Rightarrow E(x, Beef)]$$

Unfortunately, this statement is true as long as there is even one vegetarian person in the entire domain of discourse (which is not the originally intended semantic interpretation). As an example of how the wrong semantic interpretation could be reached, consider the case where John is vegetarian. Therefore, $N(John)$ is *False*, which is the antecedent of this instantiation of the assertion. In other words, the following statement is *True*:

$$N(John) \Rightarrow E(John, Beef)$$

Therefore, one can use the existential quantifier to infer that the statement $\exists x \ [N(x) \Rightarrow E(x, Beef)]$ is *True*. Furthermore, the use of the existential quantifier over a domain implicitly implies that the domain is non-empty. Otherwise, the resulting statement could lead to inferences that are not valid.

Quantifiers can be quite powerful, when they are used in chain of sequentially transitive relationships. Consider the case where we are trying to infer whether one person is an ancestor of another through the specification of parent-child relationships between individual people. It is convenient to specify only parent-child relationships in the knowledge base, because they are far fewer than the number of ancestral relationships. However, the key is that ancestral relationships follow as a logical consequence of parent-child relationships, and the quantifiers of first-order logic provide precisely the tools required to infer such relationships. Suppose that $P(x, y)$ is defined to be true for every pair of people in which x is a parent of y. We intend $A(x, y)$ to be true, if x is an ancestor of y. But how do we code up this intention in the form of first-order assertions? A knowledge base has no way of knowing how the relationship associated with the parent predicate can be connected with the ancestral relationship, unless rules are added to the knowledge base specifying this connection. In particular, adding the following statements to the knowledge base can help in inferring ancestral relationships from parent-child relationships available in the knowledge base:

$$\forall x, y \ [P(x, y) \Rightarrow A(x, y)]$$
$$\forall x, y [\exists z [P(x, z) \land A(z, y)] \Rightarrow A(x, y)]$$
$$\forall x, y [\exists z [A(x, z) \land A(z, y)] \Rightarrow A(x, y)]$$

The first rule above establishes the definition that if x is a parent of y, then x is an ancestor of y as well. One can view this rule as defining the base definition of a person being an ancestor of another, although the ancestral relationship is defined recursively (in general). The second and third rules provide two different ways of characterizing the recursive portion of the definition of ancestral relationships. The second rule states that if x is the parent of some z who is an ancestor of y, then x is an ancestor of y. The third rule assumes that z is some descendent of x and ancestor of y (instead of using a direct parent-child connection). Using these definitions of ancestral relationships and parent-child relationships, one can easily infer all ancestral relationships in a knowledge base. The above definition is, however, not complete, since we also need to assert the converse of the parental definition of the above statements:

$$\forall x, y \ [A(x, y) \Rightarrow P(x, y) \lor [\exists z [P(x, z) \land A(z, y)]]]$$

Furthermore, since everyone has a parent, the following will hold:

$$\forall x \exists z P(z, x)$$

Using a more complicated expression, one can even code up the fact that everyone has exactly two parents (see Exercise 16). One can already see how complicated a simple definition (such as ancestral relationships) can become in order to handle every corner case.

A knowledge base may contain a huge number of specific instantiations of parent-child relationships, and the number of ancestral relationships is exponentially related to the number of parent-child relationships, where the exponent of the relation is related to the number of generations. One can build upon this set of relationships further, by defining sibling or cousin relationships in terms of parent-child relationships, and so on. In all these cases, first-order logic provides a compact way of defining a much larger number of relationships by using recursive forms of the definition.

It is helpful to think of the quantification operators as shorthand forms of the $\land$ and $\lor$ operators over the entire discourse domain. For example, the use of the universal quantifier before an expression implies that the expression is true for each object in the discourse domain. Therefore, one is effectively performing an $\land$ over the expression as applied to all objects in the discourse domain. Similarly, the $\exists$ operator can be viewed as shorthand form of the $\lor$ operator over all objects in the discourse domain. Understanding the universal and existential quantifiers in this way is helpful in generalizing the rules of propositional logic to first-order logic when quantifiers are used. The basic idea is that first-order expressions can be converted to propositional expressions using this trick, and then converted back to first-order expressions with quantifiers after using appropriate simplifications (see the first-order analog of De Morgan laws in Section 5.2.5).

5.2.2 Functions in First-Order Logic

In addition to predicates and quantifiers, first-order logic also contains the ability to use functions, which map objects to other objects. Therefore, while predicates take as input objects and output truth values (much like propositional variables), functions output objects instead of truth values. For example, consider the function $Fav(x)$, which outputs the favorite food of person x.

The *equality operator*, denoted by "$=$," is used to indicate that two objects are the same. For example, consider the following statement:

$$\forall x \, [Fav(x) = Beef]$$

This statement indicates that the favorite food of x is beef for every person x. Note that the value on both sides of the equality operator is an object rather than a truth value. This makes the equality operator different from the equivalence operator, which is concerned with relationships between truth values. However, the use of the equality operator itself returns a truth value of *True* or *False*, depending on whether or not the statement is true. The above statement is true when everyone has beef as their favorite food. The equality operator is reflexive, symmetric, and transitive, as in the case of almost all of its uses in mathematics.

The equality operator is particularly helpful in initializing the values of functions in the knowledge base. The value of $Fav(x)$ will often be initialized to specific values in the knowledge base, such as the following:

$$Fav(John) = Beef$$
$$Fav(Mary) = Carrots$$

Technically, one is setting the value of each of these Boolean expressions to *True*, while initializing the *Fav* function.

The assertion that each person x always eats their favorite food can be expressed with a universal quantifier and the function $Fav(x)$ as follows:

$$\forall x\ E(x, Fav(x))$$

As a result, one can now infer that the following statements take on the value of *True*:

$$E(John, Beef)$$
$$E(Mary, Carrots)$$

One can often define complex properties of objects with the help of functions, which are not always possible with the use of predicates. Consider the case where $M(y)$ is a predicate indicating whether or not y is a food that is considered a meat. In such a case, if we know that $M(Fav(x))$ is *True* for person x, it can be inferred that person x is a non-vegetarian. Note that this is an indirect property of person x, by relating them to the type of food they eat. However, this property needs to be explicitly defined and coded up in the knowledge base. In the formal machinery of first-order logic, one can combine the truth of $M(Fav(x))$ with the existential statement $\exists y E(x, y) \Rightarrow N(x)$ (which is true for all x) to make the assertion that person x is a non-vegetarian. This means that the knowledge base needs to contain the following assertion:

$$\forall x\ [M(Fav(x)) \Rightarrow N(x)]$$

One can use the expression $(Fav(x) = Beef)$ as a first-order expression returning *True* or *False*, and incorporate it like any other first-order expression within a larger expression. This larger first-order expression will return truth values using the normal rules of first-order logic:

$$\forall x\ [E(x, Fish) \wedge \neg(Fav(x) = Beef)]$$

This statement takes on the value of *True* if and only if everyone eats fish and also does not have beef as their favorite food. In other words, everyone eats fish, but no one has beef as their favorite food.

5.2.3 How First-Order Logic Builds on Propositional Logic

The ability to make semantic statements of truth about the relationships between different objects/entities is critical in order to be able to use them in practical applications of artificial intelligence, such as expert systems. Propositional logic is inherently not suited to such settings, because of its inability to incorporate the concept of objects and truth values associated with them.

Much of the machinery of first-order logic is borrowed or generalized from propositional logic, and this includes the terminology and definitions in propositional logic. For example, the predicates $A(x)$, $B(x)$, and $C(x)$ are (alternatively) referred to as *atomic formulae*, and an atomic formula is the precise analog of an atomic proposition in propositional logic. A *literal* is a generalization of the concept of atomic formula, as it can either be an atomic formula or its negation. For example, both $C(x)$ and $\neg C(x)$ are literals, whereas only $C(x)$ is an atomic formula. The operators of propositional logic, such as $\wedge$, $\vee$, $\Rightarrow$, $\Leftrightarrow$, $\equiv$, and $\neg$ are directly generalized from propositional logic to first-order logic. Similarly, the laws involving the operators as well as their relative precedence are identical to that in propositional logic.

An atomic formula may have any number of arguments, including zero arguments. For example, the predicate $F(x, y, z)$ discussed above is an example of an atomic formula with three arguments (which corresponds to the arity of the predicate). An atomic formula containing zero arguments is simply an atomic propositional variable discussed in the previous chapter; this automatically implies that propositional logic is a special case of first-order logic—propositional logic is simply first-order logic with zero-argument atomic formulae! Because objects are not referenced in zero-argument atomic formulae, quantifiers are not needed either. Even so, the theorems governing quantifiers can be justified using De Morgan laws from propositional logic (see Section 5.2.5).

In first-order logic, one has to be careful not to be confused by the notation for the bound variables when combining multiple sentences. The choice of notation for a bound variable has nothing to do with how the laws of propositional logic are generalized to a first-order expression. In order to explain this point, we discuss the combining of two rules with the use of transitivity. Consider the case where the knowledge base encodes the following two sentences in first-order logic:

> Any person who fails to appear for the examination will receive a failing grade.
> Any person who receives a failing grade will not obtain a scholarship.

One can easily infer from the semantics of these two sentences that any person not appearing for an examination will receive a failing grade. However, we want to prove the correctness of this conclusion more formally using the machinery of first-order logic. In the formalism of first-order logic, the predicate $A(x)$ corresponds to the case where person x does not appear for an examination. The predicate $B(x)$ corresponds to the case where the person x receives a failing grade. Finally, the predicate $C(x)$ corresponds to the case where person x obtains a scholarship. Then, one can encode the following rules in the knowledge base in order to model the known facts:

$$\forall x \ [A(x) \Rightarrow B(x)]$$
$$\forall y [B(y) \Rightarrow \neg C(y)]$$

Note that one can bind x and y to any object in the domain of persons, such as John, Mary, and so on. The use of different variables x and y in the above rules is not significant, as long as they refer to the same domain of entities. This is because both statements are true for all objects within the domain of discourse. Therefore, the second rule is also equivalent to the following:

$$\forall x \ [B(x) \Rightarrow \neg C(x)]$$

Note that for any particular object such as *John*, we know that both $A(John) \Rightarrow B(John)$ and $B(John) \Rightarrow C(John)$ are *True*. Therefore, one can use the transitivity of implication (as in propositional logic) to infer that $A(John) \Rightarrow C(John)$. Since this fact is true for each and every object in the domain of discourse, one can infer this statement with a universal quantifier:

$$\forall x \ [A(x) \Rightarrow \neg C(x)]$$

This means that any person who does not appear for the examination will not obtain a scholarship. In other words, if we know that $A(John)$ is *True*, one can infer $\neg C(John)$, which implies that John will not receive a scholarship.

In order to understand why the above approach works, consider a domain containing just three objects corresponding to *John*, *Mary*, and *Ann*. Then, based on the use of the universal quantifier, one can infer that each of the following rules is true:

$$A(John) \Rightarrow B(John), \quad A(Mary) \Rightarrow B(Mary), \quad A(Ann) \Rightarrow B(Ann)$$
$$B(John) \Rightarrow \neg C(John), \quad B(Mary) \Rightarrow \neg C(Mary), \quad B(Ann) \Rightarrow \neg C(Ann)$$

We can then apply the transitivity of implication of each of *John*, *Mary*, and *Ann* (based on propositional logic) in order to infer the following:

$$A(John) \Rightarrow \neg C(John), \quad A(Mary) \Rightarrow \neg C(Mary), \quad A(Ann) \Rightarrow \neg C(Ann)$$

Since the rule $A(x) \Rightarrow \neg C(x)$ is true for all persons in the domain, it is evident that the following rule is true as well:

$$\forall x \, [A(x) \Rightarrow \neg C(x)]$$

This type of proof is exactly analogous to what is done in propositional logic. By instantiating to all cases, it is possible to convert a proof involving only universal quantifiers to a proof involving only propositional logic. Therefore, *proofs involving only universal quantifiers in first-order logic are relatively straightforward because we can use exactly the same proof mechanisms as we do in propositional logic.*

It is easy enough to generalize many of the proof methods of propositional logic to first-order logic, as long as these operands do not interact with quantifiers in unexpected ways. For example, transitivity does not apply to two rules bound by existential quantifiers, since the instance to which these two rules apply may not be the same. For example, if $A(x) \Rightarrow B(x)$ and $B(x) \Rightarrow C(x)$ are each true for some x, then it is not necessarily the case than $A(x) \Rightarrow C(x)$ is true for some x. For example, if the rules $\exists x(A(x) \Rightarrow B(x))$ and $\exists x(B(x) \Rightarrow C(x))$ are true, then only the following two rules may hold:

$$A(John) \Rightarrow B(John)$$
$$B(Mary) \Rightarrow C(Mary)$$

In this case, one can no longer use transitivity of implication, since the two rules apply to different objects. In general, *it is easier to generalize laws involving universal quantifiers from propositional logic than it is to generalize laws involving existential quantifiers.* As we will see later in this chapter, this principle is particularly important to keep in mind while designing systematic proofs in first-order logic.

Modus Ponens also directly generalizes from propositional logic to first-order logic, as long as it is used within the scope of a universal quantifier. For example, if $(\forall x)[A(x) \Rightarrow B(x)]$ and $A(x)$ take on the value of *True*, one can also infer that $(\forall x)B(x)$ takes on the value of *True*. However, *Modus Ponens* cannot be used in the context of an existential quantifier. This is an issue that we will discuss in the next section.

Some laws can be used within any type of quantifier. For example, the identity laws, idempotence laws, annihilator laws, double negation law, and complementarity laws are directly inherited from propositional logic to first-order logic, as long as they are used *within the quantifiers.* For example, one can make the following assertions using the identity law:

$$(\forall x)[A(x) \vee \neg A(x)] \equiv True$$
$$(\exists x)[A(x) \vee \neg A(x)] \equiv True$$

However, the expression $(\forall x)A(x) \vee (\forall x)\neg A(x)$ is not a tautology. A key point is that we have *two separate quantifiers and the full expression is not within the scope of either of two quantifiers*. In order to be able to create an expression with nested quantifiers extending over the full expression, it is important to first *standardize* the first-order expression.

5.2.4 Standardization Issues and Scope Extension

We revisit the issue of standardization, as it is particularly important in first-order proofs for extending the scopes of quantifiers to the full first-order expression. A confusion is often caused in first-order expressions when the same notation is used to refer to fundamentally different variables. For example, consider the following expression:

$$\exists x[A(x) \vee \forall x\, B(x)]$$

Here, it is important to understand that the two occurrences of x do not refer to the same variable. The predicate $A(x)$ falls within the scope of the existential quantifier, whereas the second predicate $B(x)$ falls within the scope of the universal quantifier. However, the expression seems somewhat confusing, and it is easy for a reader to get confused that the two predicates are based on the same object. Therefore, it is preferable to use standardized forms of the expressions, such as the ones on the right-hand sides of the equivalences shown below:

$$\exists x[A(x) \vee \forall x\, B(x)] \equiv \exists y[A(y) \vee \forall x\, B(x)]$$
$$\equiv \exists y\forall x\, [A(y) \wedge B(x)]$$

Changing the variable in this way is also referred to as *substitution*, which we will discuss in greater detail in a later section. Note that standardization allows the scope of the universal quantifier to be extended to the full expression, since the variable y bound by the existential quantifier does not interact with the variable x that is bound by the universal quantifier. What this example shows is that standardization not only helps improve readability, but it also helps in writing the first-order expression in a way that extends the scope of all quantifiers to the full expression. *Scope extension is particularly useful in first-order proofs, where one of the key steps is to extend the scope of all quantifiers to the full expression.* The resulting expression only contains a bunch of nested quantifiers over different variables. After scope extension, one can directly use the machinery of propositional logic within the first-order expression. We make the following observation:

Observation 5.2.1 *The first-order expression that is fully within the scope of all the (nested) quantifiers in that expression can be simplified like any propositional expression by treating each unique predicate-object combination as a propositional variable.*

However, extending the scope of quantifiers to the full expression is not immediately possible, if negations occur directly in front of quantifiers. For example, consider the expression $\exists x[A(x) \vee \neg\forall y\, B(y)]$, which is similar to the one discussed above, but a negation occurs in front of a quantifier. Suddenly, scope extension becomes more difficult, because one cannot move the $\forall$ quantifier outside the negation:

$$\exists x[A(x) \vee \neg\forall y\, B(y)] \not\equiv \exists x\forall y[A(x) \vee \neg B(y)]$$

Therefore, we need mechanisms to address the presence of negations in front of quantifiers in order to extend the scope of the quantifier to the full expression. This point will be discussed in the next section.

5.2.5   Interaction of Negation with Quantifiers

Just as the order of the universal and existential quantifier matters in first-order logic, so does the ordering of a particular type of quantifier and the negation operator. In order to understand this point, we will use an example. Let $N(x)$ be a predicate asserting the fact that x is non-vegetarian. Then, consider the following pair of statements:

$$\forall x \ [\neg N(x)]$$
$$\neg [\forall x \ N(x)]$$

The two statements are not the same. The first statement makes the claim that no one is a non-vegetarian (i.e., every single person on the planet is a vegetarian). This is because the negation is inside the scope of the quantifier. On the other hand, the second statement makes the claim that it is not true that everyone is a non-vegetarian (i.e., *at least* one person on person on the plant is a vegetarian). In this case, the negation is outside the scope of the quantifier. The use of the phrase "at least" to refer to the occurrence of an event implies that the existential quantifier comes into play. Therefore, the second statement is equivalent to the following:

$$\exists x \ \neg N(x)$$

In other words, pushing the negation into the $\forall$ quantifier changes it into an existential quantifier. We also point out that the first statement $\forall x \ \neg N(x)$ is equivalent to the following statement obtained via quantifier flipping:

$$\neg \exists x \ N(x)$$

One can, therefore, summarize the above observation as follows:

> Flipping the order of a negation and a quantifier changes a universal quantifier into an existential quantifier, and vice versa.

It is noteworthy that these negation push-through laws are equivalent to the De Morgan laws (cf. page 114) for propositional logic. This is because the De Morgan laws flip the $\wedge$ operator into the $\vee$ operator (and vice versa) by pushing the negation in and out of an expression. The key point is that the universal quantifier is really an indirect form of the $\wedge$ operator over the entire discourse domain, whereas the existential quantifier is an indirect form of the $\vee$ operator over the same discourse domain. Since De Morgan laws flip the $\wedge$ operator into a $\vee$ operator via push=through (and vice versa), it follows that pushing the negation into a quantifier changes the type of quantifier from a universal quantifier to an existential one and vice versa.

We will illustrate the nature of quantifier-negation interaction with the help of an example. Consider the domain of discourse containing the three objects $\{a, b, c\}$. Then, the statement $\forall x N(x)$ is equivalent to the following propositional expression:

$$N(a) \wedge N(b) \wedge N(c)$$

Similarly, the statement $\exists x N(x)$ is equivalent to the following propositional expression:

$$N(a) \vee N(b) \vee N(c)$$

Therefore, the machinery used in propositional logic for constructing proofs is directly inherited by first-order logic. Note that the De Morgan laws of propositional logic imply the following:

$$\underbrace{\neg[N(a) \lor N(b) \lor N(c)]}_{\neg \exists x N(x)} \equiv \underbrace{[\neg N(a) \land \neg N(b) \land \neg N(c)]}_{\forall x \neg N(x)}$$

$$\underbrace{\neg[N(a) \land N(b) \land N(c)]}_{\neg \forall x N(x)} \equiv \underbrace{[\neg N(a) \lor \neg N(b) \lor \neg N(c)]}_{\exists x \neg N(x)}$$

Therefore, the fact that the quantification operators can be flipped by negation is explained by the De Morgan laws.

5.2.6 Substitution and Skolemization

An important problem that often arises is that some rules hold for all objects in the domain, whereas other rules might hold for specific individuals like John. The expressions within universal quantifiers are particularly easy to address with the use of the rules of propositional logic, and it is also possible to apply laws that involve multiple statements. For example, if we have two universally quantified assertions such as $\forall x[E(x, Beef) \Rightarrow N(x)]$ and $\forall x E(x, Beef)$, one can use *Modus Ponens* to infer the expression $\forall x N(x)$. This is because one can express the first two assertions in terms of each object in the domain of discourse, apply Modus Ponens for each object to get the object-specific propositional expression, and then apply universal quantification back to these object-specific inferences in order to obtain a universally quantified expression in the same form. However, if we might have an instantiated assertion such as $E(John, Beef)$, the conclusions from *Modus Ponens* are specific only to John. The process of applying *Modus Ponens* in such a case can be achieved only by the method of substitution:

Observation 5.2.2 (Ground Substitution) *For any universally quantified expression* $\forall x \, A(x)$, *one can substitute* x *with any arbitrary object* o *in the domain of discourse, and the resulting expression* $A(o)$ *takes on the same value as* $\forall x \, A(x)$.

A second type of substitution, referred to as *flat substitution*, simply exchanges the notations for variables:

Observation 5.2.3 (Flat Substitution) *For any universally quantified expression* $\forall x \, A(x)$, *one can substitute* x *with any other variable* y *as long as quantification applies to the switched variable:*

$$[\forall x \, A(x)] \equiv [\forall y \, A(y)]$$

It is clear that flat substitutions are more general than ground substitutions, because they apply to larger sets of objects.

Substitution is critical in being able to perform *Modus Ponens* in first-order logic via the idea of *lifting*. The idea in lifting is that one can make a pair of formulas identical by making appropriate substitutions. For example, if we have the propositional statement $\forall x[E(x, Beef) \Rightarrow N(x)]$, which indicates that anyone who eats Beef is a non-vegetarian. However, we also know that John eats beef, and therefore the assertion $E(John, Beef)$ exists in the knowledge base. What is true for everyone must be true for John. Therefore, one can substitute *John* for x in the universally quantified expression in order to obtain the lifted version of the rule:

$$E(John, Beef) \Rightarrow N(John)$$

Therefore, we can use *Modus Ponens* with the lifted version of the rule in order in infer $N(John)$. We were able to use *Modus Ponens* because the new antecedent of the lifted rule is identical to the assertion $E(John, Beef)$. This process of making two expressions identical via appropriate substitutions is referred to as *unification*. When performing unification, the most general unification is preferred. For example, $\forall x\ E(x, Beef)$ and $\forall y E(y, Beef)$ could either unify to $\forall z\ E(z, Beef)$ (via flat substitution), or both expressions could unify to $E(John, Beef)$ (via ground substitution). The former is preferred because it is the most general expression that unifies both expressions. This approach of combining Modus Ponens with unification is referred to as *generalized* Modus Ponens. The process of unification recursively explores two expressions simultaneously to check if they match, and its complexity is quadratic in the length of the subexpression. We omit the details of the unification algorithm, and refer the readers to the bibliographic notes at the end of the chapter. Although knowing the specific algorithm for unification is not essential (as it can be treated like a black-box package), it is important for the reader to understand this concept, because it will be used extensively in this chapter.

The principle of substitution allows the proof techniques in propositional logic to be generalized easily to first-order logic, as long as one works only with universal quantifiers (rather than existential quantifiers). For example, extending the scope of a quantifier to a single expression allows us to use the rules of propositional logic to whatever is within that single expression, but a problem arises in the case of multiple quantified expressions because *substitution does not work with existential quantifiers*. Substitution is required whenever we are using a rule of propositional logic involving multiple statements (like *Modus Ponens* or transitivity of implication). For example, the two statements $\exists x(A(x) \Rightarrow B(x))$ and $\exists y(B(y) \Rightarrow C(y))$ do not necessarily imply by transitivity that $\exists x(A(x) \Rightarrow C(x))$. On the other hand, transitivity does hold for universally quantified statements. Therefore, many propositional laws such as transitivity (that involve multiple assertions) do not work with existential quantifiers, whereas they do work with universal quantifiers via the process of substitution:

Observation 5.2.4 *The rules of propositional logic can be applied to multiple quantified expressions using the principle of substitution to create matching predicate-object combinations, as long as (i) the expressions contain (possibly nested) quantifiers each of whose extends over the entirety of the corresponding expression, and (ii) the expressions include only universal quantifiers and not existential ones.*

Handing existential quantifiers requires a process referred to as *Skolemization*. First, we discuss the simple case where the existential quantifier does not occur inside the scope of a universal quantifier, where Skolem *constants* are needed:

Definition 5.2.1 (Skolem Constant) *Any object variable that is existentially quantified outside the scope of a universal quantifier (such as $\exists x\ A(x)$) can be replaced by a single new constant expression like $A(t)$, where t is a Skolem constant. The existential quantification can be dropped.*

Note that if there are multiple variables that are associated with existential quantifiers (e.g., $\exists x, y$), then a separate Skolem **constant** would need to be used for each (e.g., t for x and u for y). Furthermore, it is important not to use the same notation for the Skolem constant as that representing one of the constant ground objects (e.g., John) or variable objects in the first-order expression.

Skolemization leads to an expression that is *inferentially equivalent* to the original expression, but the equivalence is good only for proofs. For example, replacing $\exists x C(x)$ with

$C(t)$ seems to imply that there is only one object t satisfying the predicate $C(\cdot)$, whereas this may not really be the case. However, this replacement does not affect the validity of proofs. For example, consider the following *inferential* equivalence between expressions with without existential quantifiers:

$$[(\exists y B(y)) \vee (\exists y C(y))] \equiv_I [B(t) \vee C(u)]$$

Here t and u are Skolem constants, and they are different constants because they "belong" to different existential quantifiers. Any first-order expression that can be proven using the statement on the left can also be proven using the statement on the right (and vice versa). This is in spite of the fact that the expression on the left could mean that there is more than one ground object satisfying the expression. Therefore, we have used the subscript 'I' in the above equivalence to show that the equivalence is inferential in nature. In practice, we do not use this type of subscript in proofs to avoid cryptic notation.

When existential quantifiers fall within the scope of a universal quantifier and the expression within the existential quantifier contains some of the universally bound variables, introducing a Skolem constant blindly might cause a problem. For example, consider the expression $\forall x \exists y E(x, y)$, which implies that everybody eats at least one thing. However, that one food is specific to the person we are talking about in the universal quantification. Therefore, the existential object in question (specific food) needs to be a function of the person we are talking about, which is achieved with the use of a Skolem *function* rather than a Skolem constant:

Definition 5.2.2 (Skolem Function) *Any existential quantification inside the scope of a universal quantifier (such as $\forall y \exists x B(x, y)$) can be replaced with a Skolem function of the universal variable (such as $\forall y B(f(u), y)$. An existential quantifier within the scope of more than one universal quantifier (such as $\forall y \forall z \exists x B(x, y, z)$) can be replaced with a multivariable Skolem function (such as $\forall y \forall z B(g(y, z), y, z)$).*

Different Skolem functions need to be introduced for different existential quantifiers, and the scopes of the universal quantifiers need to be taken into account during the process. Consider the following first-order expression:

$$\forall x [\exists y (A(x) \Rightarrow B(y)) \vee \forall w \exists z (D(x) \wedge E(w) \wedge F(z) \Rightarrow C(z))]$$

This expression contains multiple existential and universal quantifiers, and different existential quantifiers lie within the scope of different universal quantifiers, although the universal quantifier involving variable x applies to all existential quantifiers. Then, Skolemization results in the following expression:

$$\forall x ([A(x) \Rightarrow B(f(x))] \vee \forall w [D(x) \wedge E(w) \wedge F(g(x, w)) \Rightarrow C(g(x, w))])$$

In the above example, $f(x)$ and $g(x, w)$ are Skolem functions. Note that different Skolem functions are used for the different existential quantifiers, and the number of arguments of a Skolem function depends on the number of universal quantifiers in whose scope it lies. It is also noteworthy that the same Skolem function $g(x, w)$ is used for the two occurrences of the existential variable z, because both occurrences are bound by the same quantifier and must therefore refer to the same ground instance. The end goal of Skolemization is to create a first-order expression containing only universal quantifiers (because they are simpler to deal with).

We will illustrate the simultaneous use of substitution and Skolemization with the help of an example. Consider the case, where we have the following pairs of statements:

$$\forall x \ [A(x) \Rightarrow B(x)]$$
$$\exists x \ [B(x) \Rightarrow C(x)]$$

We want to use the above two results to show the following:

$$\exists x \ [A(x) \Rightarrow C(x)]$$

From the second statement $\exists x \ [B(x) \Rightarrow C(x)]$, one can show that $B(t) \Rightarrow C(t)$ is true for some object t in the domain of discourse. This conclusion directly follows by replacing the existential variable with a Skolem constant t. Furthermore, by combining the first statement $\forall x \ [A(x) \Rightarrow B(x)]$ with the principle of (ground) substitution, one can infer the fact that $A(t) \Rightarrow B(t)$. Therefore, one can infer $A(t) \Rightarrow C(t)$, and it implies that the following is true:

$$\exists x \ [A(x) \Rightarrow C(x)]$$

In general, first-order proofs are more challenging than those of propositional logic because of the presence of quantifiers and functions.

5.2.7   Why First-Order Logic Is More Expressive

The binding of variables to specific objects is what gives first-order logic its true expressive power. It gives first-order logic the ability to *reason* over objects and the relationships among them. In propositional logic, separate propositional statements are used to express statements such as *"Steven Spielberg directed Saving Private Ryan"* and *"Steven Spielberg directed The Terminal."* Unfortunately, using this approach loses the fact that *the semantic structure of* these two statements is very similar, with the main difference being that they use different objects as arguments. However, in first-order logic, both statements can be expressed as $G(StevenSpielberg, y)$, where the variable y corresponds to any movie that Steven Spielberg has directed. This allows us to use these algebraic connections during inferences. This ability to express similar statements as propositional expressions allows one to come up with more complex and rich inferences that can be leveraged to unleash the full power of first-order logic.

A key point is that propositional logic is *declarative*, whereas first-order logic is *compositional*. In compositional languages, the meaning of a sentence depends on the semantics of the objects and predicates that it is made out of. As a result, the meaning of a sentence in first-order logic depends on its context. This is one of the reasons that first-order logic can come closer to expressing the semantics of natural language, as compared to propositional logic.

As we will see in Chapter 12, there are various other ways to represent these relationships in structural form, such as the use of a *knowledge graph*. In knowledge graphs, objects correspond to nodes and the relationships among them correspond to edges. However, knowledge graphs represent a (practical and useful) simplification of these relationships, because each relationship is defined between only two objects. On the other hand, a predicate in first-order logic can be a relationship between any number of objects. For example, the relationship between *Steven Spielberg, Saving Private Ryan*, and *Best Director Award* will be represented as three separate binary relationships between pairs of entities (see Figure 12.1(b) of Chapter 12). In knowledge graphs, these relationships are explicitly represent in structural form, as the edges of a graph of objects. Furthermore, relationships between subsets of objects can be expressed using ontologies, and edges between these nodes. Therefore, knowledge graphs implicitly use some aspects of second-order logic where distinctions are allowed between classes of objects and hierarchical relationships among the different classes are used in the analysis.

5.3 Populating a Knowledge Base

In order to use first-order logic in artificial intelligence, it is first necessary to populate a knowledge base with important facts about the domain of discourse. As in the case of propositional logic, sentences are added to a knowledge base via assertions. Subsequently, the assertions for a particular domain of discourse are entered into the knowledge base. These axioms correspond to the basic relationships among objects in the database and the definitions of these relationships. An example of such a relationship is that of an ancestor relationship $A(x, y)$, based on the parent relationship $P(x, y)$. We revisit these rules that were discussed earlier on page 144:

$$\forall x, y \ [P(x, y) \Rightarrow A(x, y)]$$
$$\forall x, y [\exists z [P(x, z) \wedge A(z, y)] \Rightarrow A(x, y)]$$
$$\forall x, y \ [A(x, y) \Rightarrow P(x, y) \vee [\exists z [P(x, z) \wedge A(z, y)]]]$$

Here, $A(x, y)$ denotes the fact that x is the parent of y, and $A(x, y)$ denotes the fact that x is the ancestor of y. These can be viewed as the basic assertions upon which a knowledge base is built. Some assertions can be simple facts about particular objects; these facts form the basis on which more complex facts are inferred about specific objects in knowledge bases. For example, if one knows that Jim is the parent of Sue, and Sue is the parent of Ann, one can specify these facts with two assertions:

$$P(Jim, Ann)$$
$$P(Ann, Sue)$$

This part of the construction of the knowledge base is often a simple matter of mechanically reading in facts that we already know about the world, just as a computer program might often mechanically read in large amounts of input data. Starting with these basic assertions, one can use the other relationship-centric assertions to infer non-obvious facts. For example, the two basic assertions can be combined with the other general rules in the knowledge base to infer that Jim is an ancestor of Sue.

It is important for the axioms be specified in self-consistent fashion. Consider the specific example of Jim, Sue, and Ann discussed above. If Jim is the parent of Ann, and Ann is the parent of Sue, there is no way in which Sue could be the parent of Jim. On the other hand, if the knowledge base contains the assertion $P(Sue, Jim)$, it should lead to an inconsistent knowledge base. Therefore, one needs to add mechanisms to knowledge bases to detect and rule out such inconsistencies. There are several ways of achieving this goal, one of which is to add more rules to the knowledge base to explicitly rule out such situations. An example of such an assertion is as follows:

$$\forall x, y \ [A(x, y) \Rightarrow \neg A(y, x)]$$

In other words, x and y cannot be ancestors of one another, which rules out the possibility of cycles in the graph. Here, a key point is that knowledge bases have no way of knowing "obvious" semantic relationships that would be natural to us, unless they are explicitly coded up. It is often a challenge to ensure that a knowledge base contains a consistent and complete set of assertions about the domain at hand. In many cases, important axioms are missing from the knowledge base and are discovered only when the queries to the knowledge base do not yield expected results. As a result, even simple knowledge bases often contain a large number of assertions in order to define the useful relationships in an effective way. The most basic assertions without which the knowledge base would be incomplete are referred to as axioms. However, an assertion such as the following would be considered a theorem, since it can be derived from the earlier statement about ancestral relationships:

$$\forall x, y \ [P(x, y) \Rightarrow \neg P(y, x)]$$

Even though it suffices to only have axioms in the knowledge base, many of the rules in the knowledge base are often not axioms, and they may be derivative theorems of other axioms. This is done in order to make proofs more concise and workable. Even a human mathematician does not derive every proof only from first principles, but will often use (important) intermediate results in order to create concise proofs.

The overall task of building a knowledge base is to first assemble all the required knowledge for the task at hand, and then construct a vocabulary of objects, predicates, functions, and constants. The general knowledge about the domain is then coded up in terms of this vocabulary. The general-purpose inference procedure with the use of first-order logic (see Section 5.5) is then used in order to derive further facts and respond to queries. These general purpose procedures often use similar procedures (e.g., forward or backward chaining) as propositional logic in order to reach similar conclusions. However, some modifications have to be made to these procedures in order to account for the greater complexity of first-order logic. This greater complexity is usually associated with the object-centric nature of propositional formulas and their associated quantifiers.

An important point is that it is often very easy to miss basic facts about the relationships between objects when one is adding them to the knowledge base. A knowledge base does not have a pre-existing concept of semantic understanding that may be obvious to humans—in other words, any seemingly "obvious" inferences need to be explicitly represented in some way during the construction of the knowledge base. For example, consider the case where one is trying to set up a definition of a the sibling relationship by stating that they have the same set of two parents. An erroneous way of stating this would be as follows:

$$\forall x \forall y \exists z \exists w [P(z, x) \wedge P(z, y) \wedge P(w, x) \wedge P(w, y) \Rightarrow S(x, y)]$$

This way of defining the sibling relationship is erroneous, because it also includes half-siblings, when z and w are both chosen to be the *same* biological parent of the two half-siblings. A correct way of defining the sibling relationship would be to use the equality operator to ensure that the two objects z and w are not the same:

$$\forall x \forall y \exists z \exists w [\neg(z = w) \wedge (P(z, x) \wedge P(z, y) \wedge P(w, x) \wedge P(w, y) \Rightarrow S(x, y))]$$

This example shows how easy it is to make logical mistakes during the construction of a knowledge base. The most common problem is that of missing relationships that were often implicitly assumed by the analyst without explicitly coding them into the knowledge base. Considerable debugging of a knowledge base is often necessary to ensure that incorrect inferences are not made using the available rules. This can be achieved by repeatedly testing the knowledge base for known relationships with appropriately chosen queries. The process of building a knowledge base is a rather tedious one, and unexpected results are often returned in earlier iterations, as "obvious" but missing truths about the state of the world are discovered and added to the knowledge base.

First-order logic forms the workhorse of many existing expert systems. The main challenge in all these cases is to build the underlying knowledge base, and populate it with the appropriate rules that regulate the relationships among objects. Knowledge bases do not have any understanding of "obvious" facts about the discourse domains, unless the appropriate rules are carefully coded into the knowledge base. It is extremely common to end up with knowledge bases that are incomplete because of missing rules, and therefore the resolution procedure may not work as one expects. Creating knowledge bases in open domain settings is often an extremely challenging and unending task. This is one of the reasons that most of the successes of these methods have come in heavily restricted domain settings.

5.4 Example of Expert System with First-Order Logic

In order to understand the power of first-order logic, we will revisit the example of a medical expert system that is discussed in the previous chapter on propositional logic (cf. Section 4.4). The main problem with using propositional logic for this medical expert system is that a separate proposition needs to be created for each object in the knowledge base, which can be computationally expensive and sometimes not feasible (when one does not know the precise membership of the domain of objects up front). As we will show in this section, first-order logic provides a much more compact way of representing the knowledge base with the use of quantifiers, so that the same set of rules need not be repeated over each object in the domain of discourse (and one can also work with domains that are not explicitly defined up front in terms of their precise membership). Working through the details of this approach provides an appreciation of how one might be able to make inferences about specific objects, given generic statements about all or some objects. Although this section does not provide a systematic approach for inference, it provides an understanding of how the general rules of first-order logic can be used to make inferences about specific objects.

The medical expert system discussed in the previous chapter on propositional logic uses a number of symptomatic statements about the patient John and then attempts to infer a diagnosis. Specifically, John's symptoms are captured by the following statements:

John is running a temperature
John is coughing
John has colored phlegm

Now imagine a case where the knowledge base of the expert system contains the following subset of rules relating symptoms, diagnoses, and treatments:

IF coughing AND temperature THEN infection
IF colored phlegm AND infection THEN bacterial infection
IF bacterial infection THEN administer antibiotic

Although the previous chapter (implicitly) creates propositional expressions for each possible entity, this is not an efficient approach for inference, because one needs a separate rule for each possible entity:

IF coughing$_\text{John}$ AND temperature$_\text{John}$ THEN infection$_\text{John}$
IF coughing$_\text{Mary}$ AND temperature$_\text{Mary}$ THEN infection$_\text{Mary}$

Note that this type of approach is a reduction of first-order logic to propositional logic, which creates many repetitive rules about the entire domain of discourse. The main issue is that some of the statements (such as John having colored phlegm) are specific to the entity John, whereas other statements (such as the fact that coughing and temperature indicate an infection) relate to any person in the domain of discourse. By creating a statement for each object in the domain of discourse, it is indeed possible to perform resolution with the use of propositional logic. However, this is not a practical approach if the domain of discourse is large. Furthermore, if the domain of discourse changes over time, it will lead to challenges in continually updating the knowledge base in a rather awkward way. For domains of infinite size or in cases where we do not know the names of all the individuals up front, it is not possible to use the approach at all. It is here that using quantified predicates rather than propositional expressions is useful. After all, quantifiers provide a route to express facts about domains without explicitly worrying about the specific membership of objects. This makes the representation compact, irrespective of the domain size.

One can define the propositional variable $C(x)$ for person x coughing, the variable $T(x)$ for person x running a temperature, $F(x)$ for person x having an infection, $P(x)$ for person x having colored phlegm, $B(x)$ for person x having bacterial infection, and $A(x)$ for administering an antibiotic to person x. Unlike our analysis in Chapter 4, these rules are no longer defined only for John, but can apply to an arbitrary person x. In other words, we no longer need a domain of objects whose members are explicitly enumerated. The rules in the knowledge base can be coded up as follows:

$$\forall x \ [C(x) \wedge T(x) \Rightarrow F(x)]$$
$$\forall x \ [P(x) \wedge F(x) \Rightarrow B(x)]$$
$$\forall x \ [B(x) \Rightarrow A(x)]$$

The conditions for John can be summarized as follows:

$$T(John) \wedge C(John) \wedge P(John)$$

The process of inferencing is similar to that described on page 119 of Chapter 4, and it proceeds by using the bound variables with respect to John to derive new inferences that are specific to John. Since each of the assertions $C(John)$ and $T(John)$ are true, and we have the rule $C(x) \wedge T(x) \Rightarrow F(x)$ with x being allowed to bind to John, one can also infer $F(john)$. This follows directly from the use of substitution, where a universal quantification can bind to any value. This is essentially a use of generalized *Modus Ponens*. Furthermore, it is known that John has colored phlegm, it follows that $P(John)$ is true. In other words, $P(John) \wedge F(John)$ is true as well, which implies that $B(John)$ is true (via generalized *Modus Ponens*). One has effectively inferred from John's symptoms that he has a bacterial infection. Using the rule $B(x) \Rightarrow A(x)$ together with generalized *Modus Ponens*, one obtains $A(John)$. In other words, the expert system recommends the adminstration of an antibiotic to John.

The above example uses an ad hoc approach of reaching the goal state, which is obviously not practical for large knowledge bases. Therefore, we need some kind of systematic procedure to reach goal states, where the new facts are inferred in a particular order. This will be the topic of discussion in the next section.

5.5 Systematic Inferencing Procedures

The previous sections discuss how the individual inferencing rules in propositional logic can be generalized to first-order logic. Furthermore, the previous section provides an ad hoc inferencing method in first-order logic. In this section, we will introduce systematic ways of leveraging the assertions in the knowledge base to achieve a particular inferencing goal. This will be achieved with the use of proof by contradiction, as well as forward and backward chaining procedures; these procedures are exactly analogous to those used in propositional logic. In general, proofs in first-order logic are somewhat more difficult because of its object-centric algebra. For example, truth tables provide a way of constructing proofs in propositional logic, but they do not work in the case of first-order logic. This is because the sizes of such tables might be too large in first-order logic, or the membership of the domain of objects might not be fully known up front. In such cases, it is not possible to create propositions (and corresponding truth table values) over the objects in the domain.

The inferencing procedures in first-order logic are similar to those in propositional logic. For example, procedures such as *proof by contradiction, forward chaining*, and *backward chaining* have corresponding analogs in first-order logic (and they represent generalizations of how they are used in propositional logic). This will be the topic of discussion in this section.

5.5.1 The Method of Proof by Contradiction

The proof by contradiction in first-order logic is similar to that in propositional logic. For a given knowledge base KB, if one wants to infer q from KB, it suffices to show that $KB \land \neg q$ is not satisfiable. It is assumed that the variables are already *standardized*, so that a single variable is not used multiple times in a quantification. For example, an expression such as $\forall x A(x) \lor \exists x B(x)$ is not standardized, because the variable x is used multiple times in a quantification. However, this expression can also be expressed as $\forall x A(x) \lor \exists y B(y)$, which is standardized. It is important to work with standardized expressions in order to avoid confusion during the inference process. Once a standardized representation is obtained, the database is converted to conjunctive normal form (as in the case of propositional logic). Once the database is converted to conjunctive normal form, a similar resolution approach is used for simplifying this representation and arriving at a contradiction.

5.5.1.1 Conversion to Conjunctive Normal Form

The overall steps of converting a knowledge base to conjunctive normal form are similar to those in propositional logic, with the main difference being that existential quantifiers need to be dealt with in a special way. Therefore, the steps for conversion of $KB \land \neg q$ to conjunctive normal form are similar to those in Section 4.7 of Chapter 4. However, some changes need to be made in order to address the presence of quantifiers in first-order expressions. The steps for converting a first-order logical expression to conjunctive normal form are summarized as follows:

1. Replace each $\equiv$ in $KB \land \neg q$ with an expression containing only the operator "$\Rightarrow$". For example, the statement $A(x) \equiv B(x)$ can be replaced with the following expression:

$$(A(x) \Rightarrow B(x)) \land (B(x) \Rightarrow A(x))$$

Note that $A(x)$ and $B(x)$ might themselves be complex first-order expressions rather than literals.

2. Replace each "$\Rightarrow$" operator in $KB \wedge \neg q$ with $\vee$. For example, if we have the expression $A(x) \Rightarrow B(x)$, it can be replaced with $\neg A(x) \vee B(x)$. Note that each $A(x)$ and $B(x)$ could itself be a longer first-order expression rather than a literal.

3. When an expression is expressed in conjunctive normal form, the negation operator appears only in front of literals. For example, a subexpression such as $\neg(A(x) \vee B(x))$ would never be allowed in the CNF form of $KB \wedge \neg q$, because the negation operator appears in front of a complex expression rather than a literal. One can use De Morgan's laws in order to get rid of this problem, when it does occur in the current representation of $KB \wedge \neg q$. In this case, we push the negations into the subexpressions, so that *all negations apply only to literals*. In the case of first-order logic expressions, it is important to flip a quantifier when pushing a negation through it. Therefore, expressions like $\neg \forall x \ A(x)$ and $\neg \exists x \ B(x)$ are converted to $\exists x \ \neg A(x)$ and $\forall x \ \neg B(x)$, respectively.

4. At this point, all existential quantifiers are removed via Skolemization. As a result, the expression now contains only universal quantifiers (and constants). Since the expression is standardized the scopes of all universal quantifiers can be extended to the full expression.

5. After pushing through the negations to the level of the literals, one might have a nested expression containing both the $\wedge$ and $\vee$ operators. At this point, we use the distributive law (inherited from propositional logic), in which the operator $\vee$ is distributed over $\wedge$. For example, the expression $A(x) \vee (B(x) \wedge C(x))$ is replaced by the following expression using the distributive property of $\vee$ over $\wedge$:

$$\forall x[(A(x) \vee B(x)) \wedge (A(x) \vee C(x))]$$

After performing repeated distribution of $\vee$ over $\wedge$, one obtains an expression in conjunctive normal form.

Since universal quantifiers are assumed to bind to all matching object variables in the expression (which is standardized), one does not need to explicitly write out these quantifiers for compactness. This makes the overall expression look even more similar to the propositional expressions in conjunctive normal form. Each unique predicate-variable and predicate-constant can be treated in a manner similar to propositional variables (in terms of the fact that they will always obey the laws of propositional logic).

The main difference between this procedure and that for propositional logic is the additional step required in order to deal with quantifiers. Once $KB \wedge \neg q$ has been converted to conjunctive normal form, a resolution procedure is used to decide whether the expression is satisfiable.

5.5.1.2   Resolution Procedure

The resolution procedure in first-order logic is similar to that in propositional logic in terms of using the CNF form of the expression in order to show that it is equivalent to the truth value of *False*. However, the primary difference from propositional logic is that exact matching between a literal and its negation may not be possible; therefore, one might sometimes need to perform unification in order to match atomic formulae to atomic propositions. In other words, some amount of lifting may be needed.

The simplest situation is that the resolution procedure repeatedly looks at pairs of conjunctions that differ only in the negation in front of a single literal. An example of such a pair of conjuncts would be $(F(x) \vee \neg G(x) \vee H(x))$ and $(F(x) \vee \neg G(x) \vee \neg H(x))$. Note that the only difference between these two conjuncts is in terms of the negation in front of the final literal $H(x)$. One can use the distributive property to infer the following:

$$[\underbrace{(F(x) \vee \neg G(x)}_{D(x)} \vee H(x)) \wedge \underbrace{(F(x) \vee \neg G(x)}_{D(x)} \vee \neg H(x))] \equiv [(F(x) \vee \neg G(x)) \vee (H(x) \wedge \neg H(x))]$$

$$\equiv [(F(x) \vee \neg G(x)) \vee \textit{False}]$$
$$\equiv (F(x) \vee \neg G(x))$$

Note that the two conjuncts become consolidated, and the single literal and the differing portion of the conjunct becomes eliminated. However, it is possible to have a situation that the elimination of one of the literals via *unification*. For example, consider the following expression, which is universally quantified over x:

$$[D(x) \vee H(x)] \wedge [D(x) \vee \neg H(John)]$$

In this case, it is still possible to unify $H(x)$ and $H(John)$, and conclude that the expression is equivalent to $D(x)$. This type of unification may be repeatedly needed for simplification. In each step, a pair of clauses is identified that can be consolidated in this way in order to reduce the length of each of a pair of conjuncts. Therefore, the resolution procedure is a lifted version of the resolution procedure used in propositional logic.

Each such step of eliminating a literal and its negation leads to the addition of a new statement to the knowledge base. At some point, it is possible for the repeated reduction in the sizes of the individual conjuncts to lead to a situation where an atomic formula and its negation are present in the augmented knowledge base. This situation leads to a contradiction, and one can conclude that KB entails q. Therefore, the overall algorithm for resolution may be described using the following pair of steps repeatedly:

1. If a pair of conjuncts exists in the CNF form of $KB \wedge \neg q$ that are atomic formulas and are negations of one another, then report that KB implies q and terminate.

2. Select a pair of clauses differing in the negation of a single literal and apply the resolution procedure to create a new, shorter clause.

3. Add the new clause to KB and return to the first step.

It is noteworthy that the resolution procedure can also be used to construct a proof, by tracing how one obtains the atomic formulas that are negations of one another.

A natural question arises as to whether the resolution process in first-order logic is both sound and complete, as in the case of propositional logic. Completeness refers to the fact that any fact that is entailed by the knowledge base can indeed be inferred in a finite number of steps. Soundness refers to the fact that if a fact is not entailed by the knowledge base, then it can also be shown in a finite number of steps. It turns out that the resolution procedure of first-order logic is complete but not sound. The procedure is complete, because any sentence that is truly entailed by the knowledge base can be arrived at by the resolution procedure in a finite number of steps. However, if the sentence is not entailed by the knowledge base, the algorithm may not be able to arrive at this conclusion in a finite number of steps. For any sentence that can indeed be entailed, it is possible to convert the problem to that of propositional entailment over a knowledge base of finite size. This result is referred to as *Herbrand's theorem*. However, this is not possible to achieve for sentences that are not entailed by the knowledge base. Therefore, one difference from propositional logic is that this procedure is not guaranteed to terminate in first-order logic. As a result, first-order logic is considered *semi-decidable*.

5.5.2   Forward Chaining

The previous chapter discusses forward and backward chaining methods in propositional logic. It is also possible to generalize the forward and backward chaining algorithms of propositional logic to first-order logic. To achieve this goal, it is necessary to create first-order definite clauses, which are analogous to the positive definite clauses in propositional logic. The main difference is that first-order definite clauses may also contain variables, which are assumed to be universally quantified. As in propositional logic, first-order definite clauses typically have a set of positive literals in the antecedent, and a single positive literal in the consequent. Sometimes the antecedent might be missing, in which case the clause is in the form a single positive literal (assertion). The variables are assumed to be standardized, in that two separate variables do not use the same symbol, unless they refer to the same variable. Since the universal quantifier is implicit, it is often omitted in the exposition (for brevity). The following are examples are positive definite clauses:

$$A(x)$$
$$A(John)$$
$$A(x) \wedge B(x) \Rightarrow C(x)$$

Although it is not guaranteed that an arbitrary knowledge base can be converted to first-order definite form, the reality is that most knowledge bases can be converted to this form. Once the first-order definite clauses have been created, the methods for forward chaining in propositional logic can be generalized to first-order logic. However, since one is dealing with variables, unification becomes more important in this case. Therefore, in each iteration, the unsatisfied antecedents of each rule are checked, and unified with existing statements in the knowledge base. The unification may require the substitution of variables in the antecedents of the rule. The same rule may be substituted in multiple ways for different substitutions, and all possible substitutions that result in a unification are attempted. This process is continued iteratively until the knowledge base does not expand any more, or a generalization of the goal clause is discovered. If a generalization of the goal clause is discovered, then the algorithm terminates with a Boolean value of *True*. On the other hand, in the case where the goal clause is not discovered (but the knowledge base does not expand further), the situation is referred to as the *fixed point* of the knowledge base. In such a case, the approach terminates with a Boolean value of *False*. The overall structure of the approach is similar to that of the forward chaining algorithm discussed in Figure 4.1 of Chapter 4. The main difference is that unification must be used during the process of finding matching rules.

When the goal clause is entailed by the knowledge base, the approach is guaranteed to terminate. However, if the goal clause is not entailed by the knowledge base, the procedure is not guaranteed to terminate, because infinitely many new facts could be generated by function symbols. Therefore, the approach is semi-decidable, even when first-order definite clauses are used. Nevertheless, in most practical cases, a reasonable termination point can be found, at which it can be concluded whether or not the goal clause is entailed by the knowledge base. The aforementioned discussion provides only a basic discussion of the forward chaining algorithm, whereas there are many ways of making it more efficient.

5.5.3 Backward Chaining

Backward chaining is a procedure that works backward from the desired conclusion to the known facts, just as forward chaining works from the known facts to the desired conclusion. Therefore, in this case, the unification is performed in the backwards direction. The algorithm first checks if the goal clause can be obtained by substitution of appropriate variables into known facts of the knowledge base. If this is the case, then the algorithm terminates with a success. Otherwise, all rules are identified whose consequents are generalizations of the goal clause. The antecedent of at least one of these rules must be shown to be true. In other words, one must unify the right-hand side of as many rules as possible with the goal clause, and then try to prove all conjuncts of the left-hand side as new goals (of at least one of these rules). Therefore, a recursive depth-first search is initiated with each of these conjuncts as the new goal and the Boolean results of all of these clauses are then combined with the $\land$ operator. This procedure is repeated in the backwards direction, until termination. The approach is similar to that used in propositional logic, except that it is combined with the process of unification. The overall structure of the algorithm is similar to that in Figure 4.2 if Chapter 4, except that one must take care to match goals with rule consequents (leveraging the process of unification) in order to discover new goals in the antecedents. As in the case of forward chaining, the approach is semi-decidable; it is not guaranteed to terminate if the goal clause is not implied by the knowledge base.

5.6 Summary

Although propositional logic provides the mathematical basis for symbolic artificial intelligence, it is too weak in its own right to support the needs of large-scale inferencing systems. The reason is that propositional logic is unable to reason over entire domains of objects and also instantiate these objects to specific cases. On the other hand, first-order logic is able to reason over facts about objects by leveraging the concepts of predicates, functions, and quantifiers. The most important innovation of first-order logic over propositional logic is the introduction of the notion of object variables, and the use of predicates in order to construct atomic formulae that take on Boolean values. It then becomes possible to make statements about domains of objects by using the notion of quantifiers.

First-order logic builds upon propositional logic by adding concepts to the existing tools of propositional logic, so that it can be viewed as a strict superset of propositional logic. The proof systems of first-order logic are also similar to those of propositional logic. First-order logic has been used in order to build out large expert systems that are often used in real-world settings. However, all existing systems using first-order logic are used in relatively narrow problem domains, and so far not been applied to generalized forms of artificial intelligence. One of the reasons for this is that it requires very large knowledge bases in order to make inferences about real-world settings. In such cases, computational complexity continues to be a problem. Furthermore, first-order logic is restricted to settings where inferences can be derived from known facts in an interpretable way. This is often not the case in general settings of artificial intelligence.

5.7 Further Reading

The basics of first-order logic are discussed in the classic textbook by Hurley and Watson [88]. This book contains broader discussions of different types of methods in logic, including propositional logic. However, this book is written from a mathematician's point of view, rather than from the point of view of a person in the field of artificial intelligence. Detailed discussions of first-order logic in the context of artificial intelligence may be found in [11, 153]. A pseudo-code for the unification algorithm may be found in [153]. More detailed mathematical discussions and methods for automated theorem proving are discussed in [57, 171].

5.8 Exercises

1. Is the statement $\forall x P(x) \Rightarrow \exists y P(y)$ a tautology?

2. Consider a knowledge case containing two sentences. The first sentence is $\exists x \, [A(x) \Rightarrow B(x)]$, and the second is $A(John)$. Does this knowledge base entail $B(John)$?

3. Argue using the connections of first-order logic to propositional logic as to why the following statements are tautologies:

$$(\forall x)(A(x) \wedge B(x)) \equiv [(\forall y)A(y)] \wedge [(\forall z)B(z)]$$
$$(\exists x)(A(x) \wedge B(x)) \Rightarrow [(\exists y)A(y)] \wedge [(\exists z)B(z)]$$

   Argue by counterexample, why the converse of the unidirectional implication in the second assertion does not hold. You may again use the connections between first-order logic and propositional logic.

4. Consider the following knowledge base:

$$\forall x[A(x) \Rightarrow B(x)]$$
$$\forall x[B(x) \wedge C(x) \Rightarrow D(x)]$$
$$A(John)$$

   Does this knowledge base entail $D(John)$?

5. Consider the following knowledge base:

$$\forall x[A(x) \wedge \neg A(John) \Rightarrow B(x)]$$

   Is this statement a tautology? Is the following statement a tautology?

$$\exists x[A(x) \wedge \neg A(John) \Rightarrow B(x)]$$

6. Consider the following statement in first-order logic:

$$\forall x P(x) \vee \forall y Q(y)$$

   Suppose that the domain of discourse for each of the atomic formulas $P(\cdot)$ and $Q(\cdot)$ is $\{a_1, a_2, a_3\}$. Write the statement using propositional logic.

7. Let $E(x, y)$ denote the statement that x eats y. Which of the following is a way of stating that Tom eats fish or beef? (a) $E(Tom, Fish \lor Beef)$, (b) $E(Tom, Fish) \lor E(Tom, Beef)$.

8. Consider the following expression:

$$(A(x) \lor B(x)) \land (A(x) \lor \neg B(John)) \land \neg A(Tom)$$

This expression is in conjunctive normal form, and the variable x is universally quantified. Use unification to show that the expression always evaluates to *False*.

9. Find a way to express the statement $\forall x \; [\neg A(x) \Rightarrow \neg B(x)]$ in first-order definite form.

10. Consider the following first-order definite rules, which are all universally quantified in terms of the variable x:

$$A(x) \Rightarrow B(x)$$
$$B(x) \Rightarrow C(x)$$
$$B(x) \land C(x) \Rightarrow D(x)$$

Suppose that the statement $A(John)$ is true. Use forward chaining to show that $D(John)$ is also true.

11. Suppose that the following statements are true:

$$(\forall x) \; (A(x) \land B(x) \Rightarrow C(x))$$
$$(\exists x) \; (B(x) \land \neg C(x))$$

Show using the laws of first-order logic that the following statement is true:

$$(\exists x) \neg A(x)$$

12. Convert the following set of sentences into symbolic form and construct a proof for the final sentence based on the first two sentences::

> Wherever there are deer, there are also lions. There is at least one deer in Serengeti. Therefore, there must also be at least one lion in Serengeti.

Use $D(x)$ and $L(x)$ respectively for x being a deer or lion, and $P(x, y)$ for x living in place y. Use the constant s for the place Serengeti.

13. Convert the following set of sentences into symbolic form and then construct a proof for the final sentence based on the first two sentences::

> Anyone who studies both artificial intelligence and botany is cool. There is at least one person who studies botany but is not cool. Therefore, there must be at least one person who does not study artificial intelligence.

Use $A(x)$ and $B(x)$ respectively for x studying artificial intelligence and botany. Use $C(x)$ for x being cool.

14. Suppose that the following statements are true:

$$(\exists x)(A(x) \Rightarrow (\forall y)(F(y) \Rightarrow C(y)))$$
$$(\exists x)(B(x) \Rightarrow (\forall y)(E(y) \Rightarrow \neg F(y)))$$
$$\neg(\exists x)(C(x) \wedge \neg E(x))$$

Then show using the laws of first-order logic that the following statement is true:

$$(\exists x)(A(x) \wedge B(x)) \Rightarrow \neg(\exists x)F(x)$$

15. Suppose that the following statements are true:

$$(\exists x)(A(x) \wedge B(x)) \Rightarrow (\forall x)(C(x) \wedge D(x))$$
$$(\exists x)A(x) \Rightarrow (\forall x)(B(x) \wedge C(x))$$
$$A(c)$$

Show that the following statement is true:

$$(\forall x)(B(x) \wedge D(x))$$

16. Let $P(x,y)$ denote the fact that x is the parent of y. Use the equality operator to write a first-order expression asserting that everyone has exactly two parents.

Chapter 6

Machine Learning: The Inductive View

Truth is much too complicated to allow anything but approximations. — John von Neumann

6.1 Introduction

In the deductive view of artificial intelligence, one starts with initial hypotheses (e.g., statements in the knowledge base), and then reasons in order to infer further conclusions. Unfortunately, this view of artificial intelligence is rather limited, since one cannot infer facts other than those that can be related to what is already present in the knowledge base, or can be enunciated as concrete sentences from these known facts. In the inductive view, one words in the reverse order, where disparate facts are used to create hypotheses, which are then used to make predictions.

In general, the humans approach to cognition is based on *evidentiary experiences*, which have an element of *generalization* that cannot be fully justified using logic-based methods. For example, a human (typically) does not learn to recognize a truck by being provided a description of what a truck looks like (as in a knowledge base). Rather, the human typically encounters examples of trucks during their childhood, and eventually learns how to classify an object as a truck, even when the truck looks somewhat different from other trucks the human has seen before. In other words, the human has been able to create a *generalized hypothesis* from the examples, although this hypothesis is often implicit, which cannot be easily described in words (like the statements in a knowledge base). A similar observation applies to the learning of a language, wherein it is often more efficient for a human to learn a language by speaking and listening to it rather than by being taught the grammar of the language in a classroom setting. Even in concretely defined fields such as mathematical/logical proofs, humans learn better by practice rather than by trying to peruse the rules of a formally defined system. In many cases, it may be possible for a human to learn faster by working out a few examples, rather than by being schooled in the formal methods for doing so. The examples need not be comprehensive in their scope in covering

© Springer Nature Switzerland AG 2021
C. C. Aggarwal, *Artificial Intelligence*, https://doi.org/10.1007/978-3-030-72357-6_6

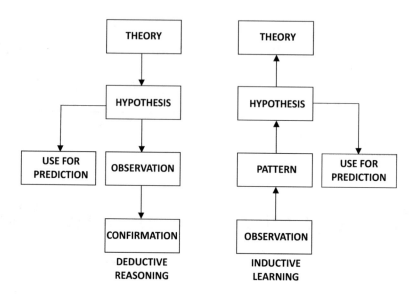

Figure 6.1: Revisiting Figure 1.1: the two schools of thought in artificial intelligence

all possible cases, and human have a natural ability to *generalize* their limited experiences to broader settings. In other words, human naturally create hypotheses from examples (in some implicit form), which they then use to address new problems in similar settings. Such an approach can be naturally considered inductive learning. In other words, *humans are inherently inductive learners*, and they often learn more efficiently from examples than they do by assimilating rigorous logical rules (even in situations well suited to such methods). At the same time, humans do retain the ability to perform deductive reasoning from known facts.

This natural dichotomy in human ability is also reflected in the abilities of methods designed using artificial intelligence. We revisit the two schools of thought in artificial intelligence in Figure 6.1. Note that inductive learning starts with observations (e.g., labeled data) rather than hypotheses (e.g., knowledge base with statements), and it is therefore a form of evidence-driven learning (or *statistical learning*). This is quite different from logic-based methods, which are inherently designed to discover absolute truths via a logical chain of inferences from known facts.

Inductive learning is arguably more powerful, especially when a lot of data is available for training. In cases, where the predictions of an inductive learning methods cannot be explained easily in terms of the underlying attributes, the quality of the predictions are sometimes hard to achieve using deductive reasoning methods. For example, when a human learns to drive a car over time, many decisions are made using "hypotheses" learned from experience, which cannot be enunciated in an interpretable way. Providing knowledge to a deductive reasoning system in well-formed sentences (as in a knowledge base) is inherently limiting, when we do not even know the important characteristics of observations that are useful for prediction. In such cases, humans combine their prior experience with intuition in order to make inferences about the optimal choice in any particular scenario. Most of the strength of humans comes from the ability to translate real-world observations into predictive power and decision making (in a way that can be expressed in a concrete way from a mathematical point of view).

Logic-based methods are based on absolute truths, because the laws of propositional logic and first-order logic deal in *provably* correct inferences from known assertion. Machine learning methods are designed to learn from data and make predictions on similar data that have not been seen before. While the predictions can sometimes be wrong (as humans are), more powerful insights can be achieved by evidence-driven learning. For example, it is often difficult for a human to describe exactly why a specific course of action is taken during driving, other than the fact that earlier evidentiary experiences during driving have guided the human towards this path. Clearly, it would be impossible for a human to enunciate all the possible choices that one should make during driving in order to decide on various courses of actions (which would be the approach one would make using a knowledge base). A similar observation holds true for machine learning, where the predictions are not guaranteed to be true or interpretable, but are often more accurate than would be possible to make by using a system of absolute, interpretable truths. While the lack of interpretability of machine learning algorithms is often criticized, the reality is that this feature of machine learning is one of its strengths.

What types of data do inductive learning methods use? The most common type of data is *multidimensional data*, in which the input data contains vectors $\overline{X}_1 \ldots \overline{X}_n$, and each $\overline{X}_i$ might be (optionally) associated with a dependent variable y_i. The components of the vectors $\overline{X}_i$ are referred to as *attributes* or *dimensions*. While conventional vectors contain only numerical attributes, it is possible for algorithms in machine learning to use categorical attributes. Other forms of data include sequences, images, and text data. The multidimensional form of data will be primarily used in this book in order to elucidate various ideas.

Inductive learning has several different abstractions that are briefly discussed in Chapter 1. These abstractions are discussed below:

- *Supervised learning:* In supervised learning, the agent tries to learn a function of the multidimensional vector $\overline{X}_i$ (containing the set of *independent variables*), and map it to a *dependent variable* y_i:

$$y_i \approx f(\overline{X}_i)$$

  The dependent variable y_i may be numerical, or it may be categorical. When the dependent variable is numerical, the problem is referred to as that of *regression*. When the dependent variable is categorical, the problem is referred to as *classification*. The function $f(\overline{X}_i)$ may or may not be expressible in closed form, but its exact form is often controlled either by a set of parameters or algorithmic choices that are fixed during *training time*. One can view of the process of *training* as that of building hypotheses from observations, which are then used for prediction (see Figure 6.1). This chapter will primarily focus on supervised learning.

- *Unsupervised learning:* Unsupervised learning applications try to learn the interrelationships among the data attributes without being given a dependent variable in an explicit way. Examples of unsupervised learning methods include clustering and dimensionality reduction. This will be the topic of discussion in Chapter 9.

- *Reinforcement learning:* Reinforcement learning methods represent general forms of learning in which an optimal *sequence* of actions need to be learned, just as an agent needs to learn a sequence of actions in order to maximize its utility in search-based methods. Therefore, reinforcement learning methods provide a data-driven alternative for many of the deductive methods discussed in the previous chapters. For example, a game like chess can be addressed using either a search-based methodology (and a domain-specific utility function), or a computer can be trained to play chess using reinforcement learning methods. The former can be viewed as a deductive reasoning method, whereas the latter can be considered an inductive learning method.

This chapter is organized as follows. The next section will introduce the problem of linear regression. Section 6.3 discusses the problem of least-squares classification. Many of these models are binary classifiers, which work with only two classes. The generalization of these binary classification to the multiclass case is discussed in section 6.6. The Bayes classifier is introduced in section 6.7. Nearest neighbor classifiers are introduced in section 6.8. Decision tree classifiers are discussed in section 6.9. Rule-based classifiers are discussed in section 6.10. A summary is given in section 6.12.

6.2   Linear Regression

In the problem of linear regression, we have n pairs of observations $(\overline{X}_i, y_i)$ for $i \in \{1 \ldots n\}$. As discussed in the previous section, the (row) vector $\overline{X}_i$ contains d numerical values corresponding to the properties of a data point. The target y_i is predicted using the following relationship:

$$\hat{y}_i \approx f(\overline{X}_i) = \overline{W} \cdot \overline{X}_i^T$$

Note the circumflex on top of $\hat{y}_i$ to indicate that it is a predicted value. Here, $\overline{W} = [w_1 \ldots w_d]^T$ is a d-dimensional column vector, which needs to be learned in a data-driven manner. The values in each vector $\overline{X}_i$ are referred to as independent variables or *regressors*, whereas each y_i is referred to as a dependent variable or *regressand*. Each $\overline{X}_i$ is a row vector, because it is common for data points to be represented as rows of data matrices in machine learning. Therefore, the row vector $\overline{X}_i$ needs to be transposed before performing a dot product with the column vector $\overline{W}$. The vector $\overline{W}$ defines a set of *parameters* that need to be learned in a data-driven manner. The goal is to find the vector $\overline{W}$ so that each $\overline{W} \cdot \overline{X}_i^T$ is as close to y_i as possible over the training data.

This learned parameter vector is useful for making predictions on unseen *test instances*. Once the vector $\overline{W}$ has been learned from the training data by optimizing the aforementioned objective function, the numerical value of the target variable of an unseen test instance $\overline{Z}$ (which is a d-dimensional row vector) can be predicted as $\overline{W} \cdot \overline{Z}^T$. Note that test instances show *similar* behavior as training instances, but the model is not directly trained on them. Therefore, the accuracy of these methods on the test data is almost always lower than that on the training data.

Linear regression is one of the oldest problems in machine learning, and it precedes the broader field of machine learning by several years. It is often used in many applications such as *forecasting* or *recommender systems*. For example, the vectors $\overline{X}_i$ might contain the attributes of different items, and the variables y_i might contain the numerical rating given by a particular user to item i. The training data can be used in order to predict the rating of this user for an item $\overline{Z}$, based on its attributes (contained in vector $\overline{Z}$). This new item was not seen either by the user or during training, and therefore the predicted rating can be used to decide whether or not to show an advertisement of this item to the user.

How can one learn the parameter vector $\overline{W}$ in order to ensure that $\overline{W} \cdot \overline{X}_i$ predicts y_i as closely as possible? To achieve this goal, we compute the loss $(y_i - \overline{W} \cdot \overline{X}_i^T)^2$ for each training data point, and then add up these losses over all points in order to create the objective function:

$$J = \frac{1}{2}\sum_{i=1}^{n}(y_i - \overline{W} \cdot \overline{X}_i^T)^2 \tag{6.1}$$

In most cases, a *regularization term* $\lambda\|\overline{W}\|^2/2$ is added to the objective function in order to reduce overfitting:

$$J = \frac{1}{2}\sum_{i=1}^{n}(y_i - \overline{W} \cdot \overline{X}_i^T)^2 + \frac{\lambda}{2}\|\overline{W}\|^2 \tag{6.2}$$

Here, $\lambda > 0$ is the regularization parameter. The purpose of regularization is to favor solution vectors $\overline{W}$ that are small in absolute magnitude. This type of approach avoids overfitting, wherein the weight vector predicts well on the training data, but performs poorly on the test data. The importance of regularization is discussed in section 6.2.4.

Such continuous optimization formulations are solved by combining differential calculus with computational methods such as *gradient descent*. One can compute the derivative of the loss function with respect to the weight vector $\overline{W}$ in order to perform optimization. The derivative of J with respect to the weight $\overline{W}$ is as follows:

$$\frac{\partial J}{\partial \overline{W}} = \left[\frac{\partial J}{\partial w_1} \cdots \frac{\partial J}{\partial w_d}\right]^T = -\sum_{i=1}^{n}(y_i - \overline{W} \cdot \overline{X}_i^T)\overline{X}_i^T + \lambda\overline{W} \tag{6.3}$$

The aforementioned matrix calculus notation of the derivative of a scalar J with respect to column vector $\overline{W}$ leads to a column vector in the *denominator layout* [8] of matrix calculus. The specific derivative is obtained using an identity from matrix calculus on the derivative of a quadratic function with respect to a vector [8].

In gradient descent, one updates the weight vector in the negative direction of the derivative in order to perform optimization. Specifically, the updates are as follows:

$$\overline{W} \Leftarrow \overline{W} - \alpha\frac{\partial J}{\partial \overline{W}} \tag{6.4}$$

$$= \overline{W}(1 - \alpha\lambda) + \alpha\sum_{i=1}^{n}\underbrace{(y_i - \overline{W} \cdot \overline{X}_i^T)}_{\text{Error on } \overline{X}_i}\overline{X}_i^T \tag{6.5}$$

The value of $\alpha > 0$ is referred to as the step-size or the *learning rate*. In most cases, the objective function reduces over the different steps (albeit not monotonically), and *converges* to a solution that is close to optimal. Larger step sizes lead to faster termination of the algorithm, but if the step size is chosen to be too large, it can lead to instability of the algorithm. This type of instability is often manifested by *divergence behavior* in which the weight vectors get successively larger, leading to numerical overflows. One can initialize the weight vector randomly, and then perform the updates to convergence.

6.2.1 Stochastic Gradient Descent

The above approach uses the *entire* data set in order to set up the objective function J. In practice, it might be possible to set up the objective function $J(S)$ with a subset of training instances:

$$J(S) = \frac{1}{2}\sum_{i \in S}(y_i - \overline{W} \cdot \overline{X}_i^T)^2 + \frac{\lambda}{2}\|\overline{W}\|^2 \tag{6.6}$$

It is noteworthy that the regularization parameter λ needs to be proportionally adjusted to a smaller value, as we are using a smaller number of training instances. The corresponding updates with the use of this modified objective function are as follows:

$$\overline{W} \Leftarrow \overline{W} - \alpha \frac{\partial J(S)}{\partial \overline{W}}$$

$$= \overline{W}(1 - \alpha\lambda) + \alpha \sum_{i \in S} \underbrace{(y_i - \overline{W} \cdot \overline{X}_i^T)}_{\text{Error on } \overline{X}_i} \overline{X}_i^T$$

All training instances are permuted in some random order, and then batches of k training instances are extracted from this permutation in order to perform the update. An entire cycle of n/k updates (so that each training instance is seen exactly once) is referred to as an *epoch*.

   This type of update is referred to as *mini-batch stochastic gradient descent*. The set S is referred to as the *mini-batch* of training instances used for the update. The basic idea behind this type of mini-batch stochastic gradient descent is that a subset of training instances is sufficient to estimate the direction of the gradient very well in most cases. A very accurate estimation can be performed using a set S that is much less than the number of data instances n; as a result, each update requires a tiny fraction of the computational effort without much loss in the accuracy of the individual update. For example, a gradient that is computed using a sample of 1000 training points will often be almost the same as that obtained using a full data set of a million points. In such cases, convergence is much faster because the number of steps does not increase significantly because of the approximation.

   A limiting case of this approach is one in which a mini-batch size of 1 is used. In other words, the set S contains a single element. This setting is referred to as *stochastic gradient descent*. In such a case, the update for training instances $(\overline{X}_i, y_i)$ is as follows:

$$\overline{W} \Leftarrow \overline{W}(1 - \alpha\lambda) + \alpha \underbrace{(y_i - \overline{W} \cdot \overline{X}_i^T)}_{\text{Error on } \overline{X}_i} \overline{X}_i^T$$

One typically cycles through all the training instances in random order to perform the updates. A single cycle of updates over all training instances is referred to as an epoch.

6.2.2   Matrix-Based Solution

While gradient descent is the natural approach to solve most machine learning problems, it is also possible to find closed-form solutions in the special case of least-squares regression. Consider the gradient of J with respect to weight vector $\overline{W}$, which we replicate below from Equation 6.3:

$$\frac{\partial J}{\partial \overline{W}} = -\sum_{i=1}^{n}(y_i - \overline{W} \cdot \overline{X}_i^T)\overline{X}_i^T + \lambda\overline{W} \tag{6.7}$$

One can rewrite this gradient in matrix form by defining an $n \times d$ data matrix D, which contains the data points $\overline{X}_1 \ldots \overline{X}_n$ its rows, and an n-dimensional column vector $\overline{y} = [y_1, y_2, \ldots, y_n]^T$ containing the regressands:

$$\frac{\partial J}{\partial \overline{W}} = D^T(D\overline{W} - \overline{y}) + \lambda\overline{W}$$

The reader should take a moment to verify that the aforementioned matrix form simplifies to Equation 6.7. Note that this gradient can also be derived by using matrix calculus techniques on the matrix form of the objective function:

$$J = \frac{1}{2}\|D\overline{W} - \overline{y}\|^2 + \frac{\lambda}{2}\|\overline{W}\|^2 \tag{6.8}$$

The matrix calculus method is described in [8]. One can also implement the updates using the matrix form of the equation:

$$\overline{W} \Leftarrow \overline{W}(1 - \alpha\lambda) + \alpha D^T \underbrace{(\overline{y} - D\overline{W})}_{\text{Error vector}}$$

One advantage of the matrix form is that it allows us to specify a closed form of the solution to linear regression. This is achieved by setting the gradient of the objective function to 0:

$$D^T(D\overline{W} - \overline{y}) + \lambda\overline{W} = 0$$

One can simplify this condition as follows:

$$(D^T D + \lambda I)\overline{W} = D^T\overline{y}$$

Here, the matrix I is the $d \times d$ identity matrix. As a result, one obtains the following weight vector:

$$\overline{W} = (D^T D + \lambda I)^{-1} D^T\overline{y}$$

This is the closed-form solution to the problem of linear regression. All machine learning problems do not have such closed-form solutions. Linear regression is a special case because of the simplicity of its objective function.

6.2.3 Use of Bias

It is noteworthy that the prediction $y_i = \overline{W} \cdot \overline{X}_i^T$ always passes through the origin. In other words, the prediction at $\overline{X}_i = \overline{0}$ must always be $y_i = 0$. However, for some problem domains, this is not the case, and one would end up with large errors. Therefore, one often adds a constant parameter b, referred to as the *bias*, which needs to be learned in a data-driven manner:

$$y_i = \overline{W} \cdot \overline{X}_i^T + b$$

The parameter b is the offset that defines the prediction y_i, when all values of $\overline{X}_i$ are set to 0. Although one can create an objective function $\sum(y_i - \overline{W} \cdot \overline{X}_i^T - b)^2$ with this new prediction function in order to evaluate the gradient algebraically, it is more common to indirectly incorporate the bias with the use of a feature engineering trick. Instead of explicitly using a bias b, one adds an additional column to D containing only 1s. Therefore, each feature vector now becomes $\overline{V}_i = (\overline{X}_i, 1)$. Now one uses the same prediction function as before with the modified weight vector $\overline{W}'$, which is obtained by adding the $(d+1)$th element w_{d+1} to the end of $\overline{W}$ as a $(d+1)$-dimensional parameter vector. and simply treats the problem as a $(d+1)$-dimensional linear regression $\overline{y}_i = \overline{W}' \cdot \overline{V}_i^T$. The same updates and closed-form solutions apply as in the previous case, but with an augmented data matrix containing $(d+1)$ dimensions. The parameter w_{d+1} yields the bias. This type of feature engineering trick can be used in most machine learning problems. Therefore, we will not explicitly introduce a bias in most of our subsequent discussion, although it is extremely important to use the bias in practice (in order to account for constant effects).

6.2.4 Why Is Regularization Important?

As discussed early, a norm-penalty $\lambda \|\overline{W}\|^2$ is added to the objective function in linear regression. This penalty might seem somewhat odd, because it does not seem to be related to the prediction error $\|y_i - \overline{W} \cdot \overline{X}_i\|^2$ or even to the characteristics of the training data in any way. Why should such a seemingly unrelated part of the objective function improve prediction error at all?

Here, it is important to understand that machine learning models are built on top of training data, but the prediction is done on test data. It is almost always the case that the prediction accuracy on the training data is better than that on the test data. This is particularly evident when the training data is small, and the gap between training and test accuracy increases. In such cases, regularization tends to favor parameter vectors with small magnitudes of the parameters. An important point is that *parameter vectors with smaller magnitudes of the components tend to have lower gaps between training and test error performance.* Therefore, even though regularization tends to worsen the error performance on the training data, it improves the error performance on the test data. In order to understand this point, we will use an example.

Consider a situation, where we have two training points and three attributes. The training points, denoted by the attributes $[x_1, x_2, x_3]$, are $[1, -1, 2]$ and $[5, 2, 3]$, respectively. It is known from domain knowledge that the target variable value is always twice the value of x_1, whereas the other two variables x_2 and x_3 are completely irrelevant for predicting the target variable. Therefore, the target variable values in this case for the two vectors $[1, -1, 2]$ and $[5, 2, 3]$ are 2 and 10, respectively. Now consider the case where we try to perform linear regression without access to the domain knowledge about the relevance of x_1. In such a case, we might learn three parameters w_1, w_2, and w_3, so that the following holds:

$$y \approx w_1 x_1 + w_2 x_2 + w_3 x_3$$

One can see that setting $w_1 = 2$, $w_2 = 0$, and $w_3 = 0$ yields perfect prediction, which is consistent with known domain knowledge. However, the paucity of the number of training examples leads to an unfortunate consequence — there are many other choices of w_1, w_2, and w_3 that would also yield perfect prediction. To understand this point, note that we have two equations with three unknowns:

$$2 \approx w_1 - w_2 + 2w_3$$
$$10 \approx 5w_1 + 2w_2 + 3w_3$$

As a result, there are an infinite number of possible solutions that provide perfect prediction. For example, we could set the only relevant parameter w_1 to 0, and set each of w_2 and w_3 to 2 to yield perfect prediction. This "perfect" prediction is caused by the random nuances of the training data, and the "luckily perfect" prediction is possible only because of the paucity of training data. Had one added more training instances, the prediction would (most likely) no longer continue to hold over the new training instances, whereas the original solution of setting $w_1 = 2$ would work well. These problems arise even in cases where the number of training instances is modestly greater than the number of attributes. In general, one needs a very large training data set to generalize predictions from training to test data.

A key point is that an increased number of attributes relative to training points provides additional degrees of freedom to the optimization problem, as a result of which irrelevant solutions become more likely. Therefore, a natural solution is to add a penalty for using additional features. Specifically we can add a penalty for each parameter w_i, which is non-zero. One can express this penalty using the L_0-norm $\|\overline{W}\|_0$ of the vector $\overline{W}$. Unfortunately, doing so creates a discrete optimization problem which cannot be optimized with the use of differential calculus. Therefore, one uses the squared L_2-norm instead, which is denoted by $\|\overline{W}\|^2 = w_1^2 + w_2^2 + w_3^2$. The parameter λ controls the weight of regularization, which is typically set by holding out of a part of the training data. The held out training data is not used in gradient descent, but is used for estimating the error at a particular value of λ. This held out part of the data is referred to as the *validation data*, and it is used to decide an appropriate value of λ that minimizes the error. It is noteworthy that adding the regularization in the previous example will cause the algorithm to prefer the solution $w_1 = 2$, rather than $w_2 = w_3 = 2$, because the former is a more *concise* solution. In general, one does not want the optimization problem to use unnecessary degrees of freedom to create a complicated solution that works well on the training data (because of random nuances), but *generalizes* poorly to test data.

It is noteworthy that regularization is closely related to the deductive school of thought in artificial intelligence, where the hypotheses are constructed by the analyst for prediction (based on known theories or domain knowledge), rather than performing predictions in a completely data-driven manner. Here, an important hypothesis is that *given two solutions, a more concise solution is likely to be better.* Adding this hypothesis to the optimization formulation is similar to adding domain knowledge to the problem in order to improve the underlying solution. There are many other indirect ways in which regularization is often used. In many machine learning problems, we might have additional domain knowledge about the relationships between different parameters. This information is used in order to modify the objective function appropriately. For example, one might be aware that an attribute such as number of previous defaults is negatively related to a person's credit score. Therefore, if one were trying to model the credit score in a regression problem with the credit score as the ith attribute and regression parameters $[w_1 \ldots w_d]$, one might constrain w_i to be non-positive in the optimization model. This type of constraint will often worsen the loss value on the training data (because the optimization space is more constrained), but it will improve performance on the test data. Therefore, some amount of domain knowledge is used in many inductive settings in order to reduce the data requirements.

6.3   Least-Squares Classification

The previous section discusses regression in which the target variable y_i is numerical. However, in many real-world settings, the target variable is discrete in nature, wherein the variable could take on one of a number of possible values that are *unordered*. For example, if one has the attributes of a set of animals, and we are trying to classify them as "Bird," "Reptile," or "Mammal," there is no ordering among these values. Given a set of feature vectors in $\overline{X}_i$, one needs to place the data point in one of these categories.

The aforementioned setting with more than two unordered labels represents *multiclass classification*, which is the most general case of the classification setting. An important special case of classification is one in which the class variable is binary. In such a case, we can impose an ordering among the classes by assuming that the two labels are drawn from $\{-1, +1\}$. Imposing an ordering among the classes is helpful in generalizing the methods used in least-squares regression to classification. The instances with label $+1$ are referred to as *positive class instances*, whereas the instances with label -1 are referred to as *negative class instances*. In this section, we will primarily discuss binary classification because of its importance and ubiquity in machine learning. Furthermore, multilabel classification problems can be reduced to repeated applications of binary classification (see section 6.6).

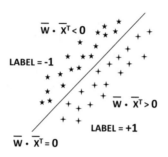

Figure 6.2: An example of linear separation between two classes

Before discussing the details of the learning model, we introduce the notations, which are similar to the case of linear regression. The d-dimensional row vectors containing the features values are stored in $\overline{X}_1, \ldots \overline{X}_n$. The target variables drawn from $\{-1, +1\}$ are stored in the n-dimensional vector $\overline{y} = [y_1, \ldots, y_n]^T$.

Least-squares classification directly adapts linear regression to classification by pretending that the binary targets are real valued. This is possible only in the binary case, where one can create an arbitrary ordering between the two classes and setting the labels from $\{-1, +1\}$. Therefore, we model each target as $y_i \approx \overline{W} \cdot \overline{X}_i^T$, where $\overline{W} = [w_1, \ldots, w_d]^T$ is a column vector containing the weights. The same squared loss function as linear regression is used:

$$J = \frac{1}{2} \sum_{i=1}^{n} (y_i - \overline{W} \cdot \overline{X}_i^T)^2 + \frac{\lambda}{2} \|\overline{W}\|^2 = \frac{1}{2} \|D\overline{W} - \overline{y}\|^2 + \frac{\lambda}{2} \|\overline{W}\|^2 \qquad (6.9)$$

Since the objective function is the same as linear regression, it results in the same closed-form solution for $\overline{W}$:

$$\overline{W} = (D^T D + \lambda I)^{-1} \overline{y} \qquad (6.10)$$

Even though $\overline{W} \cdot \overline{X}_i^T$ yields a real-valued prediction for instance $\overline{X}_i$ (like regression), it makes more sense to view the hyperplane $\overline{W} \cdot \overline{X}^T = 0$ as a *separator* or *modeled decision boundary*, where any instance $\overline{X}_i$ with label $+1$ will satisfy $\overline{W} \cdot \overline{X}_i^T > 0$, and any instance with label -1 will satisfy $\overline{W} \cdot \overline{X}_i^T < 0$. Because of the way in which the model has been trained, most *training* points will align themselves on the two sides of the separator, so that the sign of the training label y_i matches the sign of $\overline{W} \cdot \overline{X}_i^T$. An example of a two-class data set in two dimensions is illustrated in Figure 6.2 in which the two classes are denoted by '+' and '*', respectively. In this case, it is evident that the value of $\overline{W} \cdot \overline{X}_i^T = 0$ is true only for points on the separator. The training points on the two sides of the separator satisfy either $\overline{W} \cdot \overline{X}_i^T < 0$ or $\overline{W} \cdot \overline{X}_i^T > 0$. The separator $\overline{W} \cdot \overline{X}^T = 0$ between the two classes is the modeled decision boundary. Note that some data distributions might not have the kind of neat separability as shown in Figure 6.2. In such cases, one either needs to live with errors or use feature transformation techniques to create linear separability.

The aligning of positive and negative class training instances on the two sides of the separator is then generalized to an unseen test instance $\overline{Z}$ (once training is completed and $\overline{W}$ has been computed). Note that the test instance $\overline{Z}$ is a row vector, whereas $\overline{W}$ is a column vector. The dot product between the column vectors $\overline{W}$ and $\overline{Z}^T$ yields a real-valued prediction, which is converted to a binary prediction with the use of the sign function:

$$\hat{y} = \text{sign}\{\overline{W} \cdot \overline{Z}^T\} \tag{6.11}$$

Therefore, the model learns a linear hyperplane $\overline{W} \cdot \overline{X}^T = 0$ separating the positive and negative classes. All test instances for which $\overline{W} \cdot \overline{Z}^T > 0$ are predicted to belong to the positive class, and all instances for which $\overline{W} \cdot \overline{Z}^T < 0$ are predicted to belong to the negative class. The linear hyperplane separating the two classes (see Figure 6.2) is also referred to as the *decision boundary*. All classifiers (directly or indirectly) create decision boundaries to partition the different classes.

The closed-form method is not the only way to solve this problem. As in the case of real-valued targets, one can also use mini-batch stochastic gradient-descent for regression on binary targets. Let S be a mini-batch of pairs $(\overline{X}_i, y_i)$ of feature variables and targets. Each $\overline{X}_i$ is a row of the data matrix D and y_i is a target value drawn from $\{-1, +1\}$. Then, the mini-batch update for least-squares classification is identical to that of least-squares regression:

$$\overline{W} \Leftarrow \overline{W}(1 - \alpha\lambda) - \alpha \sum_{(\overline{X}_i, y_i) \in S} \overline{X}_i^T (\overline{W} \cdot \overline{X}_i^T - y_i) \tag{6.12}$$

Here, $\alpha > 0$ is the learning rate, and $\lambda > 0$ is the regularization parameter. Note that this update is *identical* to that in least-squares regression. However, since each target y_i is drawn from $\{-1, +1\}$, an alternative approach also exists for writing the targets by using the fact that $y_i^2 = 1$. This alternative form of the update is as follows:

$$\overline{W} \Leftarrow \overline{W}(1 - \alpha\lambda) - \alpha \sum_{(\overline{X}_i, y_i) \in S} \underbrace{y_i^2}_{1} \overline{X}_i^T (\overline{W} \cdot \overline{X}_i^T - y_i)$$

$$= \overline{W}(1 - \alpha\lambda) - \alpha \sum_{(\overline{X}_i, y_i) \in S} y_i \overline{X}_i^T (y_i[\overline{W} \cdot \overline{X}_i^T] - y_i^2)$$

Setting $y_i^2 = 1$, we obtain the following:

$$\overline{W} \Leftarrow \overline{W}(1 - \alpha\lambda) + \alpha \sum_{(\overline{X}_i, y_i) \in S} y_i \overline{X}_i^T (1 - y_i[\overline{W} \cdot \overline{X}_i^T]) \tag{6.13}$$

We emphasize that this form of the update is valid only for a particular type of coding of the target variable to be drawn from $\{-1, +1\}$. This form of the update is more closely related to updates of closely related models like the support vector machine and logistic regression (discussed later in this chapter). The loss function can also be converted to a more convenient representation for binary targets drawn from $\{-1, +1\}$.

Alternative Representation of Loss Function

The alternative form of the aforementioned updates can also be derived from an alternative form of the loss function. The loss function of (regularized) least-squares classification can be written as follows:

$$J = \frac{1}{2} \sum_{i=1}^{n} (y_i - \overline{W} \cdot \overline{X}_i^T)^2 + \frac{\lambda}{2} \|\overline{W}\|^2 \tag{6.14}$$

Using the fact that $y_i^2 = 1$ for binary targets, we can modify the objective function as follows:

$$J = \frac{1}{2} \sum_{i=1}^{n} y_i^2 (y_i - \overline{W} \cdot \overline{X}_i^T)^2 + \frac{\lambda}{2} \|\overline{W}\|^2$$

$$= \frac{1}{2} \sum_{i=1}^{n} (y_i^2 - y_i [\overline{W} \cdot \overline{X}_i^T])^2 + \frac{\lambda}{2} \|\overline{W}\|^2$$

Setting $y_i^2 = 1$, we obtain the following loss function:

$$J = \frac{1}{2} \sum_{i=1}^{n} (1 - y_i [\overline{W} \cdot \overline{X}_i^T])^2 + \frac{\lambda}{2} \|\overline{W}\|^2 \tag{6.15}$$

Differentiating this loss function directly leads to Equation 6.13. However, it is important to note that the loss function/updates of least-squares classification are identical to the loss function/updates of least-squares regression, even though one might use the binary nature of the targets in the former case in order to make them *look* superficially different.

A good way to perform heuristic initialization is to determine the mean $\overline{\mu}_0$ and $\overline{\mu}_1$ of the points belonging to the negative and positive classes, respectively. The difference between the two means is $\overline{w}_0 = \overline{\mu}_1^T - \overline{\mu}_0^T$ is a d-dimensional column vector, which satisfies $\overline{w}_0 \cdot \overline{\mu}_1^T \geq \overline{w}_0 \cdot \overline{\mu}_0^T$. The choice $\overline{W} = \overline{w}_0$ is a good starting point, because positive-class instances will have larger dot products with $\overline{w}_0$ than will negative-class instances (on the average). In many real applications, the classes are roughly separable with a linear hyperplane, and the normal hyperplane to the line joining the class centroids provides a good initial separator.

Least-squares classification has also been studied independently in the field of neural network learning. The updates of least-squares classification are also referred to as Widrow-Hoff updates [202]. The Widrow-Hoff updates were proposed independently of the classical literature on least-squares regression; yet, the updates turn out to be identical.

6.3.1 Problems with Least-Squares Classification

There are several challenges associated with least-squares classification, which are inherent to its loss function. We replicate the objective function of least-squares classification below:

$$J = \frac{1}{2} \sum_{i=1}^{n} (1 - y_i [\overline{W} \cdot \overline{X}_i^T])^2 + \frac{\lambda}{2} \|\overline{W}\|^2$$

An important issue is that a point is penalized, when $\overline{W} \cdot \overline{X}_i^T$ is *different* from -1 or $+1$, and the direction of error does not seem to matter. Consider a positive class instance for which $\overline{W} \cdot \overline{X}_i^T = 100$ is highly positive. This is obviously an desirable situation at least from a predictive point of view because the training instance is on the correct side of the separator in a "confident" way. However, the loss function in the training model treats this prediction as a large loss contribution of $(1 - y_i [\overline{W} \cdot \overline{X}_i^T])^2 = (1 - (1)(100))^2 = 99^2 = 9801$. Therefore, a large gradient descent update will be performed for a training instance that is located at a large distance from the hyperplane $\overline{W} \cdot \overline{X}^T = 0$ on the correct side. In fact, this update will often be larger than the update to an instance $\overline{X}_j$ for which $\overline{W} \cdot \overline{X}_j^T$ is on the wrong side of the separator. Such a situation is undesirable because it tends to confuse least-squares classification; the updates from these points on the correct side of the hyperplane

$\overline{W} \cdot \overline{X}^T = 0$ tend to push the hyperplane in the same direction as some of the incorrectly classified points. In order to address this issue, many machine learning algorithms treat such correctly classified points (which are far from the separator) in a special way. This has led to modern machine learning models such as the support vector machine.

6.4 The Support Vector Machine

The support vector machine (SVM) addresses some of the weaknesses of the least-squares classification model. We start by introducing the notations and definitions. We assume that we have n training pairs of the form $(\overline{X}_i, y_i)$ for $i \in \{1 \dots n\}$. Each $\overline{X}_i$ is a d-dimensional row vector, and each $y_i \in \{-1, +1\}$ is the label. We would like to find a d-dimensional column vector $\overline{W}$ so that the sign of $\overline{W} \cdot \overline{X}_i^T$ yields the class label.

The main difference between the least-squares classification model and the support vector machine is the way in which *well separated points* are treated. We start by formally defining a well-separated point. A point is correctly classified by the least-squares classification model when $y_i[\overline{W} \cdot \overline{X}_i^T] > 0$. In other words, y_i has the same sign as $\overline{W} \cdot \overline{X}_i^T$. Furthermore, the point is well-separated when $y_i[\overline{W} \cdot \overline{X}_i^T] > 1$. Therefore, a well-separated point is not only correctly classified, but the correct classification occurs in a confident way. The loss function of least-squares classification can be modified by setting the loss to 0, when this condition is satisfied. This can be achieved by modifying the least-squares loss to SVM loss:

$$J = \frac{1}{2} \sum_{i=1}^{n} \max \left\{0, \left(1 - y_i[\overline{W} \cdot \overline{X}_i^T]\right)\right\}^2 + \frac{\lambda}{2} \|\overline{W}\|^2 \quad [L_2\text{-loss SVM}]$$

Note that the *only* difference from the least-squares classification model is the use of the maximization term in order to set the loss of well-separated points to 0. Although this objective function is directly related to the linear regression and classification loss, it is more common to use L_1-loss in the support vector machine, which is how it was proposed in Cortes and Vapnik's seminal paper [41]. This loss is referred to as the *hinge-loss*, which is defined as follows:

$$J = \sum_{i=1}^{n} \max\{0, (1 - y_i[\overline{W} \cdot \overline{X}_i^T])\} + \frac{\lambda}{2} \|\overline{W}\|^2 \quad [\text{Hinge-loss SVM}] \tag{6.16}$$

Throughout this section, we will work with the hinge loss because of its popularity in the machine learning community.

It is noteworthy that the loss function of the L_2-SVM was proposed [77] by Hinton much earlier than the Cortes and Vapnik [41] work on the hinge-loss SVM. Interestingly, Hinton proposed the L_2-loss as a way to repair the Widrow-Hoff loss (i.e., least-squares classification loss) in order to treat well-separated points in a more effective way. Furthermore, since the SVM had not yet been proposed, Hinton did not propose the work as an example of the SVM. Rather, the loss function was proposed in the context of neural networks. Eventually, the relationship between Hinton's work and the SVM was discovered in the previous decade.

The objective function is optimized with gradient descent of either the objective function or a *dual formulation*. The use of dual formulations is an optimization methodology used in many machine learning settings. In this book, we focus only on the original formulation (even though dual formulations of SVMs are extremely popular as well). Once the vector $\overline{W}$

has been learned with the use of gradient descent, the classification process for an unseen test instance is similar to that of least-squares classification. For an unseen test instance $\overline{Z}$, the sign of $\overline{W} \cdot \overline{Z}^T$ yields the class label.

6.4.1 Mini-Batch Stochastic Gradient Descent

In this section, we will compute the gradient of the hinge-loss SVM. The objective functions for the L_1-loss (hinge loss) and L_2-loss SVM are both in the form $J = \sum_i J_i + \lambda \|\overline{W}\|^2 / 2$, where $J_i = \max\{0, (1 - y_i[\overline{W} \cdot \overline{X}_i^T])\}$ is the point-specific loss. The gradient of J_i with respect to $\overline{W}$ is either $-y_i \overline{X}_i^T$ or the zero vector, depending on whether or not $y_i[\overline{W} \cdot \overline{X}_i^T] < 1$. The gradient of the regularization term is $\lambda \overline{W}$.

Consider the case of mini-batch stochastic gradient descent, in which a set S of training instances contain feature-label pairs of the form $(\overline{X}_i, y_i)$. For the hinge-loss SVM, we first determine the set $S^+ \subseteq S$ of training instances in which $y_i[\overline{W} \cdot \overline{X}_i^T] < 1$.

$$S^+ = \{(\overline{X}_i, y_i) : (\overline{X}_i, y_i) \in S, y_i[\overline{W} \cdot \overline{X}_i^T] < 1\} \tag{6.17}$$

The subset of instances in S^+ correspond to those which are either on the wrong side of the decision boundary, or they are uncomfortably close to the decision boundary (on the correct side). Both these types of instances trigger updates in the SVM. By using the gradient of the loss function, the updates in the L_1-loss SVM can be shown to be the following:

$$\overline{W} \Leftarrow \overline{W}(1 - \alpha\lambda) + \sum_{(\overline{X}_i, y_i) \in S^+} \alpha y_i \overline{X}_i^T \tag{6.18}$$

This algorithm is referred to as the primal support vector machine algorithm. It is also possible to come up with a similar update for the L_2-loss SVM (see Exercise 2).

6.5 Logistic Regression

Unlike the hinge loss, logistic regression uses a smooth loss function. However, the shape of the two loss functions is very similar. Using the same notations as the previous section, the loss function of logistic regression is formulated as follows:

$$J = \sum_{i=1}^{n} \underbrace{\log(1 + \exp(-y_i[\overline{W} \cdot \overline{X}_i^T]))}_{J_i} + \frac{\lambda}{2}\|\overline{W}\|^2 \quad \text{[Logistic Regression]} \tag{6.19}$$

The logarithms in this section are all natural logarithms. When $\overline{W} \cdot \overline{X}_i^T$ is large in absolute magnitude and has the same sign as y_i, the point-specific loss J_i is close to $\log(1 + \exp(-\infty)) = 0$. On the other hand, the loss is larger than $\log(2)$ when the signs of y_i and $\overline{W} \cdot \overline{X}_i^T$ disagree. Furthermore, like hinge loss, the loss function increases almost linearly with the magnitude of $\overline{W} \cdot \overline{X}_i^T$ for large magnitudes of $\overline{W} \cdot \overline{X}_i^T$, when the signs of y_i and $\overline{W} \cdot \overline{X}_i^T$ disagree.

6.5.1 Computing Gradients

As in the case of SVMs, the objective function for logistic regression is in the form $J = \sum_i J_i + \lambda\|\overline{W}\|^2/2$, where J_i is defined as follows:

$$J_i = \log(1 + \exp(-y_i[\overline{W} \cdot \overline{X}_i^T]))$$

One can use the chain rule of differential calculus to compute the gradient of J_i with respect to $\overline{W}$:

$$\frac{\partial J_i}{\partial \overline{W}} = \frac{-y_i \overline{X}_i^T}{(1 + \exp(y_i[\overline{W} \cdot \overline{X}_i^T]))}$$

Given a mini-batch of S of feature-target pairs $(\overline{X}, y)$, one can define an objective function $J(S)$, which uses the loss of only the training instances in S. The regularization term remains unchanged, as one can simply re-scale the regularization parameter by $|S|/n$. It is relatively easy to compute the gradient $\nabla J(S)$ based on mini-batch S as follows:

$$\nabla J(S) = \lambda \overline{W} - \sum_{(\overline{X}_i, y_i) \in S} \frac{y_i \overline{X}_i^T}{(1 + \exp(y_i[\overline{W} \cdot \overline{X}_i^T]))} \qquad (6.20)$$

Therefore, the mini-batch stochastic gradient-descent method can be implemented as follows:

$$\overline{W} \Leftarrow \overline{W}(1 - \alpha\lambda) + \sum_{(\overline{X}_i, y_i) \in S} \frac{\alpha y_i \overline{X}_i^T}{(1 + \exp(y_i[\overline{W} \cdot \overline{X}_i^T]))} \qquad (6.21)$$

Logistic regression makes similar updates as the hinge-loss SVM. The main difference is in terms of the treatment of well-separated points, where SVM does not make any updates and logistic regression makes (small) updates.

6.5.2 Comparing the SVM and Logistic Regression

The performance of the SVM is surprisingly similar to that of logistic regression, especially when a point $\overline{X}_i$ is incorrectly classified with a large magnitude of $\overline{W} \cdot \overline{X}_i$. Therefore, we will consider a test instance for which $x_i = y_i[\overline{W} \cdot \overline{X}_i]$ is large in absolute magnitude and negative. Let $J_l(z_i)$ be the loss function of logistic regression for this instance, and let $J_s(z_i)$ be the loss function of the hinge-loss SVM for this instance. In this case, we will show that $J_s(z_i) - J_l(z_i)$ goes to the constant value of 1, as $z_i \Rightarrow -\infty$. Therefore, the gradients for grossly misclassified points will also be similar (since a constant difference in loss functions implies a zero difference in gradients). Since grossly misclassified points result in the largest updates, it means that the updates for the two loss functions will be similar at least in the initial stages when many points are grossly misclassified. First, we express the loss functions in terms of z_i *under the assumption that z_i is grossly negative*. The loss function for the SVM is as follows:

$$J_s(z_i) = \max\{0, 1 - z_i\} = 1 - z_i \qquad \text{[For } z_i < 0\text{]}$$

The loss function for logistic regression can be expressed as follows:

$$J_l(z_i) = \log(1 + \exp(-z_i))$$

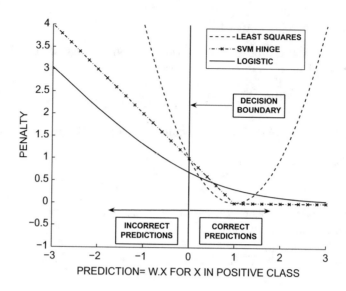

Figure 6.3: Showing the differences among the loss functions of least-squares classification, SVM, and logistic regression

Therefore, one can express the difference in the loss functions as follows:

$$\lim_{z \to -\infty} [J_s(z_i) - J_l(z_i)] = \lim_{z \to -\infty} [1 - z_i - \log(1 + \exp(-z_i))]$$
$$= \lim_{z \to -\infty} [1 - \log(\exp(z_i)) - \log(1 + \exp(-z_i))]$$
$$= 1 - \lim_{z \to -\infty} \log(1 + \exp(z_i)) = 1 - \log(1) = 1$$

Therefore, for grossly misclassified instances, the loss functions in two cases become very similar in that they differ only by a constant offset of 1. A difference of a constant value between the two loss functions will cause zero difference in the gradients, since the slopes in the two cases will be the same:

$$\frac{\partial J_s(z_i)}{\partial z_i} - \frac{\partial J_l(z_i)}{\partial z_i} = 0$$

We plot the two objective functions for varying values of z_i in Figure 6.3. The value of z_i (which is the same as $\overline{W} \cdot \overline{X}$ for a positive-class instance) is illustrated on the X-axis, whereas the loss function is illustrated on the Y-axis. The objective function of least-squares classification is shown in the same figure. It is evident that the shapes of the two loss functions of SVM and logistic regression are very similar. The main differences arise for the case of well-separated points in which the hinge-loss SVM does not make any updates because of a loss value of 0, whereas logistic regression does make updates because of slightly non-zero loss. As a result, the hinge loss SVM tends to converge relatively quickly when the points are linearly separable. Nevertheless, the accuracy performance of both models is very similar in most cases. The loss function of least-squares classification is quite different because it allows the objective function to worsen for increasingly correct classification (see Figure 6.3). This is evident from the rising portion of the corresponding plot with increasing values on the X-axis.

6.5.3 Logistic Regression as a Probabilistic Classifier

So far, we have presented logistic regression without any interpretation of its loss function. In this section, we provide a probabilistic interpretation of the logistic regression classifier. One can interpret the prediction of test instance $\overline{Z}$ in logistic regression either from a deterministic point of view or from a probabilistic point of view.

$$F(\overline{Z}) = \text{sign}\{\overline{W} \cdot \overline{Z}\} \text{ [Deterministic Prediction]}$$

$$P(F(\overline{Z}) = 1) = \frac{1}{1 + \exp(-\overline{W} \cdot \overline{Z})} \text{ [Probabilistic Prediction]}$$

It is noteworthy that points on the decision boundary satisfying $\overline{W} \cdot \overline{Z} = 0$ will be predicted to a probability of $1/(1 + \exp(0)) = 0.5$, which is a reasonable prediction.

The probabilistic point of view is critical in terms of being able to design an interpretable loss function. In essence, logistic regression assumes that the target variable $y_i \in \{-1, +1\}$ is the observed value generated from a hidden Bernoulli probability distribution that is parameterized by $\overline{W} \cdot \overline{X_i}$. Since $\overline{W} \cdot \overline{X_i}$ might be an arbitrary quantity (unlike the parameters of a Bernoulli distribution), we need to apply some type of function to it in order to bring it to the range $(0, 1)$. The specific function chosen is the *sigmoid function*. In other words, we have:

$$y_i \sim \text{Bernoulli distribution parametrized by sigmoid of } \overline{W} \cdot \overline{X_i}$$

It is this probabilistic interpretation because of which we get our prediction function $F(\overline{Z})$ for a given data point $\overline{Z}$:

$$P(F(\overline{Z}) = 1) = \frac{1}{1 + \exp(-\overline{W} \cdot \overline{Z})}$$

One can write this prediction function more generally for any target $y \in \{-1, +1\}$.

$$P(F(\overline{Z}) = y) = \frac{1}{1 + \exp(-y(\overline{W} \cdot \overline{Z}))} \tag{6.22}$$

It is easy to verify that the sum of the probabilities over both outcomes of y is 1.

Probabilistic models learn the parameters of the probabilistic process in order to maximize the probability that each training instance is classified to the correct class. The *likelihood* of the entire training data set with n pairs of the form $(\overline{X_i}, y_i)$ is defined as the product of these probabilities:

$$\mathcal{L}(\text{Training Data}|\overline{W}) = \prod_{i=1}^{n} P(F(\overline{X_i}) = y_i) = \prod_{i=1}^{n} \frac{1}{1 + \exp(-y_i(\overline{W} \cdot \overline{X_i}))}$$

Maximizing the likelihood $\mathcal{L}$ is the same as minimizing the negative logarithm of $\mathcal{L}$; this quantity is referred to as the *log-likelihood* of the training data. Log-likelihood is numerically more stable because it can be expressed as the sums of quantities, rather than as the product of many values less than 1 (which can cause underflow). Therefore, the minimization objective function $\mathcal{LL}$ can be expressed as follows:

$$\mathcal{LL} = \sum_{i=1}^{n} \log[1 + \exp\{-y_i(\overline{W} \cdot \overline{X_i})\}] \tag{6.23}$$

After adding the regularization term, this (negative) log-likelihood function is *identical* to the objective function introduced earlier for logistic regression. Therefore, logistic regression is essentially a (negative) log-likelihood minimization algorithm.

One can even interpret the updates on logistic regression in terms of the *probabilities* of errors on training points:

$$\overline{W} \Leftarrow \overline{W}(1 - \alpha\lambda) + \sum_{(\overline{X}_i, y_i) \in S} \frac{\alpha y_i \overline{X}_i^T}{(1 + \exp(y_i[\overline{W} \cdot \overline{X}_i^T]))}$$
$$= \overline{W}(1 - \alpha\lambda) + \alpha y_i \overline{X}_i^T P(F(\overline{X}_i) = -y_i)$$
$$= \overline{W}(1 - \alpha\lambda) + \alpha y_i \overline{X}_i^T P(\text{Error on } \overline{X}_i)$$

Note that the updates in least-squares regression are proportional to the *magnitudes* of the errors (see Equation 6.5). On the other hand, in the case of logistic regression, the *probabilities* of the errors are used to regulate the updates.

6.6 Multiclass Setting

In the multiclass setting, we have multiple classes associated with each data point. Associated with each training point $\overline{X}_i$, we have one of k *unordered classes*, which are indexed by $\{1, \ldots, k\}$. Note that these k indices are not ordered; for example, the indices could represent colors such as "Green," "Blue," or "Red." In this case, there are two separate approaches that can be used in order to perform the learning:

- One can decompose the problem into multiple binary class problems by testing the classes against one another and then vote on the result.

- One can *simultaneously* learn k separators for the different classes (each with a positive and negative side), and select the separator for which the class lies as far as possible on the positive side of the separator.

The latter approach is more effective, because one is learning to separate between the various classes in an integrated way rather than in a decoupled way. Nevertheless, the first approach has the advantage that there are many binary classifiers available to use as subroutines, which makes implementation particularly simple. We will discuss both approaches.

6.6.1 One-Against-Rest and One-Against-One Voting

In each of the two voting methods, a binary classifier $\mathcal{A}$ is treated as a subroutine. Subsequently, a meta-algorithm is wrapped around this binary classifier in order to create an *ensemble* of the different approaches. Subsequently, a post-processing approach is used, which is also referred to as the *voting phase*. The winner of the voting phase is used to decide the class label. Both approaches require a modification of the training data either in terms of data labeling or in terms of pre-selection of specific instances.

We first discuss the *one-against-rest* approach, which is also referred to as the *one-against-all* approach. In this approach, k different binary classification problems are created, such that one problem corresponds to each class. In the ith problem, the ith class is considered the set of positive examples whereas all the remaining examples are considered negative examples. The binary classifier $\mathcal{A}$ is applied to each of these training data sets.

This creates a total of k models. Then each of these models is applied to the test instance $\overline{Z}$. If the positive class is predicted in the ith problem for test instance $\overline{Z}$, the ith class is rewarded with a vote that is proportional to the confidence of prediction. One may also use the numeric output of a classifier (e.g., probability of positive class in logistic regression) to weight the corresponding vote. The highest numeric score for a particular class is selected to predict the label. Note that the choice of the numeric score for weighting the votes depends on the classifier at hand.

The second strategy is the *one-against-one* approach. In this strategy, a training data set is constructed for each of the $\binom{k}{2}$ pairs of classes. The algorithm $\mathcal{A}$ is applied to each training data set. This results in a total of $k(k-1)/2$ models. For each model, the prediction provides a vote to the winner. One may also weight the vote with a numeric score based on the classifier at hand. The class label with the most votes is declared as the winner at the end. At first glance, it seems that this approach is computationally more expensive, because it requires us to train $k(k-1)/2$ classifiers, rather than training k classifiers, as in the one-against-rest approach. However, the computational cost is ameliorated by the smaller size of the training data in the one-against-one approach. Specifically, the training data size in the latter case is approximately $2/k$ of the training data size used in the one-against-rest approach on the average. If the running time of each individual classifier scales super-linearly with the number of training points, then the overall running time of this approach may actually be lower than the first approach (which requires us to train only k classifiers). This can be the case with many classifier that use complex feature engineering methods.

6.6.2 Multinomial Logistic Regression

Multinomial logistic regression is a direct approach to learning multiple separators $\overline{W}_1 \ldots \overline{W}_k$, one for each class. The broad idea is to generalize logistic regression to multiple classes. We assume that the ith training instance is denoted by $(\overline{X}_i, c(i))$. The training instance contains a d-dimensional feature vector $\overline{X}_i$ (which is a row vector) and the index $c(i) \in \{1 \ldots k\}$ of its observed class. In this case, k different separators are learned whose parameter vectors are $\overline{W}_1 \ldots \overline{W}_k$, and the class j with the largest dot product $\overline{W}_j \cdot \overline{Z}^T$ is predicted as the class of test instance $\overline{Z}$.

Multinomial logistic regression is a natural generalization of binary logistic regression, which models the probability of a point belonging to the rth class. The probability of training point $\overline{X}_i$ belonging to class r is defined as follows:

$$P(r|\overline{X}_i) = \frac{\exp(\overline{W}_r \cdot \overline{X}_i^T)}{\sum_{j=1}^{k} \exp(\overline{W}_j \cdot \overline{X}_i^T)} \tag{6.24}$$

One would like to learn $\overline{W}_1 \ldots \overline{W}_k$, so that the probability $P(c(i)|\overline{X}_i)$ is as high as possible for the class $c(i)$, given training instance $\overline{X}_i$. This is achieved by using the *cross-entropy loss*, which is as natural generalization of the loss function in logistic regression. This loss is defined as the negative logarithm of the probability of the instance $\overline{X}_i$ belonging to the correct class $c(i)$:

$$J = -\sum_{i=1}^{n} \underbrace{\log[P(c(i)|\overline{X}_i)]}_{J_i} + \frac{\lambda}{2}\sum_{r=1}^{k} \|\overline{W}_r\|^2$$

6.6.2.1   Stochastic Gradient Descent

Since each of the separators needs to be updated during gradient descent, we need to evaluate the gradient of J with respect to each $\overline{W}_r$. In addition to the regularization term, the point-specific portion of the loss function is denoted by $J_i = -\log[P(c(i)|\overline{X}_i)]$. Therefore, the gradient can also be decomposed into the sum of point-specific gradients, along with the gradient of the regularization term. The point-specific gradient is denoted by $\frac{\partial J_i}{\partial W_r}$. Let v_{ji} denote the quantity $\overline{W}_j \cdot \overline{X}_i^T$. Then, the value of $\frac{\partial J_i}{\partial W_r}$ is computed as follows:

$$\frac{\partial J_i}{\partial \overline{W}_r} = \sum_j \left(\frac{\partial J_i}{\partial v_{ji}} \right) \frac{\partial v_{ji}}{\partial \overline{W}_r} = \frac{\partial J_i}{\partial v_{ri}} \underbrace{\frac{\partial v_{ri}}{\partial \overline{W}_r}}_{\overline{X}_i^T} = \overline{X}_i^T \frac{\partial J_i}{\partial v_{ri}} \qquad (6.25)$$

Several terms are dropped in the above summation, because v_{ji} has a zero gradient with respect to $\overline{W}_r$ for $j \neq r$. One only needs to compute the partial derivative of J_i with respect to v_{ri}. In order to achieve this goal, the point-specific loss J_i is expressed directly as a function of $v_{1i}, v_{2i}, \ldots, v_{ki}$ as follows:

$$J_i = -\log[P(c(i)|\overline{X}_i)] = -\overline{W}_{c(i)} \cdot \overline{X}_i^T + \log[\sum_{j=1}^{k} \exp(\overline{W}_j \cdot \overline{X}_i^T)] \quad \text{[Using Equation 6.24]}$$

$$= -v_{c(i),i} + \log[\sum_{j=1}^{k} \exp(v_{ji})]$$

Therefore, we can compute the partial derivative of J_i with respect to v_{ri} as follows:

$$\frac{\partial J_i}{\partial v_{ri}} = \begin{cases} -\left(1 - \frac{\exp(v_{ri})}{\sum_{j=1}^{k} \exp(v_{ji})} \right) & \text{if } r = c(i) \\ \left(\frac{\exp(v_{ri})}{\sum_{j=1}^{k} \exp(v_{ji})} \right) & \text{if } r \neq c(i) \end{cases}$$

$$= \begin{cases} -(1 - P(r|\overline{X}_i)) & \text{if } r = c(i) \\ P(r|\overline{X}_i) & \text{if } r \neq c(i) \end{cases}$$

By substituting the value of the partial derivative $\frac{\partial J_i}{\partial v_{ri}}$ in Equation 6.25, we obtain the following:

$$\frac{\partial J_i}{\partial \overline{W}_r} = \begin{cases} -\overline{X}_i^T (1 - P(r|\overline{X}_i)) & \text{if } r = c(i) \\ \overline{X}_i^T P(r|\overline{X}_i) & \text{if } r \neq c(i) \end{cases} \qquad (6.26)$$

One can then use this point-specific gradient to compute the stochastic gradient descent updates:

$$\overline{W}_r \Leftarrow \overline{W}_r(1 - \alpha\lambda) + \alpha \begin{cases} \overline{X}_i^T (1 - P(r|\overline{X}_i)) & \text{if } r = c(i) \\ -\overline{X}_i^T P(r|\overline{X}_i) & \text{if } r \neq c(i) \end{cases} \quad \forall r \in \{1 \ldots k\} \qquad (6.27)$$

The probabilities in the above update can be substituted using Equation 6.24.

6.7 The Naïve Bayes Model

The previous sections introduce various forms of logistic regression, which models the probability that an instance belongs to particular class. This approach of modeling the *probability of the classes directly as a function of given feature instances* is referred to as a *discriminative model*. Another type of model is the *generative model*, which models the *probability distribution of the feature vectors as a function of the class labels*. In other words, instead of modeling the class probabilities (given the features), we model the feature probability distributions (given the classes). This reversed approach allows us to generate a sample data set in a probabilistic way by first selecting a class and then generating the feature values from the probability distribution. This is why the approach is referred to as a generative model. The generative approach requires us to use a fundamental theorem in probability, referred to as the *Bayes theorem*, which is how the method derives its name. The naïve Bayes classifier assumes that each point in the data set is generated using a probabilistic process of first sampling the class and then generating the feature vector using a probability distribution that is specific to the class. We will work with the multiway classification setting in which there are a total of k class. The mixture component associated with the rth class is denoted by $\mathcal{C}_r$, where $r \in \{1, \ldots, k\}$. The generative process for each data point $\overline{X}_i$ is as follows:

1. Select the rth class (mixture component) $\mathcal{C}_r$ with prior probability $\alpha_r = P(\mathcal{C}_r)$.

2. Generate the data point $\overline{X}_i$ from the probability distribution for $\mathcal{C}_r$. For simplicity, we discuss the case where the features in $\overline{X}_i$ are drawn from $\{0, 1\}$, and therefore a Bernoulli model is used.

The observed (training and test) data are assumed to be outcomes of this generative process, and the parameters of this generating process are estimated (using the training data set) so that the likelihood of this data set being created by the generative process is maximized. Subsequently, these parameters are used to estimate the probability of each class for a test instance. The aforementioned model is referred to as a *mixture model*, which is also used in probabilistic forms of unsupervised learning.

In the Bernoulli model, it is assumed that each feature of $\overline{X}_i$ is drawn from $\{0, 1\}$. Although this assumption might seem restrictive at first glance, it is possible to apply the model to other types of data sets by simply changing the probability distribution of each class. For example, if we have a data set containing continuous attributes, we can simply use a Gaussian distribution to model each class. The Bernoulli model assumes that the jth attribute value of a data point is set to 1 in the rth class (mixture component) with probability $p_j^{(r)}$. Now, consider a test instance $\overline{Z}$ with binary values $[z_1, z_2, \ldots, z_d]$ in its d attributes. Then, the probability $P(\overline{Z}|\mathcal{C}_r)$ of the generation of the data point $\overline{Z}$ from mixture component $\mathcal{C}_r$ is given by the product of the d different Bernoulli probabilities corresponding to the values of the binary attributes being set to 1 and 0, respectively:

$$P(\overline{Z}|\mathcal{C}_r) = \prod_{j:z_j=1} p_j^{(r)} \prod_{j:z_j=0} (1 - p_j^{(r)}) \qquad (6.28)$$

This model makes the *naïve Bayes assumption* that the values of the binary attributes are conditionally independent, given the choice of class. This is the reason that the method is referred to as a *naïve Bayes classifier*. The assumption is convenient because it allows one to express the joint probability of the attributes in $\overline{Z}$ as the product of the corresponding values on individual attributes.

A maximum likelihood model is used to estimate the parameters of the model in the training phase. These are then used in the prediction phase as follows:

- **Training phase:** Estimate the maximum-likelihood values of the parameters $p_j^{(r)}$ and α_r using only the training data.

- **Prediction phase:** Use the estimated values of the parameters to predict the class of each unlabeled test instance.

The training phase is executed first, which is followed by the prediction phase. However, we present the prediction phase first, since this phase is the key to understanding a naïve Bayes classifier. The following section assumes that the model parameters have already been learned in the training phase.

Prediction Phase

The prediction phase uses the Bayes rule of posterior probabilities to predict the class of an instance. According to the Bayes rule of posterior probabilities, the posterior probability of $\overline{Z}$ being generated by the mixture component $\mathcal{C}_r$ (i.e., generating component of the rth class) can be estimated as follows:

$$P(\mathcal{C}_r|\overline{Z}) = \frac{P(\mathcal{C}_r) \cdot P(\overline{Z}|\mathcal{C}_r)}{P(\overline{Z})} \propto P(\mathcal{C}_r) \cdot P(\overline{Z}|\mathcal{C}_r) \qquad (6.29)$$

A constant of proportionality is used instead of the $P(\overline{Z})$ in the denominator, because the estimated probability is only compared between multiple classes to determine the predicted class, and $P(\overline{Z})$ is independent of the class at hand. We further expand the relationship in Equation 6.29 using the Bernoulli distribution of Equation 6.28 as follows:

$$P(\mathcal{C}_r|\overline{Z}) \propto P(\mathcal{C}_r) \cdot P(\overline{Z}|\mathcal{C}_r) = \alpha_r \prod_{j:z_j=1} p_j^{(r)} \prod_{j:z_j=0} (1 - p_j^{(r)}) \qquad (6.30)$$

All the parameters on the right-hand side are estimated during the training phase discussed below. Therefore, one now has an estimated probability of each class being predicted up to a constant factor of proportionality. The class with the highest posterior probability is predicted as the relevant one.

Training Phase

The training phase of the Bayes classifier uses the labeled training data to estimate the maximum likelihood values of the parameters in Equation 6.30. There are two key sets of parameters that need to be estimated; these are the prior probabilities α_r and the Bernoulli generative parameters, $p_j^{(r)}$, for each mixture component. The statistics available for parameter estimation include the number of labeled data points n_r belonging to the rth class $\mathcal{C}_r$, and the number, $m_j^{(r)}$, of the data points belonging to class $\mathcal{C}_r$ that contain term t_j. The maximum likelihood estimates of these parameters can be shown to be the following:

1. *Estimation of prior probabilities:* Since the training data contains n_r data points for the rth class in a corpus size of n, the natural estimate for the prior probability of the class is as follows:

$$\alpha_r = \frac{n_r}{n} \qquad (6.31)$$

If the corpus size is small, it is helpful to perform Laplacian smoothing by adding a small value $\beta > 0$ to the numerator and $\beta \cdot k$ to the denominator:

$$\alpha_r = \frac{n_r + \beta}{n + k \cdot \beta} \tag{6.32}$$

The precise value of β contains the amount of smoothing, and it is often set to 1 in practice. When the amount of data is very small, this results in the prior probabilities being estimated closer to $1/k$, which is a sensible assumption in the absence of sufficient data.

2. *Estimation of class-conditioned mixture parameters:* The class-conditioned mixture parameters, $p_j^{(r)}$, are estimated as follows:

$$p_j^{(r)} = \frac{m_j^{(r)}}{n_r} \tag{6.33}$$

The estimations can be poor when the number of training data points is small. For example, it is possible for the training data to contain no data points belonging to the rth class for which the jth attribute takes on the value of 1. In such a case, one would estimate the corresponding value of $p_j^{(r)}$ to 0. As a result of the multiplicative nature of Equation 6.30, the estimated probability of the rth class will be 0. Such predictions are often erroneous, and are caused by overfitting to the small training data size.

Laplacian smoothing of class-conditioned probability estimation can alleviate this problem. Let d_a be the average number of 1s in each row of the binary data and d be the dimensionality. The basic idea is to add a Laplacian smoothing parameter $\gamma > 0$ to the numerator of Equation 6.33 and $d\gamma/d_a$ to the denominator:

$$p_j^{(r)} = \frac{m_j^{(r)} + \gamma}{n_r + d\gamma/d_a} \tag{6.34}$$

The value of γ is often set to 1 in practice. When the amount of training data is very small, this choice leads to a default value of d_a/d for $p_j^{(r)}$, which reflects the level of sparsity in the data.

This probabilistic model is extremely popular because of its simplicity and interpretability. It can be generalized to any type of data (and not just binary data) by changing the nature of the generative model for each mixture component.

6.8 Nearest Neighbor Classifier

As in the case of the nearest neighbor classifier, we work with the (more generic) multiway setting in which we have $k \geq 2$ different classes. Nearest-neighbor classifiers use the following principle:

Similar instances have similar labels.

A natural way of implementing this principle is to use a κ-nearest-neighbor[1] classifier. The basic idea is to identify the κ-nearest neighbors of a test point, and compute the number

---

[1]We use κ instead of the more common use of the variable k, since the number of classes is k.

of points that belong to each class. The class with the largest number of points is reported as the relevant one. A variety of distance functions can be used in order to implement the nearest neighbor classifier. Therefore, the approach can be used for any type of data, as long as an appropriate distance function is available. Nearest-neighbor classification can be used for both binary classes and multi-way classification, as long as the class with the largest vote is used. If the dependent variable is numeric, the average value of the dependent variable among the nearest neighbors can be reported. Therefore, an important advantage of nearest neighbor classifiers is that they can be used for virtually any type of data, and the complexity of the approach is nicely restricted to the design of a distance (or similarity) function.

Nearest-neighbor classifiers are also referred to as *lazy learners, memory-based learners*, and *instance-based learners*. They are referred to as lazy learners because most of the work of classification is postponed to the very end. In a sense, these methods *memorize* all the training examples, and use the best matching ones to the *instance* at hand. Unlike model-based methods like the support vector machine, less generalization and learning is done up front, and most of the work of classification is left to the very end in a *lazy* way. A straightforward implementation of the nearest-neighbor method requires no training, but it requires $O(n)$ similarity computations for classifying *each test instance*. It is possible to speed up the nearest neighbor classifier by using a variety of index structures.

The number of nearest neighbors, κ, is a parameter for the algorithm. Its value can be set by trying different values of κ on the training data. The value of κ at which the highest accuracy is achieved on the training data is used. While computing accuracy on the training data, a *leave-one-out* approach is used, in which the point to which the κ-nearest neighbors are computed is not included among the nearest neighbors. For example, if we did not take this precaution, every point with be its own nearest neighbor, and a value of $\kappa = 1$ would always be deemed as optimal. This is a manifestation of overfitting, which is avoided with the leave-one-out approach. The classification accuracy is computed by using a *validation sample* of size s. For each point in the sample, the similarities with respect to the entire data are computed (without including the point itself among the nearest neighbors). These computed similarities are used to rank the $n - 1$ training points for each sample, and test various values of κ. This process requires $O(n \cdot s)$ similarity computations and $O(n \cdot s \cdot \log(n))$ time for sorting the points. For a validation sample size of s, the time required is $O(s \cdot n \cdot (T + \log(n)))$ for tuning the parameter κ. Here, T is the time required for each similarity computation.

Nearest neighbor classifiers can be extremely powerful when a large amount of data is available. With an infinite amount of data, the decision boundary between a pair of classes can be learned with a large level of accuracy. However, in practice, the data is often limited, and nearest neighbor classifiers often provide poor performance. It is possible to improve the accuracy of nearest neighbor classifiers by adding some supervision to the neighbor selection process. In fact, it can be shown that many other classifiers, such as *decision trees* and the support vector machine can be viewed as special cases of a supervised nearest neighbor classifier. A detailed discussion of these connections may be found in [7].

6.9   Decision Trees

A decision tree is a hierarchical partitioning of the data space, in which the partitioning is achieved with a series of split conditions (i.e., decisions) on the attributes. The idea is to partition the data space into attribute regions that are heavily biased towards a particular class during the training phase. Therefore, partitions are associated with their favored (i.e., majority) class labels. During the testing phase, the relevant partition of the data space is identified for the test instance, and the label of the partition is returned. Note that each *node* in the decision tree corresponds to a region of the data space defined by the split conditions at its ancestor nodes, and the root node corresponds to the entire data space.

6.9.1 Training Phase of Decision Tree Construction

Decision trees partition the data space recursively in top-down fashion using *split conditions* or *predicates*. The basic idea is to choose the split conditions in such a way that the subdivided portions are dominated by one or more classes. The evaluation criteria for such split predicates are often similar to feature selection criteria in classification. The split criteria typically correspond to constraints on the frequencies of one or more words. A split that uses a single attribute is referred to as a *univariate split*, whereas a split using multiple attributes is referred to as a *multivariate split*. It is common for each node in the decision tree to have only two children. For example, if the split predicate corresponds to an attribute such as *Age* being less than 30, then all individuals with age less than 30 will lie in one branch, whereas individuals with age greater than 30 will lie in the other branch. The splits are applied recursively in top-down fashion, until each node in the tree contains a single class. These nodes are the leaf nodes, and are labeled with the classes of their instances. In order to classify a test instance for which the label is unknown, the split predicates are used in top-down fashion over various nodes of the tree in order to identify the branch to follow down the tree until the leaf node is reached. For example, if a split predicate corresponds to the age of an individual being less than 30, it is checked whether the test point corresponds to the age attribute being less than 30. This process is repeated until the relevant leaf node is identified, and its label is reported as the prediction of the test instance.

This type of extreme way of creating a tree until each leaf contains instances of only a single class is referred to as growing a tree to full height. Such a fully-grown tree will provide 100% accuracy on the *training data* even for a data set in which class labels are generated randomly and independently of the features in the training instances. This is clearly the result of overfitting, because one cannot expect to learn anything from a data set with random labels. A fully-grown tree will often misinterpret random nuances in the training data as indicative of discriminative power, and these types of overfitted choices will cause the predictions of the same test instance to vary significantly between trees constructed on different training samples. This type of variability is usually a sign of a poor classifier *in expectation*, because at least some of these diverse predictions are bound to be incorrect. As a result, the performance on the test data of such a tree will be poor even for those data sets in which the feature values are related to the class labels. This problem is addressed by *pruning* the nodes at the lower levels of the tree that do not contribute in a positive way to the generalization power on unseen test instances. As a result, the leaves of the pruned tree may no longer contain a single class, and are therefore labeled with the majority class (or *dominant* class for k-way classification).

Pruning is accomplished by holding out a part of the training data, which is not used in the (initial) decision-tree construction. For each internal node, it is tested whether or not the accuracy improves on the held out data by removing the subtree rooted at that node (and converting that internal node to a leaf). Depending on whether or not the accuracy improves, the pruning is performed. Internal nodes are selected for testing in bottom-up order, until all of them have been tested once. The overall procedure of decision-tree construction is shown in Figure 6.4. Note that the specific split criterion is not spelled out in these generic pseudo-

Algorithm *ConstructDecisionTree* (Labeled Training Data Set: D_y)
begin
    Hold out a subset H from D_y to create $D'_y = D_y - H$;
    Initialize decision tree T to a single root node containing D'_y;
    { **Tree Construction Phase** }
    repeat
      Select any eligible leaf node from T with data set L;
      Use split criteria of section 6.9.2 to partition L into subsets L_1 and L_2;
      Store split condition at L and make $\{L_1, L_2\}$ children of L in T;
    until no more eligible nodes in T;
    { **Tree Pruning Phase** }
    repeat
      Select an untested internal node N in T in bottom-up order;
      Create T_n obtained by pruning subtree of T at N;
      Compare accuracy of T and T_n on held out set H;
      if T_n has better accuracy **then** replace T with T_n;
    until no untested internal nodes remain in T;
    Label each leaf node of T with its dominant class;
    return T;
end

Figure 6.4: Training process in a decision tree

code. This is an issue that will be discussed in the next section. The notion of eligibility of a node to be split is also not specified in the pseudo-code. Since bottom nodes are pruned anyway, it is possible to stop early using other criteria than growing the tree to full height. Various stopping criteria make nodes ineligible for splitting, such as a maximum threshold on the number of instances, or a minimum percentage threshold on the dominant class. The simplest possible criterion for splitting is one in which a node is split when all data points in the node belong to the same label. In such a case, no additional benefit can be obtained by splitting further. This approach is referred to as growing the tree to *full height*.

To illustrate the basic idea of decision-tree construction, an illustrative example will be used. In Table 6.1, a snapshot of a hypothetical charitable donation data set has been illustrated. The two feature variables represent the age and salary attributes. Both attributes are related to the donation propensity, which is also the class label. Specifically, the likelihood of an individual to donate is positively correlated with his or her age and salary. However, the best separation of the classes may be achieved only by combining the two attributes. The goal in the decision-tree construction process is to perform a sequence of splits in top-down fashion to create nodes at the leaf level in which the donors and non-donors are separated well. One way of achieving this goal is depicted in Figure 6.5(a). The figure illustrates a hierarchical arrangement of the training examples in a tree-like structure. The first-level split uses the age attribute, whereas the second-level split for both branches uses the salary attribute. Note that different splits at the same decision-tree level need not be on the same attribute. Furthermore, the decision tree of Figure 6.5(a) has two branches at each node, but this need not always be the case. In this case, the training examples in all leaf nodes belong to the same class, and, therefore, there is no point in growing the decision tree beyond the leaf nodes. The splits shown in Figure 6.5(a) are referred to as *univariate* splits because they use a single attribute. To classify a test instance, a single relevant path in the tree is traversed in top-down fashion by using the split criteria to decide which branch to follow at each node of the tree. The dominant class label in the leaf node is reported as

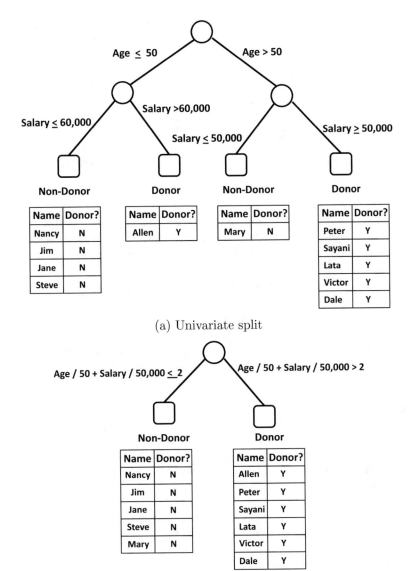

(a) Univariate split

(b) Multivariate split

Figure 6.5: Illustration of univariate and multivariate splits for decision tree construction

Name	Age	Salary	Donor?
Nancy	21	37000	N
Jim	27	41000	N
Allen	43	61000	Y
Jane	38	55000	N
Steve	44	30000	N
Peter	51	56000	Y
Sayani	53	70000	Y
Lata	56	74000	Y
Mary	59	25000	N
Victor	61	68000	Y
Dale	63	51000	Y

Table 6.1: Training data snapshot relating the salary and age features to charitable donation propensity

the relevant class. For example, a test instance with age less than 50 and salary less than 60,000 will traverse the leftmost path of the tree in Figure 6.5(a). Because the leaf node of this path contains only non-donor training examples, the test instance will also be classified as a non-donor.

Multivariate splits use more than one attribute in the split criteria. An example is illustrated in Figure 6.5(b). In this particular case, a single split leads to full separation of the classes. This suggests that multivariate criteria are more powerful because they lead to shallower trees. For the same level of class separation in the training data, shallower trees are generally more desirable because the leaf nodes contain more examples and, therefore, are statistically less likely to overfit the noise in the training data.

6.9.2   Splitting a Node

The goal of the split criterion is to maximize the separation of the different classes among the children nodes. In the following, only univariate criteria will be discussed. Assume that a quality criterion for evaluating a split is available. The design of the split criterion depends on the nature of the underlying attribute:

1. *Binary attribute:* Only one type of split is possible, and the tree is always binary. Each branch corresponds to one of the binary values.

2. *Categorical attribute:* If a categorical attribute has r different values, there are multiple ways to split it. One possibility is to use an r-way split, in which each branch of the split corresponds to a particular attribute value. The other possibility is to use a binary split by testing each of the $2^r - 1$ combinations (or groupings) of categorical attributes, and selecting the best one. This is obviously not a feasible option when the value of r is large. A simple approach that is sometimes used is to convert categorical data to binary data is to create binary attributes, in which one binary variable corresponds to each possible outcome of the categorical attribute. Therefore, only one binary variable will take on the value of 1, whereas other attributes take on the value of 0. In this case, the approach for binary attributes may be used.

3. *Numeric attribute:* If the numeric attribute contains a small number r of ordered values (e.g., integers in a small range $[1, r]$), it is possible to create an r-way split for each distinct value. However, for continuous numeric attributes, the split is typically performed by using a binary condition, such as $x \leq a$, for attribute value x and constant a.

   Consider the case where a node contains m data points. Therefore, there are m possible split points for the attribute, and the corresponding values of a may be determined by sorting the data in the node along this attribute. One possibility is to test all the possible values of a for a split and select the best one. A faster alternative is to test only a smaller set of possibilities for a, based on equi-depth division of the range.

Many of the aforementioned methods requires the determination of the "best" split from a set of choices. Specifically, it is needed to choose from multiple attributes and from the various alternatives available for splitting each attribute. Therefore, quantifications of split quality are required. Some examples of such quantifications are as follows:

1. *Error rate:* Let p be the fraction of the instances in a set of data points S belonging to the dominant class. Then, the error rate is simply $1 - p$. For an r-way split of set S into sets $S_1 \ldots S_r$, the overall error rate of the split may be quantified as the weighted average of the error rates of the individual sets S_i, where the weight of S_i is $|S_i|$. The split with the lowest error rate is selected from the alternatives.

2. *Gini index:* The Gini index $G(S)$ for a set S of data points may be computed on the class distribution $p_1 \ldots p_k$ of the training data points in S.

$$G(S) = 1 - \sum_{j=1}^{k} p_j^2 \tag{6.35}$$

   The overall Gini index for an r-way split of set S into sets $S_1 \ldots S_r$ may be quantified as the weighted average of the Gini index values $G(S_i)$ of each S_i, where the weight of S_i is $|S_i|$.

$$\text{Gini-Split}(S \Rightarrow S_1 \ldots S_r) = \sum_{i=1}^{r} \frac{|S_i|}{|S|} G(S_i) \tag{6.36}$$

   The split with the lowest Gini index is selected from the alternatives. The *CART* algorithm uses the Gini index as the split criterion.

3. *Entropy:* The entropy measure is used in one of the earliest classification algorithms, referred to as *ID3*. The entropy $E(S)$ for a set S may be computed on the class distribution $p_1 \ldots p_k$ of the training data points in the node.

$$E(S) = - \sum_{j=1}^{k} p_j \log_2(p_j) \tag{6.37}$$

   As in the case of the Gini index, the overall entropy for an r-way split of set S into sets $S_1 \ldots S_r$ may be computed as the weighted average of the Gini index values $G(S_i)$ of each S_i, where the weight of S_i is $|S_i|$.

$$\text{Entropy-Split}(S \Rightarrow S_1 \ldots S_r) = \sum_{i=1}^{r} \frac{|S_i|}{|S|} E(S_i) \tag{6.38}$$

Lower values of the entropy are more desirable. The entropy measure is used by the *ID3* and *C4.5* algorithms.

The information gain is closely related to entropy, and is equal to the *reduction* in the entropy $E(S)$ − Entropy-Split($S \Rightarrow S_1 \ldots S_r$) as a result of the split. Large values of the reduction are desirable. At a conceptual level, there is no difference between using either of the two for a split although a normalization for the degree of the split is possible in the case of information gain. Note that the entropy and information gain measures should be used only to compare two splits of the same degree because both measures are naturally biased in favor of splits with larger degree. For example, if a categorical attribute has many values, attributes with many values will be preferred. It has been shown by the *C4.5* algorithm that dividing the overall information gain with the normalization factor of $-\sum_{i=1}^{r} \frac{|S_i|}{|S|} \log_2(\frac{|S_i|}{|S|})$ helps in adjusting for the varying number of categorical values.

The aforementioned criteria are used to select the choice of the split attribute and the precise criterion on the attribute. For example, in the case of a numeric database, different split points are tested for each numeric attribute, and the best split is selected.

It is relatively easy to generalize decision trees to numeric target variables in order to perform regression. The main difference is that the splits need to use the variance of the target variables to choose attributes for splitting (rather than using measures such as the Gini index). Such trees are referred to as *regression trees*. One advantage of regression trees is that they can learn the relationship between the feature variables and the target variable, even when this relationship is nonlinear. This is not the case for linear regression, which tends to model such nonlinear relationships rather poorly.

Prediction

Once the decision tree has been set up, it is relatively easy to use it for prediction. The split criterion associated with each node is always stored with that node during decision tree construction. For a test instance, the split criterion at the root node is tested to decide which branch to follow. This process is repeated recursively until the leaf node is reached. The label of the leaf node is returned as the prediction. A confidence is associated with the prediction, corresponding to the fraction of the labels belonging to the predicted class in the relevant leaf node.

Strengths and Weaknesses of Decision Trees

Decision trees are very similar to nearest neighbor classifiers in that they use the class distribution in local regions of the data in order to make predictions. Like a nearest neighbor classifier, a decision tree can learn arbitrarily complex decision boundaries in the underlying data, provided that an infinite amount of data is available. Unfortunately, this is never the case. With a limited amount of data, a decision tree provides a very rough approximation of the decision boundary. Such an approximation can overfit the data, and the problem can be ameliorated by using random forests.

6.9.3   Generalizing Decision Trees to Random Forests

Even though decision trees can capture arbitrary decision boundaries with an infinite amount of data, they can capture only piecewise linear approximations of these boundaries with a finite amount of data. These approximations are particularly inaccurate in

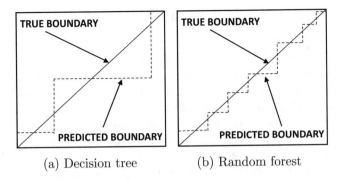

(a) Decision tree   (b) Random forest

Figure 6.6: Decision boundaries in decision trees and random forests

smaller data sets. An effective approach to deal with this problem is to randomize the tree construction process by allowing the splits at the higher levels of the tree to use the best feature selected out of a restricted subset of features. In other words, r features are randomly selected at each node, and the best splitting feature is selected only out of these features. Furthermore, different nodes use different subsets of randomly selected features. Using smaller values of r results in an increasing amount of randomization in tree construction. At first sight, it would seem that using such a randomized tree construction should impact the prediction in a detrimental way. However, the key is that multiple such randomized trees are grown, and the predictions of each test point over different trees are averaged to yield the final result. By averaging, we mean that the number of times a class is predicted by a randomized tree for a test instance is counted. The class receiving the most number of votes is predicted for the test instance. This averaging process improves the quality of the predictions significantly over a single tree by effectively using diverse choices of features at higher levels of the different trees in various ensemble components. This results in more robust predictions. The individual trees are grown to full height without pruning because the averaged predictions do not have the overfitting problem of the predictions of individual trees. The overall approach is referred to as an *ensemble-centric method*, which reduces the propensity of the classifier to create jagged decision boundaries of the underlying data.

Examples of the decision boundaries created by decision trees and random forests are shown in Figure 6.6. It is evident that the decision boundary in the case of the decision tree is quite jagged, which occurs frequently in the presence of limited data. In order to understand this point, note that a decision tree constructed on a data set with only two points of different class will be a straight line dividing the two points. However, this kind of decision tree clearly overfits the data, and does not work well for arbitrary test instances. By creating multiple trees and averaging over the predictions of the different trees, one obtains a smoother decision boundary as shown in Figure 6.6(b).

6.10 Rule-Based Classifiers

Rule-based classifiers use a set of "if then" rules $\mathcal{R} = \{R_1 \ldots R_m\}$ to match conditions on features on the left-hand side of the rule to the class labels on the right-hand side. As in the case of logical rules, the expression on the left-hand side of the rule is referred to as the *antecedent* and that on the right-hand side of the rule is referred to as the *consequent*. A rule is typically expressed in the following form:

IF *Condition* THEN *Conclusion*

In order to effectively implement rule-based classifiers, it is common to discretize the data points into categorical values. Then, for a discretized feature vector $\overline{X} = [x_1, x_2, \ldots x_d]$, the antecedent contains conditions of the form $(x_j = a)$ AND $(x_l = b)$ AND $(\ldots)$. Here, a and b are choices of categorical values. The matching of the condition in the antecedent with a data point causes the rule to be triggered. Each condition $(t_j \in \overline{X})$ is referred to as a conjunct, as in the case of propositional logic. The right-hand side of the rule is referred to as the consequent, and it contains the class variable. Therefore, a rule R_i is of the form $Q_i \Rightarrow c$ where Q_i is the antecedent, and c is the class variable. The "$\Rightarrow$" symbol denotes the "THEN" condition. In other words, the rules relate the presence of particular categorical values in the data record to the class variable c.

As in all inductive classifiers, rule-based methods have a training phase and a prediction phase. The training phase of a rule-based algorithm creates a set of rules. The prediction phase for a test instance discovers some or all rules that are *triggered* or *fired* by the test instance. A rule is said to be triggered by a training or test instance when the logical condition in the antecedent is satisfied by the features in the instance. Alternatively, for the specific case of training instances, it is said that such a rule *covers* the training instance. In some algorithms, the rules are *ordered* by priority and therefore, the first rule fired by the test instance is used to predict the class label in the consequent. In some algorithms, the rules are unordered, and multiple rules with (possibly) conflicting consequent values are triggered by the test instance. In such cases, methods are required to resolve the conflicts in class label prediction. Rules generated from *sequential covering algorithms* are ordered, although there are other algorithms that generate unordered rules.

6.10.1   Sequential Covering Algorithms

The basic idea in sequential covering algorithms is to generate the rules for each class at one time, by treating the class of interest as the positive class, and the union of all other classes as the negative class. Each generated rule always contains the positive class as the consequent. In each iteration, a single rule is generated using a *Learn-One-Rule* procedure and training examples that are covered by the class are removed. The generated rule is added to the bottom of the rule list. This procedure is continued until at least a certain minimum fraction of the instances of that class have been covered. Other termination criteria are often used. For example, the procedure can be terminated when the error of the next generated rule exceeds a certain pre-determined threshold on a separate validation set. A minimum description length (MDL) criterion is sometimes used when further addition of a rule increases the minimum description length of the model by more than a certain amount. The procedure is repeated for all classes. Note that less prioritized classes start with a smaller training data set because many instances have already been removed in the rule generation of higher priority classes. The *RIPPER* algorithm orders the rules belonging to the rare classes before those of more frequent classes, although other criteria are used by other algorithms, whereas *C4.5rules* uses various accuracy and information-theoretic measures to order the classes. The broad framework of the sequential covering algorithm is as follows:

for each class c in a particular order **do**
  repeat
    Extract the next rule $R \Rightarrow c$ using *Learn-One-Rule* on training data V;
    Remove examples covered by $R \Rightarrow c$ from training data V;
    Add extracted rule to bottom of rule list;
  until class c has been sufficiently covered

The procedure for learning a single rule is described in section 6.10.1.1. Only rules for $(k-1)$ classes are grown, and the final class is assumed to be a default catch-all class. One can also view the final rule for the remaining class c_l as the catch-all rule $\{\} \Rightarrow c_l$. This rule is added to the very bottom of the entire rule list. This type of ordered approach to rule generation makes the prediction process a relatively simple matter. For any test instance, the first triggered rule is identified. The consequent of that rule is reported as the class label. Note that the catch-all rule is guaranteed to be triggered when no other rule is triggered. One criticism of this approach is that the ordered rule generation mechanism might favor some classes more than others. However, since multiple criteria exist to order the different classes, it is possible to repeat the entire learning process with these different orderings, and report an averaged prediction.

6.10.1.1   Learn-One-Rule

It remains to be explained how the rule for a single class is generated. When the rules for class c are generated, each conjunct is sequentially added to the antecedent. The approach starts with the empty rule $\{\} \Rightarrow c$ for the class c, and then adds conjuncts such as $x_j = a$ one by one to the antecedent. What should be the criterion for adding a term to the antecedent of the current rule $R \Rightarrow c$?

1. The simplest criterion is to add the term to the antecedent that increases the accuracy of the rule as much as possible. In other words, if n_* is the number of training examples covered by the rule (after addition of a conjunct to the antecedent), and n_+ is the number of positive examples among these instances, then the accuracy of the rule is defined by n_+/n_*. To reduce overfitting, the smoothed accuracy A is sometimes used:

$$A = \frac{n_+ + 1}{n_* + k} \tag{6.39}$$

Here, k is the total number of classes.

2. Another criterion is *FOIL's information gain*. The term "FOIL" stands for *First Order Inductive Learner*. Consider the case where a rule covers n_1^+ positive examples and n_1^- negative examples, where positive examples are defined as training examples matching the class in the consequent. Furthermore, assume that the addition of a term to the antecedent changes the number of positive examples and negative examples to n_2^+ and n_2^-, respectively. Then, FOIL's information gain FG is defined as follows:

$$FG = n_2^+ \left(\log_2 \frac{n_2^+}{n_2^+ + n_2^-} - \log_2 \frac{n_1^+}{n_1^+ + n_1^-} \right) \tag{6.40}$$

This measure tends to select rules with high coverage because n_2^+ is a multiplicative factor in FG. At the same time, the information gain increases with higher accuracy because of the term inside the parentheses. This particular measure is used by the *RIPPER* algorithm.

Several other measures are often used, such as the *likelihood ratio* and *entropy*. Conjuncts can be successively added to the antecedent of the rule, until 100% accuracy is achieved by the rule on the training data or when the addition of a term cannot improve the accuracy of a rule. In many cases, this point of termination leads to overfitting. Just as node pruning is done in a decision tree, antecedent pruning is necessary in rule-based learners to avoid overfitting. Another modification to improve generalization power is to grow the r best rules simultaneously at a given time, and only select one of them at the very end based on the performance on a held-out set. This approach is also referred to as *beam search*.

Rule Pruning

Overfitting may result from the presence of too many conjuncts. As in decision-tree pruning, the Minimum Description Length principle can be used for pruning. For example, for each conjunct in the rule, one can add a penalty term δ to the quality criterion in the rule-growth phase. This will result in a *pessimistic error rate*. Rules with many conjuncts will therefore have larger aggregate penalties to account for their greater model complexity. A simpler approach for computing pessimistic error rates is to use a separate holdout validation set that is used for computing the error rate (without a penalty). However, this type of approach is not used by Learn-One-Rule.

The conjuncts successively added during rule growth (in sequential covering) are then tested for pruning in reverse order. If pruning reduces the pessimistic error rate on the training examples covered by the rule, then the generalized rule is used. While some algorithms such as *RIPPER* test the most recently added conjunct first for rule pruning, it is not a strict requirement to do so. It is possible to test the conjuncts for removal in any order, or in greedy fashion, to reduce the pessimistic error rate as much as possible. Rule pruning may result in some of the rules becoming identical. Duplicate rules are removed from the rule set before classification.

6.10.2   Comparing Rule-Based Classifiers to Logical Rules in Expert Systems

The earlier sections of this book introduce rule-based reasoning methods in the way that they are used in expert systems. It is, therefore, natural to compare these rule-based classifiers to the rule-based reasoning systems. Expert systems create logical rules in order to execute inferences from domain knowledge. Therefore, the rules often represent the understanding of the expert, and are contained in the knowledge base. On the other hand, the rule-based classifiers in this section are purely inductive and data-driven systems. One important point is that expert systems cannot grow beyond what the domain expert already knows about a particular setting. On the other hand, provided that sufficient data is available, the rule-based methods in this section can often infer novel insights from the underlying data. However, if the amount of data is small, it makes more sense to work with domain knowledge. In many cases, such rule-based classifiers can be integrated with domain knowledge in order to create an integrated system that can provide robust predictions without losing the insights available in data-driven analysis.

6.11   Evaluation of Classification

Evaluation algorithms are important not only from the perspective of understanding the performance characteristics of a learning algorithm, but also from the point of view of optimizing algorithm performance via *model selection*. Given a particular data set, how can we know which algorithm to use? Should we use a support vector machine or a random forest? Therefore, the notions of model evaluation and model selection are closely intertwined.

Given a labeled data set, one cannot use all of it for model building. This is because the main goal of classification is to *generalize* a model of labeled data to unseen test instances. Therefore, using the same data set for both model building and testing grossly overestimates the accuracy. Furthermore, the portion of the data set used for *model selection* and *parameter tuning* also needs to be different from that used for model building. A common mistake is to use the same data set for both parameter tuning and final evaluation (testing). Such an approach partially mixes the training and test data, and the resulting accuracy is overly optimistic. Given a data set, it should always be divided into three parts.

1. *Training data:* This part of the data is used to build the training model such as a decision tree or a support vector machine. The training data may be used multiple times over different choices of the parameters or completely different algorithms to build the models in multiple ways. This process sets up the stage for *model selection*, in which the best algorithm is selected out of these different models. However, the actual *evaluation* of these algorithms for selecting the best model is not done on the training data but on a separate validation data set to avoid favoring overfitted models.

2. *Validation data:* This part of the data is used for model selection and parameter tuning. For example, the choice of the kernel bandwidth and the regularization parameters may be tuned by constructing the model multiple times on the first part of the data set (i.e., training data), and then using the validation set to estimate the accuracy of these different models. The best choice of the parameters is determined by using this accuracy. In a sense, validation data should be viewed as a kind of test data set to tune the parameters of the algorithm, or to select the best choice of the algorithm (e.g., decision tree versus support vector machine).

3. *Testing data:* This part of the data is used to test the accuracy of the final (tuned) model. It is important that the testing data are not even looked at during the process of parameter tuning and model selection to prevent overfitting. The testing data are *used only once at the very end of the process*. Furthermore, if the analyst uses the results on the test data to adjust the model in some way, then the results will be contaminated with knowledge from the testing data. The idea that one is allowed to look at a test data set only once is an extraordinarily strict requirement (and an important one). Yet, it is frequently violated in real-life benchmarks. The temptation to use what one has learned from the final accuracy evaluation is simply too high.

The division of the labeled data set into training data, validation data, and test data is shown in Figure 6.7. Strictly speaking, the validation data is also a part of the training data, because it influences the final model (although only the model building portion is often referred to as the training data). The division in the ratio of 2:1:1 is quite common. However, it should not be viewed as a strict rule. For very large labeled data sets, one needs only a modest number of examples to estimate accuracy. When a very large data set is available, it makes sense to use as much of it for model building as possible, because the estimation error induced by the validation and evaluation stage is often quite low. A constant number of examples (e.g., less than a few thousand) in the validation and test data sets are sufficient to provide accurate estimates.

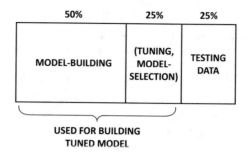

Figure 6.7: Partitioning a labeled data set for evaluation design

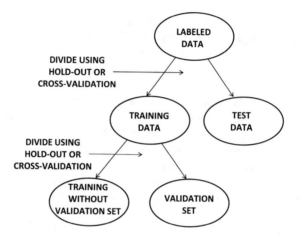

Figure 6.8: Hierarchical division into training, validation, and testing portions

6.11.1   Segmenting into Training and Testing Portions

The aforementioned description of partitioning the labeled data into three segments is an implicit description of a method referred to as *hold-out* for segmenting the labeled data into various portions. However, the division into *three* parts is not done in one shot. Rather, the training data is first divided into *two* parts for training and testing. The testing part is then carefully hidden away from any further analysis *until the very end where it can be used only once.* The remainder of the data set is then divided again into the training and validation portions. This type of recursive division is shown in Figure 6.8.

A key point is that the types of division at both levels of the hierarchy are conceptually identical. In the following, we will consistently use the terminology of the first level of division in Figure 6.8 into "training" and "testing" data, even though the same approach can also be used for the second-level division into model building and validation portions. This consistency in terminology allows us to provide a common description for both levels of the division.

6.11.1.1   Hold-Out

In the hold-out method, a fraction of the instances are used to build the training model. The remaining instances, which are also referred to as the *held out* instances, are used for testing. The accuracy of predicting the labels of the held out instances is then reported as the overall accuracy. Such an approach ensures that the reported accuracy is not a result of overfitting to the specific data set, because different instances are used for training and testing. The approach, however, underestimates the true accuracy. Consider the case where the held-out examples have a higher presence of a particular class than the labeled data set. This means that the held-in examples have a lower average presence of the same class, which will cause a mismatch between the training and test data. Furthermore, the class-wise frequency of the held-in examples will always be inversely related to that of the held-out examples. This will lead to a consistent pessimistic bias in the evaluation.

6.11.1.2 Cross-Validation

In the cross-validation method, the labeled data is divided into q equal segments. One of the q segments is used for testing, and the remaining $(q-1)$ segments are used for training. This process is repeated q times by using each of the q segments as the test set. The average accuracy over the q different test sets is reported. Note that this approach can closely estimate the true accuracy when the value of q is large. A special case is one where q is chosen to be equal to the number of labeled data points and therefore a single data point is used for testing. Since this single data point is left out from the training data, this approach is referred to as *leave-one-out cross-validation*. Although such an approach can closely approximate the accuracy, it is usually too expensive to train the model a large number of times. Nevertheless, leave-one-out cross-validation is the method of choice for lazy learning algorithms like nearest neighbor classifiers.

6.11.2 Absolute Accuracy Measures

Once the data have been segmented between training and testing, a natural question arises about the type of accuracy measure that one can use in classification and regression.

6.11.2.1 Accuracy of Classification

When the output is presented in the form of class labels, the ground-truth labels are compared to the predicted labels to yield the following measures:

1. *Accuracy:* The accuracy is the fraction of test instances in which the predicted value matches the ground-truth value.

2. *Cost-sensitive accuracy:* Not all classes are equally important in all scenarios, while comparing the accuracy. This is particularly important in imbalanced class problems, in which one of the classes is much rarer than the other. For example, consider an application in which it is desirable to classify tumors as *malignant* or *non-malignant* where the former is much rarer than the latter. In such cases, the misclassification of the former is often much less desirable than misclassification of the latter. This is frequently quantified by imposing differential costs $c_1 \ldots c_k$ on the misclassification of the different classes. Let $n_1 \ldots n_k$ be the number of test instances belonging to each class. Furthermore, let $a_1 \ldots a_k$ be the accuracies (expressed as a fraction) on the subset of test instances belonging to each class. Then, the overall accuracy A can be computed as a weighted combination of the accuracies over the individual labels.

$$A = \frac{\sum_{i=1}^{k} c_i n_i a_i}{\sum_{i=1}^{k} c_i n_i} \tag{6.41}$$

The cost sensitive accuracy is the same as the unweighted accuracy when all costs $c_1 \ldots c_k$ are the same.

Aside from the accuracy, the statistical robustness of a model is also an important issue. For example, if two classifiers are trained over a small number of test instances and compared, the difference in accuracy may be a result of random variations, rather than a truly *statistically significant* difference between the two classifiers. This measure is related to that of the variance of a classifier that was discussed earlier in this chapter. When the variance of two classifiers is high, it is often difficult to assess whether one is truly better than the other. One way of testing the robustness is to repeat the aforementioned process of cross-validation (or hold-out) in many different ways (or *trials*) by repeating the randomized process of creating the folds in many different ways. The difference δa_i in accuracy between the ith pair of classifiers (constructed on the same folds) is computed, and the standard deviation σ of this difference is computed as well. The overall difference in accuracy over s trials is computed as follows:

$$\Delta A = \frac{\sum_{i=1}^{s} \delta a_i}{s} \tag{6.42}$$

Note that ΔA might be positive or negative, depending on which classifier is winning. The standard deviation is computed as follows:

$$\sigma = \sqrt{\frac{\sum_{i=1}^{s}(\delta a_i - \Delta A)^2}{s-1}} \tag{6.43}$$

Then, the overall statistical level of significance by which one classifier wins over the other is given by the following:

$$Z = \frac{\Delta A \sqrt{s}}{\sigma} \tag{6.44}$$

The factor $\sqrt{s}$ accounts for the fact that we are using the sample mean ΔA, which is more stable that the individual accuracy differences δa_i. The standard deviation of ΔA is a factor $1/\sqrt{s}$ of the standard deviation of individual accuracy differences. Values of Z that are significantly greater than 3, are strongly indicative of one classifier being better than the other in a statistically significant way.

6.11.2.2   Accuracy of Regression

The effectiveness of linear regression models can be evaluated with a measure known as the Mean Squared Error (MSE), or the Root Mean Squared Error, which is the RMSE. Let $y_1 \ldots y_r$ be the observed values over r test instances, and let $\hat{y}_1 \ldots \hat{y}_r$ be the predicted values. Then, the mean-squared error, denoted by MSE is defined as follows:

$$MSE = \frac{\sum_{i=1}^{r}(y_i - \hat{y}_i)^2}{r} \tag{6.45}$$

The Root-Mean-Squared Error (RMSE) is defined as the square root of this value:

$$RMSE = \sqrt{\frac{\sum_{i=1}^{r}(y_i - \hat{y}_i)^2}{r}} \tag{6.46}$$

Another measure is the R^2-*statistic*, or the *coefficient of determination*, which provides a better idea of the *relative performance* of a particular model. In order to compute the R^2-statistic, we first compute the variance σ^2 of the observed values. Let $\mu = \sum_{j=1}^{r} y_j / r$ be the mean of the dependent variable. Then, the variance σ^2 of the r observed values of the test instances is computed as follows:

$$\sigma^2 = \frac{\sum_{i=1}^{r}(y_i - \mu)^2}{r} \tag{6.47}$$

Then, the R^2-statistic is as follows:

$$R^2 = 1 - \frac{MSE}{\sigma^2} \tag{6.48}$$

Larger values of the R^2 statistic are desirable, and the maximum value of 1 corresponds to an MSE of 0. It is possible for the R^2-statistic to be negative, when it is applied on an out-of-sample test data set, or even when it is used in conjunction with a nonlinear model. Although we have described the computation of the R^2-statistic for the test data, this measure is often used on the training data in order to compute the fraction of unexplained variance in the model. In such cases, linear regression models always return an R^2-statistic in the range $(0, 1)$. This is because the mean value μ of the dependent variable in the training data can be predicted by a linear regression model, when the coefficients of the features are set to 0 and only the bias term (or coefficient of dummy column) is set to the mean. Since the linear regression model will always provide a solution with a lower objective function value on the training data, it follows that the value of MSE is no larger than σ^2. As a result, the value of the R^2-statistic on the training data always lies in the range $(0, 1)$. In other words, a training data set can never be predicted better using its mean than by using the predictions of linear regression. However, an out-of-sample test data set *can* be modeled better by using its mean than by using the predictions of linear regression.

One can increase the R^2-statistic on the training data simply by increasing the number of regressors, as the MSE reduces with increased overfitting. When the dimensionality is large, and it is desirable to compute the R^2-statistic on the training data, the adjusted R^2-statistic provides a more accurate measure. In such cases, the use of a larger number of features for regression is penalized. The adjusted R^2-statistic for a training data set with n data points and d dimensions is computed as follows:

$$R^2 = 1 - \frac{(n-d)}{(n-1)}\frac{MSE}{\sigma^2} \tag{6.49}$$

The R^2-statistic is generally used only for linear models. For nonlinear models, it is more common to use the MSE as a measure of the error.

6.11.3 Ranking Measures

The classification problem is posed in different ways, depending on the setting in which it is used. The absolute accuracy measures discussed in the previous section are useful in cases where the labels or numerical dependent variables are predicted as the final output. However, in some settings, a particular *target* class is of special interest, and all the test instances are *ranked* in order of their propensity to belong to the target class. A particular example is that of classifying email as "*spam*" or "*not spam.*" When one has a large number of data points with a high imbalance in relative proportion of classes, it makes little sense to directly return binary predictions. In such cases, only the top-ranked emails will be returned based on the probability of belonging to the "*spam*" category, which is the target class. Ranking-based evaluation measures are often used in imbalanced class settings in

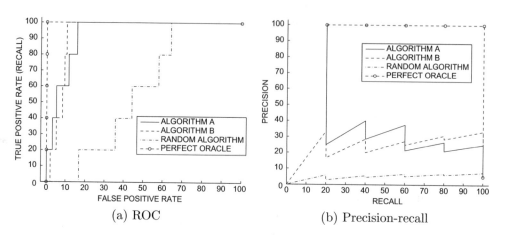

(a) ROC   (b) Precision-recall

Figure 6.9: ROC curve and precision-recall curves

Table 6.2: Rank of ground-truth positive instances

Algorithm	Rank of ground-truth positives (ground-truth positives)
Algorithm A	1, 5, 8, 15, 20
Algorithm B	3, 7, 11, 13, 15
Random Algorithm	17, 36, 45, 59, 66
Perfect Oracle	1, 2, 3, 4, 5

which one of the classes (i.e., the rare class) is considered more relevant from a detection point of view. Discussions of some of these different ranking measures is also provided in different contexts [5, 7].

6.11.3.1   Receiver Operating Characteristic

Ranking methods are used frequently in cases where a ranked list of a particular class of interest is returned. The ground-truth is assumed to be binary in which the class of interest corresponds to the positive class, and the remaining data points belong to the negative class. In most such settings, the relative frequencies of the two classes are heavily imbalanced, so that the discovery of (rare) positive class instances is more desirable.

The instances that belong to the positive class in the *observed data* are *ground-truth positives* or *true positives*. It is noteworthy that when information retrieval, search, or classification applications are used, the algorithm can *predict* any number of instances as positives, which might be different from the number of *observed* positives (i.e., true positives). When a larger number of instances are predicted as positives, one would recover a larger number of the true positives, but a smaller percentage of the predicted list would be correct. This type of trade-off can be visualized with the use of a precision-recall or a *receiver operating characteristic (ROC)* curve. Such trade-off plots are commonly used in rare class detection, outlier analysis evaluation, recommender systems, and information retrieval. In fact, such trade-off plots can be used in any application where a binary ground truth is compared to a ranked list discovered by an algorithm.

The basic assumption is that it is possible to rank all the test instances using a numerical score, which is the output of the algorithm at hand. This numerical score is often available from classification algorithms in the form of a probability of belonging to the positive class in methods like the naïve Bayes classifier or logistic regression. For methods like SVMs, one can report the (signed) distance of a point from the separating class instead of converting it into a binary prediction. A threshold on the numerical score creates a predicted list of positives. By varying the threshold (i.e., size of predicted list), one can quantify the fraction of relevant (ground-truth positive) instances in the list, and the fraction of relevant instances that are missed by the list. If the predicted list is too small, the algorithm will miss relevant instances (false-negatives). On the other hand, if a very large list is recommended, there will be too many spuriously predicted instances (i.e., false-positives). This leads to a trade-off between the false-positives and false-negatives, which can be visualized with the *precision-recall* curve or the *receiver operating characteristic (ROC)* curve.

Assume that one selects the top-t set of ranked instances and predicted them to belong to the positive class. For any given value t of the size of the positively predicted list, the set of instances predicted to belong to the positive class is denoted by $S(t)$. Note that $|S(t)| = t$. Therefore, as t changes, the size of $S(t)$ changes as well. Let G represent the true set of relevant data points (ground-truth positives). Then, for any given size t of the predicted list, the *precision* is defined as the percentage of percentage of instances predicted to belong to the positive class that truly turn out to belong to the positive class in the predicted labels:

$$Precision(t) = 100 \cdot \frac{|S(t) \cap G|}{|S(t)|}$$

The value of $Precision(t)$ is *not* necessarily monotonic in t because both the numerator and denominator may change with t differently. The *recall* is correspondingly defined as the percentage of *ground-truth* positives that have been recommended as positive for a list of size t.

$$Recall(t) = 100 \cdot \frac{|S(t) \cap G|}{|G|}$$

While a natural trade-off exists between precision and recall, this trade-off is not necessarily monotonic. In other words, an increase in recall does not always lead to a reduction in precision. One way of creating a single measure that summarizes both precision and recall is the F_1-measure, which is the harmonic mean between the precision and the recall.

$$F_1(t) = \frac{2 \cdot Precision(t) \cdot Recall(t)}{Precision(t) + Recall(t)} \tag{6.50}$$

While the $F_1(t)$ measure provides a better quantification than either precision or recall, it is still dependent on the size t of the number of instances predicted to belong to the positive class, and is therefore still not a complete representation of the trade-off between precision and recall. It is possible to visually examine the entire trade-off between precision and recall by varying the value of t and plotting the precision versus the recall. The lack of monotonicity of the precision makes the results hard to interpret.

A second way of generating the trade-off in a more intuitive way is through the use of the ROC curve. The *true-positive rate*, which is the same as the recall, is defined as the percentage of ground-truth positives that have been included in the predicted list of size t.

$$TPR(t) = Recall(t) = 100 \cdot \frac{|S(t) \cap G|}{|G|}$$

The false-positive rate $FPR(t)$ is the percentage of the falsely reported positives in the predicted list out of the ground-truth negatives (i.e., irrelevant data points belonging to the negative class in the observed labels). Therefore, if $\mathcal{U}$ represents the universe of all test instances, the ground-truth negative set is given by $(\mathcal{U} - \mathcal{G})$, and the falsely reported part in the predicted list is $(\mathcal{S}(t) - \mathcal{G})$. Therefore, the false-positive rate is defined as follows:

$$FPR(t) = 100 \cdot \frac{|\mathcal{S}(t) - \mathcal{G}|}{|\mathcal{U} - \mathcal{G}|} \qquad (6.51)$$

The false-positive rate can be viewed as a kind of "bad" recall, in which the fraction of the ground-truth negatives (i.e., test instances with observed labels in the negative class), which are incorrectly captured in the predicted list $\mathcal{S}(t)$, is reported. The ROC curve is defined by plotting the $FPR(t)$ on the X-axis and $TPR(t)$ on the Y-axis for varying values of t. In other words, the ROC curve plots the "good" recall against the "bad" recall. Note that both forms of recall will be at 100% when $\mathcal{S}(t)$ is set to the entire universe of test data points (or entire universe of data points to return in response to a query). Therefore, the end points of the ROC curve are always at $(0,0)$ and $(100, 100)$, and a random method is expected to exhibit performance along the diagonal line connecting these points. The *lift* obtained above this diagonal line provides an idea of the accuracy of the approach. The area under the ROC curve provides a concrete quantitative evaluation of the effectiveness of a particular method. Although one can directly use the area shown in Figure 6.9(a), the staircase-like ROC curve is often modified to use local linear segments which are not parallel to either the X-axis or the Y-axis. The area of the resulting trapezoids [52] is then used to compute the area slightly more accurately. From a practical point of view, this change often makes very little difference to the final computation.

To illustrate the insights gained from these different graphical representations, consider an example of a scenario with 100 test instances, in which 5 data points truly belong to the positive class. Two algorithms A and B are applied to this data set that rank all test instances from 1 to 100 to belong to the positive class, with lower ranks being selected first in the predicted list. Thus, the true-positive rate and false-positive rate values can be generated from the ranks of the 5 test instances in the positive class. In Table 6.2, some hypothetical ranks for the 5 truly positive instances have been illustrated for the different algorithms. In addition, the ranks of the ground-truth positive instances for a random algorithm have been indicated. This algorithm ranks all the test instances randomly. Similarly, the ranks for a "perfect oracle" algorithm are such that the correct positive instances are placed as the top 5 instances in the ranked list. The resulting ROC curves are illustrated in Figure 6.9(a). The corresponding precision-recall curves are illustrated in Figure 6.9(b). Note that the ROC curves are always increasing monotonically, whereas the precision-recall curves are not monotonic. While the precision-recall curves are not quite as nicely interpretable as the ROC curves, it is easy to see that the *relative trends* between different algorithms are the same in both cases. In general, ROC curves are used more frequently because of greater ease in interpretability.

What do these curves really tell us? For cases in which one curve strictly dominates another, it is clear that the algorithm for the former curve is superior. For example, it is immediately evident that the oracle algorithm is superior to all algorithms and that the random algorithm is inferior to all the other algorithms. On the other hand, algorithms A and B show domination at different parts of the ROC curve. In such cases, it is hard to say that one algorithm is strictly superior. From Table 6.2, it is clear that Algorithm A ranks three positive instances very highly, but the remaining two positive instances are ranked poorly. In the case of Algorithm B, the highest ranked positive instances are not as well ranked as Algorithm A, though all 5 positive instances are determined much earlier in terms of rank threshold. Correspondingly, Algorithm A dominates on the earlier part of the ROC curve, whereas Algorithm B dominates on the later part. It is possible to use the area under the ROC curve as a proxy for the overall effectiveness of the algorithm. However, not all parts of the ROC curve are equally important because there are usually practical limits on the size of the predicted list.

Interpretation of Area Under ROC Curve

The area under the ROC curve has a neat interpretation. Consider a situation where one has two classes, and we sample one instance randomly from each class. Then, a perfect classifier will always score the truly positive instance to belong to the positive class higher than the score of the truly negative instance to belong to the negative score. The area under the ROC curve is simply the fraction of the time that the correct ordering of the instances is maintained by the algorithm. In a sense, the area under the ROC curve is an indirect measure of the classification accuracy, although it is done after a particular type of pairwise sampling of instances from the two classes (irrespective of the relative presence of the instances belonging to each class).

6.12 Summary

This chapter provides the inductive learning view of artificial intelligence, wherein data-driven methods are used in order to learn models and perform prediction. Inductive learning methods have an advantage over deductive reasoning methods, because they can often learn non-obvious conclusions from the data, which are not easily expressible in an interpretable way. Deductive reasoning methods often require the incorporation of interpretable statements in the knowledge base in order to perform inferences. Not all predictions in machine learning can be performed in an interpretable way, simply because many of the choices that humans make (e.g., driving a car) are not easily expressible in terms of interpretable choices. Numerous machine learning methods have been proposed in the literature for different types of data, such as numerical data and categorical data. This chapter discusses a number of optimization methods such as linear regression, least-squares classification, support vector machine, logistic regression, Bayes classifier, nearest neighbor methods, decision trees, and rule-based methods. In addition, various techniques for evaluation of classification and regression-based methods were discussed. In the subsequent chapters, recent advancements such as deep learning methods, will be discussed as well.

6.13 Further Reading

The design of classification methods for machine learning is discussed in several books [3, 4, 20, 71]. A classical discussion of machine learning algorithms may be found in [43, 48]. A discussion of linear algebra and optimization methods for machine learning is provided in [8]. The artificial intelligence book by [153] also provides a discussion of inductive learning methods from the point of view of artificial intelligence. Discussions of nearest neighbor classifiers may be found in [48]. The C4.5 algorithm was introduced in [144], and random forests are discussed in [29]. The sequential covering algorithm of this chapter is discussed in the context of the RIPPER algorithm [38, 39].

6.14 Exercises

1. Discuss why any linear classifier is a special case of a rule-based classifier.

2. The text of the chapter introduces the loss function of the L_2-loss SVM, but it does not discuss the update used by stochastic gradient descent. Derive the stochastic gradient descent update for the L_2-loss SVM.

3. The goal of this exercise is to show that the stochastic gradient-descent updates for various machine learning models are closely related. The updates for the least-squares classification, SVM, and logistic-regression models can be expressed in a unified way in terms of a model-specific *mistake function* $\delta(\overline{X}_i, y_i)$ for the training pair $(\overline{X}_i, y_i)$ at hand. In particular, show that the stochastic gradient-descent updates of all three algorithms are of the following form:

$$\overline{W} \Leftarrow \overline{W}(1 - \alpha\lambda) + \alpha y_i[\delta(\overline{X}_i, y_i)]\overline{X}_i^T \tag{6.52}$$

Derive the form of the mistake function $\delta(\overline{X}_i, y_i)$ for least-squares classification, hinge-loss SVMs, and logistic regression.

4. The L_1-loss regression model uses a modified loss function in which the L_1-norm of the error is used to create the objective function (rather than the squared norm). Derive the stochastic gradient-descent updates of L_1-loss regression.

5. **Hinge-loss without margin:** Suppose that we modified the hinge-loss on page 179 by removing the constant value within the maximization function as follows:

$$J = \sum_{i=1}^{n} \max\{0, (-y_i[\overline{W} \cdot \overline{X}_i^T])\} + \frac{\lambda}{2}\|\overline{W}\|^2$$

This loss function is referred to as the *perceptron criterion*. Derive the stochastic gradient descent updates for this loss function.

6. Discuss how you can generate rules from a decision tree. Comment on the relationship between decision trees and nearest neighbor classifiers.

Chapter 7

Neural Networks

"When we talk mathematics, we may be discussing a secondary language built on the primary language of the nervous system."– John von Neumann

7.1 Introduction

The previous chapter discusses a number of machine learning algorithms that learn functions by first deciding a general form of a prediction function and then parameterizing them with weights. The goal is to *learn* the target variable y_i as a function of the ith data point $\overline{X}_i$ (which is a d-dimensional vector):

$$y_i \approx f(\overline{X}_i)$$

The target variable may be either categorical or numerical It is possible for the function $f(\cdot)$ to take on a highly complicated and difficult-to-interpret form, especially in the case of neural network and deep learning models. The function $f(\overline{X}_i)$ is often parameterized with a weight vector $\overline{W}$. In traditional machine learning, the nature of the function $f(\overline{X}_i)$ is relatively simple and easy to understand. An example is the problem of linear regression, in which we create a linear prediction function of $\overline{X}_i$ in order to predict the numeric variable y_i:

$$y_i \approx f_{\overline{W}}(\overline{X}_i) = \overline{W} \cdot \overline{X}_i \tag{7.1}$$

The nature of the prediction depends on the representation $\overline{X}_i$ and the target variable y_i. Another example of a prediction function is that of binary classification of feature vector $\overline{X}_i$ into the labels $\{-1, +1\}$:

$$y_i \approx f_{\overline{W}}(\overline{X}_i) = \text{sign}\{\overline{W} \cdot \overline{X}_i\} \tag{7.2}$$

In each of the cases of regression and classification, we have added a subscript to the function to indicate its parametrization. The parameter vector $\overline{W}$ heavily controls the nature of the prediction function, and the parameter vector $\overline{W}$ needs to be learned in order to penalize any kind of mismatching between the *observed value* y_i and the predicted value $f(\overline{X}_i)$ with

© Springer Nature Switzerland AG 2021
C. C. Aggarwal, *Artificial Intelligence*, https://doi.org/10.1007/978-3-030-72357-6_7

the use of a carefully constructed *loss function*. Therefore, many of the machine learning models reduce to the following optimization problem:

$$\text{Minimize}_{\overline{W}} \sum_i \text{Mismatching between } y_i \text{ and } f_{\overline{W}}(\overline{X}_i)$$

Once the weight vector $\overline{W}$ has been computed by solving the optimization model, it is used to predict the value of the target variable y_i for instances in which the class variable is not known. In the case of classification, the loss function is often applied on a continuous relaxation of $f(\overline{W} \cdot \overline{X}_i)$ in order to enable to use of differential calculus for optimization. In other words, the sign function is not used in the loss function. Examples of such loss functions include the least-squares classification loss, the SVM hinge loss, and the logistic loss (for logistic regression). A gradient descent algorithm is then used in order to perform the optimization. Several examples of such gradient-descent algorithms, such as SVM gradient descent and logistic regression, are discussed in the previous chapter.

Neural networks represent a natural generalization of the optimization ideas that we have already seen in the previous chapters. In these cases, the modeled functions $f_{\overline{W}}(\cdot)$ are represented as *computational graphs*, in which the input nodes contain the argument of the function and the output nodes contain the output of the function. However, intermediate nodes might also be available to compute intermediate values in cases where the overall function is a complex composition of simpler functions, such as the following:

$$f_{\overline{W}}(x_1, x_2, x_3) = F_{\overline{w}}(G_{\overline{u}}(x_1, x_2, x_3), H_{\overline{v}}(x_1, x_2, x_3))$$

Here, the functions $G(\cdot)$ and $H(\cdot)$ are computed at intermediate nodes, and the overall parameter vector $\overline{W}$ is obtained as the concatenation of node-specific parameters $\overline{u}$, $\overline{v}$, and $\overline{w}$. It is also immediately noticeable that this type of prediction function is much more complex than any of the prediction functions we have seen in earlier chapters. The functions computed in intermediate nodes are often nonlinear functions, which create more complex features for downstream nodes. The ability to create such complex prediction functions with the use of the computational graph abstraction is how neural networks gain their power. In fact, neural networks are referred to as *universal function approximators*, which can model any prediction function accurately, given sufficient data.

The parameter vector $\overline{W}$ corresponds to parameters associated with edges, and they have a direct effect on the functions computed at nodes connected to these edges. These parameters are learned in a *data-driven* manner so that variables in the nodes mirror relationships among attribute values in data instances. Each data instance contains both *input* and *target* attributes. The variables in a subset of the *input* nodes are fixed to input attribute values in data instances, whereas the variables in all other nodes are *computed* using the node-specific functions. The variables in some of the computed nodes are compared to observed *target* values in data instances, and edge-specific parameters are modified to match the observed and computed values as closely as possible. By learning the parameters along the edges in a data-driven manner, one can learn a function relating the input and target attributes in the data.

A *feed-forward neural network* is an important special case of this type of computational graph. The inputs often correspond to the features in each data point, whereas the output nodes might correspond to the target variables (e.g., class variable or regressand). The optimization problem is defined over the edge parameters so that the predicted variables match the observed values in the corresponding nodes as closely as possible. This is similar to the process of learning the parameter vector $\overline{W}$ in the prediction function $f_{\overline{W}}(\cdot)$ in linear regression. In order to achieve the goal of learning $\overline{W}$, the *loss function* of a computational graph might penalize differences between predicted and observed values at the output nodes of the computational graph. In computational graphs with continuous variables, one can use gradient descent for optimization. *Many machine learning algorithms based on continuous optimization, such as linear regression, logistic regression, and SVMs, can be modeled as directed acyclic computational graphs with continuous variables.*

This chapter is organized as follows. The next section will introduce the basics of computational graphs.

7.2 An Introduction to Computational Graphs

Since neural networks build upon computational graphs, we will start by providing an introduction to the basics of computational graphs. A computational graph is typically a directed acyclic graph (i.e., graph without cycles) in which each node performs a computation on its inputs in order to create outputs, which feed forward into down stream nodes. This creates a successive composition of functions, resulting in a more complicated composition function being computed by the graph. A directed acyclic computational graph is defined as follows:

Definition 7.2.1 (Directed Acyclic Computational Graph) *A directed acyclic computational graph contains nodes, so that each node is associated with a variable. A set of directed edges connect nodes, which indicate functional relationships among nodes. Edges* **might** *be associated with learnable parameters. A variable in a node is either fixed externally (for input nodes with no incoming edges), or it is computed as a function of the variables in the tail ends of edges incoming into the node and the learnable parameters on the incoming edges.*

A directed acyclic computational graph contains three types of nodes, which are the input, output, and hidden nodes. The input nodes contain the external inputs to the computational graph, and the output node(s) contain(s) the final output(s). The hidden nodes contain intermediate values. Each hidden and output node computes a relatively simple **local** function of its incoming node variables. When there are many input nodes (e.g., regressors) and a single output node (e.g., regressand), one is computing a vector-to-scalar function. On the other hand, if there are multiple input and output nodes, one is computing a vector-to-vector function using the computational graph. Such a situation arises often in multiclass or multilabel learning applications. The cascading effect of the computations over the whole graph implicitly defines a **global** function from input to output nodes. The variable in each input node is fixed to an externally specified input value. Therefore, no function is computed at an input node. The node-specific functions also use parameters associated with their incoming edges, and the inputs along those edges are scaled with the weights. By choosing weights appropriately, one can control the (global) function defined by the computational graph. This global function is often *learned* by feeding the computational graph input-output pairs (training data) and adjusting the weights so that predicted outputs matched observed outputs. Interestingly, all the continuous optimization models discussed in the previous chapter can be modeled by using an appropriate choice of a computational graph.

An example of a computational graph with two weighted edges is provided in Figure 7.1. This graph has three inputs, denoted by x_1, x_2, and x_3. Two of the edges have weights w_2 and w_3. Other than the input nodes, all nodes perform a computation such as addition, multiplication, or evaluating a function like the logarithm. In the case of weighted edges,

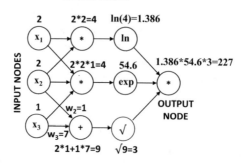

Figure 7.1: Examples of computational graph with two weighted edges

the values at the tail of the edge are scaled with the weights before computing the node-specific function. The graph has a single output node, and computations are cascaded in the forward direction from the input to the output. For example, if the weights w_2 and w_3 are chosen to be 1 and 7, respectively, the global function $f(x_1, x_2, x_3)$ is as follows:

$$f(x_1, x_2, x_2) = \ln(x_1 x_2) \cdot \exp(x_1 x_2 x_3) \cdot \sqrt{x_2 + 7 x_3}$$

For $[x_1, x_2, x_3] = [2, 2, 1]$, the cascading sequence of computations is shown in the figure with a final output value of approximately 227.1. However, if the *observed* value of the output is only 100, it means that the weights need to be readjusted to change the computed function. In this case, one can observe from inspection of the computational graph that reducing either w_2 or w_3 will help reduce the output value. For example, if we change the weight w_3 to -1, while keeping $w_2 = 1$, the computed function becomes the following:

$$f(x_1, x_2, x_2) = \ln(x_1 x_2) \cdot \exp(x_1 x_2 x_3) \cdot \sqrt{x_2 - x_3}$$

In this case, for the same set of inputs $[x_1, x_2, x_3] = [2, 2, 1]$, the computed output becomes 75.7, which is much closer to the true output value of 100. Therefore, it is clear that one must use the mismatch of predicted values with observed outputs to adjust the computational function, so that there is a better matching between predicted and observed outputs across the data set. Although we adjusted w_3 here by inspection, such an approach will not work in very large computational graphs containing millions of weights.

The goal in machine learning is to learn parameters (like weights) using examples of input-output pairs, while adjusting weights with the help of the observed data. The key point is to convert the problem of adjusting weights into an optimization problem. The computational graph may be associated with a loss function, which typically penalizes the differences in the *predicted* outputs from *observed* outputs, and adjusts weights accordingly. Since the outputs are functions of inputs and edge-specific parameters, the loss function can also be viewed as a complex function of the inputs and edge-specific parameters. The goal of learning the parameters is to minimize the loss, so that the input-output pairs in the computational graph mimic the input-output pairs in the observed data. It should be immediately evident that the problem of learning the weights is likely to challenging, if the underlying computational graph is large with a complex topology.

The choice of loss function depends on the application at hand. For example, one can model least-squares regression by using as many input nodes as the number of input variables (regressors), and a single output node containing the predicted regressand. Directed edges exist from each input node to this output node, and the parameter on each such edge

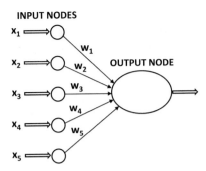

Figure 7.2: A single-layer computational graph that can perform linear regression

corresponds to the weight associated with that input variable (cf. Figure 7.2). The output node computes the following function of the variables $x_1 \ldots x_d$ in the d input nodes:

$$\hat{o} = f(x_1, x_2, \ldots, x_d) = \sum_{i=1}^{d} w_i x_i$$

If the observed regressand is o, then the loss function simply computes $(o - \hat{o})^2$, and adjusts the weights $w_1 \ldots w_d$ so as to reduce this value. Typically, the derivative of the loss is computed with respect to each weight in the computational graph, and the weights are updated by using this derivative. One processes each training point one-by-one and updates the weights. *The resulting algorithm is identical to using stochastic gradient descent in the linear regression problem.* In fact, by changing the nature of the loss function at the output node, it is possible to model both logistic regression and the support vector machine. In particular, the computational graphs and the loss functions in the two cases are as follows:

- **Logistic Regression with Computational Graph:** We assume that o is an observed **binary** class label drawn from $\{-1, +1\}$ and $\hat{o}$ be the predicted **real** value by the neural architecture of Figure 7.2. Then, the loss function $\log(1 + \exp(-o\hat{o}))$ yields the same loss function for each data instance as the logistic regression model in Equation 6.19 of Chapter 6. The only difference is that the regularization term is not used in this case.

- **SVM with Computational Graph:** We assume that o is an observed **binary** class label drawn from $\{-1, +1\}$ and $\hat{o}$ is the predicted **real** value by the neural architecture of Figure 7.2. Then, the loss function $\max\{0, 1 - o\hat{o}\}$ yields the same loss function for each data instance as the L_1-loss SVM in Equation 6.16 of Chapter 6.

In the particular case of Figure 7.2, the choice of a computational graph for model representation does not seem to be useful because a single computational node is rather rudimentary for model representation. As shown in Chapter 6, one can directly compute gradients of the loss function with respect to the weights without worrying about computational graphs. The main usefulness of computational graphs is realized when the topology of computation is more complex. In such cases, the loss function becomes more complex, and computing gradients in a straightforward way becomes more difficult.

The nodes in the directed acyclic graph of Figure 7.2 are arranged in *layers*, because all paths from an input node to any node in the network have the same length. This type of architecture is common in computational graphs. Nodes that are reachable by a path of a particular length i from input nodes are assumed to belong to layer i. At first glance, Figure 7.2 looks like a two-layer network. However, such networks are considered single-layer networks, because the non-computational input layer is not counted among the number of layers. Any computational graph with two or more layers is referred to as a *multilayer network*, and the nature of the function computed by it is always a composition of the (simpler) functions computed in individual nodes. In such cases, it no longer makes sense to compute the gradients with respect to weights with the use of closed-form expressions (as in Chapter 6). Rather, the structure of the computational graph is used in a systematic way in order to compute gradients. This is the essence of the idea behind the backpropagation algorithm, which is discussed in later sections.

7.2.1   Neural Networks as Directed Computational Graphs

The real power of computational graphs is realized when one uses multiple layers of nodes. Neural networks represent the most common use case of a multi-layer computational graph. The nodes are (typically) arranged in layerwise fashion, so that all nodes in layer-i are connected to nodes in layer-$(i + 1)$ (and no other layer). The vector of variables in each layer can be written as a vector-to-vector function of the variables in the previous layer. A pictorial illustration of a multilayer neural network is shown in Figure 7.3(a). In this case, the network contains three computational layers in addition to the input layer. The input layer only transmits values and does not perform any computation. The final layer whose outputs are visible to the user and can be compared to an observed (target) value is referred to as the *output layer*. For example, consider the first hidden layer with output values $h_{11} \ldots h_{1r} \ldots h_{1,p_1}$, which can be computed as a function of the input nodes with variables $x_1 \ldots x_d$ in the input layer as follows:

$$h_{1r} = \Phi(\sum_{i=1}^{d} w_{ir} x_i) \quad \forall r \in \{1, \ldots, p_1\}$$

The value p_1 represents the number of nodes in the first hidden layer. Here, the function $\Phi(\cdot)$ is referred to as an *activation* function. The final numerical value of the variable in a particular node (i.e., h_{1r} in this case) for a particular input is also sometimes referred to as its *activation* for that input. In the case of linear regression, the activation function is missing, which is also referred to as using the *identity* activation function or *linear* activation function. However, computational graphs primarily gain better expressive power by using nonlinear activation functions such as the following:

$$\Phi(v) = \frac{1}{1 + e^{-v}} \qquad\qquad\qquad \text{[Sigmoid function]}$$
$$\Phi(v) = \frac{e^{2v} - 1}{e^{2v} + 1} \qquad\qquad\qquad \text{[Tanh function]}$$
$$\Phi(v) = \max\{v, 0\} \qquad\qquad \text{[ReLU: Rectified Linear Unit]}$$
$$\Phi(v) = \max\{\min[v, 1], -1\} \qquad\qquad \text{[Hard tanh]}$$

It is noteworthy that these functions are nonlinear, and nonlinearity is essential for greater expressive power of networks with increased depth. Networks containing only linear activation functions are not any more powerful than single-layer networks.

In order to understand this point, consider a two-layer computational graph (not counting the input layer) with 4-dimensional input vector $\overline{x}$, 3-dimensional hidden-layer vector

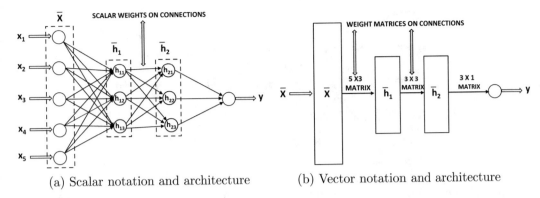

(a) Scalar notation and architecture   (b) Vector notation and architecture

Figure 7.3: A feed-forward network with two hidden layers and a single output layer.

$\overline{h}$, and 2-dimensional output-layer vector $\overline{o}$. Note that we are creating a column vector from the node variables in each layer. Let W_1 and W_2 be two matrices of sizes 3×4 and 2×3 so that $\overline{h} = W_1\overline{x}$ and $\overline{o} = W_2\overline{h}$. The matrices W_1 and W_2 contain the weight parameters of each layer. Note that one can express $\overline{o}$ directly in terms of $\overline{x}$ without using $\overline{h}$ as $\overline{o} = W_2W_1\overline{x} = (W_2W_1)\overline{x}$. One can replace the matrix W_2W_1 with a single 2×4 matrix W without any loss of expressive power. In other words, this is a single-layer network! It is not possible to use this type of approach to (easily) eliminate the hidden layer in the case of nonlinear activation functions without creating extremely complex functions at individual nodes (thereby increasing node-specific complexity). This means that increased depth results in increased complexity only when using nonlinear activation functions.

In the case of Figure 7.3(a), the neural network contains three layers. Note that the input layer is often not counted, because it simply transmits the data and no computation is performed in that layer. If a neural network contains $p_1 \ldots p_k$ units in each of its k layers, then the (column) vector representations of these outputs, denoted by $\overline{h}_1 \ldots \overline{h}_k$ have dimensionalities $p_1 \ldots p_k$. Therefore, the number of units in each layer is referred to as the *dimensionality* of that layer. It is also possible to create a computational graph in which the variables in nodes are vectors, and the connections represent vector-to-vector functions. Figure 7.3(b) creates a computational graph in which the nodes are represented by *rectangles* rather than *circles*. Rectangular representations of nodes correspond to nodes containing vectors. The connections now contain matrices. The sizes of the corresponding *connection* matrices are shown in Figure 7.3(b). For example, if the input layer contains 5 nodes and the first hidden layer contains 3 nodes, the *connection matrix* is of size 5×3. However, as we will see later, the *weight* matrix has size that is the transpose of the connection matrix (i.e., 3×5) in order to facilitate matrix operations. Note that the computational graph in the vector notation has a simpler structure, where the entire network contains only a single path. The weights of the connections between the input layer and the first hidden layer are contained in a *matrix* W_1 with size $p_1 \times d$, whereas the weights between the rth hidden layer and the $(r+1)$th hidden layer are denoted by the $p_{r+1} \times p_r$ matrix denoted by W_r. If the output layer contains s nodes, then the final matrix W_{k+1} is of size $s \times p_k$. The final layer of nodes is referred to as the *output layer*. Note that the weight matrix has transposed dimensions with respect to the connection matrix. The d-dimensional input

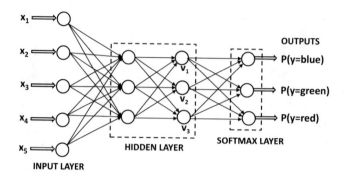

Figure 7.4: An example of multiple outputs for categorical classification with the use of a softmax layer

vector $\overline{x}$ is transformed into the outputs using the following recursive equations:

$$\overline{h}_1 = \Phi(W_1\overline{x}) \qquad\qquad \text{[Input to Hidden Layer]}$$
$$\overline{h}_{p+1} = \Phi(W_{p+1}\overline{h}_p) \ \forall p \in \{1\ldots k-1\} \qquad \text{[Hidden to Hidden Layer]}$$
$$\overline{o} = \Phi(W_{k+1}\overline{h}_k) \qquad\qquad \text{[Hidden to Output Layer]}$$

Here, the activation functions are applied in *element-wise* fashion to their vector arguments. Here, it is noteworthy that the final output is a **recursively nested composition function of the inputs**, which is as follows:

$$\overline{o} = \Phi(W_{k+1}(\Phi(W_k\Phi(W_{k-1}\ldots))))$$

This type of neural network is harder to train than single-layer networks because one must compute the derivative of a **nested** composition function with respect to each weight. In particular, the weights of earlier layers lie inside the recursive nesting, and are harder to learn with gradient descent, because the methodology for computation of the gradient of weights in the inner portions of the nesting (i.e., earlier layers) is not obvious, especially when the computational graph has a complex topology. It is also noticeable that the **global** input-to-output function computed by the neural network is harder to express in closed form neatly. The recursive nesting makes the closed-form representation look extremely cumbersome. A cumbersome closed-form representation causes challenges in derivative computation for parameter learning.

7.2.2   Softmax Activation Function

The softmax activation function is unique in that it is *almost always used in the output layer* to map k real values into k probabilities of discrete events. For example, consider the k-way classification problem in which each data record needs to be mapped to one of k unordered class labels. In such cases, k output values can be used, with a *softmax activation function* with respect to k real-valued outputs $\overline{v} = [v_1, \ldots, v_k]$ at the nodes in a given layer. This activation function maps real values to probabilities that sum to 1. Specifically, the activation function for the ith output is defined as follows:

$$\Phi(\overline{v})_i = \frac{\exp(v_i)}{\sum_{j=1}^{k} \exp(v_j)} \quad \forall i \in \{1, \ldots, k\} \tag{7.3}$$

An example of the softmax function with three outputs is illustrated in Figure 7.4, and the values v_1, v_2, and v_3 are also shown in the same figure. Note that the three outputs correspond to the probabilities of the three classes, and they convert the three outputs of the final hidden layer into probabilities with the softmax function. The final hidden layer often uses linear (identity) activations, when it is input into the softmax layer. Furthermore, there are no weights associated with the softmax layer, since it is only converting real-valued outputs into probabilities. Each output is the probability of a particular class. Note that using a single hidden layer with as many units as the number of classes (and the cross-entropy loss discussed in the next section) is exactly equivalent to the multinomial logistic regression model (cf. Section 6.6.2 of Chapter 6). In fact, many of the machine learning models discussed in this book can be easily simulated with appropriately chosen neural architectures.

7.2.3 Common Loss Functions

The choice of the loss function is critical in defining the outputs in a way that is sensitive to the application at hand. For example, least-squares regression with numeric outputs requires a simple squared loss of the form $(y - \hat{y})^2$ for a single training instance with target y and prediction $\hat{y}$. For probabilistic predictions of categorical data, two types of loss functions are used, depending on whether the prediction is binary or whether it is multiway:

1. **Binary targets (logistic regression):** In this case, it is assumed that the observed value y is drawn from $\{-1, +1\}$, and the prediction $\hat{y}$ uses a sigmoid activation function to output $\hat{y} \in (0, 1)$, which indicates the probability that the observed value y is 1. Then, the negative logarithm of $|y/2 - 0.5 + \hat{y}|$ provides the loss. This is because $|y/2 - 0.5 + \hat{y}|$ indicates the probability that the prediction is correct.

2. **Categorical targets:** In this case, if $\hat{y}_1 \ldots \hat{y}_k$ are the probabilities of the k classes (using the softmax activation of Equation 7.4), and the rth class is the ground-truth class, then the loss function for a single instance is defined as follows:

$$L = -\log(\hat{y}_r) \qquad (7.4)$$

This type of loss function implements multinomial logistic regression, and it is referred to as the *cross-entropy loss*. Note that binary logistic regression is identical to multinomial logistic regression, when the value of k is set to 2 in the latter.

The key point to remember is that the nature of the output nodes, the activation function, and the loss function depend on the application at hand.

7.2.4 How Nonlinearity Increases Expressive Power

The previous section provides a concrete proof of the fact that a neural network with only linear activations does not gain from increasing the number of layers in it. For example, consider the two-class data set illustrated in Figure 7.5, which is represented in two dimensions denoted by x_1 and x_2. There are two instances, A and B, of the class denoted by '*' with coordinates $(1, 1)$ and $(-1, 1)$, respectively. There is also a single instance B of the class denoted by '+' with coordinates $(0, 1)$, A neural network with only linear activations will never be able to classify the training data perfectly because the points are not linearly separable.

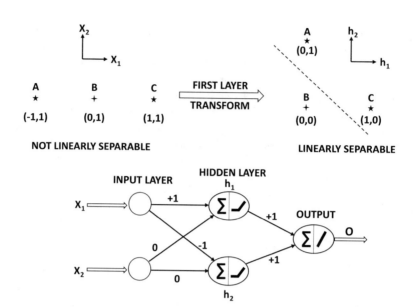

Figure 7.5: The power of nonlinear activation functions in transforming a data set to linear separability

On the other hand, consider a situation in which the hidden units have ReLU activation, and they learn the two new features h_1 and h_2, which are as follows:

$$h_1 = \max\{x_1, 0\}$$
$$h_2 = \max\{-x_1, 0\}$$

Note that these goals can be achieved by using appropriate weights from the input to hidden layer, and also applying a ReLU activation unit. The latter achieves the goal of thresholding negative values to 0. We have indicated the corresponding weights in the neural network shown in Figure 7.5. We have shown a plot of the data in terms of h_1 and h_2 in the same figure. The coordinates of the three points in the 2-dimensional hidden layer are $\{(1, 0), (0, 1), (0, 0)\}$. It is immediately evident that the two classes become linearly separable in terms of the new hidden representation. In a sense, the task of the first layer was *representation learning* to enable the solution of the problem with a linear classifier. Therefore, if we add a single linear output layer to the neural network, it will be able to classify these training instances perfectly. The key point is that the use of the nonlinear ReLU function is crucial in ensuring this linear separability. *Activation functions enable nonlinear mappings of the data, so that the embedded points can become linearly separable.* In fact, if both the weights from hidden to output layer are set to 1 with a linear activation function, the output O will be defined as follows:

$$O = h_1 + h_2 \tag{7.5}$$

This simple linear function separates the two classes because it always takes on the value of 1 for the two points labeled '*' and takes on 0 for the point labeled '+'. Therefore, much of the power of neural networks is hidden in the use of activation functions. The weights shown in Figure 7.5 will need to be *learned* in a data-driven manner, although there are many alternative choices of weights that can make the hidden representation linearly separable. Therefore, the learned weights may be different than the ones shown in Figure 7.5 if actual training is performed. Nevertheless, in the case of neural classifiers like (linear) logistic regression, there is no choice of weights at which one could hope to classify this training data set perfectly because the data set is not linearly separable in the original space. In other words, the activation functions enable nonlinear transformations of the data, that become increasingly powerful with multiple layers. A sequence of nonlinear activations imposes a specific type of structure on the learned model, whose power increases with the depth of the sequence (i.e., number of layers in the neural network).

Another classical example is the XOR function in which the two points $\{(0,0),(1,1)\}$ belong to one class, and the other two points $\{(1,0),(0,1)\}$ belong to the other class. It is possible to use ReLU activation to separate these two classes as well, although bias neurons will be needed in this case (see Exercise 1). The original backpropagation paper [151] discusses the XOR function, because this function was one of the motivating factors for designing multilayer networks and the ability to train them. The XOR function is considered a litmus test to determine the basic feasibility of a particular family of neural networks to properly predict nonlinearly separable classes. Although we have used the ReLU activation function above for simplicity, it is possible to use most of the other nonlinear activation functions to achieve the same goals. There are several types of neural architectures that are used commonly in various machine learning applications.

7.3   Optimization in Directed Acyclic Graphs

The optimization of loss functions in computational graphs requires the computation of gradients of the loss functions with respect to the network weights. This computation is done using *dynamic programming*. Dynamic programming is a technique from optimization that can be used to compute all types of path-centric functions in *directed acyclic graphs*.

In order to train computational graphs, it is assumed that we have training data corresponding to input-output pairs. The number of input nodes is equal to the number of input attributes and the number of output nodes is equal to the number of output attributes. The computational graph can predict the outputs using the inputs, and compare them to the observed outputs in order to check whether the function computed by the graph is consistent with the training data. If this is not the case, the weights of the computational graph need to be modified.

7.3.1   The Challenge of Computational Graphs

A computational graph naturally evaluates compositions of functions. Consider a variable x at a node in a computational graph with only three nodes containing a path of length 2. The first node applies the function $g(x)$, whereas the second node applies the function $f(\cdot)$ to the result. Such a graph computes the function $f(g(x))$, and it is shown in Figure 7.6. The example shown in Figure 7.6 uses the case when $f(x) = \cos(x)$ and $g(x) = x^2$. Therefore, the overall function is $\cos(x^2)$. Now, consider another setting in which both $f(x)$ and $g(x)$ are set to the same function, which is the sigmoid function:

$$f(x) = g(x) = \frac{1}{1 + \exp(-x)}$$

Then, the global function evaluated by the computational graph is as follows:

$$f(g(x)) = \frac{1}{1 + \exp\left[-\frac{1}{1+\exp(-x)}\right]} \tag{7.6}$$

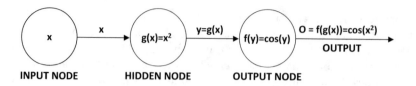

Figure 7.6: A simple computational graph with an input node and two computational nodes

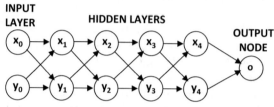

TOP NODE IN EACH LAYER COMPUTES BIVARIATE FUNCTION $F(x_{i-1}, y_{i-1})$
BOTTOM NODE IN EACH LAYER COMPUTES BIVARIATE FUNCTION $G(x_{i-1}, y_{i-1})$

Figure 7.7: The awkwardness of recursive nesting caused by a computational graph

This simple graph already computes a rather awkward composition function. Trying to find the derivative of this composition function algebraically becomes increasingly tedious with increasing complexity of the graph.

Consider a case in which the functions $g_1(\cdot)$, $g_2(\cdot) \ldots g_k(\cdot)$ are the functions computed in layer m, and they feed into a particular layer-$(m + 1)$ node that computes the multivariate function $f(\cdot)$ that uses the values computed in the previous layer as arguments. Therefore, the layer-$(m+1)$ function computes $f(g_1(\cdot), \ldots g_k(\cdot))$. This type of multivariate composition function already appears rather awkward. As we increase the number of layers, a function that is computed several edges downstream will have as many layers of nesting as the length of the path from the source to the final output. For example, if we have a computational graph which has 10 layers, and 2 nodes per layer, the overall composition function would have 2^{10} nested "terms". This makes the handling of closed-form functions of deep networks unwieldy and impractical.

In order to understand this point, consider the function in Figure 7.7. In this case, we have two nodes in each layer other than the output layer. The output layer simply sums its inputs. Each hidden layer contains two nodes. The variables in the ith layer are denoted by x_i and y_i, respectively. The input nodes (variables) use subscript 0, and therefore they are denoted by x_0 and y_0 in Figure 7.7. The two computed functions in the ith layer are $F(x_{i-1}, y_{i-1})$ and $G(x_{i-1}, y_{i-1})$, respectively.

In the following, we will write the expression for the variable in each node in order to show the increasing complexity with increasing number of layers:

$$x_1 = F(x_0, y_0)$$
$$y_1 = G(x_0, y_0)$$
$$x_2 = F(x_1, y_1) = F(F(x_0, y_0), G(x_0, y_0))$$
$$y_2 = G(x_1, y_1) = G(F(x_0, y_0), G(x_0, y_0))$$

We can already see that the expressions have already started looking unwieldy. On computing the values in the next layer, this becomes even more obvious:

$$x_3 = F(x_2, y_2) = F(F(F(x_0, y_0), G(x_0, y_0)), G(F(x_0, y_0), G(x_0, y_0)))$$
$$y_3 = G(x_2, y_2) = G(F(F(x_0, y_0), G(x_0, y_0)), G(F(x_0, y_0), G(x_0, y_0)))$$

An immediate observation is that the complexity and length of the closed-form function increases *exponentially* with the path lengths in the computational graphs. This type of complexity further increases in the case when optimization parameters are associated with the edges, and one tries to express the outputs/losses in terms of the inputs and the parameters on the edges. This is obviously a problem, if we try to use the boilerplate approach of first expressing the loss function in closed form in terms of the optimization parameters on the edges (in order to compute the derivative of the closed-form loss function).

7.3.2 The Broad Framework for Gradient Computation

The previous section makes it evident that differentiating closed-form expressions is not practical in the case of computational graphs. Therefore, one must somehow *algorithmically* compute gradients with respect to edges by using the topology of the computational graph. The purpose of this section is to introduce this broad algorithmic framework, and later sections will expand on the specific details of individual steps.

To learn the weights of a computational graph, an input-output pair is selected from the training data and the error of trying to predict the observed output with the observed input with the current values of the weights in the computational graph is quantified. When the errors are large, the weights need to be modified because the current computational graph does not reflect the observed data. Therefore, a loss function is computed as a function of this error, and the weights are updated so as to reduce the loss. This is achieved by computing the gradient of the loss with respect to the weights and performing a gradient-descent update. The overall approach for training a computational graph is as follows:

1. Use the attribute values from the input portion of a training data point to fix the values in the input nodes. Repeatedly select a node for which the values in all incoming nodes have already been computed and apply the node-specific function to also compute its variable. Such a node can be found in a directed acyclic graph by processing the nodes in order of increasing distance from input nodes. Repeat the process until the values in all nodes (including the output nodes) have been computed. If the values on the output nodes do not match the observed values of the output in the training point, compute the loss value. This phase is referred to as the *forward phase*.

2. Compute the gradient of the loss with respect to the weights on the edges. This phase is referred to as the *backwards phase*. The rationale for calling it a "backwards phase" will become clear later, when we introduce an algorithm that works backwards along the topology of the (directed acyclic) computational graph from the outputs to the inputs.

3. Update the weights in the negative direction of the gradient.

As in any stochastic gradient descent procedure, one cycles through the training points repeatedly until convergence is reached. A single cycle through all the training points is referred to as an *epoch*.

The main challenge is in computing the gradient of the loss function with respect to the weights in a computational graph. It turns out that *the derivatives of the node variables with respect to one another can be easily used to compute the derivative of the loss function with respect to the weights on the edges.* Therefore, in this discussion, we will focus on the computation of the derivatives of the variables with respect to one another. Later, we will show how these derivatives can be converted into gradients of loss functions with respect to weights.

7.3.3 Computing Node-to-Node Derivatives Using Brute Force

As discussed in an earlier section, one can express the function in a computational graph in terms of the nodes in early layers using an awkward closed-form expression that uses nested compositions of functions. If one were to indeed compute the derivative of this closed-form expression, it would require the use of the *chain rule of differential calculus* in order to deal with the repeated composition of functions. However, a blind application of the chain rule is rather wasteful in this case because many of the expressions in different portions of the inner nesting are identical, and one would be repeatedly computing the same derivative. The key idea in *automatic differentiation over computational graphs* is to recognize the fact that structure of the computational graph already provides all the information about which terms are repeated. We can avoid repeating the differentiation of these terms by using the structure of the computational graph itself to store intermediate results (by working backwards starting from output nodes to compute derivatives)! This is a well-known idea from dynamic programming, which has been used frequently in control theory [32, 99]. In the neural network community, this same algorithm is referred to as *backpropagation* (cf. Section 7.4). It is noteworthy that the applications of this idea in control theory were well-known to the traditional optimization community in 1960 [32, 99], although they remained unknown to researchers in the field of artificial intelligence for a while (who coined the term "backpropagation" in the 1980s to independently propose and describe this idea in the context of neural networks).

The simplest version of the chain rule is defined for a univariate composition of functions:

$$\frac{\partial f(g(x))}{\partial x} = \frac{\partial f(g(x))}{\partial g(x)} \cdot \frac{\partial g(x)}{\partial x} \tag{7.7}$$

This variant is referred to as the *univariate chain rule*. Note that each term on the right-hand side is a *local gradient* because it computes the derivative of a *local* function with respect to its immediate argument rather than a recursively derived argument. The basic idea is that a composition of functions is applied on the input x to yield the final output, and the gradient of the final output is given by the product of the local gradients along that path. Each local gradient only needs to worry about its specific input and output, which simplifies the computation. An example is shown in Figure 7.6 in which the function $f(y)$ is $\cos(y)$ and $g(x) = x^2$. Therefore, the composition function is $\cos(x^2)$. On using the univariate chain rule, we obtain the following:

$$\frac{\partial f(g(x))}{\partial x} = \underbrace{\frac{\partial f(g(x))}{\partial g(x)}}_{-\sin(g(x))} \cdot \underbrace{\frac{\partial g(x)}{\partial x}}_{2x} = -2x \cdot \sin(x^2)$$

Note that we can annotate each of the above two multiplicative components on the two *connections* in the graph, and simply compute the product of these values. Therefore, *for a computational graph containing a single path, the derivative of one node with respect to another is simply the product of these annotated values on the connections between the*

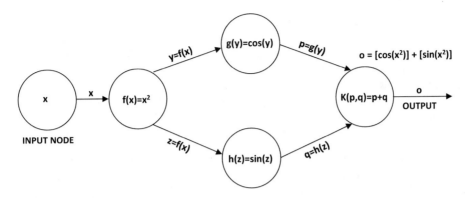

Figure 7.8: A simple computational function that illustrates the chain rule.

two nodes. The example of Figure 7.6 is a rather simple case in which the computational graph is a single path. In general, a computational graph with good expressive power will not be a single path. Rather, a single node may feed its output to multiple nodes. For example, consider the case in which we have a single input x, and we have k independent computational nodes that compute the functions $g_1(x), g_2(x), \ldots g_k(x)$. If these nodes are connected to a single output node computing the function $f()$ with k arguments, then the resulting function that is computed is $f(g_1(x), \ldots g_k(x))$. In such cases, the *multivariate chain rule* needs to be used. The multivariate chain rule is defined as follows:

$$\frac{\partial f(g_1(x), \ldots g_k(x))}{\partial x} = \sum_{i=1}^{k} \frac{\partial f(g_1(x), \ldots g_k(x))}{\partial g_i(x)} \cdot \frac{\partial g_i(x)}{\partial x} \qquad (7.8)$$

It is easy to see that the multivariate chain rule of Equation 7.8 is a simple generalization of that in Equation 7.7.

One can also view the multivariate chain rule in a path-centric fashion rather than a node-centric fashion. *For any pair of source-sink nodes, the derivative of the variable in the sink node with respect to the variable in the source node is simply the sum of the expressions arising from the univariate chain rule being applied to all paths existing between that pair of nodes.* This view leads to a direct expression for the derivative between any pair of nodes (rather than the recursive multivariate rule). However, it leads to an excessive computation, because the number of paths between a pair of nodes is exponentially related to the path length. In order to show the repetitive nature of the operations, we work with a very simple closed-form function with a single input x:

$$o = \sin(x^2) + \cos(x^2) \qquad (7.9)$$

The resulting computational graph is shown in Figure 7.8. In this case, the multivariate chain rule is applied to compute the derivative of the output o with respect to x. This is achieved by summing the results of the univariate chain rule for each of the two paths from x to o in Figure 7.8:

$$\frac{\partial o}{\partial x} = \underbrace{\frac{\partial K(p,q)}{\partial p}}_{1} \cdot \underbrace{g'(y)}_{-\sin(y)} \cdot \underbrace{f'(x)}_{2x} + \underbrace{\frac{\partial K(p,q)}{\partial q}}_{1} \cdot \underbrace{h'(z)}_{\cos(z)} \cdot \underbrace{f'(x)}_{2x}$$

$$= -2x \cdot \sin(y) + 2x \cdot \cos(z)$$

$$= -2x \cdot \sin(x^2) + 2x \cdot \cos(x^2)$$

In this simple example, there are two paths, both of which compute the function $f(x) = x^2$. As a result, the function $f(x)$ is differentiated *twice*, once for each path. This type of repetition can have severe effects for large multilayer networks containing many shared nodes, where the same function might be differentiated hundreds of thousands of times as a portion of the nested recursion. It is this *repeated and wasteful* approach to the computation of the derivative, that it is impractical to express the global function of a computational graph in closed form and explicitly differentiating it.

One can summarize the path-centric view of the multivariate chain rule as follows:

Lemma 7.3.1 (Pathwise Aggregation Lemma) *Consider a directed acyclic computational graph in which the ith node contains variable $y(i)$. The local derivative $z(i, j)$ of the directed edge (i, j) in the graph is defined as $z(i, j) = \frac{\partial y(j)}{\partial y(i)}$. Let a non-null set of paths $\mathcal{P}$ exist from a node s in the graph to node t. Then, the value of $\frac{\partial y(t)}{\partial y(s)}$ is given by computing the product of the local gradients along each path in $\mathcal{P}$, and summing these products over all paths in $\mathcal{P}$.*

$$\frac{\partial y(t)}{\partial y(s)} = \sum_{P \in \mathcal{P}} \prod_{(i,j) \in P} z(i, j) \tag{7.10}$$

This lemma can be easily shown by applying the multivariate chain rule (Equation 7.8) recursively over the computational graph. Although the use of the pathwise aggregation lemma is a wasteful approach for computing the derivative of $y(t)$ with respect to $y(s)$, it enables a simple and intuitive exponential-time algorithm for derivative computation.

An Exponential-Time Algorithm

The pathwise aggregation lemma provides a natural exponential-time algorithm, which is roughly similar to the steps one would go through by expressing the computational function in closed form with respect to a particular variable and then differentiating it. Specifically, the pathwise aggregation lemma leads to the following exponential-time algorithm to compute the derivative of the output o with respect to a variable x in the graph:

1. Use computational graph to compute the value $y(i)$ of each node i in a forward phase.

2. Compute the local partial derivatives $z(i, j) = \frac{\partial y(j)}{\partial y(i)}$ on each edge in the computational graph.

3. Let $\mathcal{P}$ be the set of all paths from an input node with value x to the output o. For each path $P \in \mathcal{P}$ compute the product of each local derivative $z(i, j)$ on that path.

4. Add up these values over all paths in $\mathcal{P}$.

In general, a computational graph will have an exponentially increasing number of paths with depth and one must add the product of the local derivatives over all paths. An example is shown in Figure 7.9, in which we have five layers, each of which has only two units. Therefore, the number of paths between the input and output is $2^5 = 32$. The jth hidden unit of the ith layer is denoted by $h(i, j)$. Each hidden unit is defined as the product of its inputs:

$$h(i, j) = h(i - 1, 1) \cdot h(i - 1, 2) \quad \forall j \in \{1, 2\} \tag{7.11}$$

In this case, the output is x^{32}, which is expressible in closed form, and can be differentiated easily with respect to x. In other words, we do not really need computational graphs in

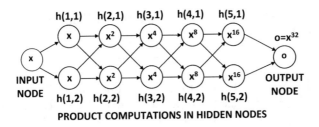

h(1,1) h(2,1) h(3,1) h(4,1) h(5,1)

PRODUCT COMPUTATIONS IN HIDDEN NODES

Figure 7.9: The chain rule aggregates the product of local derivatives along $2^5 = 32$ paths.

order to perform the differentiation. However, we will use the exponential-time algorithm to elucidate its workings. The derivatives of each $h(i, j)$ with respect to its two inputs are the values of the complementary inputs, because the partial derivative of the multiplication of two variables is the complementary variable:

$$\frac{\partial h(i, j)}{\partial h(i - 1, 1)} = h(i - 1, 2), \quad \frac{\partial h(i, j)}{\partial h(i - 1, 2)} = h(i - 1, 1)$$

The pathwise aggregation lemma implies that the value of $\frac{\partial o}{\partial x}$ is the product of the local derivatives (which are the complementary input values in this particular case) along all 32 paths from the input to the output:

$$\frac{\partial o}{\partial x} = \sum_{j_1, j_2, j_3, j_4, j_5 \in \{1,2\}^5} \prod \underbrace{h(1, j_1)}_{x} \underbrace{h(2, j_2)}_{x^2} \underbrace{h(3, j_3)}_{x^4} \underbrace{h(4, j_4)}_{x^8} \underbrace{h(5, j_5)}_{x^{16}}$$

$$= \sum_{\text{All 32 paths}} x^{31} = 32x^{31}$$

This result is, of course, consistent with what one would obtain on differentiating x^{32} directly with respect to x. However, an important observation is that it requires 2^5 aggregations to compute the derivative in this way for a relatively simple graph. More importantly, *we repeatedly differentiate the same function computed in a node* for aggregation. For example, the differentiation of the variable $h(3, 1)$ is performed 16 times because it appears in 16 paths from x to o.

Obviously, this is an inefficient approach to compute gradients. For a network with 100 nodes in each layer and three layers, we will have a million paths. *Nevertheless, this is exactly what we do in traditional machine learning when our prediction function is a complex composition function.* Manually working out the details of a complex composition function is tedious and impractical beyond a certain level of complexity. It is here that one can apply dynamic programming (which is guided by the structure of the computational graph) in order to store important intermediate results. By using such an approach, one can minimize repeated computations, and achieve polynomial complexity.

7.3.4 Dynamic Programming for Computing Node-to-Node Derivatives

In graph theory, computing all types of path-aggregative values over directed acyclic graphs is done using dynamic programming. Consider a directed acyclic graph in which the value $z(i, j)$ (interpreted as local partial derivative of variable in node j with respect to variable in node i) is associated with edge (i, j). In other words, if $y(p)$ is the variable in the node p, we have the following:

$$z(i, j) = \frac{\partial y(j)}{\partial y(i)} \tag{7.12}$$

An example of such a computational graph is shown in Figure 7.10. In this case, we have associated the edge $(2, 4)$ with the corresponding partial derivative. We would like to compute the product of $z(i, j)$ over each path $P \in \mathcal{P}$ from source node s to output node t and then add them in order to obtain the partial derivative $S(s, t) = \frac{\partial y(t)}{\partial y(s)}$:

$$S(s, t) = \sum_{P \in \mathcal{P}} \prod_{(i,j) \in P} z(i, j) \tag{7.13}$$

Let $A(i)$ be the set of nodes at the end points of outgoing edges from node i. We can compute the aggregated value $S(i, t)$ for each intermediate node i (between source node s and output node t) using the following well-known dynamic programming update:

$$S(i, t) \Leftarrow \sum_{j \in A(i)} S(j, t) z(i, j) \tag{7.14}$$

This computation can be performed backwards starting from the nodes directly incident on o, since $S(t, t) = \frac{\partial y(t)}{\partial y(t)}$ is already known to be 1. This is because the partial derivative of a variable with respect to itself is always 1. Therefore one can describe the pseudocode of this algorithm as follows:

> Initialize $S(t, t) = 1$;
> **repeat**
>   Select an unprocessed node i such that the values of $S(j, t)$ all of its outgoing
>   nodes $j \in A(i)$ are available;
>   Update $S(i, t) \Leftarrow \sum_{j \in A(i)} S(j, t) z(i, j)$;
> **until** all nodes have been selected;

Note that the above algorithm always selects a node i for which the value of $S(j, t)$ is available for all nodes $j \in A(i)$. Such a node is always available in directed acyclic graphs, and the node selection order will always be in the backwards direction starting from node t. Therefore, the above algorithm will work only when the computational graph does not have cycles, and is referred to as the *backpropagation algorithm*.

The algorithm discussed above is used by the network optimization community for computing all types of path-centric functions between *source-sink* node pairs (s, t) on directed acyclic graphs, which would otherwise require exponential time. For example, one can even use a variation of the above algorithm to find the longest path in a directed acyclic graph [12].

Interestingly, *the aforementioned dynamic programming update is exactly the multivariate chain rule of Equation 7.8, which is repeated in the backwards direction starting at the output node where the local gradient is known.* This is because we derived the path-aggregative form of the loss gradient (Lemma 7.3.1) using this chain rule in the first place. The main difference is that we apply the rule in a particular order in order to minimize computations. We emphasize this important point below:

> Using dynamic programming to efficiently aggregate the product of local gradients along the exponentially many paths in a computational graph results in a dynamic programming update that is identical to the multivariate chain rule of

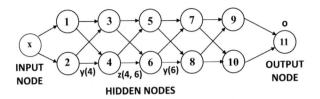

Figure 7.10: Edges are labeled with local partial derivatives such as $z(4,6) = \frac{\partial y(6)}{\partial y(4)}$.

differential calculus. The main point of dynamic programming is to apply this rule in a particular order, so that the derivative computations at different nodes are not repeated.

This approach is the backbone of the backpropagation algorithm used in neural networks. We will discuss more details of neural network-specific enhancements in Section 7.4. In the case where we have multiple output nodes $t_1, \ldots t_p$, one can initialize each $S(t_r, t_r)$ to 1, and then apply the same approach for each t_r.

7.3.4.1 Example of Computing Node-to-Node Derivatives

In order to show how the backpropagation approach works, we will provide an example of computation of node-to-node derivatives in a graph containing 10 nodes (see Figure 7.11). A variety of functions are computed in various nodes, such as the sum function (denoted by '+'), the product function (denoted by '*'), and the trigonometric sine/cosine functions. The variables in the 10 nodes are denoted by $y(1) \ldots y(10)$, where the variable $y(i)$ belongs to the ith node in the figure. Two of the edges incoming into node 6 also have the weights w_2 and w_3 associated with them. Other edges do not have weights associated with them. The functions computed in the various layers are as follows:

Layer 1: $y(4) = y(1) \cdot y(2), \quad y(5) = y(1) \cdot y(2) \cdot y(3), \quad y(6) = w_2 \cdot y(2) + w_3 \cdot y(3)$

Layer 2: $y(7) = \sin(y(4)), \quad y(8) = \cos(y(5)), \quad y(9) = \sin(y(6))$

Layer 3: $y(10) = y(7) \cdot y(8) \cdot y(9)$

We would like to compute the derivative of $y(10)$ with respect to each of the inputs $y(1)$, $y(2)$, and $y(3)$. One possibility is to simply express the $y(10)$ in closed form in terms of the inputs $y(1)$, $y(2)$ and $y(3)$, and then compute the derivative. By recursively using the above relationships, it is easy to show that $y(10)$ can be expressed in terms of $y(1)$, $y(2)$, and $y(3)$ as follows:

$$y(10) = \sin(y(1) \cdot y(2)) \cdot \cos(y(1) \cdot y(2) \cdot y(3)) \cdot \sin(w_2 \cdot y(2) + w_3 \cdot y(3))$$

As discussed earlier computing the closed-form derivative is not practical for larger networks. Furthermore, since one needs to compute the derivative of the output with respect to each and every node in the network, such an approach would also required closed-form expressions in terms of upstream nodes like $y(4)$, $y(5)$ and $y(6)$. All this tends to increase the amount of repeated computation. Luckily, backpropagation frees us from this repeated computation, since the derivative in $y(10)$ with respect to each and every node is computed by the backwards phase. The algorithm starts, by initializing the derivative of the output $y(10)$ with respect to itself, which is 1:

$$S(10, 10) = \frac{\partial y(10)}{\partial y(10)} = 1$$

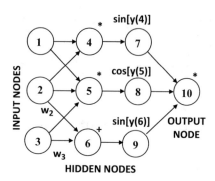

Figure 7.11: Example of node-to-node derivative computation

Subsequently, the derivatives of $y(10)$ with respect to all the variables on its incoming nodes are computed. Since $y(10)$ is expressed in terms of the variables $y(7)$, $y(8)$, and $y(9)$ incoming into it, this is easy to do, and the results are denoted by $z(7, 10)$, $z(8, 10)$, and $z(9, 10)$ (which is consistent with the notations used earlier in this chapter). Therefore, we have the following:

$$z(7, 10) = \frac{\partial y(10)}{\partial y(7)} = y(8) \cdot y(9)$$

$$z(8, 10) = \frac{\partial y(10)}{\partial y(8)} = y(7) \cdot y(9)$$

$$z(9, 10) = \frac{\partial y(10)}{\partial y(9)} = y(7) \cdot y(8)$$

Subsequently, we can use these values in order to compute $S(7, 10)$, $S(8, 10)$, and $S(9, 10)$ using the recursive backpropagation update:

$$S(7, 10) = \frac{\partial y(10)}{\partial y(7)} = S(10, 10) \cdot z(7, 10) = y(8) \cdot y(9)$$

$$S(8, 10) = \frac{\partial y(10)}{\partial y(8)} = S(10, 10) \cdot z(8, 10) = y(7) \cdot y(9)$$

$$S(9, 10) = \frac{\partial y(10)}{\partial y(9)} = S(10, 10) \cdot z(9, 10) = y(7) \cdot y(8)$$

Next, we compute the derivatives $z(4, 7)$, $z(5, 8)$, and $z(6, 9)$ associated with all the edges incoming into nodes 7, 8, and 9:

$$z(4, 7) = \frac{\partial y(7)}{\partial y(4)} = \cos[y(4)]$$

$$z(5, 8) = \frac{\partial y(8)}{\partial y(5)} = -\sin[y(5)]$$

$$z(6, 9) = \frac{\partial y(9)}{\partial y(6)} = \cos[y(6)]$$

These values can be used to compute $S(4, 10)$, $S(5, 10)$, and $S(6, 10)$:

$$S(4, 10) = \frac{\partial y(10)}{\partial y(4)} = S(7, 10) \cdot z(4, 7) = y(8) \cdot y(9) \cdot \cos[y(4)]$$

$$S(5, 10) = \frac{\partial y(10)}{\partial y(5)} = S(8, 10) \cdot z(5, 8) = -y(7) \cdot y(9) \cdot \sin[y(5)]$$

$$S(6, 10) = \frac{\partial y(10)}{\partial y(6)} = S(9, 10) \cdot z(6, 9) = y(7) \cdot y(8) \cdot \cos[y(6)]$$

In order to compute the derivatives with respect to the input values, one now needs to compute the values of $z(1, 3)$, $z(1, 4)$, $z(2, 4)$, $z(2, 5)$, $z(2, 6)$, $z(3, 5)$, and $z(3, 6)$:

$$z(1, 4) = \frac{\partial y(4)}{\partial y(1)} = y(2)$$

$$z(2, 4) = \frac{\partial y(4)}{\partial y(2)} = y(1)$$

$$z(1, 5) = \frac{\partial y(5)}{\partial y(1)} = y(2) \cdot y(3)$$

$$z(2, 5) = \frac{\partial y(5)}{\partial y(2)} = y(1) \cdot y(3)$$

$$z(3, 5) = \frac{\partial y(5)}{\partial y(3)} = y(1) \cdot y(2)$$

$$z(2, 6) = \frac{\partial y(6)}{\partial y(2)} = w_2$$

$$z(3, 6) = \frac{\partial y(6)}{\partial y(3)} = w_3$$

These partial derivatives can be backpropagated to compute $S(1, 10)$, $S(2, 10)$, and $S(3, 10)$:

$$S(1, 10) = \frac{\partial y(10)}{\partial y(1)} = S(4, 10) \cdot z(1, 4) + S(5, 10) \cdot z(1, 5)$$
$$= y(8) \cdot y(9) \cdot \cos[y(4)] \cdot y(2) - y(7) \cdot y(9) \cdot \sin[y(5)] \cdot y(2) \cdot y(3)$$

$$S(2, 10) = \frac{\partial y(10)}{\partial y(2)} = S(4, 10) \cdot z(2, 4) + S(5, 10) \cdot z(2, 5) + S(6, 10) \cdot z(2, 6)$$
$$= y(8) \cdot y(9) \cdot \cos[y(4)] \cdot y(1) - y(7) \cdot y(9) \cdot \sin[y(5)] \cdot y(1) \cdot y(3) +$$
$$+ y(7) \cdot y(8) \cdot \cos[y(6)] \cdot w_2$$

$$S(3, 10) = \frac{\partial y(10)}{\partial y(3)} = S(5, 10) \cdot z(3, 5) + S(6, 10) \cdot z(3, 6)$$
$$= -y(7) \cdot y(9) \cdot \sin[y(5)] \cdot y(1) \cdot y(2) + y(7) \cdot y(8) \cdot \cos[y(6)] \cdot w_3$$

Note that the use of a backward phase has the advantage of computing the derivative of $y(10)$ (output node variable) with respect to all the hidden and input node variables. These different derivatives have many sub-expressions in common, although the derivative computation of these sub-expressions is not repeated. This is the advantage of using the backwards phase for derivative computation as opposed to the use of closed-form expressions.

Because of the tedious nature of the closed-form expressions for outputs, the algebraic expressions for derivatives are also very long and awkward (no matter how we compute them). One can see that this is true even for the simple, ten-node computational graph of this section. For example, if one examines the derivative of $y(10)$ with respect to each of nodes $y(1)$, $y(2)$ and $y(3)$, the algebraic expression wraps into multiple lines. Furthermore, one cannot avoid the presence of repeated subexpressions within the algebraic derivative. This is counter-productive because our original goal in the backwards algorithm was to avoid the repeated computation endemic to traditional derivative evaluation with closed-form expressions. Therefore, one does not *algebraically* compute these types of expressions in real-world networks. One would first *numerically* compute all the node variables for a *specific* set of numerical inputs from the training data. Subsequently, one would *numerically* carry the derivatives backward, so that one does not have to carry the large algebraic expressions (with many repeated sub-expressions) in the backwards direction. The advantage of carrying numerical expressions is that multiple terms get consolidated into a single numerical value, which is specific to a particular input. By making the numerical choice, one must repeat the backwards computation algorithm *for each training point*, but it is still a better choice than computing the (massive) symbolic derivative in one shot and substituting the values in different training points. This is the reason that such an approach is referred to as *numerical differentiation* rather than *symbolic differentiation*. In much of machine learning, one first computes the algebraic derivative (which is symbolic differentiation) before substituting numerical values of the variables in the expression (for the derivative) to perform gradient-descent updates. This is different from the case of computational graphs, where the backwards algorithm is *numerically* applied to each training point.

7.3.5 Converting Node-to-Node Derivatives into Loss-to-Weight Derivatives

Most computational graphs define loss functions with respect to output node variables. One needs to compute the derivatives with respect to weights on *edges* rather than the node variables (in order to update the weights). In general, the node-to-node derivatives can be converted into loss-to-weight derivatives with a few additional applications of the univariate and multivariate chain rule.

Consider the case in which we have computed the node-to-node derivative of output variables in nodes indexed by $t_1, t_2, \ldots t_p$ with respect to the variable in node i using the dynamic programming approach in the previous section. Therefore, the computational graph has p output nodes in which the corresponding variable values are $y(t_1) \ldots y(t_p)$ (since the indices of the output nodes are $t_1 \ldots t_p$). The loss function is denoted by $L(y(t_1), \ldots y(t_p))$. We would like to compute the derivative of this loss function with respect to *the weights in the incoming edges of i*. For the purpose of this discussion, let w_{ji} be the weight of an edge from node index j to node index i. Therefore, we want to compute the derivative of the loss function with respect to w_{ji}. In the following, we will abbreviate $L(y_{t_1}, \ldots y_{t_p})$ with L for compactness of notation:

$$\frac{\partial L}{\partial w_{ji}} = \left[\frac{\partial L}{\partial y(i)} \right] \frac{\partial y(i)}{\partial w_{ji}} \qquad \text{[Univariate chain rule]}$$

$$= \left[\sum_{k=1}^{p} \frac{\partial L}{\partial y(t_k)} \frac{\partial y(t_k)}{\partial y(i)} \right] \frac{\partial y(i)}{\partial w_{ji}} \qquad \text{[Multivariate chain rule]}$$

Here, it is noteworthy that the loss function is typically a closed-form function of the variables in the node indices $t_1 \ldots t_p$, which is often either is least-squares function or a logarithmic loss function (like the logistic loss function in Chapter 6). Therefore, each derivative of the loss L with respect to $y(t_i)$ is easy to compute. Furthermore, the value of each $\frac{\partial y(t_k)}{\partial y(i)}$ for $k \in \{1 \ldots p\}$ can be computed using the dynamic programming algorithm of the previous section. The value of $\frac{\partial y_i}{\partial w_{ji}}$ is a derivative of the **local** function at each node, which usually has a simple form. Therefore, the loss-to-weight derivatives can be computed relatively easily, once the node-to-node derivatives have been computed using dynamic programming.

Although one can apply the pseudocode of page 228 to compute $\frac{\partial y(t_k)}{\partial y(i)}$ for each $k \in \{1 \ldots p\}$, it is more efficient to collapse all these computations into a single backwards algorithm. In practice, one initializes the derivatives at the output nodes to the loss derivatives $\frac{\partial L}{\partial y(t_k)}$ for each $k \in \{1 \ldots p\}$ rather than the value of 1 (as shown in the pseudocode of page 228). Subsequently, the entire loss derivative $\Delta(i) = \frac{\partial L}{\partial y(i)}$ is propagated backwards. Therefore, the modified algorithm for computing the loss derivative with respect to the *node variables* as well as the *edge variables* is as follows:

Initialize $\Delta(t_r) = \frac{\partial L}{\partial y(t_k)}$ for each $k \in \{1 \ldots p\}$;
repeat
    Select an unprocessed node i such that the values of $\Delta(j)$ all of its outgoing
        nodes $j \in A(i)$ are available;
    Update $\Delta(i) \Leftarrow \sum_{j \in A(i)} \Delta(j) z(i, j)$;
until all nodes have been selected;
for each edge (j, i) with weight w_{ji} **do** compute $\frac{\partial L}{\partial w_{ji}} = \Delta(i) \frac{\partial y(i)}{\partial w_{ji}}$;

In the above algorithm, $y(i)$ denotes the variable at node i. The key difference of this algorithm from the algorithm on page 228 is in the nature of the initialization and the addition of a final step computing the edge-wise derivatives. However, the core algorithm for computing the node-to-node derivatives remains an integral part of this algorithm. In fact, one can convert all the weights on the edges into additional "input" nodes containing weight parameters, and also add computational nodes that multiply the weights with the corresponding variables at the tail nodes of the edges. Furthermore, a computational node can be added that computes the loss from the output node(s). For example, the architecture of Figure 7.11 can be converted to that in Figure 7.12. Therefore, a computational graph with *learnable weights* can be converted into an unweighted graph with *learnable node variables* (on a subset of nodes). Performing only node-to-node derivative computation in Figure 7.12 from the loss node to the weight nodes is equivalent to loss-to-weight derivative computation. In other words, *loss-to-weight derivative computation in a weighted graph is equivalent to node-to-node derivative computation in a modified computational graph.* The derivative of the loss with respect to each weight can be denoted by the vector $\frac{\partial L}{\partial W}$ (in matrix calculus notation), where $\overline{W}$ denotes the weight vector. Subsequently, the standard gradient descent update can be performed:

$$\overline{W} \Leftarrow \overline{W} - \alpha \frac{\partial L}{\partial \overline{W}} \qquad (7.15)$$

Here, α is the learning rate. This type of update is performed to convergence by repeating the process with different inputs in order to learn the weights of the computational graph.

7.3.5.1 Example of Computing Loss-to-Weight Derivatives

Consider the case of Figure 7.11 in which the loss function is defined by $L = \log[y(10)^2]$, and we wish to compute the derivative of the loss with respect to the weights w_2 and w_3. In

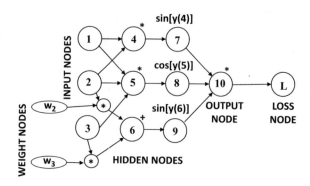

Figure 7.12: Converting loss-to-weight derivatives into node-to-node derivative computation based on Figure 7.11. Note the extra weight nodes and an extra loss node.

such a case, the derivative of the loss with respect to the weights is given by the following:

$$\frac{\partial L}{\partial w_2} = \frac{\partial L}{\partial y(10)} \frac{\partial y(10)}{\partial y(6)} \frac{\partial y(6)}{\partial w_2} = \left[\frac{2}{y(10)} \right] [y(7) \cdot y(8) \cdot \cos[y(6)]] \, y(2)$$

$$\frac{\partial L}{\partial w_3} = \frac{\partial L}{\partial y(10)} \frac{\partial y(10)}{\partial y(6)} \frac{\partial y(6)}{\partial w_3} = \left[\frac{2}{y(10)} \right] [y(7) \cdot y(8) \cdot \cos[y(6)]] \, y(3)$$

Note that the quantity $\frac{\partial y(10)}{\partial y(6)}$ has been obtained using the example in the previous section on node-to-node derivatives. In practice, these quantities are not computed algebraically. This is because the aforementioned algebraic expressions can be extremely awkward for large networks. Rather, for each numerical input set $\{y(1), y(2), y(3)\}$, one computes the different values of $y(i)$ in a forward phase. Subsequently, the derivatives of the loss with respect to each node variable (and incoming weights) are computed in a backwards phase. Again, these values are computed numerically for a specific input set $\{y(1), y(2), y(3)\}$. The numerical gradients can be used in order to update the weights for learning purposes.

7.3.6 Computational Graphs with Vector Variables

The previous section discuss the simple case in which each node of a computational graph contains a single scalar variable, whereas this section allows vector variables. In other words, the ith node contains the vector variable $\overline{y}_i$. Therefore, the **local** functions applied at the computational nodes are also vector-to-vector functions. For any node i, its local function uses an argument which corresponds to all the vector components of all its incoming nodes. From the input perspective of this local function, this situation is not too different from the previous case, where the argument is a vector corresponding to all the scalar inputs. However, the main difference is that the *output* of this function is a vector rather than a scalar. One example of such a vector-to-vector function is the *softmax* function (cf. Equation 6.24 of Chapter 6), which takes k real values as inputs and outputs k probabilities. Specifically, for inputs, $v_1 \ldots v_k$, the outputs $p_1 \ldots p_k$ of the softmax function are as follows:

$$p_r = \frac{\exp(v_r)}{\sum_{j=1}^{k} \exp(v_j)} \quad \forall r \in \{1 \ldots k\} \tag{7.16}$$

Note that this equation is the same of Equation 6.24 of Chapter 6, except that we are using $v_j = \overline{W}_j \cdot \overline{X}_i^T$. In general, the number of inputs of the function need not be the same as the number of outputs in a vector-to-vector function.

A vector-to-vector derivative is a *matrix*. Consider two vectors, $\overline{v} = [v_1 \ldots v_d]^T$ and $\overline{h} = [h_1 \ldots h_m]^T$, which occur somewhere in the computational graph shown in Figure 7.13(a). There might be nodes incoming into $\overline{v}$ as well as a loss L computed in a later layer. Then, using the denominator layout of matrix calculus, the vector-to-vector derivative is the transpose of the *Jacobian* matrix:

$$\frac{\partial \overline{h}}{\partial \overline{v}} = \text{Jacobian}(\overline{h}, \overline{v})^T$$

The (i, j)th entry of the above vector-to-vector derivative is simply $\frac{\partial h_j}{\partial v_i}$. Since $\overline{h}$ is an m-dimensional vector and $\overline{v}$ is a d-dimensional vector, the vector derivative is a $d \times m$ matrix. The chain rule over a single vector-centric path looks almost identical to the univariate chain rule over scalars, when one substitutes local partial derivatives with Jacobians. In the univariate case with scalars, the rule is quite simple. For example, consider the case where the scalar objective J is a function of the scalar w as follows:

$$J = f(g(h(w))) \tag{7.17}$$

All of $f(\cdot)$, $g(\cdot)$, and $h(\cdot)$ are assumed to be scalar functions. In such a case, the derivative of J with respect to the scalar w is simply $f'(g(h(w)))g'(h(w))h'(w)$. This rule is referred to as the univariate chain rule of differential calculus. Note that the order of multiplication does not matter because scalar multiplication is commutative.

Similarly, consider the case where you have the following functions, where one of the functions is a vector-to-scalar function:

$$J = f(g_1(w), g_2(w), \ldots, g_k(w))$$

In such a case, the *multivariate chain rule* states that one can compute the derivative of J with respect to w as the sum of the products of the partial derivatives using all arguments of the function:

$$\frac{\partial J}{\partial w} = \sum_{i=1}^{k} \left[\frac{\partial J}{\partial g_i(w)} \right] \left[\frac{\partial g_i(w)}{\partial w} \right]$$

One can generalize *both* of the above results into a single form by considering the case where the functions are vector-to-vector functions. Note that vector-to-vector derivatives are matrices, and therefore we will be multiplying matrices together instead of scalars. *Unlike the case of the scalar chain rule, the order of multiplication is important when dealing with matrices and vectors.* In a composition function, the derivative of the argument (inner level variable) is always pre-multiplied with the derivative of the function (outer level variable). In many cases, the order of multiplication is self-evident because of the size constraints associated with matrix multiplication. We formally define the vectored chain rule as follows:

Theorem 7.3.1 (Vectored Chain Rule) *Consider a composition function of the following form:*

$$\overline{o} = F_k(F_{k-1}(\ldots F_1(\overline{x})))$$

Assume that each $F_i(\cdot)$ takes as input an n_i-dimensional column vector and outputs an n_{i+1}-dimensional column vector. Therefore, the input $\overline{x}$ is an n_1-dimensional vector and the final output $\overline{o}$ is an n_{k+1}-dimensional vector. For brevity, denote the vector output of $F_i(\cdot)$ by $\overline{h}_i$. Then, the vectored chain rule asserts the following:

$$\underbrace{\left[\frac{\partial \overline{o}}{\partial \overline{x}} \right]}_{n_1 \times n_{k+1}} = \underbrace{\left[\frac{\partial \overline{h}_1}{\partial \overline{x}} \right]}_{n_1 \times n_2} \underbrace{\left[\frac{\partial \overline{h}_2}{\partial \overline{h}_1} \right]}_{n_2 \times n_3} \cdots \underbrace{\left[\frac{\partial \overline{h}_{k-1}}{\partial \overline{h}_{k-2}} \right]}_{n_{k-1} \times n_k} \underbrace{\left[\frac{\partial \overline{o}}{\partial \overline{h}_{k-1}} \right]}_{n_k \times n_{k+1}}$$

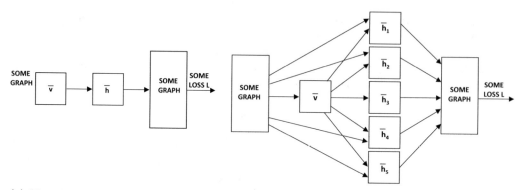

(a) Vector-centric graph with single path   (b) Vector-centric graph with multiple paths

Figure 7.13: Examples of vector-centric computational graphs

It is easy to see that the size constraints of matrix multiplication are respected in this case.

In other words, we can derive the following vector-valued chain rule for the single path of Figure 7.13(a):

$$\frac{\partial L}{\partial \overline{v}} = \underbrace{\frac{\partial \overline{h}}{\partial \overline{v}}}_{d \times m} \underbrace{\frac{\partial L}{\partial \overline{h}}}_{m \times 1} = \text{Jacobian}(\overline{h}, \overline{v})^T \frac{\partial L}{\partial \overline{h}}$$

Therefore, once the gradient of the loss is available with respect to a layer, it can be backpropagated by multiplying it with the transpose of a Jacobian! Here the *ordering of the matrices is important*, since matrix multiplication is not commutative.

The above provides the chain rule only for the case where the computational graph is a single path. What happens when the computational graph has an arbitrary structure? In such a case, we might have a situation where we have multiple nodes $\overline{h}_1 \ldots \overline{h}_s$ between node $\overline{v}$ and a network in later layers, as shown in Figure 7.13(b). Furthermore, there are connections between alternate layers, which are referred to as *skip connections*. Assume that the vector $\overline{h}_i$ has dimensionality m_i. In such a case, the partial derivative turns out to be a simple generalization of the previous case:

$$\frac{\partial L}{\partial \overline{v}} = \sum_{i=1}^{s} \underbrace{\frac{\partial \overline{h}_i}{\partial \overline{v}}}_{d \times m_i} \underbrace{\frac{\partial L}{\partial \overline{h}_i}}_{m_i \times 1} = \sum_{i=1}^{s} \text{Jacobian}(\overline{h}_i, \overline{v})^T \frac{\partial L}{\partial \overline{h}_i}$$

In most layered neural networks, we only have a single path and we rarely have to deal with the case of branches. Such branches might, however, arise in the case of neural networks with skip connections [see Figures 7.13(b) and 7.15(b)]. However, even in complicated network architectures like Figures 7.13(b) and 7.15(b), *each node only has to worry about its local outgoing edges during backpropagation*. Therefore, we provide a very general vector-based algorithm below that can work even in the presence of skip connections.

Consider the case where we have p output nodes containing vector-valued variables, which have indices denoted by $t_1 \ldots t_p$, and the variables in it are $\overline{y}(t_1) \ldots \overline{y}(t_p)$. In such a case, the loss function L might be function of all the components in these vectors. Assume that the ith node contains a column vector of variables denoted by $\overline{y}(i)$. Furthermore, in

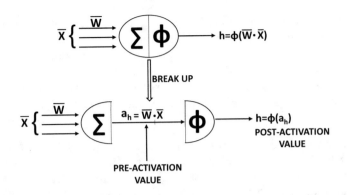

Figure 7.14: Pre- and post-activation values within a neuron

the denominator layout of matrix calculus, each $\overline{\Delta}(i) = \frac{\partial L}{\partial \overline{y}(i)}$ is a column vector with dimensionality equal to that of $\overline{y}(i)$. It is this *vector* of loss derivatives that will be propagated backwards. The vector-centric algorithm for computing derivatives is as follows:

Initialize $\overline{\Delta}(t_k) = \frac{\partial L}{\partial \overline{y}(t_k)}$ for each output node t_k for $k \in \{1 \ldots p\}$;
repeat
    Select an unprocessed node i such that the values of $\overline{\Delta}(j)$ all of its outgoing
        nodes $j \in A(i)$ are available;
    Update $\overline{\Delta}(i) \Leftarrow \sum_{j \in A(i)} \text{Jacobian}(\overline{y}(j), \overline{y}(i))^T \overline{\Delta}(j)$;
until all nodes have been selected;
for the vector $\overline{w}_i$ of edges incoming to each node i **do** compute $\frac{\partial L}{\partial \overline{w}_i} = \frac{\partial \overline{y}(i)}{\partial \overline{w}_i} \overline{\Delta}(i)$;

In the final step of the above pseudocode, the derivative of vector $\overline{y}(i)$ with respect to the vector $\overline{w}_i$ is computed, which is itself the transpose of a Jacobian matrix. This final step converts a vector of partial derivatives with respect to node variables into a vector of partial derivatives with respect to weights incoming at a node.

7.4 Application: Backpropagation in Neural Networks

In this section, we will describe how the generic algorithm based on computational graphs can be used in order to perform the backpropagation algorithm in neural networks. The key idea is that specific variables in the neural networks need to be defined as nodes of the computational-graph abstraction. The same neural network can be represented by different types of computational graphs, depending on which variables in the neural network are used to create computational graph nodes. The precise methodology for performing the backpropagation updates depends heavily on this design choice.

Consider the case of a neural network that first applies a linear function with weights w_{ij} on its inputs to create the pre-activation value $a(i)$, and then applies the activation function $\Phi(\cdot)$ in order to create the output $h(i)$:

$$h(i) = \Phi(a(i))$$

The variables $h(i)$ and $a(i)$ are shown in Figure 7.14. In this case, it is noteworthy that there are several ways in which the computational graph can be created. For example, one might create a computational graph in which each node contains the post-activation value $h(i)$, and therefore we are implicitly setting $y(i) = h(i)$. A second choice is to create a computational graph in which each node contains the pre-activation variable $a(i)$ and therefore we are setting $y(i) = a(i)$. It is even possible to create a decoupled computational graph containing both $a(i)$ and $h(i)$; in the last case, the computational graph will have twice as many nodes as the neural network. In all these cases, a relatively straightforward special-case/simplification of the pseudocodes in the previous section can be used for learning the gradient:

1. The post-activation value $y(i) = h(i)$ could represent the variable in the ith computational node in the graph. Therefore, each computational node in such a graph *first* applies the linear function, *and then* applies the activation function. The post-activation value is shown in Figure 7.14. In such a case, the value of $z(i, j) = \frac{\partial y(j)}{\partial y(i)} = \frac{\partial h(j)}{\partial h(i)}$ in the pseudocode of page 233 is $w_{ij}\Phi'_j$. Here, w_{ij} is the weight of the edge from i to j and $\Phi'_j = \frac{\partial \Phi(a(j))}{\partial a(j)}$ is the local derivative of the activation function at node j with respect to its argument. The value of each $\Delta(t_r)$ at output node t_r is simply the derivative of the loss function with respect to $h(t_r)$. The final derivative with respect to the weight w_{ji} (in the final line of the pseudocode on page 233) is equal to $\Delta(i)\frac{\partial h(i)}{\partial w_{ji}} = \Delta(i)h(j)\Phi'_i$.

2. The pre-activation value (after applying the linear function), which is denoted by $a(i)$, could represent the variable in each computational node i in the graph. Note the subtle distinction between the work performed in computational nodes and neural network nodes. Each *computational* node *first* applies the activation function to each of its inputs before applying a linear function, whereas these operations are performed in the reverse order in a neural network. The structure of the computational graph is roughly similar to the neural network, except that the first layer of computational nodes do not contain an activation. In such a case, the value of $z(i, j) = \frac{\partial y(j)}{\partial y(i)} = \frac{\partial a(j)}{\partial a(i)}$ in the pseudocode of page 233 is $\Phi'_i w_{ij}$. Note that $\Phi(a(i))$ is being differentiated with respect to its argument in this case, rather than $\Phi(a(j))$ as in the case of the post-activation variables. The value of the loss derivative with respect to the pre-activation variable $a(t_r)$ in the rth output node t_r needs to account for the fact that it is a pre-activation value, and therefore, we cannot directly use the loss derivative with respect to post-activation values. Rather the post-activation loss derivative needs to be *multiplied with the derivative Φ'_{t_r} of the activation function at that node*. The final derivative with respect to the weight w_{ji} (final line of pseudocode on page 233) is equal to $\Delta(i)\frac{\partial a(i)}{\partial w_{ji}} = \Delta(i)h(j)$.

The use of pre-activation variables for backpropagation is more common than the use of post-activation variables. Therefore, we present the backpropagation algorithm in a crisp pseudocode with the use of pre-activation variables. Let t_r be the index of the rth output node. Then, the backpropagation algorithm with pre-activation variables may be presented as follows:

Initialize $\Delta(t_r) = \frac{\partial L}{\partial y(t_r)} = \Phi'(a(t_r))\frac{\partial L}{\partial h(t_r)}$ for each output node t_r with $r \in \{1 \ldots k\}$;
repeat
    Select an unprocessed node i such that the values of $\Delta(j)$ all of its outgoing
        nodes $j \in A(i)$ are available;
    Update $\Delta(i) \Leftarrow \Phi'_i \sum_{j \in A(i)} w_{ij}\Delta(j)$;
until all nodes have been selected;
for each edge (j, i) with weight w_{ji} **do** compute $\frac{\partial L}{\partial w_{ji}} = \Delta(i)h(j)$;

It is also possible to use both pre-activation and post-activation variables as separate nodes of the computational graph. In the next section, we will combine this approach with a vector-centric representation.

7.4.1 Derivatives of Common Activation Functions

It is evident from the discussion in the previous section that backpropagation requires the computation of derivatives of activation functions. Therefore, we discuss the computation of the derivatives of common activation functions in this section:

1. *Sigmoid activation:* The derivative of sigmoid activation is particularly simple, when it is expressed in terms of the *output* of the sigmoid, rather than the input. Let o be the output of the sigmoid function with argument v:

$$o = \frac{1}{1 + \exp(-v)} \tag{7.18}$$

Then, one can write the derivative of the activation as follows:

$$\frac{\partial o}{\partial v} = \frac{\exp(-v)}{(1 + \exp(-v))^2} \tag{7.19}$$

The key point is that this sigmoid can be written more conveniently in terms of the outputs:

$$\frac{\partial o}{\partial v} = o(1 - o) \tag{7.20}$$

The derivative of the sigmoid is often used as a function of the output rather than the input.

2. *Tanh activation:* As in the case of the sigmoid activation, the tanh activation is often used as a function of the output o rather than the input v:

$$o = \frac{\exp(2v) - 1}{\exp(2v) + 1} \tag{7.21}$$

One can then compute the derivative as follows:

$$\frac{\partial o}{\partial v} = \frac{4 \cdot \exp(2v)}{(\exp(2v) + 1)^2} \tag{7.22}$$

One can also write this derivative in terms of the output o:

$$\frac{\partial o}{\partial v} = 1 - o^2 \tag{7.23}$$

3. *ReLU and hard tanh activations:* The ReLU takes on a partial derivative value of 1 for non-negative values of its argument, and 0, otherwise. The hard tanh function takes on a partial derivative value of 1 for values of the argument in $[-1, +1]$ and 0, otherwise.

7.4.2   The Special Case of Softmax

Softmax activation is a special case because the function is not computed with respect to one input, but with respect to multiple inputs. Therefore, one cannot use exactly the same type of update, as with other activation functions. The softmax activation function converts k real-valued predictions $v_1 \ldots v_k$ into output probabilities $o_1 \ldots o_k$ using the following relationship:

$$o_i = \frac{\exp(v_i)}{\sum_{j=1}^{k} \exp(v_j)} \quad \forall i \in \{1, \ldots, k\} \tag{7.24}$$

Note that if we try to use the chain rule to backpropagate the derivative of the loss L with respect to $v_1 \ldots v_k$, then one has to compute each $\frac{\partial L}{\partial o_i}$ and also each $\frac{\partial o_i}{\partial v_j}$. This backpropagation of the softmax is greatly simplified, when we take two facts into account:

1. The softmax is almost always used in the output layer.

2. The softmax is almost always paired with the *cross-entropy loss*. If $y_1 \ldots y_k \in \{0, 1\}$, be the one-hot encoded (observed) outputs for the k mutually exclusive classes, then the cross-entropy loss is defined as follows:

$$L = -\sum_{i=1}^{k} y_i \log(o_i) \tag{7.25}$$

The key point is that the value of $\frac{\partial L}{\partial v_i}$ has a particularly simple form in the case of the softmax:

$$\frac{\partial L}{\partial v_i} = \sum_{j=1}^{k} \frac{\partial L}{\partial o_j} \cdot \frac{\partial o_j}{\partial v_i} = o_i - y_i \tag{7.26}$$

Note that this derivative has already been shown in Section 6.6.2 of Chapter 6. In this case, we have decoupled the backpropagation update of the softmax activation from the updates of the weighted layers. In general, it is helpful to create a view of backpropagation in which the linear matrix multiplications and activation layers are decoupled because it greatly simplifies the updates. This view will be discussed in the next section.

7.4.3   Vector-Centric Backpropagation

As illustrated in Figure 7.3, any *layer-wise* neural architecture can be represented as a computational graph of vector variables *with a single path*. We repeat the vector-centric illustration of Figure 7.3(b) in Figure 7.15(a). Note that the architecture corresponds to a single path of vector variables, which can be further decoupled into linear layers and activation layers. Although it is possible for the neural network to have an arbitrary architecture (with paths of varying length), this situation is not so common. Some variations of this idea have been explored recently in the context of a specialized[1] neural network for image data, referred to as *ResNet* [6, 72]. We illustrate this situation in Figure 7.15(b), where there is a shortcut between alternate layers.

Since the layerwise situation of Figure 7.15(a) is more common, we discuss the approach used for performing backpropagation in this case. As discussed earlier, a node in a neural network performs a combination of a linear operation and a nonlinear activation function. In

---

[1] *ResNet* is a convolutional neural network in which the structure of the layer is spatial, and the operations correspond to convolutions.

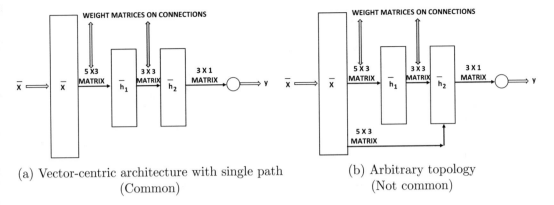

(a) Vector-centric architecture with single path
(Common)

(b) Arbitrary topology
(Not common)

Figure 7.15: Most neural networks have a layer-wise architecture, and therefore the vector-centric architecture has a single path. However, if there are shortcuts across layers, it is possible for the topology of the vector-centric architecture to be arbitrary.

order to simplify the gradient evaluations, the linear computations and the activation computations are decoupled as separate "layers," and one separately backpropagates through the two layers. Therefore, one can create a neural network in which activation layers are alternately arranged with linear layers, as shown in Figure 7.16. Activation layers (usually) perform one-to-one, elementwise computations on the vector components with the activation function $\Phi(\cdot)$, whereas linear layers perform all-to-all computations by multiplying with the coefficient matrix W. Then, if $\overline{g}_i$ and $\overline{g}_{i+1}$ be the loss gradients in the ith and $(i+1)$th layers, and J_i be the Jacobian matrix between the ith and $(i+1)$th layers, the update is as follows. Let J be the matrix whose elements are J_{kr}. Then, it is easy to see that the backpropagation update from layer to layer can be written as follows:

$$\overline{g}_i = J_i^T \overline{g}_{i+1} \tag{7.27}$$

Writing backpropagation equations as matrix multiplications is often beneficial from an implementation-centric point of view, such as acceleration with Graphics Processor Units, which work particularly well with vector and matrix operations.

First, the forward phase is performed on the inputs in order to compute the activations in each layer. Subsequently, the gradients are computed in the backwards phase. For each pair of matrix multiplication and activation function layers, the following forward and backward steps need to be performed:

1. Let $\overline{z}_i$ and $\overline{z}_{i+1}$ be the column vectors of activations in the forward direction when the matrix of linear transformations from the ith to the $(i+1)$th layer is denoted[2] by W. Each element of the gradient $\overline{g}_i$ is the partial derivative of the loss function with respect to a hidden variable in the ith layer. Then, we have the following:

$$\overline{z}_{i+1} = W\overline{z}_i \qquad \text{[Forward Propagation]}$$
$$\overline{g}_i = W^T \overline{g}_{i+1} \qquad \text{[Backward Propagation]}$$

---

[2]Strictly speaking, we should use the notation W_i instead of W, although we omit the subscripts here for simplicity, since the entire discussion in this section is devoted to the linear transforms between a single pair of layers.

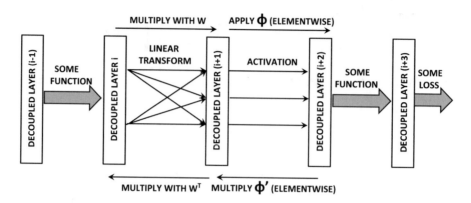

Figure 7.16: A decoupled view of backpropagation

Table 7.1: Examples of different functions and their backpropagation updates between layers i and $(i+1)$. The hidden values and gradients in layer i are denoted by $\overline{z}_i$ and $\overline{g}_i$. Some of these computations use $I(\cdot)$ as the binary indicator function

Function	Type	Forward	Backward
Linear	Many-Many	$\overline{z}_{i+1} = W \overline{z}_i$	$\overline{g}_i = W^T \overline{g}_{i+1}$
Sigmoid	One-One	$\overline{z}_{i+1} = \text{sigmoid}(\overline{z}_i)$	$\overline{g}_i = \overline{g}_{i+1} \odot \overline{z}_{i+1} \odot (\overline{1} - \overline{z}_{i+1})$
Tanh	One-One	$\overline{z}_{i+1} = \tanh(\overline{z}_i)$	$\overline{g}_i = \overline{g}_{i+1} \odot (\overline{1} - \overline{z}_{i+1} \odot \overline{z}_{i+1})$
ReLU	One-One	$\overline{z}_{i+1} = \overline{z}_i \odot I(\overline{z}_i > 0)$	$\overline{g}_i = \overline{g}_{i+1} \odot I(\overline{z}_i > 0)$
Hard Tanh	One-One	Set to ± 1 ($\notin [-1, +1]$) Copy ($\in [-1, +1]$)	Set to 0 ($\notin [-1, +1]$) Copy ($\in [-1, +1]$)
Max	Many-One	Maximum of inputs	Set to 0 (non-maximal inputs) Copy (maximal input)
Arbitrary function $f_k(\cdot)$	Anything	$\overline{z}_{i+1}^{(k)} = f_k(\overline{z}_i)$	$\overline{g}_i = J_i^T \overline{g}_{i+1}$ J_i is Jacobian$(\overline{z}_{i+1}, \overline{z}_i)$

2. Now consider a situation where the activation function $\Phi(\cdot)$ is applied to each node in layer $(i+1)$ to obtain the activations in layer $(i+2)$. Then, we have the following:

$$\overline{z}_{i+2} = \Phi(\overline{z}_{i+1}) \qquad \text{[Forward Propagation]}$$
$$\overline{g}_{i+1} = \overline{g}_{i+2} \odot \Phi'(\overline{z}_{i+1}) \quad \text{[Backward Propagation]}$$

Here, $\Phi(\cdot)$ and its derivative $\Phi'(\cdot)$ are applied in element-wise fashion to vector arguments. The symbol $\odot$ indicates elementwise multiplication.

Note the extraordinary simplicity once the activation is decoupled from the matrix multiplication in a layer. The forward and backward computations are shown in Figure 7.16. Examples of different types of backpropagation updates for various forward functions are shown in Table 7.1. Therefore, the backward propagation operation is just like forward propagation. Given the vector of gradients in a layer, one only has to apply the operations shown in the final column of Table 7.1 to obtain the gradients of the loss with respect to the previous layer. In the table, the vector indicator function $I(\overline{x} > 0)$ is an *element-wise* indicator function that returns a binary vector of the same size as $\overline{x}$; the ith output component is set to 1 when the ith component of $\overline{x}$ is larger than 0. The notation $\overline{1}$ denotes a column vector of 1s.

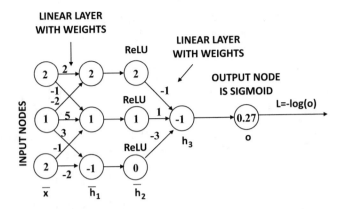

Figure 7.17: Example of decoupled neural network with vector layers $\overline{x}$, $\overline{h}_1$, $\overline{h}_2$, h_3, and o: variable values are shown within the nodes

Converting to Weight-Centric Derivatives

Upon performing the backpropagation, one only obtains the loss-to-node derivatives but not the loss-to-weight derivatives. Note that the elements in $\overline{g}_i$ represent gradients of the loss with respect to the *activations* in the ith layer, and therefore an additional step is needed to compute gradients with respect to the *weights*. The gradient of the loss with respect to a weight between the pth unit of the $(i-1)$th layer and the qth unit of ith layer is obtained by multiplying the pth element of $\overline{z}_{i-1}$ with the qth element of $\overline{g}_i$. One can also achieve this goal using a vector-centric approach, by simply computing the *outer product* of $\overline{g}_i$ and $\overline{z}_{i-1}$. In other words, the entire matrix M of derivatives of the loss with respect to the weights in the $(i-1)$th layer and the ith layer is given by the following:

$$M = \overline{g}_i \overline{z}_{i-1}^T$$

Since M is given by the product of a column vector and a row vector of sizes equal to two successive layers, it is a matrix of exactly the same size as the weight matrix between the two layers. The (q, p)th element of M yields the derivative of the loss with respect to the weight between the pth element of $\overline{z}_{i-1}$ and qth element of $\overline{z}_i$.

7.4.4 Example of Vector-Centric Backpropagation

In order to explain vector-specific backpropagation, we will use an example in which the linear layers and activation layers have been decoupled. Figure 7.17 shows an example of a neural network with two computational layers, but they appear as four layers, since the activation layers have been decoupled as separated layers from the linear layers. The vector for the input layer is denoted by the 3-dimensional column vector $\overline{x}$, and the vectors for the computational layers are $\overline{h}_1$ (3-dimensional), $\overline{h}_2$ (3-dimensional), h_3 (1-dimensional), and output layer o (1-dimensional). The loss function is $L = -\log(o)$. These notations are annotated in Figure 7.17. The input vector $\overline{x}$ is $[2, 1, 2]^T$, and the weights of the edges in the two linear layers are annotated in Figure 7.17. Missing edges between $\overline{x}$ and $\overline{h}_1$ are assumed to have zero weight. In the following, we will provide the details of both the forward and the backwards phase.

Forward phase: The first hidden layer $\overline{h}_1$ is related to the input vector $\overline{x}$ with the weight matrix W as $\overline{h}_1 = W\overline{x}$. We can reconstruct the weights matrix W and then compute $\overline{h}_1$ for forward propagation as follows:

$$W = \begin{bmatrix} 2 & -2 & 0 \\ -1 & 5 & -1 \\ 0 & 3 & -2 \end{bmatrix} ; \quad \overline{h}_1 = W\overline{x} = \begin{bmatrix} 2 & -2 & 0 \\ -1 & 5 & -1 \\ 0 & 3 & -2 \end{bmatrix} \begin{bmatrix} 2 \\ 1 \\ 2 \end{bmatrix} = \begin{bmatrix} 2 \\ 1 \\ -1 \end{bmatrix}$$

The hidden layer $\overline{h}_2$ is obtained by applying the ReLU function in element-wise fashion to $\overline{h}_1$ during the forward phase. Therefore, we obtain the following:

$$\overline{h}_2 = \text{ReLU}(\overline{h}_1) = \text{ReLU} \begin{bmatrix} 2 \\ 1 \\ -1 \end{bmatrix} = \begin{bmatrix} 2 \\ 1 \\ 0 \end{bmatrix}$$

Subsequently, the 1×3 weight matrix $W_2 = [-1, 1, -3]$ is used to transform the 3-dimensional vector $\overline{h}_2$ to the 1-dimensional "vector" h_3 as follows:

$$h_3 = W_2 \overline{h}_2 = [-1, 1, -3] \begin{bmatrix} 2 \\ 1 \\ 0 \end{bmatrix} = -1$$

The output o is obtained by applying the sigmoid function to h_3. In other words, we have the following:

$$o = \frac{1}{1 + \exp(-h_3)} = \frac{1}{1 + e} \approx 0.27$$

The point-specific loss is $L = -\log_e(0.27) \approx 1.3$.

Backwards phase: In the backward phase, we first start by initializing $\frac{\partial L}{\partial o}$ to $-1/o$, which is $-1/0.27$. Then, the 1-dimensional "gradient" g_3 of the hidden layer h_3 is obtained by using the backpropagation formula for the sigmoid function in Table 7.1:

$$g_3 = o(1-o) \underbrace{\frac{\partial L}{\partial o}}_{-1/o} = o - 1 = 0.27 - 1 = -0.73 \tag{7.28}$$

The gradient $\overline{g}_2$ of the hidden layer $\overline{h}_2$ is obtained by multiplying g_3 with the transpose of the weight matrix $W_2 = [-1, 1, -3]$:

$$\overline{g}_2 = W_2^T g_3 = \begin{bmatrix} -1 \\ 1 \\ -3 \end{bmatrix} (-0.73) = \begin{bmatrix} 0.73 \\ -0.73 \\ 2.19 \end{bmatrix}$$

Based on the entry in Table 7.1 for the ReLU layer, the gradient $\overline{g}_2$ can be propagated backwards to $\overline{g}_1 = \frac{\partial L}{\partial \overline{h}_1}$ by copying the components of $\overline{g}_2$ to $\overline{g}_1$, when the corresponding components in $\overline{h}_1$ are positive; otherwise, the components of $\overline{g}_1$ are set to zero. Therefore, the gradient $\overline{g}_1 = \frac{\partial L}{\partial \overline{h}_1}$ can be obtained by simply copying the first and second components of $\overline{g}_2$ to the first and second components of $\overline{g}_1$, and setting the third component of $\overline{g}_1$ to 0. In other words, we have the following:

$$\overline{g}_1 = \begin{bmatrix} 0.73 \\ -0.73 \\ 0 \end{bmatrix}$$

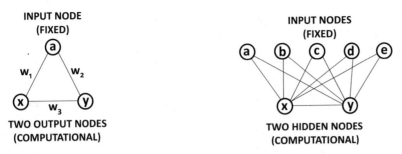

(a) An undirected computational graph   (b) More input states than hidden states

Figure 7.18: Examples of undirected computational graphs

Note that we can also compute the gradient $\overline{g}_0 = \frac{\partial L}{\partial \overline{x}}$ of the loss with respect to the input layer $\overline{x}$ by simply computing $\overline{g}_0 = W^T \overline{g}_1$. However, this is not really needed for computing loss-to-weight derivatives.

Computing loss-to-weight derivatives: So far, we have only shown how to compute loss-to-node derivatives in this particular example. These need to be converted to loss-to-weight derivatives with the additional step of multiplying with a hidden layer. Let M be the loss-to-weight derivatives for the weight matrix W between the two layers. Note that there is a one-to-one correspondence between the positions of the elements of M and W. Then, the matrix M is defined as follows:

$$M = \overline{g}_1 \overline{x}^T = \begin{bmatrix} 0.73 \\ -0.73 \\ 0 \end{bmatrix} [2,1,2] = \begin{bmatrix} 1.46 & 0.73 & 1.46 \\ -1.46 & -0.73 & -1.46 \\ 0 & 0 & 0 \end{bmatrix}$$

Similarly, one can compute the loss-to-weight derivative matrix M_2 for the 1×3 matrix W_2 between $\overline{h}_2$ and h_3:

$$M_2 = g_3 \overline{h}_2^T = (-0.73)[2,1,0] = [-1.46, -0.73, 0]$$

Note that the size of the matrix M_2 is identical to that of W_2, although the weights of the missing edges should not be updated.

7.5 A General View of Computational Graphs

Although the use of directed acyclic graphs on continuous-valued data is extremely common in machine learning (with neural networks being a prominent use case), other variations of such graphs exist. For example, it is possible for a computational graph to define probabilistic functions on edges, to have discrete-valued variables, and also to have cycles in the graph. In fact, the entire field of probabilistic graphical models is devoted to these types of computational graphs. Although the use of cycles in computational graphs is not common in feed-forward neural networks, they are extremely common in many advanced variations of neural networks like Kohonen self-organizing maps, Hopfield networks, and Boltzmann machines. Furthermore, these neural networks use discrete and probabilistic data types as variables within their nodes (implicitly or explicitly).

Another important variation is the use of undirected computational graphs. In undirected computational graphs, each node computes a function of the variables in nodes

Table 7.2: Types of computational graphs for different machine learning problems. The properties of the computational graph vary according to the application at hand

Model	Cycles?	Variable	Function	Methodology
SVM Logistic Regression Linear Regression SVD Matrix Factorization	No	Continuous	Deterministic	Gradient Descent
Feedforward Neural Networks	No	Continuous	Deterministic	Gradient Descent
Kohonen Map	Yes	Continuous	Deterministic	Gradient Descent
Hopfield Networks	Yes (Undirected)	Discrete (Binary)	Deterministic	Iterative (Hebbian Rule)
Boltzmann Machines	Yes (Undirected)	Discrete (Binary)	Probabilistic	Monte Carlo Sampling + Iterative (Hebbian)
Probabilistic Graphical Models	Varies	Varies	Probabilistic (largely)	Varies

incident on it, and there is no direction to the links. This is the only difference between an undirected computational graph and a directed computational graph. As in the case of directed computational graphs, one can define a loss function on the observed variables in the nodes. Examples of undirected computational graphs are shown in Figure 7.18. Some nodes are fixed (for observed data) whereas others are computational nodes. The computation can occur in both directions of an edge as long as the value in the node is not fixed externally.

It is harder to learn the parameters in undirected computational graphs, because the presence of cycles creates additional constraints on the values of variables in the nodes. In fact, it is not even necessary for there to be a set of variable values in nodes that satisfy all the functional constraints implied by the computational graph. For example, consider a computational graph with two nodes in which the variable of each node is obtained by adding 1 to the variable on the other node. It is impossible to find a pair of values in the two nodes that can satisfy both constraints (because both variable values cannot be larger than the other by 1). Therefore, one would have to be satisfied with a *best-fit* solution in many cases. This situation is different from a directed acyclic graph, where appropriate variables values can always be defined over all values of the inputs and parameters (as long as the function in each node is computable over its inputs).

Undirected computational graphs are often used in all types of unsupervised algorithms, because the cycles in these graphs help in relating other hidden nodes to the input nodes. For example, if the variables x and y are assumed to be hidden variables in Figure 7.18(b), this approach learns weights so that the two hidden variables correspond to the compressed representations of 5-dimensional data. The weights are often learned to minimize a loss function (or energy function) that rewards large weights when connected nodes are highly correlated in a positive way. For example, if variable x is heavily correlated with input a in a positive way, then the weight between these two nodes should be large. By learning these weights, one can compute the hidden representation of any 5-dimensional point by providing it as an input to the network.

The level of difficulty in learning the parameters of a computational graph is regulated by three characteristics of the graph. The first characteristic is the structure of the graph itself. It is generally much easier to learn the parameters of computational graphs without cycles (which are always directed). The second characteristic is whether the variable in a node is continuous or discrete. It is much easier to optimize the parameters of a computational graph with continuous variables with the use of differential calculus. Finally, the function computed at a node of can be either probabilistic or deterministic. The parameters of deterministic computational graphs are almost always much easier to optimize with observed data. All these variations are important, and they arise in different types of machine learning applications. Some examples of different types of computational graphs in machine learning are as follows:

1. **Hopfield networks:** Hopfield networks are *undirected* computational graphs, in which the nodes always contain discrete, binary values. Since the graph is undirected, it contains cycles. The discrete nature of the variables makes the problem harder to optimize, because it precludes the use of simple techniques from calculus. In many cases, the optimal solutions to undirected graphs with discrete-valued variables are known to be NP-hard [61]. For example, a special case of the Hopfield network can be used to solve the *traveling salesman problem*, which is known to be NP-hard. Most of the algorithms for such types of optimization problems are iterative heuristics.

2. **Probabilistic graphical models:** Probabilistic graphical models [104] are graphs representing the structural dependencies among random variables. Such dependencies may be either undirected or directed; directed dependencies may or may not contain cycles. The main distinguishing characteristic of a probabilistic graphical model from other types of graphical computational models is that the variables are probabilistic in nature. In other words, a variable in the computational graph corresponds to the *outcome* resulting from sampling from a *probability distributions that is conditioned on the variables in the incoming nodes*. Among all classes of models, probabilistic graphical models are the hardest to solve, and often require computationally intensive procedures like *Markov chain Monte Carlo sampling*. Interestingly, a generalization of Hopfield networks, referred to as *Boltzmann machines*, represents an important class of probabilistic graphical models.

3. **Kohonen self-organizing map:** A Kohonen self-organizing map uses a 2-dimensional *lattice-structured graph* on the hidden nodes. The activations on hidden nodes are analogous to the centroids in a k-means algorithm. This type of approach is a *competitive learning algorithm*. The lattice structure ensures that hidden nodes that are close to one another in the graph have similar values. As a result, by associating data points with their closest hidden nodes, one is able to obtain a 2-dimensional visualization of the data.

Table 7.2 shows several variations of the computational graph paradigm in machine learning, and their specific properties. It is evident that the methodology used for a particular problem is highly dependent on the structure of the computational graph, its variables, and the nature of the node-specific function. We refer the reader to [6] for the neural architectures of the basic machine learning models discussed in this book (like linear regression, logistic regression, matrix factorization, and SVMs).

7.6 Summary

This chapter introduces the basics of computational graphs for machine learning applications. Computational graphs often have parameters associated with their edges, which need to be learned. Learning the parameters of a computational graph from observed data provides a route to learning a function from observed data (whether it can be expressed in closed form or not). The most commonly used type of computational graph is a directed acyclic graph. Traditional neural networks represent a class of models that is a special case of this type of graph. However, other types of undirected and cyclic graphs are used to represent other models like Hopfield networks and restricted Boltzmann machines.

7.7   Further Reading

Computational graphs represent a fundamental way of defining the computations associated with many machine learning models such as neural networks or probabilistic models. Detailed discussions of neural networks may be found in [6, 67], whereas detailed discussions of probabilistic graphical models may be found in [104]. Automatic differentiation in computational graphs has historically been used extensively in control theory [32, 99]. The backpropagation algorithm was first proposed in the context of neural networks by Werbos [200], although it was forgotten. Eventually, the algorithm was popularized in the paper by Rumelhart *et al.* [150]. The Hopfield network and the Boltzmann machine are both discussed in [6]. A discussion of Kohonen self-organizing maps may also be found in [6].

7.8   Exercises

1. The discussion on page 215 proposes a loss function for the L_1-SVM in the context of a computational graph. How would you change this loss function, so that the same computational graph results in an L_2-SVM?

2. Repeat Exercise 1 with the changed setting that you want to simulate Widrow-Hoff learning (least-squares classification) with the same computational graph. What will be the loss function associated with the single output node?

3. The book discusses a vector-centric view of backpropagation in which backpropagation in linear layers can be implemented with matrix-to-vector multiplications. Discuss how you can deal with *batches* of training instances at a time (i.e., mini-batch stochastic gradient descent) by using matrix-to-matrix multiplications.

4. Let $f(x)$ be defined as follows:

$$f(x) = \sin(x) + \cos(x)$$

Consider the function $f(f(f(f(x))))$. Write this function in closed form to obtain an appreciation of the awkwardly long function. Evaluate the derivative of this function at $x = \pi/3$ radians by using a computational graph abstraction.

5. Suppose that you have a computational graph with the constraint that specific sets of weights are always constrained to be at the same value. Discuss how you can compute the derivative of the loss function with respect to these weights. [Note that this trick is used frequently in the neural network literature to handle shared weights.]

6. Consider a computational graph in which you are told that the variables on the edges satisfy k linear equality constraints. Discuss how you would train the weights of such a graph. How would your answer change, if the variables satisfied box constraints. [The

reader is advised to refer to the chapter on constrained optimization for answering this question.]

7. Discuss why the dynamic programming algorithm for computing the gradients will not work in the case where the computational graph contains cycles.

8. Consider the neural architecture with connections between alternate layers, as shown in Figure 7.15(b). Suppose that the recurrence equations of this neural network are as follows:

$$\overline{h}_1 = \text{ReLU}(W_1 \overline{x})$$
$$\overline{h}_2 = \text{ReLU}(W_2 \overline{x} + W_3 \overline{h}_1)$$
$$y = W_4 \overline{h}_2$$

Here, W_1, W_2, W_3, and W_4 are matrices of appropriate size. Use the vector-centric backpropagation algorithm to derive the expressions for $\frac{\partial y}{\partial \overline{h}_2}$, $\frac{\partial y}{\partial \overline{h}_1}$, and $\frac{\partial y}{\partial \overline{x}}$ in terms of the matrices and activation values in intermediate layers.

9. Consider a neural network that has hidden layers $\overline{h}_1 \ldots \overline{h}_t$, inputs $\overline{x}_1 \ldots \overline{x}_t$ into each layer, and outputs $\overline{o}$ from the final layer $\overline{h}_t$. The recurrence equation for the pth layer is as follows:

$$\overline{o} = U \overline{h}_t$$
$$\overline{h}_p = \tanh(W \overline{h}_{p-1} + V \overline{x}_p) \quad \forall p \in \{1 \ldots t\}$$

The vector output $\overline{o}$ has dimensionality k, each $\overline{h}_p$ has dimensionality m, and each $\overline{x}_p$ has dimensionality d. The "tanh" function is applied in element-wise fashion. The notations U, V, and W are matrices of sizes $k \times m$, $m \times d$, and $m \times m$, respectively. The vector $\overline{h}_0$ is set to the zero vector. Start by drawing a (vectored) computational graph for this system. Show that node-to-node backpropagation uses the following recurrence:

$$\frac{\partial \overline{o}}{\partial \overline{h}_t} = U^T$$
$$\frac{\partial \overline{o}}{\partial \overline{h}_{p-1}} = W^T \Delta_{p-1} \frac{\partial \overline{o}}{\partial \overline{h}_p} \quad \forall p \in \{2 \ldots t\}$$

Here, Δ_p is a diagonal matrix in which the diagonal entries contain the components of the vector $\overline{1} - \overline{h}_p \odot \overline{h}_p$. What you have just derived contains the node-to-node backpropagation equations of a recurrent neural network. What is the size of each matrix $\frac{\partial \overline{o}}{\partial \overline{h}_p}$?

10. Show that if we use the loss function $L(\overline{o})$ in Exercise 9, then the loss-to-node gradient can be computed for the final layer $\overline{h}_t$ as follows:

$$\frac{\partial L(\overline{o})}{\partial \overline{h}_t} = U^T \frac{\partial L(\overline{o})}{\partial \overline{o}}$$

The updates in earlier layers remain similar to Exercise 9, except that each $\overline{o}$ is replaced by $L(\overline{o})$. What is the size of each matrix $\frac{\partial L(\overline{o})}{\partial \overline{h}_p}$?

11. Suppose that the output structure of the neural network in Exercise 9 is changed so that there are k-dimensional outputs $\overline{o}_1 \ldots \overline{o}_t$ in each layer, and the overall loss is $L = \sum_{i=1}^{t} L(\overline{o}_i)$. The output recurrence is $\overline{o}_p = U\overline{h}_p$. All other recurrences remain the same. Show that the backpropagation recurrence of the hidden layers changes as follows:

$$\frac{\partial L}{\partial \overline{h}_t} = U^T \frac{\partial L(\overline{o}_t)}{\partial \overline{o}_t}$$

$$\frac{\partial L}{\partial \overline{h}_{p-1}} = W^T \Delta_{p-1} \frac{\partial L}{\partial \overline{h}_p} + U^T \frac{\partial L(\overline{o}_{p-1})}{\partial \overline{o}_{p-1}} \quad \forall p \in \{2 \ldots t\}$$

12. For Exercise 11, show the following loss-to-weight derivatives:

$$\frac{\partial L}{\partial U} = \sum_{p=1}^{t} \frac{\partial L(\overline{o}_p)}{\partial \overline{o}_p} \overline{h}_p^T, \qquad \frac{\partial L}{\partial W} = \sum_{p=2}^{t} \Delta_{p-1} \frac{\partial L}{\partial \overline{h}_p} \overline{h}_{p-1}^T, \qquad \frac{\partial L}{\partial V} = \sum_{p=1}^{t} \Delta_p \frac{\partial L}{\partial \overline{h}_p} \overline{x}_p^T$$

What are the sizes and ranks of these matrices?

13. Consider a neural network in which a vectored node $\overline{v}$ feeds into two distinct vectored nodes $\overline{h}_1$ and $\overline{h}_2$ computing different functions. The functions computed at the nodes are $\overline{h}_1 = \text{ReLU}(W_1\overline{v})$ and $\overline{h}_2 = \text{sigmoid}(W_2\overline{v})$. We do not know anything about the values of the variables in other parts of the network, but we know that $\overline{h}_1 = [2, -1, 3]^T$ and $\overline{h}_2 = [0.2, 0.5, 0.3]^T$, that are connected to the node $\overline{v} = [2, 3, 5, 1]^T$. Furthermore, the loss gradients are $\frac{\partial L}{\partial \overline{h}_1} = [-2, 1, 4]^T$ and $\frac{\partial L}{\partial \overline{h}_2} = [1, 3, -2]^T$, respectively. Show that the backpropagated loss gradient $\frac{\partial L}{\partial \overline{v}}$ can be computed in terms of W_1 and W_2 as follows:

$$\frac{\partial L}{\partial \overline{v}} = W_1^T \begin{bmatrix} -2 \\ 0 \\ 4 \end{bmatrix} + W_2^T \begin{bmatrix} 0.16 \\ 0.75 \\ -0.42 \end{bmatrix}$$

What are the sizes of W_1, W_2, and $\frac{\partial L}{\partial \overline{v}}$?

14. **Forward Mode Differentiation:** The backpropagation algorithm needs to compute node-to-node derivatives of *output* nodes with respect to all other nodes, and therefore computing gradients in the backwards direction makes sense. Consequently, the pseudocode on page 228 propagates gradients in the backward direction. However, consider the case where we want to compute the node-to-node derivatives of all nodes with respect to *source* (input) nodes $s_1 \ldots s_k$. In other words, we want to compute $\frac{\partial x}{\partial s_i}$ for each non-input node variable x and each input node s_i in the network. Propose a variation of the pseudocode of page 228 that computes node-to-node gradients in the forward direction.

15. **All-pairs node-to-node derivatives:** Let $y(i)$ be the variable in node i in a directed acyclic computational graph containing n nodes and m edges. Consider the case where one wants to compute $S(i,j) = \frac{\partial y(j)}{\partial y(i)}$ for all pairs of nodes in a computational graph, so that at least one directed path exists from node i to node j. Propose an algorithm for all-pairs derivative computation that requires at most $O(n^2 m)$ time. [Hint: The pathwise aggregation lemma is helpful. First compute $S(i,j,t)$, which is the portion of $S(i,j)$ in the lemma belonging to paths of length exactly t. How can $S(i,k,t+1)$ be expressed in terms of the different $S(i,j,t)$?]

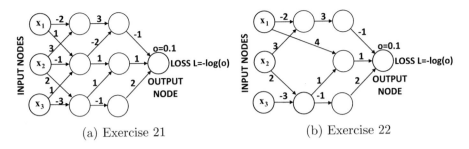

(a) Exercise 21   (b) Exercise 22

Figure 7.19: Computational graphs for Exercises 21 and 22

16. Use the pathwise aggregation lemma to compute the derivative of $y(10)$ with respect to each of $y(1)$, $y(2)$, and $y(3)$ as an algebraic expression (cf. Figure 7.11). You should get the same derivative as obtained using the backpropagation algorithm in the text of the chapter.

17. Consider the computational graph of Figure 7.10. For a particular numerical input $x = a$, you find the unusual situation that the value $\frac{\partial y(j)}{\partial y(i)}$ is 0.3 for each and every edge (i, j) in the network. Compute the numerical value of the partial derivative of the output with respect to the input x (at $x = a$). Show the computations using both the pathwise aggregation lemma and the backpropagation algorithm.

18. Consider the computational graph of Figure 7.10. The upper node in each layer computes $\sin(x + y)$ and the lower node in each layer computes $\cos(x + y)$ with respect to its two inputs. For the first hidden layer, there is only a single input x, and therefore the values $\sin(x)$ and $\cos(x)$ are computed. The final output node computes the product of its two inputs. The single input x is 1 radian. Compute the numerical value of the partial derivative of the output with respect to the input x (at $x = 1$ radian). Show the computations using both the pathwise aggregation lemma and the backpropagation algorithm.

19. Consider the computational graph shown in Figure 7.19(a), in which the local derivative $\frac{\partial y(j)}{\partial y(i)}$ is shown for each edge (i, j), where $y(k)$ denotes the activation of node k. The output o is 0.1, and the loss L is given by $-\log(o)$. Compute the value of $\frac{\partial L}{\partial x_i}$ for each input x_i using both the path-wise aggregation lemma, and the backpropagation algorithm.

20. Consider the computational graph shown in Figure 7.19(b), in which the local derivative $\frac{\partial y(j)}{\partial y(i)}$ is shown for each edge (i, j), where $y(k)$ denotes the activation of node k. The output o is 0.1, and the loss L is given by $-\log(o)$. Compute the value of $\frac{\partial L}{\partial x_i}$ for each input x_i using both the path-wise aggregation lemma, and the backpropagation algorithm.

21. Convert the weighted computational graph of Figure 7.2 into an unweighted graph by defining additional nodes containing $w_1 \ldots w_5$ along with appropriately defined hidden nodes.

22. **Multinomial logistic regression with neural networks:** Propose a neural network architecture using the softmax activation function and an appropriate loss function that can perform multinomial logistic regression. You may refer to Chapter 6 for details of multinomial logistic regression.

Chapter 8

Domain-Specific Neural Architectures

"All research in the cultural sciences in an age of specialization, once it is oriented towards a given subject matter through particular settings of problems and has established its methodological principles, will consider the analysis of the data as an end in itself."– Max Weber

8.1 Introduction

The discussion in the previous chapter introduces generic forms of neural architectures. These architectures are *fully connected* and layered, in the sense that the computational units are layered and each unit in a particular layer is connected to a unit in the next layer. However, these types of architectures are not well suited to domain-specific settings, where there are known relationships among the attributes. Some examples of such known relationships are as follows:

- In an image data set, the attributes correspond to the intensities of the pixels in a particular image. The value of the intensity in a particular pixel is often the same as the value of an adjacent pixel. An image is usually contains a deeply structured pattern of changes in values of adjacent pixels. For example, a straight line or a curve in a particular image is caused by this type of structured change of values. The architecture of the neural network should be constructed in order to capture such changes. Irrespective of the type of setting that an image is drawn from, there are certain common patterns that are repeated across different settings. For example, an arbitrary image can often be constructed from some basic geometric shapes, which can be viewed as critical *features* for the image domain.

- In a text data set, the attributes correspond to the words in a particular sentence. The identity of the word at a particular position in a sentence is closely related to words in adjacent sentences. Therefore, the architecture of the neural network should be constructed to capture these types of sequential relationships.

© Springer Nature Switzerland AG 2021
C. C. Aggarwal, *Artificial Intelligence*, https://doi.org/10.1007/978-3-030-72357-6_8

In this chapter, we will focus on two important types of neural architectures that can capture such structural relationships. In particular, we will focus on convolutional neural networks and recurrent neural networks. The former is designed to model image data, whereas the latter is designed to model sequence data.

This chapter is organized as follows. The next section introduces the basic inspiration behind the convolutional neural network architecture. Section 8.3 introduces the basics of a convolutional neural network, the various operations, and the way in which they are organized. Case studies with some typical convolutional neural networks are discussed in Section 8.4. Section 8.5 introduces the principles behind recurrent neural networks. Section 8.6 discusses the basic architecture of the recurrent neural network along with the associated training algorithm. Long short-term memory networks are discussed in Section 8.7. Applications of convolutional and recurrent neural networks are discussed in Section 8.8. A summary is given in Section 8.9.

8.2   Principles Underlying Convolutional Neural Networks

Convolutional neural networks are designed to work with grid-structured inputs, which have strong spatial dependencies in local regions of the grid. The most obvious example of grid-structured data is a 2-dimensional image. This type of data also exhibits spatial dependencies, because adjacent spatial locations in an image often have similar color values of the individual pixels. An additional dimension captures the different colors, which creates a 3-dimensional input *volume*. Therefore, the features in a convolutional neural network have dependencies among one another based on spatial distances. Other forms of sequential data like text, time-series, and sequences can also be considered special cases of grid-structured data with various types of relationships among adjacent items. This is because a sequence or time-series data set can be viewed as a 1-dimensional data set with adjacent (temporal) dependencies, whereas an image data set can be considered as a 2-dimensional data set with adjacent (spatial) dependencies. In both cases, the close relationships among adjacent values make the use of convolutional neural networks feasible. The vast majority of applications of convolutional neural networks focus on image data, although one can also use these networks for all types of temporal, spatial, and spatiotemporal data.

An important defining characteristic of convolutional neural networks is the *convolution* operation. A convolution operation is a dot-product operation between a grid-structured set of weights and similar grid-structured inputs drawn from different spatial localities in the input volume. This type of operation is useful for data with a high level of spatial or other locality, such as image data. Therefore, convolutional neural networks are defined as networks that use the convolutional operation in at least one layer, although most convolutional neural networks use this operation in multiple layers.

Biological Inspirations and Domain-Specific Properties

Convolutional neural networks were one of the first success stories of deep learning, well before recent advancements in training techniques led to improved performance in other types of architectures. In fact, the eye-catching successes of some convolutional neural network architectures in image-classification contests after 2011 led to broader attention to the field of deep learning. Convolutional neural networks are well suited to the process of hierarchical feature engineering with depth; this is reflected in the fact that the deepest neural networks in all domains are drawn from the field of convolutional networks. Furthermore, these networks also represent excellent examples of how biologically inspired neural networks can sometimes provide ground-breaking results.

The early motivation for convolutional neural networks was derived from experiments by Hubel and Wiesel on a cat's visual cortex [87]. The visual cortex has small regions of cells that are sensitive to specific regions in the visual field. In other words, if specific areas of the visual field are excited, then those cells in the visual cortex will be activated as well. Furthermore, the excited cells also depend on the shape and orientation of the objects in the visual field. For example, vertical edges cause some neuronal cells to be excited, whereas horizontal edges cause other neuronal cells to be excited. The cells are connected using a layered architecture, and this discovery led to the conjecture that mammals use these different layers to construct portions of images at different levels of abstraction. From a machine learning point of view, this principle is similar to that of hierarchical feature extraction. As we will see later, convolutional neural networks achieve something similar by encoding primitive shapes in earlier layers, and more complex shapes in later layers.

Based on these biological inspirations, the earliest neural model was the *neocognitron* [58]. However, there were several differences between this model and the modern convolutional neural network. The most prominent of these differences was that the notion of weight sharing was not used. Based on this architecture, one of the first fully convolutional architectures, referred to as *LeNet-5* [110], was developed. This network was used by banks to identify hand-written numbers on checks. Since then, the convolutional neural network has not evolved much; the main difference is in terms of using more layers and stable activation functions like the ReLU. Furthermore, numerous training tricks and powerful hardware options are available to achieve better success in training when working with deep networks and large data sets.

A factor that has played an important role in increasing the prominence of convolutional neural networks has been the annual *ImageNet* competition [218] (also referred to as "*ImageNet Large Scale Visual Recognition Challenge [ILSVRC]*"). The ILSVRC competition uses the *ImageNet* data set [217]. One of the earliest methods that achieved success in the 2012 *ImageNet* competition by a large margin was *AlexNet* [107]. Furthermore, the improvements in accuracy have been so extraordinarily large in the last few years that it has changed the landscape of research in the area.

The secret to the success of any neural architecture lies in designing the architecture of the network in a way that is sensitive to the understanding of the domain at hand. Convolutional neural networks are heavily based on this principle, because they use sparse connections with a high level of parameter-sharing in a domain-sensitive way. In other words, not all states in a particular layer are connected to those in the previous layer in an indiscriminate way. Rather, the value of a feature in a particular layer is connected only to a local spatial region in the previous layer with a consistent set of shared parameters across the full spatial footprint of the image. This type of architecture can be viewed as a domain-aware regularization, which was derived from the biological insights in Hubel and Wiesel's early work. In general, the success of the convolutional neural network has important lessons for other data domains. A carefully designed architecture, in which the relationships and dependencies among the data items are used in order to reduce the parameter footprint, provides the key to results of high accuracy.

A significant level of domain-aware regularization is also available in recurrent neural networks, which share the parameters from different temporal periods. This sharing is based on the assumption that temporal dependencies remain invariant with time. Recurrent neural networks are based on intuitive understanding of temporal relationships, whereas convolutional neural networks are based on an intuitive understanding of spatial relationships. The latter intuition was directly extracted from the organization of biological neurons in a cat's visual cortex. This outstanding success provides a motivation to explore how neuroscience may be leveraged to design neural networks in clever ways. Even though artificial neural networks are only caricatures of the true complexity of the biological brain, one should not underestimate the intuition that one can obtain by studying the basic principles of neuroscience [70].

8.3   The Basic Structure of a Convolutional Network

In convolutional neural networks, the states in each layer are arranged according to a spatial grid structure. These spatial relationships are inherited from one layer to the next because each feature value is based on a small local spatial region in the previous layer. It is important to maintain these spatial relationships among the grid cells, because the convolution operation and the transformation to the next layer is critically dependent on these relationships. Each layer in the convolutional network is a 3-dimensional grid structure, which has a *height*, *width*, and *depth*. The depth of a layer in a convolutional neural network should not be confused with the depth of the network itself. The word "depth" (when used in the context of a single layer) refers to the number of *channels* in each layer, such as the number of primary color channels (e.g., blue, green, and red) in the input image or the number of feature maps in the hidden layers. The use of the word "depth" to refer to both the number of feature maps in each layer as well as the number of layers is an unfortunate overloading of terminology used in convolutional networks, but we will be careful while using this term, so that it is clear from its context.

The convolutional neural network functions much like a traditional feed-forward neural network, except that the operations in its layers are spatially organized with sparse (and carefully designed) connections between layers. The three types of layers that are commonly present in a convolutional neural network are *convolution*, *pooling*, and *ReLU*. The ReLU activation is no different from a traditional neural network. In addition, a final set of layers is often fully connected and maps in an application-specific way to a set of output nodes. In the following, we will describe each of the different types of operations and layers, and the typical way in which these layers are interleaved in a convolutional neural network.

Why do we need depth in each layer of a convolutional neural network? To understand this point, let us examine how the input to the convolutional neural network is organized. The input data to the convolutional neural network is organized into a 2-dimensional grid structure, and the values of the individual grid points are referred to as *pixels*. Each pixel, therefore, corresponds to a spatial location within the image. However, in order to encode the precise color of the pixel, we need a multidimensional array of values at each grid location. In the RGB color scheme, we have an intensity of the three primary colors, corresponding to red, green, and blue, respectively. Therefore, if the spatial dimensions of an image are 32×32 pixels and the depth is 3 (corresponding to the red, green, and blue color channels), then the overall number of pixels in the image is $32 \times 32 \times 3$. This particular image size is quite common, and also occurs in a popularly used data set for benchmarking, known as CIFAR-10 [220]. An example of this organization is shown in Figure 8.1(a). It is natural to represent the input layer in this 3-dimensional structure because two dimensions are devoted to spatial relationships and a third dimension is devoted to the independent properties along these channels. For example, the intensities of the primary colors are the independent properties in the first layer. In the hidden layers, these independent properties correspond to various types of shapes extracted from local regions of the image. For the purpose of discussion, assume that the input in the qth layer is of size $L_q \times B_q \times d_q$. Here, L_q refers to the *height* (or length), B_q refers to the width (or breadth), and d_q is the depth. In almost all image-centric applications, the values of L_q and B_q are the same. However, we will work with separate notations for height and width in order to retain generality in presentation.

For the first (input) layer, these values are decided by the nature of the input data and its preprocessing. In the above example, the values are $L_1 = 32$, $B_1 = 32$, and $d_1 = 3$. Later layers have exactly the same 3-dimensional organization, except that each of the d_q 2-dimensional grid of values for a particular input can no longer be considered a grid of raw pixels. Furthermore, the value of d_q is much larger than three for the hidden layers because the number of independent properties of a given local region that are relevant to classification can be quite significant. For $q > 1$, these grids of values are referred to as *feature maps* or *activation maps*. These values are analogous to the values in the hidden layers in a feed-forward network.

In the convolutional neural network, the parameters are organized into sets of 3-dimensional structural units, known as *filters* or *kernels*. The filter is usually square in terms of its spatial dimensions, which are typically much smaller than those of the layer the filter is applied to. On the other hand, *the depth of a filter is always same is the same as that of the layer to which it is applied.* Assume that the dimensions of the filter in the qth layer are $F_q \times F_q \times d_q$. An example of a filter with $F_1 = 5$ and $d_1 = 3$ is shown in Figure 8.1(a). It is common for the value of F_q to be small and odd. Examples of commonly used values of F_q are 3 and 5, although there are some interesting cases in which it is possible to use $F_q = 1$.

The *convolution operation* places the filter at each possible position in the image (or hidden layer) so that the filter fully overlaps with the image, and performs a dot product between the $F_q \times F_q \times d_q$ parameters in the filter and the matching grid in the input volume (with same size $F_q \times F_q \times d_q$). The dot product is performed by treating the entries in the relevant 3-dimensional region of the input volume and the filter as vectors of size $F_q \times F_q \times d_q$, so that the elements in both vectors are ordered based on their corresponding positions in the grid-structured volume. How many possible positions are there for placing the filter? This question is important, because each such position therefore defines a spatial "pixel" (or, more accurately, a *feature*) in the next layer. In other words, the number of alignments between the filter and image defines the spatial height and width of the next hidden layer. The relative spatial positions of the features in the next layer are defined based on the relative positions of the upper left corners of the corresponding spatial grids in the previous layer. When performing convolutions in the qth layer, one can align the filter at $L_{q+1} = (L_q - F_q + 1)$ positions along the height and $B_{q+1} = (B_q - F_q + 1)$ along the width of the image (without having a portion of the filter "sticking out" from the borders of the image). This results in a total of $L_{q+1} \times B_{q+1}$ possible dot products, which defines the size of the next hidden layer. In the previous example, the values of L_2 and B_2 are therefore defined as follows:

$$L_2 = 32 - 5 + 1 = 28$$
$$B_2 = 32 - 5 + 1 = 28$$

The next hidden layer of size 28×28 is shown in Figure 8.1(a). However, this hidden layer also has a depth of size $d_2 = 5$. Where does this depth come from? This is achieved by using 5 different filters with their own independent sets of parameters. Each of these 5 sets

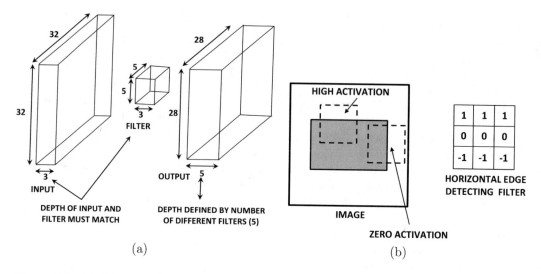

Figure 8.1: (a) The convolution between an input layer of size $32 \times 32 \times 3$ and a filter of size $5 \times 5 \times 3$ produces an output layer with spatial dimensions 28×28. The depth of the resulting output depends on the number of distinct filters and not on the dimensions of the input layer or filter. (b) Sliding a filter around the image tries to look for a particular feature in various windows of the image

of spatially arranged features obtained from the output of a single filter is referred to as a *feature map*. Clearly, an increased number of feature maps is a result of a larger number of filters (i.e., parameter footprint), which is $F_q^2 \cdot d_q \cdot d_{q+1}$ for the qth layer. *The number of filters used in each layer controls the capacity of the model because it directly controls the number of parameters.* Furthermore, increasing the number of filters in a particular layer increases the number of feature maps (i.e., depth) of the next layer. It is possible for different layers to have very different numbers of feature maps, depending on the number of filters we use for the convolution operation in the previous layer. For example, the input layer typically only has three color channels, but it is possible for the each of the later hidden layers to have depths (i.e., number of feature maps) of more than 500. The idea here is that each filter tries to identify a particular type of spatial pattern in a small rectangular region of the image, and therefore a large number of filters is required to capture a broad variety of the possible shapes that are combined to create the final image (unlike the case of the input layer, in which three RGB channels are sufficient). Typically, the later layers tend to have a smaller spatial footprint, but greater depth in terms of the number of feature maps. For example, the filter shown in Figure 8.1(b) represents a horizontal edge detector on a grayscale image with one channel. As shown in Figure 8.1(b), the resulting feature will have high activation at each position where a horizontal edge is seen. A perfectly vertical edge will give zero activation, whereas a slanted edge might give intermediate activation. Therefore, sliding the filter everywhere in the image will already detect several key outlines of the image in a single feature map of the output volume. Multiple filters are used to create an output volume with more than one feature map. For example, a different filter might create a spatial feature map of vertical edge activations.

We are now ready to formally define the convolution operation. The pth filter in the qth layer has parameters denoted by the 3-dimensional tensor $W^{(p,q)} = [w_{ijk}^{(p,q)}]$. The indices i, j, k indicate the positions along the height, width, and depth of the filter. The feature maps in the qth layer are represented by the 3-dimensional tensor $H^{(q)} = [h_{ijk}^{(q)}]$. When the value of q is 1, the special case corresponding to the notation $H^{(1)}$ simply represents the input layer (which is not hidden). Then, the convolutional operations from the qth layer to the $(q+1)$th layer are defined as follows:

$$h_{ijp}^{(q+1)} = \sum_{r=1}^{F_q} \sum_{s=1}^{F_q} \sum_{k=1}^{d_q} w_{rsk}^{(p,q)} h_{i+r-1,j+s-1,k}^{(q)} \qquad \forall i \in \{1 \ldots, L_q - F_q + 1\}$$

$$\forall j \in \{1 \ldots B_q - F_q + 1\}$$
$$\forall p \in \{1 \ldots d_{q+1}\}$$

The expression above seems notationally complex, although the underlying convolutional operation is really a simple dot product over the entire volume of the filter, which is repeated over all valid spatial positions (i, j) and filters (indexed by p). It is intuitively helpful to understand a convolution operation by placing the filter at each of the 28×28 possible spatial positions in the first layer of Figure 8.1(a) and performing a dot product between the vector of $5 \times 5 \times 3 = 75$ values in the filter and the corresponding 75 values in $H^{(1)}$. Even though the size of the input layer in Figure 8.1(a) is 32×32, there are only $(32-5+1) \times (32-5+1)$ possible spatial alignments between an input volume of size 32×32 and a filter of size 5×5.

The convolution operation brings to mind Hubel and Wiesel's experiments that use the activations in small regions of the visual field to activate particular neurons. In the case of convolutional neural networks, this visual field is defined by the filter, which is applied to all locations of the image in order to detect the presence of a shape at each spatial location. Furthermore, the filters in earlier layers tend to detect more primitive shapes, whereas the filters in later layers create more complex compositions of these primitive shapes. This is not particularly surprising because most deep neural networks are good at hierarchical feature engineering.

One property of convolution is that it shows *equivariance to translation*. In other words, if we shifted the pixel values in the input in any direction by one unit and then applied convolution, the corresponding feature values will shift with the input values. This is because of the shared parameters of the filter across the entire convolution. The reason for sharing parameters across the entire convolution is that the presence of a particular shape in any part of the image should be processed in the same way irrespective of its specific spatial location.

In the following, we provide an example of the convolution operation. In Figure 8.2, we have shown an example of an input layer and a filter with depth 1 for simplicity (which does occur in the case of grayscale images with a single color channel). Note that the depth of a layer must exactly match that of its filter/kernel, and the contributions of the dot products over all the feature maps in the corresponding grid region of a particular layer will need to be added (in the general case) to create a single output feature value in the next layer. Figure 8.2 depicts two specific examples of the convolution operations with a layer of size $7 \times 7 \times 1$ and a $3 \times 3 \times 1$ filter in the bottom row. Furthermore, the entire feature map of the next layer is shown on the upper right-hand side of Figure 8.2. Examples of two convolution

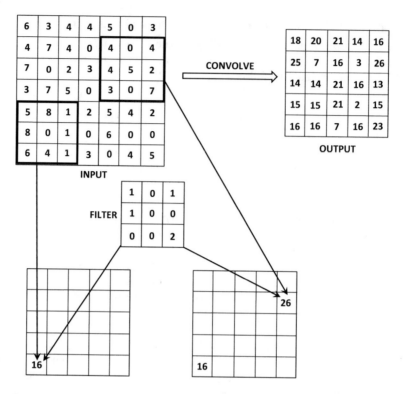

Figure 8.2: An example of a convolution between a $7 \times 7 \times 1$ input and a $3 \times 3 \times 1$ filter with stride of 1. A depth of 1 has been chosen for the filter/input for simplicity. For depths larger than 1, the contributions of each input feature map will be added to create a single value in the feature map. A single filter will always create a single feature map irrespective of its depth

operations are shown in which the outputs are 16 and 26, respectively. These values are arrived at by using the following multiplication and aggregation operations:

$$5 \times 1 + 8 \times 1 + 1 \times 1 + 1 \times 2 = 16$$
$$4 \times 1 + 4 \times 1 + 4 \times 1 + 7 \times 2 = 26$$

The multiplications with zeros have been omitted in the above aggregation. In the event that the depths of the layer and its corresponding filter are greater than 1, the above operations are performed for each spatial map and then aggregated across the entire depth of the filter.

A convolution in the qth layer increases the *receptive field* of a feature from the qth layer to the $(q+1)$th layer. In other words, each feature in the next layer captures a larger spatial region in the input layer. For example, when using a 3×3 filter convolution successively in three layers, the activations in the first, second, and third hidden layers capture pixel regions of size 3×3, 5×5, and 7×7, respectively, in the *original input image*. As we will see later, other types of operations increase the receptive fields further, as they reduce the size of the spatial footprint of the layers. This is a natural consequence of the fact that features in later layers capture complex characteristics of the image over larger spatial regions, and then combine the simpler features in earlier layers.

When performing the operations from the qth layer to the $(q + 1)$th layer, the depth d_{q+1} of the computed layer depends on the *number* of filters in the qth layer, and it is independent of the *depth* of the qth layer or any of its other dimensions. In other words, the depth d_{q+1} in the $(q + 1)$th layer is always equal to the number of filters in the qth layer. For example, the depth of the second layer in Figure 8.1(a) is 5, because a total of five filters are used in the first layer for the transformation. However, in order to perform the convolutions in the second layer (to create the third layer), one must now use filters of depth 5 in order to match the new depth of this layer, even though filters of depth 3 were used in the convolutions of the first layer (to create the second layer).

8.3.1 Padding

One observation is that the convolution operation reduces the size of the $(q + 1)$th layer in comparison with the size of the qth layer. This type of reduction in size is not desirable in general, because it tends to lose some information along the borders of the image (or of the feature map, in the case of hidden layers). This problem can be resolved by using *padding*. In padding, one adds $(F_q - 1)/2$ "pixels" all around the borders of the feature map in order to maintain the spatial footprint. Note that these pixels are really feature values in the case of padding hidden layers. The value of each of these padded feature values is set to 0, irrespective of whether the input or the hidden layers are being padded. As a result, the spatial height and width of the input volume will both increase by $(F_q - 1)$, which is exactly what they reduce by (in the output volume) after the convolution is performed. The padded portions do not contribute to the final dot product because their values are set to 0. In a sense, what padding does is to allow the convolution operation with a portion of the filter "sticking out" from the borders of the layer and then performing the dot product only over the portion of the layer where the values are defined. This type of padding is referred to as *half-padding* because (almost) half the filter is sticking out from all sides of the spatial input in the case where the filter is placed in its extreme spatial position along the edges. Half-padding is designed to maintain the spatial footprint exactly.

When padding is not used, the resulting "padding" is also referred to as a *valid padding*. Valid padding generally does not work well from an experimental point of view. Using half-padding ensures that some of the critical information at the borders of the layer is represented in a standalone way. In the case of valid padding, the contributions of the pixels on the borders of the layer will be under-represented compared to the central pixels in the next hidden layer, which is undesirable. Furthermore, this under-representation will be compounded over multiple layers. Therefore, padding is typically performed in all layers, and not just in the first layer where the spatial locations correspond to input values. Consider a situation in which the layer has size $32 \times 32 \times 3$ and the filter is of size $5 \times 5 \times 3$. Therefore, $(5 - 1)/2 = 2$ zeros are padded on all sides of the image. As a result, the 32×32 spatial footprint first increases to 36×36 because of padding, and then it reduces back to 32×32 after performing the convolution. An example of the padding of a single feature map is shown in Figure 8.3, where two zeros are padded on all sides of the image (or feature map). This is a similar situation as discussed above (in terms of addition of two zeros), except that the spatial dimensions of the image are much smaller than 32×32 in order to enable illustration in a reasonable amount of space.

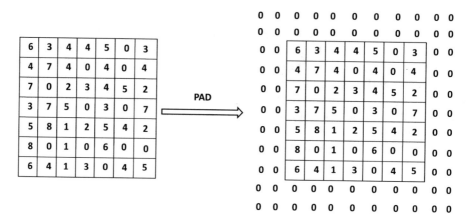

Figure 8.3: An example of padding. Each of the d_q activation maps in the entire depth of the qth layer are padded in this way

Another useful form of padding is *full-padding*. In full-padding, we allow (almost) the *full* filter to stick out from various sides of the input. In other words, a portion of the filter of size $F_q - 1$ is allowed to stick out from any side of the input with an overlap of only one spatial feature. For example, the kernel and the input image might overlap at a single pixel at an extreme corner. Therefore, the input is padded with $(F_q - 1)$ zeros on each side. In other words, each spatial dimension of the input increases by $2(F_q - 1)$. Therefore, if the input dimensions in the original image are L_q and B_q, the padded spatial dimensions in the input volume become $L_q + 2(F_q - 1)$ and $B_q + 2(F_q - 1)$. After performing the convolution, the feature-map dimensions in layer $(q+1)$ become $L_q + F_q - 1$ and $B_q + F_q - 1$, respectively. While convolution normally reduces the spatial footprint, full padding *increases* the spatial footprint. Interestingly, full-padding increases each dimension of the spatial footprint by the same value $(F_q - 1)$ that no-padding decreases it. *This relationship is not a coincidence because a "reverse" convolution operation can be implemented by applying another convolution on the fully padded output (of the original convolution) with an appropriately defined kernel of the same size.*

8.3.2 Strides

There are other ways in which convolution can reduce the spatial footprint of the image (or hidden layer). The above approach performs the convolution at every position in the spatial location of the feature map. However, it is not necessary to perform the convolution at every spatial position in the layer. One can reduce the level of granularity of the convolution by using the notion of *strides*. The description above corresponds to the case when a stride of 1 is used. When a stride of S_q is used in the qth layer, the convolution is performed at the locations 1, $S_q + 1$, $2 S_q + 1$, and so on along both spatial dimensions of the layer. The spatial size of the output on performing this convolution[1] has height of $(L_q - F_q)/S_q + 1$ and a width of $(B_q - F_q)/S_q + 1$. As a result, the use of strides will result in a reduction of each spatial dimension of the layer by a factor of approximately S_q and the area by S_q^2,

---

[1] Here, it is assumed that $(L_q - F_q)$ is exactly divisible by S_q in order to obtain a clean fit of the convolution filter with the original image. Otherwise, some ad hoc modifications are needed to handle edge effects. In general, this is not a desirable solution.

although the actual factor may vary because of edge effects. It is most common to use a stride of 1, although a stride of 2 is occasionally used as well. It is rare to use strides more than 2 in normal circumstances. Even though a stride of 4 was used in the input layer of the winning architecture [107] of the ILSVRC competition of 2012, the winning entry in the subsequent year reduced the stride to 2 [206] to improve accuracy. Larger strides can be helpful in memory-constrained settings or to reduce overfitting if the spatial resolution is unnecessarily high. Strides have the effect of rapidly increasing the receptive field of each feature in the hidden layer, while reducing the spatial footprint of the entire layer. An increased receptive field is useful in order to capture a complex feature in a larger spatial region of the image. As we will see later, the hierarchical feature engineering process of a convolutional neural network captures more complex shapes in later layers. Historically, the receptive fields have been increased with another operation, known as the *max-pooling* operation. In recent years, larger strides have been used in lieu [73, 172] of max-pooling operations.

8.3.3 Typical Settings

It is common to use stride sizes of 1 in most settings. Even when strides are used, small strides of size 2 are used. Furthermore, it is common to have $L_q = B_q$. In other words, it is desirable to work with square images. In cases where the input images are not square, preprocessing is used to enforce this property. For example, one can extract square patches of the image to create the training data. The number of filters in each layer is often a power of 2, because this often results in more efficient processing. Such an approach also leads to hidden layer depths that are powers of 2. Typical values of the spatial extent of the filter size (denoted by F_q) are 3 or 5. In general, small filter sizes often provide the best results, although some practical challenges exist in using filter sizes that are too small. Small filter sizes typically lead to deeper networks (for the same parameter footprint) and therefore tend to be more powerful. In fact, one of the top entries in an ILSVRC contest, referred to as *VGG* [169], was the first to experiment with a spatial filter dimension of only $F_q = 3$ for all layers, and the approach was found to work very well in comparison with larger filter sizes.

Use of Bias

As in all neural networks, it is also possible to add biases to the forward operations. Each unique filter in a layer is associated with its own bias. Therefore, the pth filter in the qth layer has bias $b^{(p,q)}$. When any convolution is performed with the pth filter in the qth layer, the value of $b^{(p,q)}$ is added to the dot product. The use of the bias simply increases the number of parameters in each filter by 1, and therefore it is not a significant overhead. Like all other parameters, the bias is learned during backpropagation. One can treat the bias as a weight of a connection whose input is always set to +1. This special input is used in all convolutions, irrespective of the spatial location of the convolution. Therefore, one can assume that a special pixel appears in the input whose value is always set to 1. Therefore, the number of input features in the qth layer is $1 + L_q \times B_q \times d_q$. This is a standard feature-engineering trick that is used for handling bias in all forms of machine learning.

8.3.4 The ReLU Layer

The convolution operation is interleaved with the pooling and ReLU operations. The ReLU activation is not very different from how it is applied in a traditional neural network. For each of the $L_q \times B_q \times d_q$ values in a layer, the ReLU activation function is applied to it to create $L_q \times B_q \times d_q$ thresholded values. These values are then passed on to the next layer. Therefore, applying the ReLU does not change the dimensions of a layer because it is a simple one-to-one mapping of activation values. In traditional neural networks, the activation function is combined with a linear transformation with a matrix of weights to create the next layer of activations. Similarly, a ReLU typically follows a convolution operation (which is the rough equivalent of the linear transformation in traditional neural networks), and the ReLU layer is often not explicitly shown in pictorial illustrations of the convolution neural network architectures.

It is noteworthy that the use of the ReLU activation function is a recent evolution in neural network design. In the earlier years, saturating activation functions like sigmoid and tanh were used. However, it was shown in [107] that the use of the ReLU has tremendous advantages over these activation functions both in terms of speed and accuracy. Increased speed is also connected to accuracy because it allows one to use deeper models and train them for a longer time. In recent years, the use of the ReLU activation function has replaced the other activation functions in convolutional neural network design to an extent that this chapter will simply use the ReLU as the default activation function (unless otherwise mentioned).

8.3.5 Pooling

The pooling operation is, however, quite different. The pooling operation works on small grid regions of size $P_q \times P_q$ in each layer, and produces another layer *with the same depth* (unlike filters). For each square region of size $P_q \times P_q$ in each of the d_q activation maps, the *maximum* of these values is returned. This approach is referred to as *max-pooling*. If a stride of 1 is used, then this will produce a new layer of size $(L_q - P_q + 1) \times (B_q - P_q + 1) \times d_q$. However, it is more common to use a stride $S_q > 1$ in pooling. In such cases, the length of the new layer will be $(L_q - P_q)/S_q + 1$ and the breadth will be $(B_q - P_q)/S_q + 1$. Therefore, pooling drastically reduces the spatial dimensions of each activation map.

Unlike with convolution operations, pooling is done at the level of *each* activation map. Whereas a convolution operation simultaneously uses all d_q feature maps in combination with a filter to produce a single feature value, pooling independently operates on each feature map to produce another feature map. Therefore, the operation of pooling does not change the number of feature maps. In other words, the depth of the layer created using pooling is the same as that of the layer on which the pooling operation was performed. Examples of pooling with strides of 1 and 2 are shown in Figure 8.4. Here, we use pooling over 3×3 regions. The typical size P_q of the region over which one performs pooling is 2×2. At a stride of 2, there would be no overlap among the different regions being pooled, and it is quite common to use this type of setting. However, it has sometimes been suggested that it is desirable to have at least some overlap among the spatial units at which the pooling is performed, because it makes the approach less likely to overfit.

Other types of pooling (like average-pooling) are possible but rarely used. In the earliest convolutional network, referred to as *LeNet-5*, a variant of average pooling was used and was referred[2] to as *subsampling*. In general, max-pooling remains more popular than

---

[2]In recent years, subsampling also refers to other operations that reduce the spatial footprint. Therefore,

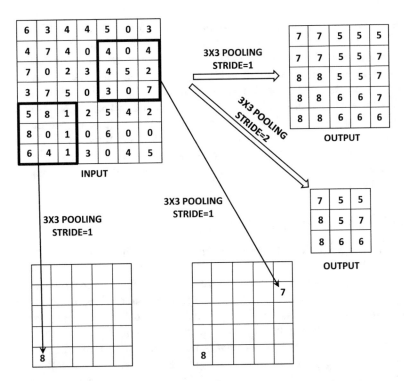

Figure 8.4: An example of a max-pooling of one activation map of size 7×7 with strides of 1 and 2. A stride of 1 creates a 5×5 activation map with heavily repeating elements because of maximization in overlapping regions. A stride of 2 creates a 3×3 activation map with less overlap. Unlike convolution, each activation map is independently processed and therefore the number of output activation maps is exactly equal to the number of input activation maps

average pooling. The max-pooling layers are interleaved with the convolutional/ReLU layers, although the former typically occurs much less frequently in deep architectures. This is because pooling drastically reduces the spatial size of the feature map, and only a few pooling operations are required to reduce the spatial map to a small constant size.

It is common to use pooling with 2×2 filters and a stride of 2, when it is desired to reduce the spatial footprint of the activation maps. Pooling results in (some) invariance to translation because shifting the image reduces the shift in the activation map significantly. This property is referred to as *translation invariance*. The idea is that similar images often have very different relative locations of the distinctive shapes within them, and translation invariance helps in being able to classify such images in a similar way. For example, one should be able to classify a bird as a bird, irrespective of where it occurs in the image.

Another important purpose of pooling is that it increases the size of the receptive field while reducing the spatial footprint of the layer because of the use of strides larger than 1. Increased sizes of receptive fields are needed to be able to capture larger regions of the image within a complex feature in later layers. Most of the rapid reductions in spatial footprints of the layers (and corresponding increases in receptive fields of the features) are caused by the pooling operations. Convolutions increase the receptive field only gently unless the

there is some difference between the classical usage of this term and modern usage.

stride is larger than 1. In recent years, it has been suggested that pooling is not always necessary. One can design a network with only convolutional and ReLU operations, and obtain the expansion of the receptive field by using larger strides within the convolutional operations [73, 172]. Therefore, there is an emerging trend in recent years to get rid of the max-pooling layers altogether. However, this trend has not been fully established and validated, as of the writing of this book. There seem to be at least some arguments in favor of max-pooling. Max-pooling introduces nonlinearity and a greater amount of translation invariance, as compared to strided convolutions. Although nonlinearity can be achieved with the ReLU activation function, the key point is that the effects of max-pooling cannot be exactly replicated by strided convolutions either. At the very least, the two operations are not fully interchangeable.

8.3.6 Fully Connected Layers

Each feature in the final spatial layer is connected to each hidden state in the first fully connected layer. This layer functions in exactly the same way as a traditional feed-forward network. In most cases, one might use more than one fully connected layer to increase the power of the computations towards the end. The connections among these layers are exactly structured like a traditional feed-forward network. Since the fully connected layers are densely connected, the vast majority of parameters lie in the fully connected layers. For example, if each of two fully connected layers has 4096 hidden units, then the connections between them have more than 16 million weights. Similarly, the connections from the last spatial layer to the first fully connected layer will have a large number of parameters. Even though the convolutional layers have a larger number of *activations* (and a larger memory footprint), the fully connected layers often have a larger number of *connections* (and parameter footprint). The reason that activations contribute to the memory footprint more significantly is that the number of activations are multiplied by mini-batch size while tracking variables in the forward and backward passes of backpropagation. These trade-offs are useful to keep in mind while choosing neural-network design based on specific types of resource constraints (e.g., data versus memory availability). It is noteworthy that the nature of the fully-connected layer can be sensitive to the application at hand. For example, the nature of the fully-connected layer for a classification application would be somewhat different from the case of a segmentation application. The aforementioned discussion is for the most common use-case of a classification application.

The output layer of a convolutional neural network is designed in an application-specific way. In the following, we will consider the representative application of classification. In such a case, the output layer is fully connected to every neuron in the penultimate layer, and has a weight associated with it. One might use the logistic, softmax, or linear activation depending on the nature of the application (e.g., classification or regression).

One alternative to using fully connected layers is to use average pooling across the whole spatial area of the final set of activation maps to create a single value. Therefore, the number of features created in the final spatial layer will be exactly equal to the number of filters. In this scenario, if the final activation maps are of size $7 \times 7 \times 256$, then 256 features will be created. Each feature will be the result of aggregating 49 values. This type of approach greatly reduces the parameter footprint of the fully connected layers, and it has some advantages in terms of generalizability. This approach was used in *GoogLeNet* [184]. In some applications like image segmentation, each pixel is associated with a class label, and one does not use fully connected layers. Fully convolutional networks with 1×1 convolutions are used in order to create an output spatial map.

8.3.7 The Interleaving between Layers

The convolution, pooling, and ReLU layers are typically interleaved in a neural network in order to increase the expressive power of the network. The ReLU layers often follow the convolutional layers, just as a nonlinear activation function typically follows the linear dot product in traditional neural networks. Therefore, the convolutional and ReLU layers are typically stuck together one after the other. Some pictorial illustrations of neural architectures like *AlexNet* [107] do not explicitly show the ReLU layers because they are assumed to be always stuck to the end of the linear convolutional layers. After two or three sets of convolutional-ReLU combinations, one might have a max-pooling layer. Examples of this basic pattern are as follows:

CRCRP

CRCRCRP

Here, the convolutional layer is denoted by C, the ReLU layer is denoted by R, and the max-pooling layer is denoted by P. This entire pattern (including the max-pooling layer) might be repeated a few times in order to create a deep neural network. For example, if the first pattern above is repeated three times and followed by a fully connected layer (denoted by F), then we have the following neural network:

CRCRPCRCRPCRCRPF

The description above is not complete because one needs to specify the number/size/padding of filters/pooling layers. The pooling layer is the key step that tends to reduce the spatial footprint of the activation maps because it uses strides that are larger than 1. It is also possible to reduce the spatial footprints with strided convolutions instead of max-pooling. These networks are often quite deep, and it is not uncommon to have convolutional networks with more than 15 layers. Recent architectures also use *skip connections* between layers, which become increasingly important as the depth of the network increases (cf. Section 8.4.3).

LeNet-5

Early networks were quite shallow. An example of one of the earliest neural networks is *LeNet-5* [110]. The input data is in grayscale, and there is only one color channel. The input is assumed to be the ASCII representation of a character. For the purpose of discussion, we will assume that there are ten types of characters (and therefore 10 outputs), although the approach can be used for any number of classes.

The network contained two convolution layers, two pooling layers, and three fully connected layers at the end. However, later layers contain multiple feature maps because of the use of multiple filters in each layer. The architecture of this network is shown in Figure 8.5. The first fully connected layer was also referred to as a convolution layer (labeled as $C5$) in the original work because the ability existed to generalize it to spatial features for larger input maps. However, the specific implementation of *LeNet-5* really used $C5$ as a fully connected layer, because the filter spatial size was the same as the input spatial size. This is why we are counting $C5$ as a fully connected layer in this exposition. It is noteworthy that two versions of *LeNet-5* are shown in Figures 8.5(a) and (b). The upper diagram of Figure 8.5(a) explicitly shows the subsampling layers, which is how the architecture was presented in the original work. However, deeper architectural diagrams like *AlexNet* [107] often do not show the subsampling or max-pooling layers explicitly in order to accommodate the large number of layers. Such a concise architecture for *LeNet-5* is illustrated in

Figure 8.5(b). The activation function layers are also not explicitly shown in either figure. In the original work in *LeNet-5*, the sigmoid activation function occurs immediately after the subsampling operations, although this ordering is relatively unusual in recent architectures. In most modern architectures, subsampling is replaced by max-pooling, and the max-pooling layers occur less frequently than the convolution layers. Furthermore, the activations are typically performed immediately after each convolution (rather than after each max-pooling).

The number of layers in the architecture is often counted in terms of the number of layers with weighted spatial filters and the number of fully connected layers. In other words, subsampling/max-pooling and activation function layers are often not counted separately. The subsampling in *LeNet-5* used 2×2 spatial regions with stride 2. Furthermore, unlike max-pooling, the values were averaged, scaled with a trainable weight and then a bias was added. In modern architectures, the linear scaling and bias addition operations have been dispensed with. The concise architectural representation of Figure 8.5(b) is sometimes confusing to beginners because it is missing details such as the size of the max-pooling/subsampling filters. In fact, there is no unique way of representing these architectural details, and many variations are used by different authors. This chapter will show several such examples in the case studies.

This network is extremely shallow by modern standards; yet the basic principles have not changed since then. The main difference is that the ReLU activation had not appeared at that point, and sigmoid activation was often used in the earlier architectures. Furthermore, the use of average pooling is extremely uncommon today compared to max-pooling. Recent years have seen a move away from both max-pooling and subsampling, with strided convolutions as the preferred choice. *LeNet-5* also used ten radial basis function (RBF) units in the final layer (cf. Chapter 9), in which the prototype of each unit was compared to its input vector and the squared Euclidean distance between them was output. This is the same as using the negative log-likelihood of the Gaussian distribution represented by that RBF unit. The parameter vectors of the RBF units were chosen by hand, and correspond to a stylized 7×12 bitmap image of the corresponding character class, which were flattened into a $7 \times 12 = 84$-dimensional representation. Note that the size of the penultimate layer is exactly 84 in order to enable the computation of the Euclidean distance between the vector corresponding to that layer and the parameter vector of the RBF unit. The ten outputs in the final layer provide the scores of the classes, and the smallest score among the ten units provides the prediction. This type of use of RBF units is now anachronistic in modern convolutional network design, and one generally tends to work with softmax units with log-likelihood loss on multinomial label outputs. *LeNet-5* was used extensively for character recognition, and was used by many banks to read checks.

8.3.8 Hierarchical Feature Engineering

It is instructive to examine the activations of the filters created by real-world images in different layers. The activations of the filters in the early layers are low-level features like edges, whereas those in later layers put together these low-level features. For example, a mid-level feature might put together edges to create a hexagon, whereas a higher-level feature might put together the mid-level hexagons to create a honeycomb. It is fairly easy to see why a low-level filter might detect edges. Consider a situation in which the color of the image changes along an edge. As a result, the difference between neighboring pixel values will be non-zero only across the edge. This can be achieved by choosing the appropriate weights in the corresponding low-level filter. Note that the filter to detect a horizontal edge

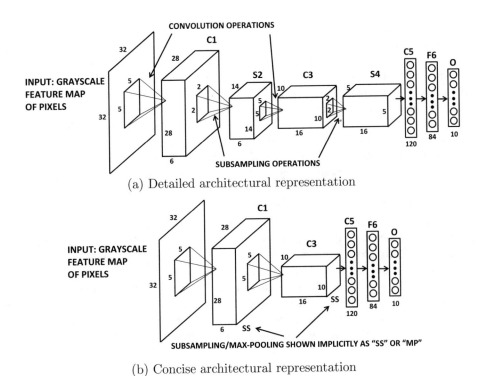

(a) Detailed architectural representation

(b) Concise architectural representation

Figure 8.5: LeNet-5: One of the earliest convolutional neural networks

will not be the same as that to detect a vertical edge. This brings us back to Hubel and Weisel's experiments in which different neurons in the cat's visual cortex were activated by different edges. Examples of filters detecting horizontal and vertical edges are illustrated in Figure 8.6. The next layer filter works on the hidden features and therefore it is harder to interpret. Nevertheless, the next layer filter is able to detect a rectangle by combining the horizontal and vertical edges.

The smaller portions of real-world image activate different hidden features, much like the biological model of Hubel and Wiesel in which different shapes seem to activate different

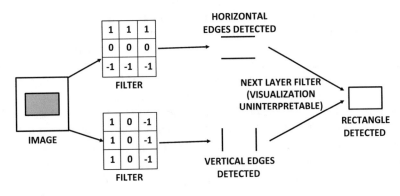

Figure 8.6: Filters detect edges and combine them to create rectangle

neurons. Therefore, the power of convolutional neural networks rests in the ability to put together these primitive shapes into more complex shapes layer by layer. Note that it is impossible for the first convolution layer to learn any feature that is larger than $F_1 \times F_1$ pixels, where the value of F_1 is typically a small number like 3 or 5. However, the next convolution layer will be able to put together many of these patches together to create a feature from an area of the image that is larger. The primitive features learned in earlier layers are put together in a semantically coherent way to learn increasingly complex and interpretable visual features. The choice of learned features is affected by how backpropagation adapts the features to the needs of the loss function at hand. For example, if an application is training to classify images as cars, the approach might learn to put together arcs to create a circle, and then it might put together circles with other shapes to create a car wheel. All this is enabled by the hierarchical features of a deep network.

Recent *ImageNet* competitions have demonstrated that much of the power in image recognition lies in increased depth of the network. Not having enough layers effectively prevents the network from learning the hierarchical regularities in the image that are combined to create its semantically relevant components. Another important observation is that the nature of the features learned will be sensitive to the specific data set at hand. For example, the features learned to recognize trucks will be different from those learned to recognize carrots. However, some data sets (like *ImageNet*) are diverse enough that the features learned by training on these data sets have general-purpose significance across many applications.

8.4 Case Studies of Convolutional Architectures

In the following, we provide some case studies of convolutional architectures. These case studies were derived from successful entries to the ILSVRC competition in recent years. These are instructive because they provide an understanding of the important factors in neural network design that can make these networks work well. Even though recent years have seen some changes in architectural design (like ReLU activation), it is striking how similar the modern architectures are to the basic design of *LeNet-5*. The main changes from *LeNet-5* to modern architectures are in terms of the explosion of depth, the use of ReLU activation, and the training efficiency enabled by modern hardware/optimization enhancements. Modern architectures are deeper, and they use a variety of computational, architectural, and hardware tricks to efficiently train these networks with large amounts of data. Hardware advancements should not be underestimated; modern GPU-based platforms are 10,000 times faster than the (similarly priced) systems available at the time *LeNet-5* was proposed. Even on these modern platforms, it often takes a week to train a convolutional neural network that is accurate enough to be competitive at ILSVRC. The hardware, data-centric, and algorithmic enhancements are connected to some extent. It is difficult to try new algorithmic tricks if enough data and computational power is not available to experiment with complex/deeper models in a reasonable amount of time. Therefore, the recent revolution in deep convolutional networks could not have been possible, had it not been for the large amounts of data and increased computational power available today.

In the following sections, we provide an overview of some of the well-known models that are often used for designing training algorithms for image classification. It is worth mentioning that some of these models are available as pretrained models over *ImageNet*, and the resulting features can be used for applications beyond classification. Such an approach is a form of transfer learning, which is discussed later in this section.

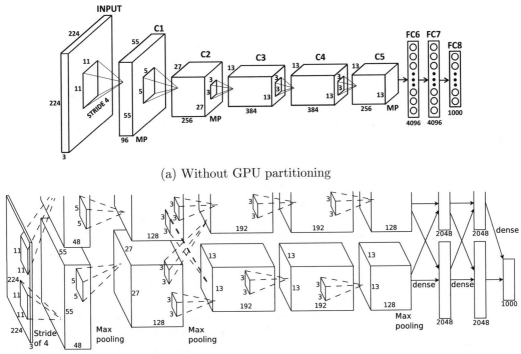

(a) Without GPU partitioning

(b) With GPU partitioning (original architecture)

(b) With GPU partitioning (original architecture)

Figure 8.7: The *AlexNet* architecture. The ReLU activations follow each convolution layer, and are not explicitly shown. Note that the max-pooling layers are labeled as MP, and they follow only a subset of the convolution-ReLU combination layers. The architectural diagram in (b) is from [A. Krizhevsky, I. Sutskever, and G. Hinton. Imagenet classification with deep convolutional neural networks. *NIPS Conference*, pp. 1097–1105. 2012.] ©2012 A. Krizhevsky, I. Sutskever, and G. Hinton

8.4.1 AlexNet

AlexNet was the winner of the 2012 ILSVRC competition. The architecture of *AlexNet* is shown in Figure 8.7(a). It is worth mentioning that there were two parallel pipelines of processing in the original architecture, which are not shown in Figure 8.7(a). These two pipelines are caused by two GPUs working together to build the training model with a faster speed and memory sharing. The network was originally trained on a GTX 580 GPU with 3 GB of memory, and it was impossible to fit the intermediate computations in this amount of space. Therefore, the network was partitioned across two GPUs. The original architecture is shown in Figure 8.7(b), in which the work is partitioned into two GPUs. We also show the architecture without the changes caused by the GPUs, so that it can be more easily compared with other convolutional neural network architectures discussed in this chapter. It is noteworthy that the GPUs are inter-connected in only a subset of the layers in Figure 8.7(b), which leads to some differences between Figure 8.7(a) and 8.7(b) in terms of the actual model constructed. Specifically, the GPU-partitioned architecture has fewer weights because not all layers have interconnections. Dropping some of the interconnections reduces the communication time between the processors and therefore helps in efficiency.

AlexNet starts with $224 \times 224 \times 3$ images and uses 96 filters of size $11 \times 11 \times 3$ in the first layer. A stride of 4 is used. This results in a first layer of size $55 \times 55 \times 96$. After the first layer has been computed, a max-pooling layer is used. This layer is denoted by 'MP' in Figure 8.7(a). Note that the architecture of Figure 8.7(a) is a simplified version of the architecture shown in Figure 8.7(b), which explicitly shows the two parallel pipelines. For example, Figure 8.7(b) shows a depth of the first convolution layer of only 48, because the 96 feature maps are divided among the GPUs for parallelization. On the other hand, Figure 8.7(a) does not assume the use of GPUs, and therefore the width is explicitly shown as 96. The ReLU activation function was applied after each convolutional layer, which was followed by response normalization and max-pooling. Although max-pooling has been annotated in the figure, it has not been assigned a block in the architecture. Furthermore, the ReLU and response normalization layers are not explicitly shown in the figure. These types of concise representations are common in pictorial depictions of neural architectures.

The second convolutional layer uses the response-normalized and pooled output of the first convolutional layer and filters it with 256 filters of size $5 \times 5 \times 96$. No intervening pooling or normalization layers are present in the third, fourth, or fifth convolutional layers. The sizes of the filters of the third, fourth, and fifth convolutional layers are $3 \times 3 \times 256$ (with 384 filters), $3 \times 3 \times 384$ (with 384 filters), and $3 \times 3 \times 384$ (with 256 filters). All max-pooling layers used 3×3 filters at stride 2. Therefore, there was some overlap among the pools. The fully connected layers have 4096 neurons. The final set of 4096 activations can be treated as a 4096-dimensional representation of the image. The final layer of AlexNet uses a 1000-way softmax in order to perform the classification. It is noteworthy that the final layer of 4096 activations (labeled by FC7 in Figure 8.7(b)) is often used to create a flat 4096 dimensional representation of an image for applications beyond classification. One can extract these features for any out-of-sample image by simply passing it through the trained neural network. These features often generalize well to other data sets and other tasks. Such features are referred to as FC7 features. In fact, the use of the extracted features from the penultimate layer as FC7 was popularized after AlexNet, even though the approach was known much earlier. As a result, such extracted features from the penultimate layer of a convolutional neural network are often referred to as FC7 features, irrespective of the number of layers in that network. It is noteworthy that the number of feature maps in middle layers is far larger than the initial depth of the volume in the input layer (which is only 3 corresponding to red, green, and blue colors) although their spatial dimensions are smaller. This is because the initial depth only contains the red, green, and blue color components, whereas the later layers capture different types of semantic features in the features maps.

Many design choices used in the architecture became standard in later architectures. A specific example is the use of ReLU activation in the architecture (instead of sigmoid or tanh units). The choice of the activation function in most convolutional neural networks today is almost exclusively focused on the ReLU, although this was not the case before AlexNet. Some other training tricks were known at the time, but their use in AlexNet popularized them. One example was the use of data augmentation, which turned out to be very useful in improving accuracy. AlexNet also underlined the importance of using specialized hardware like GPUs for training on such large data sets. Dropout was used with L_2-weight decay in order to improve generalization. The use of Dropout is common in virtually all types of architectures today because it provides an additional booster in most cases. An idea called local response normalization was also used, which was eventually found not to be useful and discarded by later architectures (including later implementations of this architecture).

We also briefly mention the parameter choices used in *AlexNet*. The interested reader can find the full code and parameter files of *AlexNet* at [219]. L_2-regularization was used with a parameter of 5×10^{-4}. *Dropout* was used by sampling units at a probability of 0.5. Momentum-based (mini-batch) stochastic gradient descent was used for training *AlexNet* with parameter value of 0.8. The batch-size was 128. The learning rate was 0.01, although it was eventually reduced a couple of times as the method began to converge. Even with the use of the GPU, the training time of *AlexNet* was of the order of a week.

The final top-5 error rate, which was defined as the fraction of cases in which the correct image was not included in the top-5 images, was about 15.4%. This error rate[3] was in comparison with the previous winners with an error rate of more than 25%. The gap with respect to the second-best performer in the contest was also similar. The use of single convolutional network provided a top-5 error rate of 18.2%, although using an ensemble of seven models provided the winning error-rate of 15.4%. Note that these types of ensemble-based tricks provide a consistent improvement of between 2% and 3% with most architectures. Furthermore, since the executions of most ensemble methods are embarrassingly parallelizable, it is relatively easy to perform them, as long as sufficient hardware resources are available. *AlexNet* is considered a fundamental advancement within the field of computer vision because of the large margin with which it won the ILSVRC contest. This success rekindled interest in deep learning in general, and convolutional neural networks in particular.

8.4.2 VGG

VGG [169] further emphasized the developing trend in terms of increased depth of networks. The tested networks were designed with various configurations with sizes between 11 and 19 layers, although the best-performing versions had 16 or more layers. *VGG* was a top-performing entry on ISLVRC in 2014, but it was not the winner. The winner was *GoogLeNet*, which had a top-5 error rate of 6.7% in comparison with the top-5 error rate of 7.3% for *VGG*. Nevertheless, *VGG* was important because it illustrated several important design principles that eventually became standard in future architectures.

An important innovation of *VGG* is that it reduced filter sizes but increased depth. It is important to understand that *reduced filter size necessitates increased depth*. This is because a small filter can capture only a small region of the image unless the network is deep. For example, a single feature that is a result of three sequential convolutions of size 3×3 will capture a region in the input of size 7×7. Note that using a single 7×7 filter directly on the input data will also capture the visual properties of a 7×7 input region. In the first case, we are using $3 \times 3 \times 3 = 27$ parameters, whereas we are using $7 \times 7 \times 1 = 49$ parameters in the second case. Therefore, the parameter footprint is smaller in the case when three sequential convolutions are used. However, three successive convolutions can often capture more interesting and complex features than a single convolution, and the resulting activations with a single convolution will look like primitive edge features. Therefore, the network with 7×7 filters will be unable to capture sophisticated shapes in smaller regions.

In general, greater depth forces more nonlinearity and greater regularization. A deeper network will have more nonlinearity because of the presence of more ReLU layers, and more regularization because the increased depth forces a structure on the layers through the use of repeated composition of convolutions. As discussed above, architectures with greater depth and reduced filter size require fewer parameters. This occurs in part because the number of parameters in each layer is given by the square of the filter size, whereas

---

[3]The top-5 error rate makes more sense in image data where a single image might contain objects of multiple classes. Throughout this chapter, we use the term "error rate" to refer to the top-5 error rate.

Table 8.1: Configurations used in *VGG*. The term C3D64 refers to the case in which convolutions are performed with 64 filters of spatial size 3×3 (and occasionally 1×1). The depth of the filter matches the corresponding layer. The padding of each filter is chosen in order to maintain the spatial footprint of the layer. All convolutions are followed by ReLU. The max-pool layer is referred to as M, and local response normalization as LRN. The softmax layer is denoted by S, and FC4096 refers to a fully connected layer with 4096 units. Other than the final set of layers, the number of filters always increases after each max-pooling. Therefore, reduced spatial footprint is often accompanied with increased depth

Name:	A	A-LRN	B	C	D	E
# Layers	11	11	13	16	16	19
	C3D64	C3D64	C3D64	C3D64	C3D64	C3D64
		LRN	C3D64	C3D64	C3D64	C3D64
	M	M	M	M	M	M
	C3D128	C3D128	C3D128	C3D128	C3D128	C3D128
			C3D128	C3D128	C3D128	C3D128
	M	M	M	M	M	M
	C3D256	C3D256	C3D256	C3D256	C3D256	C3D256
	C3D256	C3D256	C3D256	C3D256	C3D256	C3D256
				C1D256	C3D256	C3D256
						C3D256
	M	M	M	M	M	M
	C3D512	C3D512	C3D512	C3D512	C3D512	C3D512
	C3D512	C3D512	C3D512	C3D512	C3D512	C3D512
				C1D512	C3D512	C3D512
						C3D512
	M	M	M	M	M	M
	C3D512	C3D512	C3D512	C3D512	C3D512	C3D512
	C3D512	C3D512	C3D512	C3D512	C3D512	C3D512
				C1D512	C3D512	C3D512
						C3D512
	M	M	M	M	M	M
	FC4096	FC4096	FC4096	FC4096	FC4096	FC4096
	FC4096	FC4096	FC4096	FC4096	FC4096	FC4096
	FC1000	FC1000	FC1000	FC1000	FC1000	FC1000
	S	S	S	S	S	S

the number of parameters depend linearly on the depth. Therefore, one can drastically reduce the number of parameters by using smaller filter sizes, and instead "spend" these parameters by using increased depth. Increased depth also allows the use of a greater number of nonlinear activations, which increases the discriminative power of the model. Therefore *VGG* always uses filters with spatial footprint 3×3 and pooling of size 2×2. The convolution was done with stride 1, and a padding of 1 was used. The pooling was done at stride 2. Using a 3×3 filter with a padding of 1 maintains the spatial footprint of the output volume, although pooling always compresses the spatial footprint. Therefore, the pooling was done on non-overlapping spatial regions (unlike the previous two architectures), and always reduced the spatial footprint (i.e., both height and width) by a factor of 2. Another interesting design choice of *VGG* was that the number of filters was often increased by a factor of 2 after each max-pooling. The idea was to always increase the depth by a factor of 2 whenever the spatial footprint reduced by a factor of 2. This design choice results in some level of balance in the computational effort across layers, and was inherited by some of the later architectures like *ResNet*.

One issue with using deep configurations was that increased depth led to greater sensitivity with initialization, which is known to cause instability. This problem was solved by using pretraining, in which a shallower architecture was first trained, and then further layers were added. However, the pretraining was not done on a layer-by-layer basis. Rather, an 11-layer subset of the architecture was first trained. These trained layers were used to initialize a subset of the layers in the deeper architecture. *VGG* achieved a top-5 error of only 7.3% in the ISLVRC contest, which was one of the top performers but not the winner. The different configurations of *VGG* are shown in Table 8.1. Among these, the architecture denoted by column D was the winning architecture. Note that the number of filters increase by a factor of 2 after each max-pooling. Therefore, max-pooling causes the spatial height and width to reduce by a factor of 2, but this is compensated by increasing depth by a factor of 2. Performing convolutions with 3×3 filters and padding of 1 does not change the spatial footprint. Therefore, the sizes of each spatial dimension (i.e., height and width) in the regions between different max-pooling layers in column D of Table 8.1 are 224, 112, 56, 28, and 14, respectively. A final max-pooling is performed just before creating the fully connected layer, which reduces the spatial footprint further to 7. Therefore, the first fully connected layer has dense connections between 4096 neurons and a $7 \times 7 \times 512$ volume. As we will see later, most of the parameters of the neural network are hidden in these connections.

An interesting exercise has been shown in [96] about where most of the parameters and the memory of the activations is located. In particular, the vast majority of the *memory* required for storing the activations and gradients in the forward and backward phases are required by the early part of the convolutional neural network with the largest spatial footprint. This point is significant because the memory required by a mini-batch is scaled by the size of the mini-batch. For example, it is has been shown in [96] that about 93MB are required for each image. Therefore, for a mini-batch size of 128, the total memory requirement would be about 12GB. Although the early layers require the most memory because of their large spatial footprints, they do not have a large parameter footprint because of the sparse connectivity and weight sharing. In fact, most of the parameters are required by the fully connected layers at the end. The connection of the final $7 \times 7 \times 512$ spatial layer (cf. column D in Table 8.1) to the 4096 neurons required $7 \times 7 \times 512 \times 4096 = 102,760,448$ parameters. The total number of parameters in *all* layers was about $138,000,000$. Therefore, *nearly 75% of the parameters are in a single layer of connections.* Furthermore, the majority of the remaining parameters are in the final two fully connected layers. In all, dense connectivity accounts for 90% of the parameter footprint in the neural network.

It is notable that some of the architectures allow 1×1 convolutions. Although a 1×1 convolution does not combine the activations of spatially adjacent features, it does combine the feature values of different channels when the depth of a volume is greater than 1. Using a 1×1 convolution is also a way to incorporate additional nonlinearity into the architecture without making fundamental changes at the spatial level. This additional nonlinearity is incorporated via the ReLU activations attached to each layer. Refer to [169] for more details.

8.4.3 ResNet

ResNet [73] used 152 layers, which was almost an order of magnitude greater than previously used by other architectures. This architecture was the winner of the ILSVRC competition in 2015, and it achieved a top-5 error of 3.6%, which resulted in the first classifier with human-level performance. This accuracy is achieved by an ensemble of *ResNet* networks; even a single model achieves 4.5% accuracy. Training an architecture with 152 layers is generally not possible unless some important innovations are incorporated.

The main issue in training such deep networks is that the gradient flow between layers is impeded by the large number of operations in deep layers that can increase or decrease the size of the gradients. These problems are referred to as the *vanishing and exploding gradient problems,* which are caused by increased depth. However, the work in [73] suggests that the main training problem in such deep networks might not necessarily be caused by these problems, especially if batch normalization is used. The main problem is caused by the difficulty in getting the learning process to converge properly in a reasonable amount of time. Such convergence problems are common in networks with complex loss surfaces. Although some deep networks show large gaps between training and test error, the error on both the training ad test data is high in many deep networks. This implies that the optimization process has not made sufficient progress.

Although hierarchical feature engineering is the holy grail of learning with neural networks, its layer-wise implementations force all concepts in the image to require the same level of abstraction. Some concepts can be learned by using shallow networks, whereas others require fine-grained connections. For example, consider a circus elephant standing on a square frame. Some of the intricate features of the elephant might require a large number of layers to engineer, whereas the features of the square frame might require very few layers. Convergence will be unnecessarily slow when one is using a very deep network with a fixed depth across all paths to learn concepts, many of which can also be learned using shallow architectures. Why not let the neural network decide how many layers to use to learn each feature?

ResNet uses *skip connections* between layers in order to enable copying between layers and introduces an *iterative view* of feature engineering (as opposed to a hierarchical view). As discussed in a later section, long-short term memory networks leverage similar principles in sequence data by allowing portions of the states to be copied from one layer to the next with the use of adjustable *gates.* Most feed-forward networks only contain connections between layers i and $(i+1)$ (which tend to scramble the states), whereas *ResNet* contains *direct* connections between layers i and $(i+r)$ for $r > 1$. Examples of such skip connections, which form the basic unit of *ResNet,* are shown in Figure 8.8(a) with $r = 2$. This skip connection simply copies the input of layer i and adds it to the output of layer $(i+r)$. Such an approach enables effective gradient flow because the backpropagation algorithm now has a superhighway for propagating the gradients backwards using the skip connections. This basic unit is referred to as a *residual module,* and the entire network is created by putting together many of these basic modules. In most layers, an appropriately padded filter[4] is used with a stride of 1, so that the spatial size and depth of the input does not change from layer to layer. In such cases, it is easy to simply add the input of the ith layer to that of $(i + r)$. However, some layers do use strided convolutions to reduce each spatial dimension by a factor of 2. At the same time, depth is increased by a factor of 2 by using a larger number of filters. In such a case, one cannot use the identity function over the skip connection. Therefore, a linear projection matrix might need to be applied over the skip connection in order to adjust the dimensionality. This projection matrix defines a set of 1×1 convolution operations with stride of 2 in order to reduce spatial extent by factor of

---

[4]Typically, a 3×3 filter is used at a stride/padding of 1. This trend started with the principles in *VGG,* and was adopted by *ResNet.*

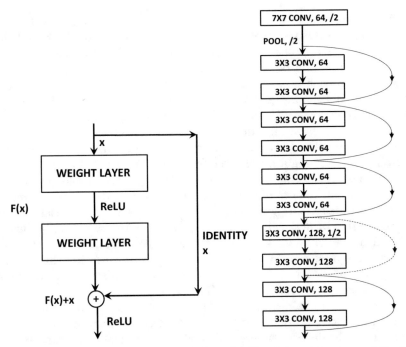

(a) Skip-connections in residual module   (b) Partial architecture of *ResNet*

Figure 8.8: The residual module and the first few layers of *ResNet*

2. The parameters of the projection matrix need to be learned during backpropagation.

In the original idea of *ResNet*, one only adds connections between layers i and $(i+r)$. For example, if we use $r = 2$, only skip connections only between successive odd layers are used. Later enhancements like *DenseNet* showed improved performance by adding connections between all pairs of layers. The basic unit of Figure 8.8(a) is repeated in *ResNet*, and therefore one can traverse the skip connections repeatedly in order to propagate input to the output after performing very few forward computations. An example of the first few layers of the architecture is shown in Figure 8.8(b). This particular snapshot is based on the first few layers of the 34-layer architecture. Most of the skip connections are shown in solid lines in Figure 8.8(b), which corresponds to the use of the identity function with an unchanged filter volume. However, in some layers, a stride of 2 is used, which causes the spatial and depth footprint to change. In these layers, a projection matrix needs to be used, which is denoted by a dashed skip connection. Four different architectures were tested in the original work [73], which contained 34, 50, 101, and 152 layers, respectively. The 152-layer architecture had the best performance, but even the 34-layer architecture performed better than did the best-performing ILSVRC entry from the previous year.

The use of skip connections provides paths of unimpeded gradient flow and therefore has important consequences for the behavior of the backpropagation algorithm. The skip connections take on the function of super-highways in enabling gradient flow, creating a situation where multiple paths of variable lengths exist from the input to the output. In such cases, the shortest paths enable the most learning, and the longer paths can be viewed as residual contributions. This gives the learning algorithm the flexibility of choosing the appropriate level of nonlinearity for a particular input. Inputs that can be classified with a small amount of nonlinearity will skip many connections. Other inputs with a more complex structure might traverse a larger number of connections in order to extract the relevant features. Therefore, the approach is also referred to as residual learning, in which learning along longer paths is a kind of fine tuning of the easier learning along shorter paths. In other words, the approach is well suited to cases in which different aspects of the image have different levels of complexity. The work in [73] shows that the residual responses from deeper layers are often relatively small, which validates the intuition that fixed depth is an impediment to proper learning. In such cases, the convergence is often not a problem, because the shorter paths enable a significant portion of the learning with unimpeded gradient flows. An interesting insight in [188] is that *ResNet* behaves like an ensemble of shallow networks because many alternative paths of shorter length are enabled by this type of architecture. Only a small amount of learning is enabled by the deeper paths, and only when it is absolutely necessary. The work in [188] in fact provides a pictorial depiction of an unraveled architecture of *ResNet* in which the different paths are explicitly shown in a parallel pipeline. This unraveled view provides a clear understanding of why *ResNet* has some similarities with ensemble-centric design principles. A consequence of this point of view is that dropping some of the layers from a trained *ResNet* at prediction time does not degrade accuracy as significantly as other networks like *VGG*.

More insights can be obtained by reading the work on *wide residual networks* [205]. This work suggests that increased depth of the residual network does not always help because most of the extremely deep paths are not used anyway. The skip connections do result in alternative paths and effectively increase the width of the network. The work in [205] suggests that better results can be obtained by limiting the total number of layers to some extent (say, 50 instead of 150), and using an increased number of filters in each layer. Note that a depth of 50 is still quite large from pre-*ResNet* standards, but is low compared to the depth used in recent experiments with residual networks. This approach also helps in parallelizing operations.

8.5 Principles Underlying Recurrent Neural Networks

The recurrent neural network is designed for sequential data types such as time-series, text, and biological data. The common characteristics of these data sets is that they contain sequential dependencies among the attributes. Examples of such dependencies are as follows:

1. In a time-series data set, the values on successive time-stamps are closely related to one another. Permuting the order of values loses the signal in the time series, and therefore one must use this ordering in the neural model.

2. Although text is often processed as a bag of words, one can obtain better semantic insights when the ordering of the words is used. In such cases, it is important to construct models that take the sequencing information into account.

3. Biological data often contains sequences, in which the symbols might correspond to amino acids or one of the nucleobases that form the building blocks of DNA.

The individual values in a sequence can be either real-valued or symbolic. Real-valued sequences are also referred to as time-series. Recurrent neural networks can be used for either type of data. In practical applications, the use of symbolic values is more common. Our exposition will primarily focus on symbolic data in general, and on text data in particular. The default assumption will be that the input to the recurrent network is a text segment

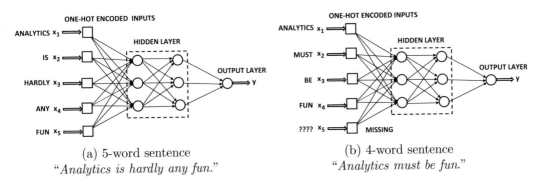

(a) 5-word sentence
"*Analytics is hardly any fun.*"

(b) 4-word sentence
"*Analytics must be fun.*"

Figure 8.9: An attempt to use a conventional neural network for sentiment analysis faces the challenge of variable-length inputs. The network architecture also does not contain any helpful information about sequential dependencies among successive words

in which the corresponding symbols of the sequence are the word identifiers of the lexicon. However, we will also examine other settings, such as cases in which the individual elements are characters or in which they are real values.

Many sequence-centric applications like text are often processed as bags of words. Such an approach ignores the ordering of words in the document, and works well for documents of reasonable size. However, in applications where the semantic interpretation of the sentence is important, or in which the size of the text segment is relatively small (e.g., a single sentence), such an approach is simply inadequate. In order to understand this point, consider the following pair of sentences:

> The lion chased the deer.
> The deer chased the lion.

The two sentences are clearly very different (and the second one is unusual). However, the bag-of-words representation would deem them identical. Hence, this type of representation works well for simpler applications (such as classification), but a greater degree of linguistic intelligence is required for more sophisticated applications such as *sentiment analysis, machine translation,* or *information extraction.*

One possible solution is to avoid the bag-of-words approach and create one input for each position in the sequence. Consider a situation in which one tried to use a conventional neural network in order to perform sentiment analysis on sentences with one input for each position in the sentence. The sentiment can be a binary label depending on whether it is positive or negative. The first problem that one would face is that the length of different sentences is different. Therefore, if we used a neural network with 5 sets of one-hot encoded word inputs (cf. Figure 8.9(a)), it would be impossible to enter a sentence with more than five words. Furthermore, any sentence with less than five words would have missing inputs (cf. Figure 8.9(b)), which needs potentially wasteful padding with dummy words. In some cases, such as Web log sequences, the length of the input sequence might run into the hundreds of thousands. More importantly, *it is important to somehow encode information about the word ordering more directly within the architecture of the network.* Therefore, the two main desiderata for the processing of sequences include (i) the ability to receive and process inputs in the same order as they are present in the sequence, and (ii) the treatment of inputs at each time-stamp in a similar manner in relation to previous history of inputs. A key challenge is that we somehow need to construct a neural network with a fixed number of parameters, but with the ability to process a variable number of inputs.

These desiderata are naturally satisfied with the use of *recurrent neural networks* *(RNNs)*. In a recurrent neural network, there is a one-to-one correspondence between the layers in the network and the specific positions in the sequence. The position in the sequence is also referred to as its *time-stamp*. Therefore, instead of a variable number of inputs in a single input layer, the network contains a variable number of layers, and each layer has a single input corresponding to that time-stamp. Therefore, the inputs are allowed to directly interact with down-stream hidden layers depending on their positions in the sequence. Each layer uses the same set of parameters to ensure similar modeling at each time stamp, and therefore the number of parameters is fixed as well. In other words, the same layer-wise architecture is repeated in time, and therefore the network is referred to as *recurrent*. Recurrent neural networks are also feed-forward networks with a specific structure based on the notion of *time layering*, so that they can take a *sequence* of inputs and produce a sequence of outputs. Each temporal layer can take in an input data point (either single attribute or multiple attributes), and optionally produce a multidimensional output. Such models are particularly useful for sequence-to-sequence learning applications like machine translation or for predicting the next element in a sequence.

There are significant challenges in learning the parameters of a recurrent neural network. As a result, a number of variants of the recurrent neural network, such as long short-term memory (LSTM) and gated recurrent unit (GRU), have been proposed. Recurrent neural networks and their variants have been used in many applications like sequence-to-sequence learning, image captioning, machine translation, and sentiment analysis.

8.6   The Architecture of Recurrent Neural Networks

In the following, the basic architecture of a recurrent network will be described. Although the recurrent neural network can be used in almost any sequential domain, its use in the text domain is both widespread and natural. We will assume the use of the text domain throughout this section in order to enable intuitively simple explanations of various concepts. Therefore, the focus of this chapter will be mostly on discrete RNNs, since that is the most popular use case. Note that exactly the same neural network can be used both for building a word-level RNN and a character-level RNN. The only difference between the two is the set of base symbols used to define the sequence. For consistency, we will stick to the word-level RNN while introducing the notations and definitions. However, variations of this setting are also discussed in this chapter.

The simplest recurrent neural network is shown in Figure 8.10(a). A key point here is the presence of the self-loop in Figure 8.10(a), which will cause the hidden state of the neural network to change after the input of each word in the sequence. In practice, one only works with sequences of finite length, and it makes sense to unfold the loop into a "time-layered" network that looks more like a feed-forward network. This network is shown in Figure 8.10(b). Note that in this case, we have a different node for the hidden state at each time-stamp and the self-loop has been unfurled into a feed-forward network. This representation is mathematically equivalent to Figure 8.10(a), but is much easier to comprehend because of its similarity to a traditional network. The weight matrices in different temporal layers *are shared* to ensure that the same function is used at each time-stamp. The annotations W_{xh}, W_{hh}, and W_{hy} of the weight matrices in Figure 8.10(b) make the sharing evident.

It is noteworthy that Figure 8.10 shows a case in which each time-stamp has an input,

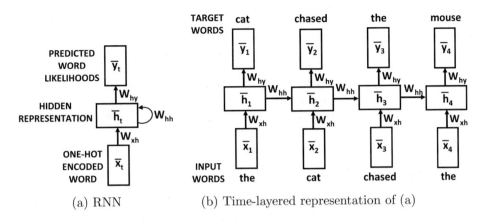

(a) RNN   (b) Time-layered representation of (a)

Figure 8.10: A recurrent neural network and its time-layered representation

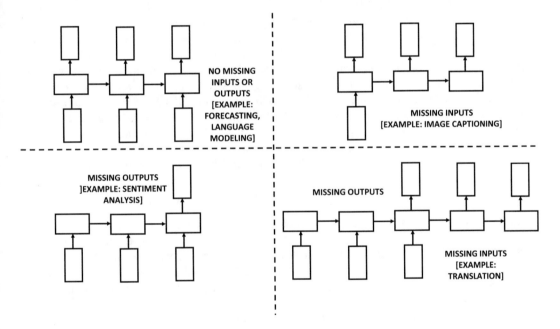

Figure 8.11: The different variations of recurrent networks with missing inputs and outputs

output, and hidden unit. In practice, it is possible for either the input or the output units to be missing at any particular time-stamp. Examples of cases with missing inputs and outputs are shown in Figure 8.11. The choice of missing inputs and outputs would depend on the specific application at hand. For example, in a time-series forecasting application, we might need outputs at each time-stamp in order to predict the next value in the time-series. On the other hand, in a sequence-classification application, we might only need a single output label at the end of the sequence corresponding to its class. In general, it is possible for any subset of inputs or outputs to be missing in a particular application. The following discussion will assume that all inputs and outputs are present, although it is easy to generalize it to the case where some of them are missing by simply removing the corresponding terms in the forward propagation equations.

The particular architecture shown in Figure 8.10 is suited to language modeling. A language model is a well-known concept in natural language processing that predicts the next word, given the previous history of words. Given a sequence of words, their one-hot encoding is fed one at a time to the neural network in Figure 8.10(a). This temporal process is equivalent to feeding the individual words to the inputs at the relevant time-stamps in Figure 8.10(b). A time-stamp corresponds to the position in the sequence, which starts at 0 (or 1), and increases by 1 by moving forward in the sequence by one unit. In the setting of language modeling, the output is a vector of probabilities predicted for the next word in the sequence. For example, consider the sentence:

The lion chased the deer.

When the word "*The*" is input, the output will be a vector of probabilities of the entire lexicon that includes the word "*lion*," and when the word "*lion*" is input, we will again get a vector of probabilities predicting the next word. This is, of course, the classical definition of a language model in which the probability of a word is estimated based on the immediate history of previous words. In general, the input vector at time t (e.g., one-hot encoded vector of the tth word) is $\overline{x}_t$, the hidden state at time t is $\overline{h}_t$, and the output vector at time t (e.g., predicted probabilities of the $(t+1)$th word) is $\overline{y}_t$. Both $\overline{x}_t$ and $\overline{y}_t$ are d-dimensional for a lexicon of size d. The hidden vector $\overline{h}_t$ is p-dimensional, where p regulates the complexity of the embedding. For the purpose of discussion, we will assume that all these vectors are column vectors. In many applications like classification, the output is not produced at each time unit but is only triggered at the last time-stamp in the end of the sentence. Although output and input units may be present only at a subset of the time-stamps, we examine the simple case in which they are present in all time-stamps. Then, the hidden state at time t is given by a function of the input vector at time t and the hidden vector at time $(t-1)$:

$$\overline{h}_t = f(\overline{h}_{t-1}, \overline{x}_t) \tag{8.1}$$

This function is defined with the use of weight matrices and activation functions (as used by all neural networks for learning), and *the same weights are used at each time-stamp.* Therefore, even though the hidden state evolves over time, the weights and the underlying function $f(\cdot, \cdot)$ remain fixed over all time-stamps (i.e., sequential elements) after the neural network has been trained. A separate function $\overline{y}_t = g(\overline{h}_t)$ is used to learn the output probabilities from the hidden states.

Next, we describe the functions $f(\cdot, \cdot)$ and $g(\cdot)$ more concretely. We define a $p \times d$ input-hidden matrix W_{xh}, a $p \times p$ hidden-hidden matrix W_{hh}, and a $d \times p$ hidden-output matrix W_{hy}. Then, one can expand Equation 8.1 and also write the condition for the outputs as follows:

$$\overline{h}_t = \tanh(W_{xh}\overline{x}_t + W_{hh}\overline{h}_{t-1})$$
$$\overline{y}_t = W_{hy}\overline{h}_t$$

Here, the "tanh" notation is used in a relaxed way, in the sense that the function is applied to the p-dimensional column vector in an element-wise fashion to create a p-dimensional vector with each element in $[-1, 1]$. Throughout this section, this relaxed notation will be used for several activation functions such as tanh and sigmoid. In the very first time-stamp, $\overline{h}_{t-1}$ is assumed to be some default constant vector (such as 0), because there is no input from the hidden layer at the beginning of a sentence. One can also learn this vector, if desired. Although the hidden states change at each time-stamp, the weight matrices stay fixed over the various time-stamps. Note that the output vector $\overline{y}_t$ is a set of continuous values with the same dimensionality as the lexicon. A softmax layer is applied on top of $\overline{y}_t$ so that the results can be interpreted as probabilities. *The p-dimensional output $\overline{h}_t$ of the hidden layer at the end of a text segment of t words yields its embedding, and the p-dimensional columns of W_{xh} yield the embeddings of individual words.*

Because of the recursive nature of Equation 8.1, the recurrent network has the *ability to compute a function of variable-length inputs.* In other words, one can expand the recurrence of Equation 8.1 to define the function for $\overline{h}_t$ in terms of t inputs. For example, starting at $\overline{h}_0$, which is typically fixed to some constant vector (such as the zero vector), we have $\overline{h}_1 = f(\overline{h}_0, \overline{x}_1)$ and $\overline{h}_2 = f(f(\overline{h}_0, \overline{x}_1), \overline{x}_2)$. Note that $\overline{h}_1$ is a function of only $\overline{x}_1$, whereas $\overline{h}_2$ is a function of both $\overline{x}_1$ and $\overline{x}_2$. In general, $\overline{h}_t$ is a function of $\overline{x}_1 \ldots \overline{x}_t$. Since the output $\overline{y}_t$ is a function of $\overline{h}_t$, these properties are inherited by $\overline{y}_t$ as well. In general, we can write the following:

$$\overline{y}_t = F_t(\overline{x}_1, \overline{x}_2, \ldots \overline{x}_t) \tag{8.2}$$

Note that the function $F_t(\cdot)$ varies with the value of t although its relationship to its immediately previous state is always the same (based on Equation 8.1). Such an approach is particularly useful for variable-length inputs. This setting occurs often in many domains like text in which the sentences are of variable length. For example, in a language modeling application, the function $F_t(\cdot)$ indicates the probability of the next word, taking into account all the previous words in the sentence.

8.6.1 Language Modeling Example of RNN

In order to illustrate the workings of the RNN, we will use a toy example of a single sequence defined on a vocabulary of four words. Consider the sentence:

The lion chased the deer.

In this case, we have a lexicon of four words, which are {*"the," "lion," "chased," "deer"*}. In Figure 8.12, we have shown the probabilistic prediction of the next word at each of time-stamps from 1 to 4. Ideally, we would like the probability of the next word to be predicted correctly from the probabilities of the previous words. Each one-hot encoded input vector $\overline{x}_t$ has length four, in which only one bit is 1 and the remaining bits are 0s. The main flexibility here is in the dimensionality p of the hidden representation, which we set to 2 in this case. As a result, the matrix W_{xh} will be a 2×4 matrix, so that it maps a one-hot encoded input vector into a hidden vector $\overline{h}_t$ vector of size 2. As a practical matter, each column of W_{xh} corresponds to one of the four words, and one of these columns is copied by the expression $W_{xh}\overline{x}_t$. Note that this expression is added to $W_{hh}\overline{h}_t$ and then transformed with the tanh function to produce the final expression. The final output $\overline{y}_t$ is defined by $W_{hy}\overline{h}_t$. Note that the matrices W_{hh} and W_{hy} are of sizes 2×2 and 4×2, respectively.

In this case, the outputs are continuous values (not probabilities) in which larger values indicate greater likelihood of presence. These continuous values are eventually converted to probabilities with the softmax function, and therefore one can treat them as substitutes to log probabilities. The word *"lion"* is predicted in the first time-stamp with a value of 1.3, although this value seems to be (incorrectly) outstripped by *"deer"* for which the corresponding value is 1.7. However, the word *"chased"* seems to be predicted correctly at the next time-stamp. As in all learning algorithms, one cannot hope to predict every value exactly, and such errors are more likely to be made in the early iterations of the backpropagation algorithm. However, as the network is repeatedly trained over multiple iterations, it makes fewer errors over the training data.

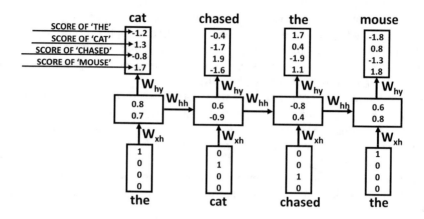

Figure 8.12: Example of language modeling with a recurrent neural network

8.6.1.1 Generating a Language Sample

Such an approach can also be used to generate an arbitrary sample of a language, once the training has been completed. How does one use such a language model at testing time, since each state requires an input word, and none is available during language generation? The likelihoods of the tokens at the first time-stamp can be generated using the <START> token as input. Since the <START> token is also available in the training data, the model will typically select a word that often starts text segments. Subsequently, the idea is to sample one of the tokens generated at each time-stamp (based on the predicted likelihood), and then use it as an input to the next time-stamp. To improve the accuracy of the sequentially predicted token, one might use beam search to expand on the most likely possibilities by always keeping track of the b best sequence prefixes of any particular length. The value of b is a user-driven parameter. By recursively applying this operation, one can generate an arbitrary sequence of text that reflects the particular training data at hand. If the <END> token is predicted, it indicates the end of that particular segment of text. Although such an approach often results in syntactically correct text, it might be nonsensical in meaning. For example, a character-level RNN[5] authored by Karpathy, Johnson, and Fei Fei [94, 222] was trained on William Shakespeare's plays. A character-level RNN requires the neural network to learn both syntax *and* spelling. After only five iterations of learning across the full data set, the following was a sample of the output:

> KING RICHARD II:
> Do cantant,-'for neight here be with hand her,-
> Eptar the home that Valy is thee.
>
> NORONCES:
> Most ma-wrow, let himself my hispeasures;
> An exmorbackion, gault, do we to do you comforr,
> Laughter's leave: mire sucintracce shall have theref-Helt.

Note that there are a large number of misspellings in this case, and a lot of the words are gibberish. However, when the training was continued to 50 iterations, the following was

---

[5]A long-short term memory network (LSTM) was used, which is a variation on the vanilla RNN discussed here.

generated as a part of the sample:

> KING RICHARD II:
> Though they good extremit if you damed;
> Made it all their fripts and look of love;
> Prince of forces to uncertained in conserve
> To thou his power kindless. A brives my knees
> In penitence and till away with redoom.
>
> GLOUCESTER:
> Between I must abide.

This generated piece of text is largely consistent with the syntax and spelling of the archaic English in William Shakespeare's plays, although there are still some obvious errors. Furthermore, the approach also indents and formats the text in a manner similar to the plays by placing new lines at reasonable locations. Continuing to train for more iterations makes the output almost error-free, and some impressive samples are also available at [95].

Of course, the semantic meaning of the text is limited, and one might wonder about the usefulness of generating such nonsensical pieces of text from the perspective of machine learning applications. The key point here is that by providing an additional *contextual* input, such as the neural representation of an image, the neural network can be made to give intelligent outputs such as a grammatically correct description (i.e., caption) of the image. In other words, language models are best used by generating *conditional* outputs.

The primary goal of the language-modeling RNN is not to create arbitrary sequences of the language, but to provide an architectural base that can be modified in various ways to incorporate the effect of the specific context. For example, applications like machine translation and image captioning learn a language model that is *conditioned* on another input such as a sentence in the source language or an image to be captioned. Therefore, the precise design of the application-dependent RNN will use the same principles as the language-modeling RNN, but will make small changes to this basic architecture in order to incorporate the specific context.

It is noteworthy that a recurrent neural network provides two different types of embeddings:

1. The *activations* of the hidden units at each time-stamp contain the multidimensional embedding of the segment sequence up to that time-stamp. Therefore, a recurrent neural network provides an embedding of each prefix of the sequence, together with the full sequence.

2. The *weight matrix* W_{xh} from the input to hidden layer contains the embedding of each word. The weight matrix W_{xh} is a $p \times d$ matrix for a lexicon of size d. This means that each of the d columns of this matrix contain a p-dimensional embedding for one of the words. These embeddings provide an alternative to the *word2vec* embeddings that are obtained using a simpler neural architecture on windows of words. Details of word2vec embeddings are provided in [6, 122, 123].

In application-dependent settings, the embeddings may be context-sensitive in nature, depend upon the other types of inputs that are used for learning. For example, an image-to-text application may take as input an embedding from a convolutional neural network, and therefore the text embedding may merge with the context of an image embedding. We will provide an example of this setting in the *image captioning application* discussed later in this chapter. In all these cases, the key is in choosing the input and output values of the recurrent units in a judicious way, so that one can backpropagate the output errors and learn the weights of the neural network in an application-dependent way.

8.6.2 Backpropagation Through Time

The negative logarithms of the softmax probability of the correct words at the various time-stamps are aggregated to create the loss function. The softmax function is introduced in Equation 7.16 of Chapter 7. If the output vector $\overline{y}_t$ can be written as $[\hat{y}_t^1 \ldots \hat{y}_t^d]$, it is first converted into a vector of d probabilities using the softmax function:

$$[\hat{p}_t^1 \ldots \hat{p}_t^d] = \text{Softmax}([\hat{y}_t^1 \ldots \hat{y}_t^d])$$

If j_t is the index of the ground-truth word at time t in the training data, then the loss function L for all T time-stamps is computed as follows:

$$L = -\sum_{t=1}^{T} \log(\hat{p}_t^{j_t}) \tag{8.3}$$

This loss function is the same as the loss function of multinomial logistic regression in Chapter 6. The derivative of the loss function with respect to the raw outputs may be computed as follows (cf. Chapter 6):

$$\frac{\partial L}{\partial \hat{y}_t^k} = \hat{p}_t^k - I(k, j_t) \tag{8.4}$$

Here, $I(k, j_t)$ is an indicator function that is 1 when k and j_t are the same, and 0, otherwise. Starting with this partial derivative, one can use the straightforward backpropagation update of Chapter 7 (on the unfurled temporal network) to compute the gradients with respect to the weights in different layers. The main problem is that the weight sharing across different temporal layers will have an effect on the update process. An important assumption in correctly using the chain rule for backpropagation (cf. Chapter 7) is that the weights in different layers are distinct from one another, which allows a relatively straightforward update process. However, it is not difficult to modify the backpropagation algorithm to handle shared weights.

The main trick for handling shared weights is to first "pretend" that the parameters in the different temporal layers are independent of one another. For this purpose, we introduce the temporal variables $W_{xh}^{(t)}$, $W_{hh}^{(t)}$ and $W_{hy}^{(t)}$ for time-stamp t. Conventional backpropagation is first performed by working under the pretense that these variables are distinct from one another. Then, the contributions of the different temporal avatars of the weight parameters to the gradient are added to create a unified update for each weight parameter. This special type of backpropagation algorithm is referred to as *backpropagation through time (BPTT)*. We summarize the BPTT algorithm as follows:

(i) Run the input sequentially in the forward direction through time and compute the errors (and the negative-log loss of softmax layer) at each time-stamp.

(ii) Compute the gradients of the edge weights in the backwards direction on the unfurled network without any regard for the fact that weights in different time layers are shared. In other words, it is assumed that the weights $W_{xh}^{(t)}$, $W_{hh}^{(t)}$ and $W_{hy}^{(t)}$ in time-stamp t are distinct from other time-stamps. As a result, one can use conventional backpropagation to compute $\frac{\partial L}{\partial W_{xh}^{(t)}}$, $\frac{\partial L}{\partial W_{hh}^{(t)}}$, and $\frac{\partial L}{\partial W_{hy}^{(t)}}$. Note that we have used matrix calculus notations where the derivative with respect to a matrix is defined by a corresponding matrix of element-wise derivatives.

(iii) Add all the derivatives with respect to different instantiations of an edge in time as follows:

$$\frac{\partial L}{\partial W_{xh}} = \sum_{t=1}^{T} \frac{\partial L}{\partial W_{xh}^{(t)}}$$

$$\frac{\partial L}{\partial W_{hh}} = \sum_{t=1}^{T} \frac{\partial L}{\partial W_{hh}^{(t)}}$$

$$\frac{\partial L}{\partial W_{hy}} = \sum_{t=1}^{T} \frac{\partial L}{\partial W_{hy}^{(t)}}$$

The above derivations follow from a straightforward application of the multivariate chain rule. Here, we are using the fact that the partial derivative of a temporal copy of each parameter (such as an element of $W_{xh}^{(t)}$) with respect to the original copy of the parameter (such as the corresponding element of W_{xh}) can be set to 1. Here, it is noteworthy that the computation of the partial derivatives with respect to the temporal copies of the weights is not different from traditional backpropagation at all. Therefore, one only needs to wrap the temporal aggregation around conventional backpropagation in order to compute the update equations. The original algorithm for backpropagation through time can be credited to Werbos's seminal work in 1990 [201], long before the use of recurrent neural networks became more popular.

Truncated Backpropagation Through Time

One of the computational problems in training recurrent networks is that the underlying sequences may be very long, as a result of which the number of layers in the network may also be very large. This can result in computational, convergence, and memory-usage problems. This problem is solved by using *truncated backpropagation through time*. This technique may be viewed as the analog of stochastic gradient descent for recurrent neural networks. In the approach, the state values are computed correctly during forward propagation, but the backpropagation updates are done only over segments of the sequence of modest length (such as 100). In other words, only the portion of the loss over the relevant segment is used to compute the gradients and update the weights. The segments are processed in the same order as they occur in the input sequence. The forward propagation does not need to be performed in a single shot, but it can also be done over the relevant segment of the sequence as long as the values in the final time-layer of the segment are used for computing the state values in the next segment of layers. The values in the final layer in the current segment are used to compute the values in the first layer of the next segment. Therefore, forward propagation is always able to accurately maintain state values, although the backpropagation uses only a small portion of the loss.

Practical Issues

The entries of each weight matrix are initialized to small values in $[-1/\sqrt{r}, 1/\sqrt{r}]$, where r is the number of columns in that matrix. One can also initialize each of the d columns of the input weight matrix W_{xh} to the *word2vec* embedding [122, 123] of the corresponding word. This approach is a form of pretraining. The specific advantage of using this type of pretraining depends on the amount of training data. It can be helpful to use this type of initialization when the amount of available training data is small.

Another detail is that the training data often contains a special <START> and an <END> token at the beginning and end of each training segment. These types of tokens help the model to recognize specific text units such as sentences, paragraphs, or the beginning of a particular module of text. The distribution of the words at the beginning of a segment of text is often very different than how it is distributed over the whole training data. Therefore, after the occurrence of <START>, the model is more likely to pick words that begin a particular segment of text.

There are other approaches that are used for deciding whether to end a segment at a particular point. A specific example is the use of a binary output that decides whether or not the sequence should continue at a particular point. Note that the binary output is in addition to other application-specific outputs. Typically, the sigmoid activation is used to model the prediction of this output, and the cross-entropy loss is used on this output. Such an approach is useful with real-valued sequences. This is because the use of <START> and <END> tokens is inherently designed for symbolic sequences. However, one disadvantage of this approach is that it changes the loss function from its application-specific formulation to one that provides a balance between end-of-sequence prediction and application-specific needs. Therefore, the weights of different components of the loss function would be yet another hyper-parameter that one would have to work with.

There are also several practical challenges in training an RNN, which make the design of various architectural enhancements of the RNN necessary. It is also noteworthy that multiple hidden layers (with long short-term memory enhancements) are used in all practical applications, which will be discussed in Section 8.6.3. However, the application-centric exposition will use the simpler single-layer model for clarity. The generalization of each of these applications to enhanced architectures is straightforward.

8.6.3   Multilayer Recurrent Networks

In all the aforementioned applications, a single-layer RNN architecture is used for ease in understanding. However, in practical applications, a multilayer architecture is used in order to build models of greater complexity. Furthermore, this multilayer architecture can be used in combination with advanced variations of the RNN, such as the LSTM architecture or the gated recurrent unit. These advanced architectures are introduced in later sections.

An example of a deep network containing three layers is shown in Figure 8.13. Note that nodes in higher-level layers receive input from those in lower-level layers. The relationships among the hidden states can be generalized directly from the single-layer network. First, we rewrite the recurrence equation of the hidden layers (for single-layer networks) in a form that can be adapted easily to multilayer networks:

$$\overline{h}_t = \tanh(W_{xh}\overline{x}_t + W_{hh}\overline{h}_{t-1})$$

$$= \tanh W \begin{bmatrix} \overline{x}_t \\ \overline{h}_{t-1} \end{bmatrix}$$

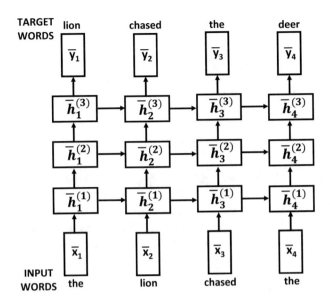

Figure 8.13: Multi-layer recurrent neural networks

Here, we have put together a larger matrix $W = [W_{xh}, W_{hh}]$ that includes the columns of W_{xh} and W_{hh}. Similarly, we have created a larger column vector that stacks up the state vector in the first hidden layer at time $t - 1$ and the input vector at time t. In order to distinguish between the hidden nodes for the upper-level layers, let us add an additional superscript to the hidden state and denote the vector for the hidden states at time-stamp t and layer k by $\overline{h}_t^{(k)}$. Similarly, let the weight matrix for the kth hidden layer be denoted by $W^{(k)}$. It is noteworthy that the weights are shared across different time-stamps (as in the single-layer recurrent network), but they are not shared across different layers. Therefore, the weights are superscripted by the layer index k in $W^{(k)}$. The first hidden layer is special because it receives inputs both from the input layer at the current time-stamp and the adjacent hidden state at the previous time-stamp. Therefore, the matrices $W^{(k)}$ will have a size of $p \times (d+p)$ only for the first layer (i.e., $k = 1$), where d is the size of the input vector $\overline{x}_t$ and p is the size of the hidden vector $\overline{h}_t$. Note that d will typically not be the same as p. The recurrence condition for the first layer is already shown above by setting $W^{(1)} = W$. Therefore, let us focus on all the hidden layers k for $k \geq 2$. It turns out that the recurrence condition for the layers with $k \geq 2$ is also in a very similar form as the equation shown above:

$$\overline{h}_t^{(k)} = \tanh W^{(k)} \begin{bmatrix} \overline{h}_t^{(k-1)} \\ \overline{h}_{t-1}^{(k)} \end{bmatrix}$$

In this case, the size of the matrix $W^{(k)}$ is $p \times (p + p) = p \times 2p$. The transformation from hidden to output layer remains the same as in single-layer networks. It is easy to see that this approach is a straightforward multilayer generalization of the case of single-layer networks. It is common to use two or three layers in practical applications. In order to use a larger number of layers, it is important to have access to more training data in order to avoid overfitting.

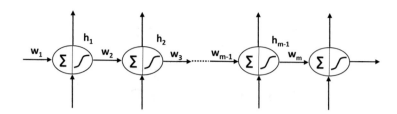

Figure 8.14: The vanishing and exploding gradient problems

8.7 Long Short-Term Memory (LSTM)

Recurrent neural networks have problems associated with vanishing and exploding gradients [84, 137, 138]. This is a common problem in neural network updates where successive multiplication by the matrix $W^{(k)}$ is inherently unstable; it either results in the gradient disappearing during backpropagation, or in it blowing up to large values in an unstable way. This type of instability is the direct result of successive multiplication with the (recurrent) weight matrix at various time-stamps.

Consider a set of T consecutive layers, in which the tanh activation function, $\Phi(\cdot)$, is applied between each pair of layers. The shared weight between a pair of hidden nodes is denoted by w. Let $h_1 \ldots h_T$ be the hidden values in the various layers. Let $\Phi'(h_t)$ be the derivative of the activation function in hidden layer t. Let the copy of the shared weight w in the tth layer be denoted by w_t so that it is possible to examine the effect of the backpropagation update. Let $\frac{\partial L}{\partial h_t}$ be the derivative of the loss function with respect to the hidden activation h_t. The neural architecture is illustrated in Figure 8.14. Then, one derives the following update equations using backpropagation:

$$\frac{\partial L}{\partial h_t} = \Phi'(w_{t+1}h_t) \cdot w_{t+1} \cdot \frac{\partial L}{\partial h_{t+1}} \tag{8.5}$$

Since the shared weights in different temporal layers are the same, the gradient is multiplied with the same quantity $w_t = w$ for each layer. Such a multiplication will have a consistent bias towards vanishing when $w < 1$, and it will have a consistent bias towards exploding when $w > 1$. However, the choice of the activation function will also play a role because the derivative $\Phi'(w_{t+1}h_t)$ is included in the product. For example, the presence of the tanh activation function, for which the derivative $\Phi'(\cdot)$ is almost always less than 1, tends to increase the chances of the vanishing gradient problem.

Although the above discussion only studies the simple case of a hidden layer with one unit, one can generalize the argument to a hidden layer with multiple units [89]. In such a case, it can be shown that the update to the gradient boils down to a repeated multiplication with the same matrix A. One can show the following result:

Lemma 8.7.1 *Let A be a square matrix, the **magnitude** of whose largest eigenvalue is λ. Then, the entries of A^t tend to 0 with increasing values of t, when we have $\lambda < 1$. On the other hand, the entries of A^t diverge to large values, when we have $\lambda > 1$.*

The proof of the above result is easy to show by diagonalizing $A = P\Delta P^{-1}$. Then, it can be shown that $A^t = P\Delta^t P^{-1}$, where Δ is a diagonal matrix. The magnitude of the largest diagonal entry of Δ^t either vanishes with increasing t or it grows to an increasingly large value (in absolute magnitude) depending on whether than eigenvalue is less than 1 or larger than 1. In the former case, the matrix A^t tends to 0, and therefore the gradient vanishes. In the latter case, the gradient explodes. Of course, this does not yet include the effect of the activation function, and one can change the threshold on the largest eigenvalue to set up the conditions for the vanishing or exploding gradients. For example, the largest possible value of the sigmoid activation derivative is 0.25, and therefore the vanishing gradient problem will definitely occur when the largest eigenvalue is less than $1/0.25 = 4$. One can, of course, combine the effect of the matrix multiplication and activation function into a single *Jacobian* matrix (cf. Table 7.1 of Chapter 7), whose eigenvalues can be tested.

One way of viewing this problem is that a neural network that uses only multiplicative updates is good only at learning over short sequences, and is therefore inherently endowed with good short-term memory but poor long-term memory [84]. To address this problem, a solution is to change the recurrence equation for the hidden vector with the use of the long short-term memory network (LSTM), which augments the hidden states with long-term memory that can easily copy values from one state to the next. The operations of the LSTM are designed to have fine-grained control over the data written into this long-term memory. This principle is similar to that used in residual convolutional neural networks like *ResNet*, where states are copied across layers in order to enable better training. In this section, we will introduce the details of long-short term memory networks.

As in the previous sections, the notation $\overline{h}_t^{(k)}$ represents the hidden states of the kth layer of a multi-layer LSTM. For notational convenience, we also assume that the input layer $\overline{x}_t$ can be denoted by $\overline{h}_t^{(0)}$ (although this layer is obviously not hidden). As in the case of the recurrent network, the input vector $\overline{x}_t$ is d-dimensional, whereas the hidden states are p-dimensional. The LSTM is an enhancement of the recurrent neural network architecture of Figure 8.13 in which we change the recurrence conditions of how the hidden states $\overline{h}_t^{(k)}$ are propagated. In order to achieve this goal, we have an additional hidden vector of p dimensions, which is denoted by $\overline{c}_t^{(k)}$ and referred to as the *cell state*. One can view the cell state as a kind of long-term memory that retains at least a part of the information in earlier states by using a combination of partial "forgetting" and "increment" operations on the previous cell states. It has been shown in [94] that the nature of the memory in $\overline{c}_t^{(k)}$ is occasionally interpretable when it is applied to text data such as literary pieces. For example, one of the p values in $\overline{c}_t^{(k)}$ might change in sign after an opening quotation and then revert back only when that quotation is closed. The upshot of this phenomenon is that the resulting neural network is able to model long-range dependencies in the language or even a specific pattern (like a quotation) extended over a large number of tokens. This is achieved by using a gentle approach to update these cell states over time, so that there is greater persistence in information storage. Persistence in state values avoids the kind of instability that occurs in the case of the vanishing and exploding gradient problems. One way of understanding this intuitively is that if the states in different temporal layers share a greater level of similarity (through long-term memory), it is harder for the gradients with respect to the incoming weights to be drastically different.

As with the multilayer recurrent network, the update matrix is denoted by $W^{(k)}$ and is used to premultiply the column vector $[\overline{h}_t^{(k-1)}, \overline{h}_{t-1}^{(k)}]^T$. However, this matrix is of size[6] $4p \times 2p$, and therefore pre-multiplying a vector of size $2p$ with $W^{(k)}$ results in a vector of size $4p$. In this case, the updates use four intermediate, p-dimensional vector variables $\overline{i}$, $\overline{f}$, $\overline{o}$, and $\overline{c}$ that correspond to the $4p$-dimensional vector. The intermediate variables $\overline{i}$, $\overline{f}$, and $\overline{o}$ are respectively referred to as *input*, *forget*, and *output* variables, because of the roles they

---

[6]In the first layer, the matrix $W^{(1)}$ is of size $4p \times (p + d)$ because it is multiplied with a vector of size $(p + d)$.

play in updating the cell states and hidden states. The determination of the hidden state vector $\overline{h}_t^{(k)}$ and the cell state vector $\overline{c}_t^{(k)}$ uses a multi-step process of first computing these intermediate variables and then computing the hidden variables from these intermediate variables. Note the difference between intermediate variable vector $\overline{c}$ and primary cell state $\overline{c}_t^{(k)}$, which have completely different roles. The updates are as follows:

$$
\begin{array}{ll}
\text{Input Gate:} \\
\text{Forget Gate:} \\
\text{Output Gate:} \\
\text{New C.-State:}
\end{array}
\left[
\begin{array}{c}
\overline{i} \\
\overline{f} \\
\overline{o} \\
\overline{c}
\end{array}
\right]
=
\left(
\begin{array}{c}
\text{sigm} \\
\text{sigm} \\
\text{sigm} \\
\text{tanh}
\end{array}
\right)
W^{(k)}
\left[
\begin{array}{c}
\overline{h}_t^{(k-1)} \\
\overline{h}_{t-1}^{(k)}
\end{array}
\right]
\quad \textbf{[Setting up intermediates]}
$$

$\overline{c}_t^{(k)} = \overline{f} \odot \overline{c}_{t-1}^{(k)} + \overline{i} \odot \overline{c}$ **[Selectively forget and add to long-term memory]**

$\overline{h}_t^{(k)} = \overline{o} \odot \tanh(\overline{c}_t^{(k)})$ **[Selectively leak long-term memory to hidden state]**

Here, the element-wise product of vectors is denoted by "$\odot$," and the notation "sigm" denotes a sigmoid operation. For the very first layer (i.e., $k = 1$), the notation $\overline{h}_t^{(k-1)}$ in the above equation should be replaced with $\overline{x}_t$ and the matrix $W^{(1)}$ is of size $4p \times (p + d)$. In practical implementations, biases are also used[7] in the above updates, although they are omitted here for simplicity. The aforementioned update seems rather cryptic, and therefore it requires further explanation.

The first step in the above sequence of equations is to set up the intermediate variable vectors $\overline{i}$, $\overline{f}$, $\overline{o}$, and $\overline{c}$, of which the first three should *conceptually* be considered binary values, although they are continuous values in $(0, 1)$. Multiplying a pair of binary values is like using an AND gate on a pair of boolean values. We will henceforth refer to this operation as gating. The vectors $\overline{i}$, $\overline{f}$, and $\overline{o}$ are referred to as input, forget, and output gates. In particular, these vectors are conceptually used as boolean gates for deciding (i) whether to add to a cell-state, (ii) whether to forget a cell state, and (iii) whether to allow leakage into a hidden state from a cell state. The use of the binary abstraction for the input, forget, and output variables helps in understanding the types of decisions being made by the updates. In practice, a continuous value in $(0, 1)$ is contained in these variables, which can enforce the effect of the binary gate in a probabilistic way if the output is seen as a probability. In the neural network setting, it is essential to work with continuous functions in order to ensure the differentiability required for gradient updates. The vector $\overline{c}$ contains the newly proposed contents of the cell state, although the input and forget gates regulate how much it is allowed to change the previous cell state (to retain long-term memory).

The four intermediate variables $\overline{i}$, $\overline{f}$, $\overline{o}$, and $\overline{c}$, are set up using the weight matrices $W^{(k)}$ for the kth layer in the first equation above. Let us now examine the second equation that updates the cell state with the use of some of these intermediate variables:

$$
\overline{c}_t^{(k)} = \underbrace{\overline{f} \odot \overline{c}_{t-1}^{(k)}}_{\text{Reset?}} + \underbrace{\overline{i} \odot \overline{c}}_{\text{Increment?}}
$$

---

[7]The bias associated with the forget gates is particularly important. The bias of the forget gate is generally initialized to values greater than 1 [93] because it seems to avoid the vanishing gradient problem at initialization.

This equation has two parts. The first part uses the p forget bits in $\overline{f}$ to decide which of the p cell states from the previous time-stamp to reset[8] to 0, and it uses the p input bits in $\overline{i}$ to decide whether to add the corresponding components from $\overline{c}$ to each of the cell states. Note that such updates of the cell states are in additive form, which is helpful in avoiding the vanishing gradient problem caused by multiplicative updates. One can view the cell-state vector as a continuously updated long-term memory, where the forget and input bits respectively decide (i) whether to reset the cell states from the previous time-stamp and forget the past, and (ii) whether to increment the cell states from the previous time-stamp to incorporate new information into long-term memory from the current word. The vector $\overline{c}$ contains the p amounts with which to increment the cell states, and these are values in $[-1, +1]$ because they are all outputs of the tanh function.

Finally, the hidden states $\overline{h}_t^{(k)}$ are updated using leakages from the cell state. The hidden state is updated as follows:

$$\overline{h}_t^{(k)} = \underbrace{\overline{o} \odot \tanh(\overline{c}_t^{(k)})}_{\text{Leak } \overline{c}_t^{(k)} \text{ to } \overline{h}_t^{(k)}}$$

Here, we are copying a functional form of each of the p cell states into each of the p hidden states, depending on whether the output gate (defined by $\overline{o}$) is 0 or 1. Of course, in the continuous setting of neural networks, partial gating occurs and only a fraction of the signal is copied from each cell state to the corresponding hidden state. It is noteworthy that the final equation does not always use the tanh activation function. The following alternative update may be used:

$$\overline{h}_t^{(k)} = \overline{o} \odot \overline{c}_t^{(k)}$$

As in the case of all neural networks, the backpropagation algorithm is used for training purposes.

In order to understand why LSTMs provide better gradient flows than vanilla RNNs, let us examine the update for a simple LSTM with a single layer and $p = 1$. In such a case, the cell update can be simplified to the following:

$$c_t = c_{t-1} * f + i * c \tag{8.6}$$

Therefore, the partial derivative c_t with respect to c_{t-1} is f, which means that the backward gradient flows for c_t are multiplied with the value of the forget gate f. Because of elementwise operations, this result generalizes to arbitrary values of the state dimensionality p. The biases of the forget gates are often set to high values initially, so that the gradient flows decay relatively slowly. The forget gate f can also be different at different time-stamps, which reduces the propensity of the vanishing gradient problem. The hidden states can be expressed in terms of the cell states as $h_t = o * \tanh(c_t)$, so that one can compute the partial derivative with respect to h_t with the use of a single tanh derivative. In other words, the long-term cell states function as gradient super-highways, which leak into hidden states.

---

[8]Here, we are treating the forget bits as a vector of binary bits, although it contains continuous values in $(0, 1)$, which can be viewed as probabilities. As discussed earlier, the binary abstraction helps us understand the conceptual nature of the operations.

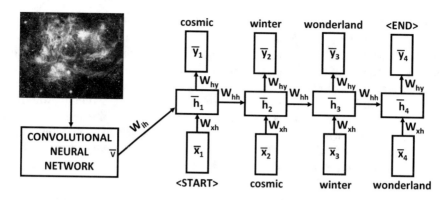

Figure 8.15: Example of image captioning with a recurrent neural network. An additional convolutional neural network is required for representational learning of the images. The image is represented by the vector $\overline{v}$, which is the output of the convolutional neural network. The inset image is by courtesy of the National Aeronautics and Space Administration (NASA)

8.8   Applications of Domain-Specific Architectures

In this section, we review some common applications of convolutional and recurrent neural networks. In particular, we will discuss diverse applications such as image captioning and machine translation.

8.8.1   Application to Automatic Image Captioning

In image captioning, the training data consists of image-caption pairs. For example, the image[9] in the left-hand side of Figure 8.15 is obtained from the National Aeronautics and Space Administration Web site. This image is captioned *"cosmic winter wonderland."* One might have hundreds of thousands of such image-caption pairs. These pairs are used to train the weights in the neural network. Once the training has been completed, the captions are predicted for unknown test instances. Therefore, one can view this approach as an instance of image-to-sequence learning.

One issue in the automatic captioning of images is that a separate neural network is required to learn the representation of the images. It is natural to use the convolutional neural network in such cases. Consider a setting in which the convolutional neural network produces the q-dimensional vector $\overline{v}$ as the output representation. This vector is then used as an input to the neural network, but only[10] at the first time-stamp. To account for this additional input, we need another $p \times q$ matrix W_{ih}, which maps the image representation to the hidden layer. Therefore, the update equations for the various layers now need to be modified as follows:

$$\overline{h}_1 = \tanh(W_{xh}\overline{x}_1 + W_{ih}\overline{v})$$
$$\overline{h}_t = \tanh(W_{xh}\overline{x}_t + W_{hh}\overline{h}_{t-1})\ \ \forall t \geq 2$$
$$\overline{y}_t = W_{hy}\overline{h}_t$$

---

[9]https://www.nasa.gov/mission_pages/chandra/cosmic-winter-wonderland.html

[10]In principle, one can also allow it to be input at all time-stamps, but it only seems to worsen performance.

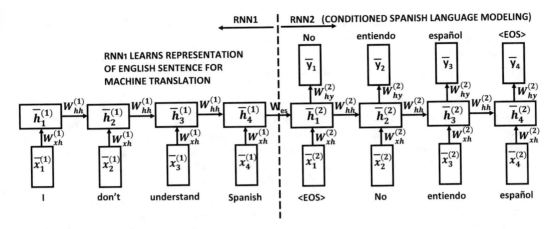

Figure 8.16: Machine translation with recurrent neural networks. Note that there are two separate recurrent networks with their own sets of shared weights. The output of $\overline{h}_4^{(1)}$ is a fixed length encoding of the 4-word English sentence

An important point here is that the convolutional neural network and the recurrent neural network are not trained in isolation. Although one might train them in isolation in order to create an initialization, the final weights are always trained jointly by running each image through the network and matching up the predicted caption with the true caption. In other words, for each image-caption pair, the weights in both networks are updated when errors are made in predicting any particular token of the caption. In practice, the errors are soft because the tokens at each point are predicted probabilistically. Such an approach ensures that the learned representation $\overline{v}$ of the images is sensitive to the specific application of predicting captions.

After all the weights have been trained, a test image is input to the entire system and passed through both the convolutional and recurrent neural network. For the recurrent network, the input at the first time-stamp is the <START> token and the representation of the image. At later time-stamps, the input is the most likely token predicted at the previous time-stamp. One can also use beam search to keep track of the b most likely sequence prefixes to expand on at each point. This approach is not very different from the language generation approach discussed in Section 8.6.1.1, except that it is conditioned on the image representation that is input to the model in the first time-stamp of the recurrent network. This results in the prediction of a relevant caption for the image.

8.8.2 Sequence-to-Sequence Learning and Machine Translation

Just as one can put together a convolutional neural network and a recurrent neural network to perform image captioning, one can put together two recurrent networks to translate one language into another. Such methods are also referred to as *sequence-to-sequence* learning because a sequence in one language is mapped to a sequence in another language. In principle, sequence-to-sequence learning can have applications beyond machine translation. For example, even question-answering (QA) systems can be viewed as sequence-to-sequence learning applications.

In the following, we provide a simple solution to machine translation with recurrent neural networks, although such applications are rarely addressed directly with the simple forms of recurrent neural networks. Rather, a variation of the recurrent neural network, referred to as the long short-term memory (LSTM) model is used. Such a model is much better in learning long-term dependencies, and can therefore work well with longer sentences. Since the general approach of using an RNN applies to an LSTM as well, we will provide the discussion of machine translation with the (simple) RNN. A discussion of the LSTM is provided in Section 8.7, and the generalization of the machine translation application to the LSTM is straightforward.

In the machine translation application, two different RNNs are hooked end-to-end, just as a convolutional neural network and a recurrent neural network are hooked together for image captioning. The first recurrent network uses the words from the source language as input. No outputs are produced at these time-stamps and the successive time-stamps accumulate knowledge about the source sentence in the hidden state. Subsequently, the end-of-sentence symbol is encountered, and the second recurrent network starts by outputting the first word of the target language. The next set of states in the second recurrent network output the words of the sentence in the target language one by one. These states also use the words of the target language as input, which is available for the case of the training instances but not for test instances (where predicted values are used instead). This architecture is shown in Figure 8.16.

The architecture of Figure 8.16 is similar to that of an *autoencoder* (see Chapter 9), and can even be used with pairs of identical sentences in the same language to create fixed-length representations of sentences. The two recurrent networks are denoted by RNN1 and RNN2, and their weights are not the same. For example, the weight matrix between two hidden nodes at successive time-stamps in RNN1 is denoted by $W_{hh}^{(1)}$, whereas the corresponding weight matrix in RNN2 is denoted by $W_{hh}^{(2)}$. The weight matrix W_{es} of the link joining the two neural networks is special, and can be independent of either of the two networks. This is necessary if the sizes of the hidden vectors in the two RNNs are different because the dimensions of the matrix W_{es} will be different from those of both $W_{hh}^{(1)}$ and $W_{hh}^{(2)}$. As a simplification, one can use[11] the same size of the hidden vector in both networks, and set $W_{es} = W_{hh}^{(1)}$. The weights in RNN1 are devoted to learning an encoding of the input in the source language, and the weights in RNN2 are devoted to using this encoding in order to create an output sentence in the target language. One can view this architecture in a similar way to the image captioning application, except that we are using two recurrent networks instead of a convolutional-recurrent pair. The output of the final hidden node of RNN1 is a fixed-length encoding of the source sentence. Therefore, irrespective of the length of the sentence, the encoding of the source sentence depends on the dimensionality of the hidden representation.

The grammar and length of the sentence in the source and target languages may not be the same. In order to provide a grammatically correct output in the target language, RNN2 needs to learn its language model. It is noteworthy that the units in RNN2 associated with the target language have both inputs and outputs arranged in the same way as a language-modeling RNN. At the same time, the output of RNN2 is conditioned on the input it receives from RNN1, which effectively causes language translation. In order to achieve this goal, training pairs in the source and target languages are used. The approach passes the source-target pairs through the architecture of Figure 8.16 and learns the model parameters with the use of the backpropagation algorithm. Since only the nodes in RNN2 have outputs, only the errors made in predicting the target language words are backpropagated to train the weights in both neural networks. The two networks are jointly trained, and therefore the weights in both networks are optimized to the errors in the translated outputs of RNN2. As a practical matter, this means that the internal representation of the source language learned by RNN1 is highly optimized to the machine translation application, and is very different

---

[11]The original work in [180] seems to use this option. In the Google Neural Machine Translation system [221], this weight is removed. This system is now used in Google Translate.

from one that would be learned if one had used RNN1 to perform language modeling of the source sentence. After the parameters have been learned, a sentence in the source language is translated by first running it through RNN1 to provide the necessary input to RNN2. Aside from this contextual input, another input to the first unit of RNN2 is the <EOS> tag, which causes RNN2 to output the likelihoods of the first token in the target language. The most likely token using beam search (cf. Section 8.6.1.1) is selected and used as the input to the recurrent network unit in the next time-stamp. This process is recursively applied until the output of a unit in RNN2 is also <EOS>. As in Section 8.6.1.1, we are generating a sentence from the target language using a language-modeling approach, except that the specific output is conditioned on the internal representation of the source sentence.

The use of neural networks for machine translation is relatively recent. Recurrent neural network models have a sophistication that greatly exceeds that of traditional machine translation models. The latter class of methods uses phrase-centric machine learning, which is often not sophisticated enough to learn the subtle differences between the grammars of the two languages. In practice, deep models with multiple layers are used to improve the performance.

One weakness of such translation models is that they tend to work poorly when the sentences are long. Numerous solutions have been proposed to solve the problem. A recent solution is that the sentence in the source language is input in the *opposite order* [180]. This approach brings the first few words of the sentences in the two languages closer in terms of their time-stamps within the recurrent neural network architecture. As a result, the first few words in the target language are more likely to be predicted correctly. The correctness in predicting the first few words is also helpful in predicting the subsequent words, which are also dependent on a neural language model in the target language.

8.9 Summary

This chapter introduces several domain-specific neural architectures, which are designed to capture the natural relationships among various attributes in domains like image ad sequence data. Convolutional neural networks and recurrent neural networks were introduced in the chapter. Convolutional neural networks are designed for image data, whereas recurrent neural networks are designed for sequence data. A number of case studies of convolutional neural networks were also provided in the chapter. In addition, the chapter also discusses a number of applications of recurrent and convolutional neural networks.

8.10 Further Reading

A discussion of various types of recurrent and convolutional neural networks may be found in [6, 67]. Video lectures and course material on these types of neural networks may be found in [96]. A tutorial on convolution arithmetic is available in [49]. A brief discussion of applications may be found in [111].

8.11 Exercises

1. Consider a 1-dimensional time-series with values 2, 1, 3, 4, 7. Perform a convolution with a 1-dimensional filter 1, 0, 1 and zero padding.

2. For a one-dimensional time series of length L and a filter of size F, what is the length of the output? How much padding would you need to keep the output size to a constant value?

3. Consider an activation volume of size $13 \times 13 \times 64$ and a filter of size $3 \times 3 \times 64$. Discuss whether it is possible to perform convolutions with strides 2, 3, 4, and 5. Justify your answer in each case.

4. Work out the sizes of the spatial convolution layers for each of the columns of Table 8.1. In each case, we start with an input image volume of $224 \times 224 \times 3$.

5. Work out the number of parameters in each spatial layer for column D of Table 8.1.

6. Download an implementation of the *AlexNet* architecture from a neural network library of your choice. Train the network on subsets of varying size from the *ImageNet* data, and plot the top-5 error with data size.

7. Compute the convolution of the input volume in the upper-left corner of Figure 8.2 with the horizontal edge detection filter of Figure 8.1(b). Use a stride of 1 without padding.

8. Perform a 4×4 pooling at stride 1 of the input volume in the upper-left corner of Figure 8.4.

9. Download the character-level RNN in [222], and train it on the *"tiny Shakespeare"* data set available at the same location. Create outputs of the language model after training for (i) 5 epochs, (ii) 50 epochs, and (iii) 500 epochs. What significant differences do you see between the three outputs?

10. Propose a neural architecture to perform binary classification of a sequence.

11. Suppose that you have a large database of biological strings containing sequences of nucleobases drawn from $\{A, C, T, G\}$. Some of these strings contain unusual mutations representing changes in the nucleobases. Propose an unsupervised method (i.e., neural architecture) using RNNs in order to detect these mutations.

12. How would your architecture for the previous question change if you were given a training database in which the mutation positions in each sequence were tagged, and the test database was untagged?

Chapter 9

Unsupervised Learning

"Education is not the filling of a pot but the lighting of a fire."– W.B. Yeats

9.1 Introduction

The supervised learning methods discussed in the previous chapters try to learn how the features in the data relate to specific target variables. Unsupervised learning methods try to learn how the features are related to one another. In other words, unsupervised learning methods do not have a specific goal in mind in order to supervise the learning process. Rather, unsupervised methods learn the key patterns in the underlying data that relate all data points and attributes to one another without a specific focus on any particular data items. In supervised learning, specific attributes (e.g., regressors or class labels) are more important, and therefore play the role of teachers (i.e., *supervisors*) to the learning process.

Unsupervised methods generally learn aggregate trends in the data matrix, and in many cases, they create a compressed model of key data characteristics. This compressed model can even be used in order to recreate examples of typical data points. From a human intelligence point of view, we experience a lot of sensory inputs on a day-to-day basis, and often store away the key aspects of these experiences. These experiences often turn out to be useful for more specific tasks. A similar analogy holds true in machine learning, where unsupervised learning is often leveraged in supervised learning tasks. The common types of relationships captured by unsupervised learning are as follows:

1. **Row-wise relationships:** In this case, we attempt to learn which rows of the data set are closely related to one another. Therefore, the problem reduces to one of finding important clusters in the data.

2. **Column-wise relationships:** In this case, the goal is to create a data set represented by a smaller set of columns, by using the interrelationships and correlations among the columns of the original data set. This problem is referred to as *dimensionality reduction*.

© Springer Nature Switzerland AG 2021
C. C. Aggarwal, *Artificial Intelligence*, https://doi.org/10.1007/978-3-030-72357-6_9

3. **Combining row-wise and column-wise relationships:** Several forms of dimensionality reduction are also well suited to clustering, because they capture the relationships among the rows as well. This is because they capture both row-wise and column-wise relationships.

In this chapter, we will discuss different types of unsupervised models like clustering and dimensionality reduction. Clustering and dimensionality reduction are both forms of data compression, and can be used to represent the data approximately in a smaller amount of space. The ability to represent the data approximately in small amount of space is a natural characteristic of unsupervised learning methods, which try to learn the broader patterns in the data. Unsupervised models build compressed models of the data, so that one can express each point $\overline{X}_i$ approximately as a function of itself:

$$\overline{h}_i = F_{compress}(\overline{X}_i), \quad \overline{X}_i \approx F_{decompress}(\overline{h}_i)$$
$$\overline{X}_i \approx G(\overline{X}_i) = F_{decompress}(F_{compress}(\overline{X}_i))$$

The compressed representation $\overline{h}_i$ can be viewed as a concise description containing the most important characteristics of $\overline{X}_i$. The "hidden" representation $\overline{h}_i$ is typically not visible to the end user. The reconstruction of a data point from this hidden representation is approximate, and might sometimes drop unusual artifacts from the point. The hidden representation can be used for various applications like generative modeling and outlier detection.

This chapter is organized as follows. The next section introduces the problem of dimensionality reduction and matrix factorization. Methods are discussed for singular value decomposition and nonnegative matrix factorization. In addition, neural networks for nonlinear dimensionality reduction are discussed. Clustering methods are discussed in Section 9.3. A discussion of various applications of unsupervised learning is provided in Section 9.4. A summary is given in Section 9.5.

9.2   Dimensionality Reduction and Matrix Factorization

Consider an $n \times d$ data matrix D in which significant correlations exist among the rows (or columns). The presence of relationships among the rows (or columns) implies that there is an inherent redundancy in the data representation. One way of capturing this data redundancy is through the use of matrix factorization. In matrix factorization, an $n \times d$ data matrix D is represented as the product of two matrices that are much smaller:

$$D \approx UV^T \tag{9.1}$$

The matrix U is of size $n \times k$, and the matrix V is of size $d \times k$, where k is much less than $\min\{n, d\}$. The value of k is referred to as the *rank* of the factorization. One can always reconstruct the matrix exactly using a factorization of rank $k = \min\{n, d\}$, and this rank is referred to as *full-rank* factorization. It is more common to use small values of k, which is referred to as a *low-rank* factorization, but exact reconstruction is not possible in such cases. Furthermore, in such cases, the total number of entries in U and V is much less than the number of entries in D:

$$(n + d)k \ll nd$$

Therefore, the matrices U and V provide a compressed representation of the data, and the data matrix D can be approximately reconstructed as UV^T. Furthermore, the matrix

UV^T will often differ significantly in those entries of D that do not naturally conform to the aggregate trends in the data. In other words, such entries are *outliers*. In fact, outlier detection is a complementary form of unsupervised learning to clustering and dimensionality reduction.

This factorization is referred to as *low-rank* because the ranks of each of U, V, and UV^T are at most $k \ll d$, whereas the rank of D might be $\min\{n, d\}$. Note that there will always be some *residual error* $(D - UV^T)$ from the factorization. In fact, the entries in U and V are often discovered by solving an optimization problem in which the sum of squares (or other aggregate function) of the residual errors in $(D - UV^T)$ are minimized. Almost all forms of dimensionality reduction and matrix factorization are special cases of the following optimization model over matrices U and V:

Maximize similarity between entries of D and UV^T

subject to:

Constraints on U and V

By varying the objective function and constraints, dimensionality reductions with different properties are obtained. The most commonly used objective function is the sum of the squares of the entries in $(D - UV^T)$, which is also defined as the (squared) *Frobenius norm* of the matrix $(D - UV^T)$. The (squared) Frobenius norm of a matrix is also referred to as its *energy*, because it is the sum of the second moments of all data points about the origin. However, some forms of factorizations with probabilistic interpretations use a *maximum-likelihood* objective function. Similarly, the constraints imposed on U and V enable different properties of the factorization. For example, if we impose orthogonality constraints on the columns of U and V, this leads to a model known as *singular value decomposition (SVD)* or *latent semantic analysis (LSA)*. The latter terminology is used in the context of document data. The orthogonality of the basis vectors is particularly helpful in mapping new data points (i.e., data points not included in the original data set on which factorization is applied) to the transformed space in a simple way. On the other hand, better semantic interpretability can be obtained by imposing nonnegativity constraints on U and V. This chapter will discuss various types of reductions and their relative advantages.

9.2.1 Symmetric Matrix Factorization

Singular value decomposition can be viewed as the generalization of a type of factorization that is performed on positive semi-definite matrices. A square and symmetric $n \times n$ matrix A is positive semi-definite, if and only if for any n-dimensional column vector $\overline{x}$, we have $\overline{x}^T A \overline{x} \geq 0$.

Positive semi-definite matrices have the property that they can be *diagonalized* to the following form:

$$A = Q\Delta Q^T$$

Here Q is a $n \times n$ matrix with orthonormal columns, and Δ is an $n \times n$ diagonal matrix with nonnegative entries. Since Δ has nonnegative entries, it is common to express Δ as the square of another $n \times n$ diagonal matrix Σ, and define the diagonalization in the following way:

$$A = Q\Sigma^2 Q^T$$

The columns of Q are referred to as *eigenvectors*, which represent an orthonormal set of n vectors. Therefore, we have $Q^T Q = I$. The diagonal entries of Δ are referred to as *eigenvalues*. What does an eigenvector and eigenvalue mean? Note that the relationship above can be expressed as follows (by postmultiplying with P and setting $Q^T Q = I$):

$$AQ = Q\Delta$$

If the ith column of Q is $\bar{q}_i$ and the ith diagonal entry of Δ is $\delta_i \geq 0$, we can express the above relationship as follows:

$$A\bar{q}_i = \delta_i \bar{q}_i$$

Note that multiplying an eigenvector with a matrix A simply scales the magnitude of the eigenvector, without changing its direction. Positive semi-definite matrix factorization is fundamental to machine learning because it is used in all sorts of machine learning applications, such as kernel methods.

One can also express positive semi-definite matrix factorization as a form of *symmetric* matrix factorization:

$$A = Q\Sigma^2 Q^T = \underbrace{(Q\Sigma)}_{U}(Q\Sigma)^T = UU^T$$

Note that this is a factorization of *full rank* in which we are not losing any accuracy of representation of A via factorization.

How does one relate positive semi-definite matrix factorization to the generic optimization formulation of the previous section? It can be shown that *truncated forms* of symmetric factorization, in which low-rank factorization is used, reduce to the following matrix factorization:

$$\text{Minimize}_U \|A - UU^T\|^2$$

subject to:

No constraints on U

Here U is an $n \times k$ matrix rather than an $n \times n$, where $k \ll n$. Therefore, this factorization is of low rank. It can be shown that the optimal solution to this problem yields $U = Q_k \Sigma_k$, where Q_k is an $n \times k$ matrix containing the top-k eigenvectors of A in its columns (i.e., eigenvectors with largest eigenvalues), and Σ_k is a diagonal matrix containing the square-root of the corresponding eigenvalues of A. Singular value decomposition is a natural generalization of this idea to asymmetric and rectangular matrices.

9.2.2   Singular Value Decomposition

Singular value decomposition (SVD) is the most common form of dimensionality reduction for multidimensional data. Consider the simplest possible factorization of the $n \times d$ matrix D into an $n \times k$ matrix $U = [u_{ij}]$ and the $d \times k$ matrix $V = [v_{ij}]$ as an *unconstrained matrix factorization problem*:

$$\text{Minimize}_{U,V} \|D - UV^T\|_F^2$$

subject to:

No constraints on U and V

Here $\|\cdot\|_F^2$ refers to the (squared) Frobenius norm of a matrix, which is the sum of squares of its entries. The matrix $(D - UV^T)$ is also referred to as the *residual matrix*, because its entries contain the residual errors obtained from a low-rank factorization of the original matrix D. This optimization problem is the most basic form of matrix factorization with a popular objective function and no constraints. This formulation has infinitely many alternative optimal solutions. However, one[1] of them is such that the columns of V are orthonormal, which allows transformations of new data points (i.e., rows not included in D) with simple axis rotations (i.e., matrix multiplication). A remarkable property of the unconstrained optimization problem above is that imposing orthogonality constraints does not worsen the optimal solution. *The following constrained optimization problem shares at least one optimal solution as the unconstrained version* [50, 175]:

$$\text{Minimize}_{U,V} \|D - UV^T\|_F^2$$

subject to:

Columns of U are mutually orthogonal

Columns of V are mutually orthonormal

In other words, one of the alternative optima to the unconstrained problem also satisfies orthogonality constraints. It is noteworthy that *only* the solution satisfying the orthogonality constraint is considered SVD because of its interesting properties, even though other optima do exist.

Another remarkable property of the solution (satisfying orthogonality) is that it can be computed using *eigen-decomposition* of either of the positive semi-definite matrices $D^T D$ or DD^T. The following properties of the solution can be shown:

1. The columns of V are defined by the top-k unit eigenvectors of the $d \times d$ positive semi-definite and symmetric matrix $D^T D$. The diagonalization of a symmetric and positive semi-definite matrix results in orthonormal eigenvectors with non-negative eigenvalues. After V has been determined, we can also compute the reduced representation U as DV, which is simply an axis rotation operation on the rows in the original data matrix. This is caused by the orthogonality of the columns of V, which results in $DV \approx U(V^T V) = U$. One can also use this approach to compute the reduced representation $\overline{X}V$ of any row-vector $\overline{X}$ that was not included in D.

2. The columns of U are also defined by the top-k *scaled* eigenvectors of the $n \times n$ *dot-product matrix* DD^T in which the (i, j)th entry is the dot-product similarity between the ith and jth data points. The scaling factor is defined so that each eigenvector is multiplied with the square-root of its eigenvalue. In other words, the *scaled eigenvectors of the dot-product matrix can be used to directly generate the reduced representation*. This fact has some interesting consequences for the nonlinear dimensionality reduction methods, which replace the dot product matrix with another similarity matrix. This approach is also efficient for linear SVD when $n \ll d$, and therefore the $n \times n$ matrix DD^T is relatively small. In such cases, U is extracted first by eigen-decomposition of DD^T, and then V is extracted as $D^T U$.

3. Even though the n eigenvectors of DD^T and d eigenvectors of $D^T D$ are different, the top $\min\{n, d\}$ eigenvalues of DD^T and $D^T D$ are the same values. All other eigenvalues are zero.

4. The total squared error of the approximate matrix factorization of SVD is equal to the sum of the eigenvalues of $D^T D$ that are *not* included among the top-k eigenvectors. If we set the rank of the factorization k to $\min\{n, d\}$, we can obtain an *exact* factorization into orthogonal basis spaces with zero error.

---

[1]This solution can be unique up to multiplication of any column of U or V with -1 under some conditions on uniqueness of retained singular values.

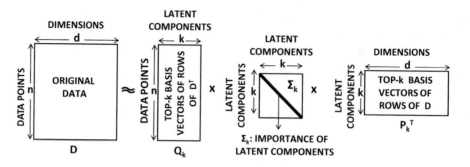

Figure 9.1: Dual interpretation of SVD in terms of the basis vectors of both D and D^T

This factorization of rank $k = \min\{n, d\}$ with zero error is of particular interest. We convert the two-way factorization (of zero error) into a three-way factorization, which results in a standard form of SVD:

$$D = Q\Sigma P^T = \underbrace{(Q\Sigma)}_{U} \underbrace{P^T}_{V^T} \tag{9.2}$$

Here, Q is an $n \times k$ matrix containing all the $k = \min\{n, d\}$ *non-zero* eigenvectors of DD^T, and P is a $d \times k$ matrix containing all the $k = \min\{n, d\}$ non-zero eigenvectors of $D^T D$. The columns of Q are referred to as the *left singular vectors*, whereas the columns of P are referred to as the *right singular vectors*. Furthermore, Σ is a (nonnegative) diagonal matrix in which the (r, r)th value is equal to the square-root of the rth largest eigenvalue of $D^T D$ (which is the same as the rth largest eigenvalue of DD^T). The diagonal entries of Σ are also referred to as *singular values*. Note that the singular values are always nonnegative by convention. The sets of columns of P and Q are each orthonormal because they are the unit eigenvectors of symmetric matrices. It is easy to verify (using Equation 9.2) that $D^T D = P\Sigma^2 P^T$ and that $DD^T = Q\Sigma^2 Q^T$, where Σ^2 is a diagonal matrix containing the top-k non-negative eigenvalues of $D^T D$ and DD^T (which are the same).

SVD is formally defined as the exact decomposition with zero error. What about the *approximate* variant of SVD, which is the primary goal of matrix factorization? In practice, one always uses values of $k \ll \min\{n, d\}$ to obtain *approximate* or *truncated* SVD:

$$D \approx Q\Sigma P^T \tag{9.3}$$

Using truncated SVD is the standard use-case in practical settings. Throughout this book, our use of the term "SVD" always refers to truncated SVD.

Just as the matrix P contains the d-dimensional basis vectors of D in its columns, the matrix Q contains the n-dimensional basis vectors of D^T in its columns. In other words, *SVD simultaneously finds approximate bases of both points and dimensions.* This ability of SVD to simultaneously find approximate bases for the row space and column space is shown in Figure 9.1. Furthermore, the diagonal entries of the matrix Σ provide a quantification of the relative dominance of the different semantic concepts.

One can express SVD as a weighted sum of rank-1 matrices. Let Q_i be the $n \times 1$ matrix corresponding to the ith column of Q and P_i be the $d \times 1$ matrix corresponding to the ith column of P. Then, the SVD product can be decomposed in *spectral form* using simple matrix-multiplication laws as follows:

$$Q\Sigma P^T = \sum_{i=1}^{k} \Sigma_{ii} Q_i P_i^T \tag{9.4}$$

Note that each $Q_i P_i$ is a rank-1 matrix of size $n \times d$ and a Frobenius norm of 1. Furthermore, it is possible to show that the Frobenius norm of $Q\Sigma P^T$ is given by $\sum_{i=1}^{k} \Sigma_{ii}^2$, which is the amount of *energy retained* in the representation. Maximizing the retained energy is the same as minimizing the loss defined by the sum of squares of the truncated singular values (which are small), because the sum of the two is always equal to $\|D\|_F^2$. The energy retained in the approximated matrix is the same as that in the transformed representation, because squared distances do not change with axis rotation. Therefore, the sum of the squares of the retained singular values provides the energy in the transformed representation DP. An important consequence of this observation is that the projection $D\bar{p}$ of D on any column $\bar{p}$ of P has an L_2-norm, which is equal to the corresponding singular value. In other words, SVD naturally selects the orthogonal directions along which the transformed data exhibits the largest scatter.

9.2.2.1 Example of SVD

An example of SVD helps in illustrating its inner workings. Consider a 6×6 matrix D defined over a document data set containing the following 6 words:

lion, tiger, cheetah, jaguar, porsche, ferrari

The data matrix D is illustrated below:

$$
D = \begin{pmatrix}
 & \text{lion} & \text{tiger} & \text{cheetah} & \text{jaguar} & \text{porsche} & \text{ferrari} \\
\text{Document-1} & 2 & 2 & 1 & 2 & 0 & 0 \\
\text{Document-2} & 2 & 3 & 3 & 3 & 0 & 0 \\
\text{Document-3} & 1 & 1 & 1 & 1 & 0 & 0 \\
\text{Document-4} & 2 & 2 & 2 & 3 & 1 & 1 \\
\text{Document-5} & 0 & 0 & 0 & 1 & 1 & 1 \\
\text{Document-6} & 0 & 0 & 0 & 2 & 1 & 2
\end{pmatrix}
$$

Note that this matrix represents topics related to both cars and cats. The first three documents are primarily related to cats, the fourth is related to both, and the last two are primarily related to cars. The word "jaguar" is *polysemous* because it could correspond to either a car or a cat. Therefore, it is often present in documents of both categories and presents itself as a confounding word. We would like to perform an SVD of rank-2 to capture the two dominant concepts corresponding to cats and cars, respectively. Then, on performing the SVD of this matrix, we obtain the following decomposition:

$$D \approx Q\Sigma P^T$$

$$
\approx \begin{pmatrix}
-0.41 & 0.17 \\
-0.65 & 0.31 \\
-0.23 & 0.13 \\
-0.56 & -0.20 \\
-0.10 & -0.46 \\
-0.19 & -0.78
\end{pmatrix}
\begin{pmatrix}
8.4 & 0 \\
0 & 3.3
\end{pmatrix}
\begin{pmatrix}
-0.41 & -0.49 & -0.44 & -0.61 & -0.10 & -0.12 \\
0.21 & 0.31 & 0.26 & -0.37 & -0.44 & -0.68
\end{pmatrix}
$$

$$
= \begin{pmatrix}
1.55 & 1.87 & 1.67 & 1.91 & 0.10 & 0.04 \\
2.46 & 2.98 & 2.66 & 2.95 & 0.10 & -0.03 \\
0.89 & 1.08 & 0.96 & 1.04 & 0.01 & -0.04 \\
1.81 & 2.11 & 1.91 & 3.14 & 0.77 & 1.03 \\
0.02 & -0.05 & -0.02 & 1.06 & 0.74 & 1.11 \\
0.10 & -0.02 & 0.04 & 1.89 & 1.28 & 1.92
\end{pmatrix}
$$

The reconstructed matrix is a very good approximation of the original document-term matrix. Furthermore, each point gets a 2-dimensional embedding corresponding to the rows of $Q\Sigma$. It is clear that the reduced representations of the first three documents are quite similar, and so are the reduced representations of the last two. The reduced representation of the fourth document seems to be somewhere in the middle of the representations of the other documents. This is logical because the fourth document corresponds to both cars and cats. From this point of view, the reduced representation seems to satisfy the basic intuitions one would expect in terms of *relative* coordinates. However, one annoying characteristic of this representation is that it is hard to get any *absolute* semantic interpretation from the embedding. For example, it is difficult to match up the two latent vectors in P with the original concepts of cats and cars. The dominant latent vector in P is $[-0.41, -0.49, -0.44, -0.61, -0.10, -0.12]$, in which all components are negative. The second latent vector contains both positive and negative components. Therefore, the correspondence between the topics and the latent vectors is not very clear. A part of the problem is that the vectors have both positive and negative components, which reduces their interpretability. The lack of interpretability of singular value decomposition is its primary weakness, as a result of which other nonnegative forms of factorization are sometimes preferred.

9.2.2.2   Alternate Optima via Gradient Descent

SVD provides one of the alternate optima to the problem of unconstrained matrix factorization. As discussed above, the problem of unconstrained matrix factorization is defined as follows:

$$\text{Minimize } _{U,V} \; J = \frac{1}{2}\|D - UV^T\|_F^2$$

Here, D, U, and V are matrices of sizes n, d, and k, respectively. The value of k is typically much smaller than the rank of the matrix D. In this section, we will investigate a method that finds a solution to the unconstrained optimization problem with the use of gradient descent. This approach does not guarantee the orthogonal solutions provided by singular value decomposition; however, the formulation is equivalent and should (ideally) lead to a solution with the same objective function value. The approach also has the advantage that it can easily adapted to more difficult settings such as the presence of missing values in the matrix. A natural application of this type of approach is that of matrix factorization in recommender systems. Recommender systems use the same optimization formulation as SVD; however, the resulting basis vectors of the factorization are not guaranteed to be orthogonal.

In order to perform gradient descent, we need to compute the derivative of the unconstrained optimization problem with respect to the parameters in the matrices $U = [u_{iq}]$ and $V = [v_{jq}]$. The simplest approach is to compute the derivative of the objective function J with respect to each parameter in the matrices U and V. First, the objective function is expressed in terms of the individual entries in the various matrices. Let the (i, j)th entry of the $n \times d$ matrix D be denoted by x_{ij}. Then, the objective function can be restated in terms of the entries of the matrices D, U, and V as follows:

$$\text{Minimize } J = \frac{1}{2}\sum_{i=1}^{n}\sum_{j=1}^{d}\left(x_{ij} - \sum_{s=1}^{k} u_{is}\cdot v_{js}\right)^2$$

The quantity $e_{ij} = x_{ij} - \sum_{s=1}^{k} u_{is}\cdot v_{js}$ is the error of the factorization for the (i, j)th entry. Note that the objective function J minimizes the sum of squares of e_{ij}. One can compute the partial derivative of the objective function with respect to the parameters in the matrices U and V as follows:

$$\frac{\partial J}{\partial u_{iq}} = \sum_{j=1}^{d} \left(x_{ij} - \sum_{s=1}^{k} u_{is} \cdot v_{js} \right) (-v_{jq}) \quad \forall i \in \{1 \dots n\}, q \in \{1 \dots k\}$$

$$= \sum_{j=1}^{d} (e_{ij})(-v_{jq}) \quad \forall i \in \{1 \dots n\}, q \in \{1 \dots k\}$$

$$\frac{\partial J}{\partial v_{jq}} = \sum_{i=1}^{n} \left(x_{ij} - \sum_{s=1}^{k} u_{is} \cdot v_{js} \right) (-u_{iq}) \quad \forall j \in \{1 \dots d\}, q \in \{1 \dots k\}$$

$$= \sum_{i=1}^{n} (e_{ij})(-u_{iq}) \quad \forall j \in \{1 \dots d\}, q \in \{1 \dots k\}$$

One can also express these derivatives in terms of matrices. Let $E = [e_{ij}]$ be the $n \times d$ matrix of errors. In the denominator layout[2] of matrix calculus, the derivatives can be expressed as follows:

$$\frac{\partial J}{\partial U} = -(D - UV^T)V = -EV$$

$$\frac{\partial J}{\partial V} = -(D - UV^T)^T U = -E^T U$$

The above matrix calculus identity can be verified by using the relatively tedious process of expanding the (i, q)th and (j, q)th entries of each of the above matrices on the right-hand side, and showing that they are equivalent to the (corresponding) scalar derivatives $\frac{\partial J}{\partial u_{iq}}$ and $\frac{\partial J}{\partial v_{jq}}$. one can also find an optimal solution by using gradient descent. The updates for gradient descent are as follows:

$$U \Leftarrow U - \alpha \frac{\partial J}{\partial U} = U + \alpha EV$$

$$V \Leftarrow V - \alpha \frac{\partial J}{\partial V} = V + \alpha E^T U$$

Here, $\alpha > 0$ is the learning rate.

The optimization model is identical to that of SVD. If the aforementioned gradient descent method is used (instead of the power iteration method of the previous chapter), one will typically obtain solutions that are equally good in terms of objective function value, but for which the columns of U (or V) are not mutually orthogonal. The power iteration methods yields solutions with orthogonal columns. Although the standardized SVD solution with orthonormal columns is typically not obtained by gradient descent, the k columns of U will span the same subspace as the columns of Q_k, and the columns of V will span the same subspace as the columns of P_k. One can also add the regularization term $\lambda(\|U\|_F^2 + \|V\|_F^2)/2$ to the objective function. Here, λ is the regularization parameter. Adding the regularization term leads to the following updates:

$$U \Leftarrow U(1 - \alpha\lambda) + \alpha EV \tag{9.5}$$

---

[2] In this layout, the (i, j)th entry of $\frac{\partial J}{\partial U}$ is $\frac{\partial J}{\partial u_{ij}}$. In the numerator layout, the (i, j)th entry of $\frac{\partial J}{\partial U}$ is $\frac{\partial J}{\partial u_{ji}}$.

$$V \Leftarrow V(I - \alpha \lambda) + \alpha E^T U \qquad (9.6)$$

The gradient-descent approach can be implemented efficiently when the matrix D is sparse by sampling entries from the matrix for making updates. This is essentially a *stochastic gradient descent* method. In other words, we sample an entry (i, j) and compute its error e_{ij}. Subsequently, we make the following updates to the ith row $\overline{u}_i$ of U and the jth row $\overline{v}_j$ of V, which are also referred to as *latent factors*:

$$\overline{u}_i \Leftarrow \overline{u}_i(1 - \alpha \lambda) + \alpha e_{ij} \overline{v}_j$$
$$\overline{v}_j \Leftarrow \overline{v}_j(1 - \alpha \lambda) + \alpha e_{ij} \overline{u}_i$$

One cycles through the sampled entries of the matrix (making the above updates) until convergence. The fact that we can sample entries of the matrix for updates means that we do not need fully specified matrices in order to learn the latent factors. This basic idea forms the foundations of recommender systems, in which partially specified matrices are used for learning the factors. The resulting factors are then used to reconstruct the entire matrix as UV^T.

9.2.3   Nonnegative Matrix Factorization

Nonnegative matrix factorization is a *highly interpretable* type of matrix factorization in which nonnegativity constraints are imposed on U and V. Therefore, this optimization problem is defined as follows:

$$\text{Minimize}_{U,V} \|D - UV^T\|_F^2$$
$$\text{subject to:}$$
$$U \geq 0, \ V \geq 0$$

As in the case of SVD, $U = [u_{ij}]$ is an $n \times k$ matrix and $V = [v_{ij}]$ is a $d \times k$ matrix of optimization parameters. Note that the optimization objective is the same but the constraints are different.

This type of constrained problem is often solved using Lagrangian relaxation. For the (i, s)th entry u_{is} in U, we introduce the Lagrange multiplier $\alpha_{is} \leq 0$, whereas for the (j, s)th entry v_{js} in V, we introduce the Lagrange multiplier $\beta_{js} \leq 0$. One can create a vector $(\overline{\alpha}, \overline{\beta})$ of dimensionality $(n+d) \cdot k$ by putting together all the Lagrangian parameters into a vector. Instead of using hard constraints on nonnegativity, Lagrangian relaxation uses penalties in order to relax the constraints into a softer version of the problem, which is defined by the augmented objective function L:

$$L = \|D - UV^T\|_F^2 + \sum_{i=1}^{n} \sum_{r=1}^{k} u_{ir} \alpha_{ir} + \sum_{j=1}^{d} \sum_{r=1}^{k} v_{jr} \beta_{jr} \qquad (9.7)$$

Note that violation of the nonnegativity constraints always lead to a positive penalty because the Lagrangian parameters cannot be positive. According to the methodology of Lagrangian optimization, this augmented problem is really a minimax problem because we need to minimize L over all U and V at any particular value of the (vector of) Lagrangian parameters, but we then need to maximize these solutions over all valid values of the Lagrangian parameters α_{is} and β_{js}. In other words, we have:

$$\text{Max}_{\overline{\alpha} \leq 0, \overline{\beta} \leq 0} \text{Min}_{U,V} L \qquad (9.8)$$

Here, $\overline{\alpha}$ and $\overline{\beta}$ represent the vectors of optimization parameters in α_{is} and β_{js}, respectively. This is a tricky optimization problem because of the way in which it is formulated with simultaneous maximization and minimization over different sets of parameters. The first step is to compute the gradient of the Lagrangian relaxation with respect to the (minimization) optimization variables u_{is} and v_{js}. Therefore, we have:

$$\frac{\partial L}{\partial u_{is}} = -(DV)_{is} + (UV^TV)_{is} + \alpha_{is} \qquad \forall i \in \{1,\ldots,n\}, s \in \{1,\ldots,k\} \qquad (9.9)$$

$$\frac{\partial L}{\partial v_{js}} = -(D^TU)_{js} + (VU^TU)_{js} + \beta_{js} \qquad \forall j \in \{1,\ldots,d\}, s \in \{1,\ldots,k\} \qquad (9.10)$$

The optimal value of the (relaxed) objective function at any particular value of the Lagrangian parameters is obtained by setting these partial derivatives to 0. As a result, we obtain the following conditions:

$$-(DV)_{is} + (UV^TV)_{is} + \alpha_{is} = 0 \qquad \forall i \in \{1,\ldots,n\}, s \in \{1,\ldots,k\} \qquad (9.11)$$

$$-(D^TU)_{js} + (VU^TU)_{js} + \beta_{js} = 0 \qquad \forall j \in \{1,\ldots,d\}, s \in \{1,\ldots,k\} \qquad (9.12)$$

We would like to eliminate the Lagrangian parameters and set up the optimization conditions purely in terms of U and V. It turns out the *Kuhn-Tucker optimality conditions* [18] are very helpful. These conditions are $u_{is}\alpha_{is} = 0$ and $v_{js}\beta_{js} = 0$ over all parameters. By multiplying Equation 9.11 with u_{is} and multiplying Equation 9.12 with v_{js}, we can use the Kuhn-Tucker conditions to get rid of these pesky Lagrangian parameters from the aforementioned equations. In other words, we have:

$$-(DV)_{is}u_{is} + (UV^TV)_{is}u_{is} + \underbrace{\alpha_{is}u_{is}}_{0} = 0 \qquad \forall i \in \{1,\ldots,n\}, s \in \{1,\ldots,k\} \qquad (9.13)$$

$$-(D^TU)_{js}v_{js} + (VU^TU)_{js}v_{js} + \underbrace{\beta_{js}v_{js}}_{0} = 0 \qquad \forall j \in \{1,\ldots,d\}, s \in \{1,\ldots,k\} \qquad (9.14)$$

One can rewrite these optimality conditions, so that a single parameter occurs on one side of the condition:

$$u_{is} = \frac{(DV)_{is}u_{is}}{(UV^TV)_{is}} \qquad \forall i \in \{1,\ldots,n\}, s \in \{1,\ldots,k\} \qquad (9.15)$$

$$v_{js} = \frac{(D^TU)_{js}v_{js}}{(VU^TU)_{js}} \qquad \forall j \in \{1,\ldots,d\}, s \in \{1,\ldots,k\} \qquad (9.16)$$

Even though these conditions are circular in nature (because the optimization parameters occur on both sides), they are natural candidates for iterative updates.

Therefore, the iterative approach starts by initializing the parameters in U and V to nonnegative random values in $(0,1)$ and then uses the following updates derived from the aforementioned optimality conditions:

$$u_{is} \Leftarrow \frac{(DV)_{is}u_{is}}{(UV^TV)_{is}} \qquad \forall i \in \{1,\ldots,n\}, s \in \{1,\ldots,k\} \qquad (9.17)$$

$$v_{js} \Leftarrow \frac{(D^TU)_{js}v_{js}}{(VU^TU)_{js}} \qquad \forall j \in \{1,\ldots,d\}, s \in \{1,\ldots,k\} \qquad (9.18)$$

These iterations are then repeated to convergence. Improved initialization provides significant advantages, and the reader is referred to [109] for such methods. Numerical stability can be improved by adding a small value $\epsilon > 0$ to the denominator during the updates:

$$u_{is} \Leftarrow \frac{(DV)_{is} u_{is}}{(UV^T V)_{is} + \epsilon} \qquad\qquad \forall i \in \{1, \dots, n\}, s \in \{1, \dots, k\} \qquad (9.19)$$

$$v_{js} \Leftarrow \frac{(D^T U)_{js} v_{js}}{(VU^T U)_{js} + \epsilon} \qquad\qquad \forall j \in \{1, \dots, d\}, s \in \{1, \dots, k\} \qquad (9.20)$$

One can also view ϵ as a type of *regularization parameter* whose primary goal is to avoid overfitting. Regularization is particularly helpful in small data sets.

As in all other forms of matrix factorization, it is possible to convert the factorization UV^T into the three-way factorization $Q\Sigma P^T$ by normalizing each column of U and V to unit norm, and storing the product of these normalization factors in a diagonal matrix Σ (i.e., the ith diagonal entry contains the product of the ith normalization factors of U and V). It is common to use L_1-normalization on each column of U and V, or that the columns of the resulting matrices Q and P each sum to 1. Interestingly, this type of normalization makes nonnegative factorization similar to a closely related factorization known as *Probabilistic Semantic Analysis (PLSA)*. The main difference between PLSA and nonnegative matrix factorization is that the former uses a maximum likelihood optimization function whereas nonnegative matrix factorization (typically) uses the Frobenius norm.

9.2.3.1 Interpreting Nonnegative Matrix Factorization

An important property of nonnegative matrix factorization is that it is highly interpretable in terms of the clusters in the underlying data. It is easiest to understand this point in terms of (semantically interpretable) document data. Consider the case in which the matrix D is an $n \times d$ matrix with n documents and d words (terms). The rth columns U_r and V_r of each of U and V respectively contain document- and word-membership information about the rth topic (or cluster) in the data. The n entries in U_r correspond to the nonnegative components (coordinates) of the n documents along the rth topic. If a document strongly belongs to topic r, then it will have a very positive coordinate in U_r. Otherwise, its coordinate will be zero or mildly positive (representing noise). Similarly, the rth column V_r of V provides the frequent vocabulary of the rth cluster. Terms that are highly related to a particular topic will have large components in V_r. The k-dimensional representation of each document is provided by the corresponding row of U. This approach allows a document to belong to multiple clusters, because a given row in U might have multiple positive coordinates. For example, if a document discusses both science and history, it will have components along latent components with science-related and history-related vocabularies. This provides a more realistic "sum-of-parts" decomposition of the corpus along various topics, which is primarily enabled by the nonnegativity of U and V. In fact, one can create a decomposition of the document-term matrix into k different rank-1 document-term matrices corresponding to the k topics captured by the decomposition. Let us treat U_r as an $n \times 1$ matrix and V_r as a $d \times 1$ matrix. If the rth component is related to science, then $U_r V_r^T$ is an $n \times d$ document-term matrix containing the science-related portion of the original corpus. Then the decomposition of the document-term matrix is defined as the sum of the following components:

$$D \approx \sum_{r=1}^{k} U_r V_r^T \qquad\qquad (9.21)$$

This decomposition is analogous to the spectral decomposition of SVD, except that its nonnegativity often gives it much better correspondence to semantically related topics.

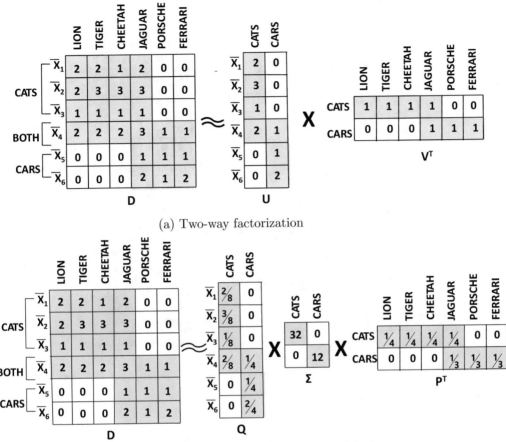

(a) Two-way factorization

(b) Three-way factorization by applying L_1-normalization to (a) above

Figure 9.2: The highly interpretable decomposition of nonnegative matrix factorization

In order to illustrate the semantic interpretability of nonnegative matrix factorization, let us revisit the same example used in Section 9.2.2.1 on singular value decomposition, and create a decomposition in terms of nonnegative matrix factorization:

$$D = \begin{pmatrix} & \text{lion} & \text{tiger} & \text{cheetah} & \text{jaguar} & \text{porsche} & \text{ferrari} \\ \text{Document-1} & 2 & 2 & 1 & 2 & 0 & 0 \\ \text{Document-2} & 2 & 3 & 3 & 3 & 0 & 0 \\ \text{Document-3} & 1 & 1 & 1 & 1 & 0 & 0 \\ \text{Document-4} & 2 & 2 & 2 & 3 & 1 & 1 \\ \text{Document-5} & 0 & 0 & 0 & 1 & 1 & 1 \\ \text{Document-6} & 0 & 0 & 0 & 2 & 1 & 2 \end{pmatrix}$$

This matrix represents topics related to both cars and cats. The first three documents are primarily related to cats, the fourth is related to both, and the last two are primarily related to cars. The word "jaguar" is polysemous because it could correspond to either a car or a cat and is present in documents of both topics.

A highly interpretable nonnegative factorization of rank-2 is shown in Figure 9.2(a). We

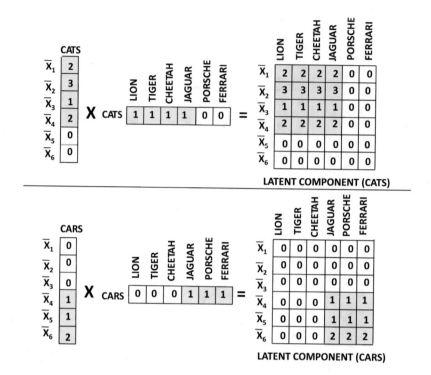

Figure 9.3: The highly interpretable "sum-of-parts" decomposition of the document-term matrix into rank-1 matrices representing different topics

have shown an approximate decomposition containing only integers for simplicity, although the optimal solution would (almost always) be dominated by floating point numbers in practice. It is clear that the first latent concept is related to cats and the second latent concept is related to cars. Furthermore, documents are represented by two non-negative coordinates indicating their affinity to the two topics. Correspondingly, the first three documents have strong positive coordinates for cats, the fourth has strong positive coordinates in both, and the last two belong only to cars. The matrix V tells us that the vocabularies of the various topics are as follows:

Cats: lion, tiger, cheetah, jaguar
Cars: jaguar, porsche, ferrari

It is noteworthy that the polysemous word "jaguar" is included in the vocabulary of both topics, and its usage is automatically inferred from its context (i.e., other words in document) during the factorization process. This fact becomes especially evident when we decompose the original matrix into two rank-1 matrices according to Equation 9.21. This decomposition is shown in Figure 9.3 in which the rank-1 matrices for cats and cars are shown. It is particularly interesting that the occurrences of the polysemous word "jaguar" are nicely divided up into the two topics, which roughly correspond with their usage in these topics.

Any two-way matrix factorization can be converted into a standardized three-way factorization by normalizing the columns of U and V, so that they add to 1, and creating a diagonal matrix from the product of these normalization factors. The three-way normalized representation is shown in Figure 9.2(b), and it tells us a little bit more about the relative frequencies of the two topics. Since the diagonal entry in Σ is 32 for cats in comparison with 12 for cars, it indicates that the topic of cats is more dominant than cars. This is consistent with the observation that more documents and terms in the collection are associated with cats as compared to cars.

9.2.4 Dimensionality Reduction with Neural Networks

Just as most of the supervised learning methods can be represented using neural networks, it is also possible to represent most of the unsupervised methods with neural networks. A key point is that all forms of learning, including unsupervised learning, can be represented in the form of input-to-output mappings, which creates a computational graph (and therefore a neural architecture as well). Autoencoders represent a fundamental architecture that is used for various types of unsupervised learning applications. In these models, the inputs and outputs are the same. In other words, the inputs are *replicated* to the outputs, with intervening layers that have fewer nodes, so that precise copying from layer to layer is not possible. In other words, the data is compressed in the intermediate layers before being replicated to the outermost layer.

The simplest autoencoders with linear layers map to well-known dimensionality reduction techniques like singular value decomposition. However, deep autoencoders with nonlinearity map to complex models that might not exist in traditional machine learning. Therefore, the goal of this section is to show two things:

1. Classical dimensionality reduction methods like singular value decomposition are special cases of shallow neural architectures.

2. By adding depth and nonlinearity to the basic architecture, one can generate sophisticated nonlinear embeddings of the data. While nonlinear embeddings are also available in traditional machine learning, the latter is limited to loss functions that can be expressed compactly in closed form. The loss functions of deep neural architectures are no longer compact; however, they provide unprecedented flexibility in controlling the properties of the unsupervised representation by making various types of architectural changes (and allowing backpropagation to take care of the complexities of differentiation).

The basic idea of an autoencoder is to have an output layer with the same dimensionality as the inputs. The idea is to try to reconstruct the input data instance. An autoencoder *replicates* the data from the input to the output, and is therefore sometimes referred to as a *replicator neural network*. Although reconstructing the data might seem like a trivial matter by simply copying the data forward from one layer to another, this is not possible when the number of units in the middle are *constricted*. In other words, the number of units in each middle layer is typically fewer than that in the input (or output). As a result, one cannot simply copy the data from one layer to another. Therefore, the activations in the constricted layers hold a reduced representation of the data; the net effect is that this type of reconstruction is inherently *lossy*. This general representation of the autoencoder is given in Figure 9.4(a), where an architecture is shown with three constricted layers. Note that the output layer has the same number of units as the input layer. The loss function of this neural network uses the sum-of-squared differences between the input and the output feature values in order to force the output to be as similar as possible to the input.

It is common (but not necessary) for an M-layer autoencoder to have a symmetric architecture between the input and output, where the number of units in the kth layer is

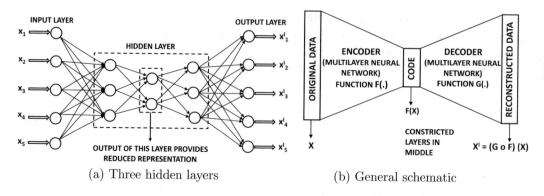

(a) Three hidden layers   (b) General schematic

Figure 9.4: The basic schematic of the autoencoder

the same as that in the $(M - k + 1)$th layer. Furthermore, the value of M is often odd, as a result of which the $(M + 1)/2$th layer is the most constricted layer. Here, we are counting the (non-computational) input layer as the first layer, and therefore the minimum number of layers in an autoencoder would be three, corresponding to the input layer, constricted layer, and the output layer.

The reduced representation of the data in the most constricted layer is also sometimes referred to as the *code*, and the number of units in this layer is the dimensionality of the reduction. The initial part of the neural architecture before the bottleneck is referred to as the *encoder* (because it creates a reduced code), and the final part of the architecture is referred to as the *decoder* (because it reconstructs from the code). The general schematic of the autoencoder is shown in Figure 9.4(b).

9.2.4.1   Linear Autoencoder with a Single Hidden Layer

A single hidden-layer autoencoder can be understood in the context of matrix factorization. In matrix factorization, we want to factorize the $n \times d$ matrix D into an $n \times k$ matrix U and a $d \times k$ matrix V:

$$D \approx UV^T \qquad (9.22)$$

Here, $k \ll n$ is the rank of the factorization. As discussed earlier in this chapter, the objective function of traditional matrix factorization is as follows:

$$\text{Minimize } J = \|D - UV^T\|_F^2$$

Here, the notation $\| \cdot \|_F$ indicates the Frobenius norm. The parameter matrices U and V need to be learned in order to optimize the aforementioned error. Although the gradient-descent steps have already been discussed in earlier sections, our goal here is to capture this optimization problem within a neural architecture. Going through this exercise helps us show that simple matrix factorization is a special case of an autoencoder architecture, which sets the stage for understanding the gains obtained with deeper autoencoders.

From the perspective of neural networks, one would need to design the architecture in such a way that the rows of D are fed to the neural network, and the reduced rows of U are obtained from the constricted layer when the original data is fed to the neural network. The single-layer autoencoder is illustrated in Figure 9.5, where the hidden layer contains k units. The rows of D are input into the autoencoder, whereas the k-dimensional rows of U are the activations of the hidden layer. Note that the $k \times d$ matrix of weights in the decoder

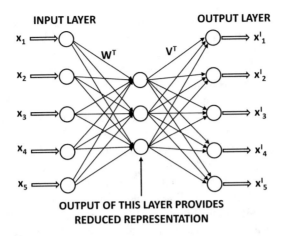

Figure 9.5: A basic autoencoder with a single layer

must be V^T, since it is necessary to be able to multiply the rows of U to reconstruct the rows of $D \approx UV^T$ according to the optimization model discussed above.

The vector of values in a particular layer of the network can be obtained by multiplying the vector of values in the previous layer with the matrix of weights connecting the two layers (with linear activation) Here, $\overline{u}_i$ is the row vector containing the ith row of U and $\overline{X}'_i$ is the reconstruction of the ith row $\overline{X}_i$ of D. One first encodes $\overline{X}_i$ with the matrix W^T, and then decodes it back using V^T:

$$\overline{u}_i = \overline{X}_i W^T \tag{9.23}$$

$$\overline{X}'_i = \overline{u}_i V^T \tag{9.24}$$

It is not difficult to see that Equation 9.24 is a row-wise variant of Equation 9.22. The autoencoder minimizes the sum-of-squared differences between the input and the output, which is equivalent to minimizing $\|D - UV^T\|_F^2$. Therefore, the neural architecture achieves precisely the goals of matrix factorization in terms of the loss value that is optimized. In fact, the basis vectors in V can be shown to span the same subspace as the top-k basis vectors of SVD. However, the optimization problem for SVD has multiple global optima, and the SVD only corresponds to an optimum in which the columns of V are orthonormal and those of U are mutually orthogonal. The neural network might find a different basis of the subspace spanned by the top-k basis vectors of SVD, and will adjust U accordingly so that the reconstruction UV^T remains unaffected.

9.2.4.2 Nonlinear Activations

So far, the discussion has focused on simulating singular value decomposition using a neural architecture. The real power of autoencoders is realized when one starts using nonlinear activations and multiple layers. For example, consider a situation in which the matrix D is binary. In such a case, one can use the same neural architecture as shown in Figure 9.5, but one can also use a sigmoid function in the final layer to predict the output. This sigmoid layer is combined with negative log loss. Therefore, for a binary matrix $B = [b_{ij}]$, the model assumes the following:

$$B \sim \text{sigmoid}(UV^T) \tag{9.25}$$

Here, the sigmoid function is applied in element-wise fashion. Note the use of $\sim$ instead of $\approx$ in the above expression, which indicates that the binary matrix B is an instantiation of random draws from Bernoulli distributions with corresponding parameters contained in sigmoid(UV^T). The resulting factorization can be shown to be equivalent to *logistic matrix factorization*. The basic idea is that the (i, j)th element of UV^T is the parameter of a Bernoulli distribution, and the binary entry b_{ij} is generated from a Bernoulli distribution with these parameters. Therefore, U and V are learned using log-likelihood loss. The log-likelihood loss implicitly tries to find parameter matrices U and V so that the probability of the matrix B being generated by these parameters is maximized.

Logistic matrix factorization has only recently been proposed [92] as a sophisticated matrix factorization method for binary data, which is useful for recommender systems with *implicit feedback* ratings. Implicit feedback refers to the binary actions of users such as buying or not buying specific items. The solution methodology of this recent work on logistic matrix factorization [92] seems to be vastly different from SVD, and it is not based on a neural network approach. However, for a neural network practitioner, the change from the SVD model to that of logistic matrix factorization is a relatively small one, where only the final layer of the neural network needs to be changed. It is this modular nature of neural networks that makes them so attractive to engineers and encourages all types of experimentation.

The real power of autoencoders in the neural network domain is realized when deeper variants with nonlinear activations in hidden layers are used. For example, an autoencoder with three hidden layers is shown in Figure 9.4(a). One can increase the number of inter-mediate layers in order to further increase the representation power of the neural network. It is noteworthy that it is essential for some of the layers of the deep autoencoder to use a nonlinear activation function to increase its representation power. As shown in Chapter 7, no additional power is gained by a multilayer network when only linear activations are used.

Deep networks with multiple layers provide an extraordinary amount of representation power. The multiple layers of this network provide *hierarchically* reduced representations of the data. For some data domains like images, hierarchically reduced representations are particularly natural. Note that there is no precise analog of this type of model in tradi-tional machine learning, and the backpropagation approach rescues us from the challenges associated in computing the complicated gradient-descent steps. A nonlinear dimensionality reduction might map a manifold of arbitrary shape into a reduced representation. Although several methods for nonlinear dimensionality reduction are known in machine learning, neural networks have some advantages over these methods:

- Many nonlinear dimensionality reduction methods have a very hard time mapping out-of-sample data points to reduced representations, unless these points are included in the training data up front. On the other hand, it is a relatively simple matter to compute the reduced representation of an out-of-sample point by passing it through the network.

- Neural networks allow more power and flexibility in the nonlinear data reduction by varying on the number and type of layers used in intermediate stages. Furthermore, by choosing specific types of activation functions in particular layers, one can engineer the nature of the reduction to the properties of the data. For example, it makes sense to use a logistic output layer with logarithmic loss for a binary data set. Indeed, for multilayer variants of the autoencoder, an exact counterpart often does not even exist in traditional machine learning. This seems to suggest that it is often more natural to

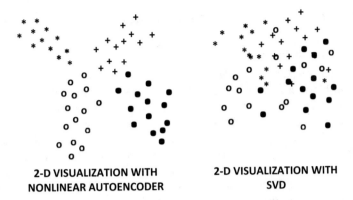

2-D VISUALIZATION WITH NONLINEAR AUTOENCODER

2-D VISUALIZATION WITH SVD

Figure 9.6: A depiction of the typical difference between the embeddings created by nonlinear autoencoders and singular value decomposition (SVD). Nonlinear and deep autoencoders are often able to separate out the entangled class structures in the underlying data, which is not possible within the constraints of linear transformations like SVD

discover sophisticated machine learning algorithms when working with the modular approach of constructing multilayer neural networks.

It is possible to achieve extraordinarily compact reductions by using this approach. Greater reduction is always achieved by using nonlinear units, which implicitly map warped manifolds into linear hyperplanes. The superior reduction in these cases is because it is easier to thread a warped surface (as opposed to a linear surface) through a larger number of points. This property of nonlinear autoencoders is often used for 2-dimensional visualizations of the data by creating a deep autoencoder in which the most compact hidden layer has only two dimensions. These two dimensions can then be mapped on a plane to visualize the points.

An illustrative example of the typical behavior of real data distributions is shown in Figure 9.6, in which the 2-dimensional mapping created by a deep autoencoder seems to clearly separate out the different classes. On the other hand, the mapping created by SVD does not seem to separate the classes well. In many cases, the data may contain heavily entangled spirals (or other shapes) that belong to different classes. Linear dimensionality reduction methods cannot attain clear separation because nonlinearly entangled shapes are not linearly separable. On the other hand, deep autoencoders with nonlinearity are far more powerful and able to disentangle such shapes.

9.3 Clustering

The problem of clustering segments data records into groups, so that similar records are placed in the same group. While dimensionality reduction methods are more deeply focused on relationships among the entries of the matrix, clustering specifically focuses on the rows. Nevertheless, some methods like nonnegative matrix factorization can be used for both clustering and dimensionality reduction. Clustering methods are either *flat* or *hierarchical.* In flat clustering, the data set is partitioned into a set of clusters in one shot, and no hierarchical relationships exist between clusters. In hierarchical clustering, the clusters are organized in tree-like fashion as a taxonomy. This section will discuss several clustering methods, some of which are flat, whereas others are hierarchical.

9.3.1   Representative-Based Algorithms

As the name implies, representative-based algorithms rely on distances (or similarities) to representative points for clustering. Such representative points serve as *prototypes* to which most cluster points are similar. Representative-based algorithms create flat clusters and they do not have hierarchical relationships among them. The partitioning representatives may either be created as a function of the data points in the clusters (e.g., the mean) or may be selected from the existing data points in the cluster. The main insight of these methods is that the discovery of high-quality clusters in the data is equivalent to discovering a high-quality set of representatives. Once the representatives have been determined, a distance function can be used to assign the data points to their closest representatives. This category of methods contains several methods such as the k-means, k-medoids, and the k-medians algorithms. In this section, we will primarily discuss the k-means algorithm.

Typically, it is assumed that the number of clusters, denoted by k, is specified by the user. Consider a data set $\mathcal{D}$ containing n data points denoted by $\overline{X}_1 \ldots \overline{X}_n$ in d-dimensional space. The goal is to determine k representatives $\overline{Y}_1 \ldots \overline{Y}_k$ that minimize the following objective function O:

$$O = \sum_{i=1}^{n} \left[\min_j Dist(\overline{X}_i, \overline{Y}_j) \right] \tag{9.26}$$

In other words, the sum of the distances of the different data points to their closest representatives needs to be minimized. Note that the assignment of data points to representatives depends on the choice of the representatives $\overline{Y}_1 \ldots \overline{Y}_k$. In some variations of representative algorithms, such as k-medoid algorithms, it is assumed that the representatives $\overline{Y}_1 \ldots \overline{Y}_k$ are drawn from the original database $\mathcal{D}$, although this will obviously not provide an optimal solution. In general, the discussion in this section will not automatically assume that the representatives are drawn from the original database $\mathcal{D}$, unless specified otherwise.

One observation about the formulation of Equation 9.26 is that the representatives $\overline{Y}_1 \ldots \overline{Y}_k$ and the optimal assignment of data points to representatives are unknown *a priori*, but they depend on each other in a circular way. For example, if the optimal representatives are known, then the optimal assignment is easy to determine, and vice versa. Such optimization problems are solved with the use of an iterative approach where candidate representatives and candidate assignments are used to improve each other. Therefore, the generic k-representatives approach starts by initializing the k representatives S with the use of a straightforward heuristic (such as random sampling from the original data), and then refines the representatives and the clustering assignment, iteratively, as follows:

- (Assign step) Assign each data point to its closest representative in S using distance function $Dist(\cdot, \cdot)$, and denote the corresponding clusters by $\mathcal{C}_1 \ldots \mathcal{C}_k$.

- (Optimize step) Determine the optimal representative $\overline{Y}_j$ for each cluster $\mathcal{C}_j$ that minimizes its *local* objective function $\sum_{\overline{X}_i \in \mathcal{C}_j} \left[Dist(\overline{X}_i, \overline{Y}_j) \right]$.

It will be evident later in this chapter that this two-step procedure is very closely related to generative models of cluster analysis in the form of *expectation-maximization* algorithms. The second step of *local* optimization is simplified by this two-step iterative approach because it no longer depends on an unknown assignment of data points to clusters as in the global optimization problem of Equation 9.26. Typically, the optimized representative can be shown to be some central measure of the data points in the jth cluster $\mathcal{C}_j$, and the precise measure depends on the choice of the distance function $Dist(\overline{X}_i, \overline{Y}_j)$. In particular, for the case of the Euclidean distance and cosine similarity functions, it can be shown that the

Algorithm *GenericRepresentative*(Database: $\mathcal{D}$, Number of Representatives: k)
begin
   Initialize representative set S;
   repeat
      Create clusters $(\mathcal{C}_1 \ldots \mathcal{C}_k)$ by assigning each
         point in $\mathcal{D}$ to closest representative in S
         using the distance function $Dist(\cdot, \cdot)$;
      Recreate set S by determining one representative $\overline{Y_j}$ for
         each $\mathcal{C}_j$ that minimizes $\sum_{\overline{X}_i \in \mathcal{C}_j} Dist(\overline{X}_i, \overline{Y}_j)$;
   until convergence;
   return $(\mathcal{C}_1 \ldots \mathcal{C}_k)$;
end

Figure 9.7: Generic representative algorithm with unspecified distance function

optimal centralized representative of each cluster is its mean. However, different distance functions may lead to a slightly different type of centralized representative, and these lead to different variations of this broader approach, such as the k-means and k-medians algorithms. Thus, the k-representative approach defines a *family* of algorithms, in which minor changes to the basic framework allow the use of different distance criteria. The generic framework for representative-based algorithms with an unspecified distance function is illustrated in the pseudocode of Figure 9.7. The idea is to improve the objective function over multiple iterations. Typically, the increase is significant in early iterations, but it slows down in later iterations. When the improvement in the objective function in an iteration is less than a user-defined threshold, the algorithm may be allowed to terminate. The primary computational bottleneck of the approach is the assignment step where the distances need to be computed between all point-representative pairs. The time complexity of each iteration is $O(k \cdot n \cdot d)$ for a data set of size n and dimensionality d. The algorithm typically terminates in a small constant number of iterations.

The k-Means Algorithm

In the k-means algorithm, the sum of the squares of the Euclidean distances of data points to their closest representatives is used to quantify the objective function of the clustering. Therefore, we have:

$$Dist(\overline{X}_i, \overline{Y}_j) = \|\overline{X}_i - \overline{Y}_j\|_2^2 \qquad (9.27)$$

Here, $\| \cdot \|_p$ represents the L_p-norm. The expression $Dist(\overline{X}_i, \overline{Y}_j)$ can be viewed as the squared error of approximating a data point with its closest representative. Thus, the overall objective minimizes the sum of square errors over different data points. This is also sometimes referred to as SSE. In such a case, it can be shown[3] that the *optimal representative $\overline{Y}_j$ for each of the "optimize" iterative steps is the mean of the data points in cluster* $\mathcal{C}_j$. Thus, the only difference between the generic pseudocode of Figure 9.7 and a k-means pseudocode is the specific instantiation of the distance function $Dist(\cdot, \cdot)$, and the choice of the representative as the local mean of its cluster.

---

[3]For a *fixed* cluster assignment $\mathcal{C}_1 \ldots \mathcal{C}_k$, the gradient of the clustering objective function $\sum_{j=1}^{k} \sum_{\overline{X}_i \in \mathcal{C}_j} \|\overline{X}_i - \overline{Y}_j\|^2$ with respect to $\overline{Y}_j$ is $2 \sum_{\overline{X}_i \in \mathcal{C}_j} (\overline{X}_i - \overline{Y}_j)$. Setting the gradient to 0 yields the mean of cluster $\mathcal{C}_j$ as the optimum value of $\overline{Y}_j$. Note that the other clusters do not contribute to the gradient, and therefore the approach effectively optimizes the local clustering objective function for $\mathcal{C}_j$.

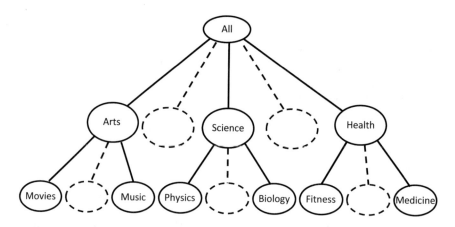

Figure 9.8: Multigranularity insights from hierarchical clustering

9.3.2   Bottom-up Agglomerative Methods

Hierarchical clustering algorithms create clusters that are arranged in the form of a hierarchical taxonomy. Although it is common to perform the clustering with the use of distances, many algorithms use other techniques, such as density- or graph-based methods, as a subroutine for constructing the hierarchy.

Hierarchical clustering algorithms are useful because different levels of clustering granularity provide different application-specific insights. This provides a *taxonomy* of clusters, which may be browsed for semantic insights. As a specific example, consider the taxonomy[4] of Web pages created by the well-known *Open Directory Project (ODP)*. In this case, the clustering has been created by a manual volunteer effort, but it nevertheless provides a good understanding of the multi-granularity insights that may be obtained with such an approach. A small portion of the hierarchical organization is illustrated in Figure 9.8. At the highest level, the Web pages are organized into topics such as arts, science, health, and so on. At the next level, the topic of science is organized into subtopics, such as biology and physics, whereas the topic of health is divided into topics such as fitness and medicine. This organization makes manual browsing very convenient for a user, especially when the content of the clusters can be described in a semantically comprehensible way. In other cases, such hierarchical organizations can be used by indexing algorithms. Furthermore, such methods can sometimes also be used for creating better "flat" clusters. There are two types of hierarchical clustering algorithms, referred to as *bottom-up hierarchical methods* and *top-down hierarchical methods*. This section will discuss bottom-up hierarchical methods, and the next section will discuss top-down methods.

In bottom-up methods, the data points are successively agglomerated into higher-level clusters. The algorithm starts with individual data points in their own clusters and successively agglomerates them into higher-level clusters. In each iteration, two clusters are selected that are deemed to be as close as possible. These clusters are merged and replaced with a newly created merged cluster. Thus, each merging step reduces the number of clusters by 1. Therefore, a method needs to be designed for measuring proximity between clusters containing multiple data points, so that they may be merged. It is in this choice of computing the distances between clusters, that most of the variations among different methods arise.

---

[4]http://www.dmoz.org

> **Algorithm** *AgglomerativeMerge* (Data: $\mathcal{D}$)
> **begin**
>   Initialize $n \times n$ distance matrix M using $\mathcal{D}$;
>   **repeat**
>   Pick closest pair of clusters i and j using M;
>   Merge clusters i and j;
>   Delete rows/columns i and j from M and create
>   a new row and column for newly merged cluster;
>   Update the entries of new row and column of M;
>   **until** termination criterion;
>   **return** current merged cluster set;
> **end**

Figure 9.9: Generic agglomerative merging algorithm with unspecified merging criterion

Let n be the number of data points in the d-dimensional database $\mathcal{D}$, and $n_t = n - t$ be the number of clusters after t agglomerations. At any given point, the method maintains an $n_t \times n_t$ distance matrix M between the current *clusters* in the data. The precise methodology for computing and maintaining this distance matrix will be described later. In any given iteration of the algorithm, the (non-diagonal) entry in the distance-matrix with the least distance is selected, and the corresponding clusters are merged. This merging will require the distance matrix to be updated to a smaller $(n_t - 1) \times (n_t - 1)$ matrix. The dimensionality reduces by 1 because the rows and columns for the two merged clusters need to be deleted, and a new row and column of distances, corresponding to the newly created cluster, needs to be added to the matrix. This corresponds to the newly created cluster in the data. The algorithm for determining the values of this newly created row and column depends on the cluster-to-cluster distance computation in the merging procedure and will be described later. The incremental update process of the distance matrix is a more efficient option than that of computing all distances from scratch. It is, of course, assumed that sufficient memory is available to maintain the distance matrix. If this is not the case, then the distance matrix will need to be fully re-computed in each iteration, and such agglomerative methods become less attractive. For termination, either a maximum threshold can be used on the distances between two merged clusters, or a minimum threshold can be used on the number of clusters at termination. The former criterion is designed to automatically determine the natural number of clusters in the data but has the disadvantage of requiring the specification of a quality threshold that is hard to guess intuitively. The latter criterion has the advantage of being intuitively interpretable in terms of the number of clusters in the data. The order of merging naturally creates a hierarchical tree-like structure illustrating the relationship between different clusters, which is referred to as a *dendrogram*. An example of a dendrogram on successive merges on six data points, denoted by A, B, C, D, E, and F, is illustrated in Figure 9.10(a).

The generic agglomerative procedure with an unspecified merging criterion is illustrated in Figure 9.9. The distances are encoded in the $n_t \times n_t$ distance matrix M. This matrix provides the pairwise cluster distances computed with the use of the merging criterion. The different choices for the merging criteria will be described later. The merging of two clusters corresponding to rows (columns) i and j in the matrix M requires the computation of some measure of distances between their constituent objects. For two clusters containing m_i and m_j objects, respectively, there are $m_i \cdot m_j$ pairs of distances between constituent objects. For example, in Figure 9.10(b), there are $2 \times 4 = 8$ pairs of distances between the constituent

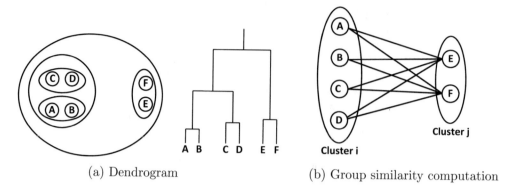

(a) Dendrogram   (b) Group similarity computation

Figure 9.10: Illustration of hierarchical clustering steps

objects, which are illustrated by the corresponding edges. The overall distance between the two clusters needs to be computed as a function of these $m_i \cdot m_j$ pairs. In the following, different ways of computing the distances will be discussed.

9.3.2.1   Group-Based Statistics

The following discussion assumes that the indices of the two clusters to be merged are denoted by i and j, respectively. In group-based criteria, the distance between two groups of objects is computed as a function of the $m_i \cdot m_j$ pairs of distances among the constituent objects. The different ways of computing distances between two groups of objects are as follows:

1. *Best (single) linkage:* In this case, the distance is equal to the minimum distance between all $m_i \cdot m_j$ pairs of objects. This corresponds to the closest pair of objects between the two groups. After performing the merge, the matrix M of pairwise distances needs to be updated. The ith and jth rows and columns are deleted and replaced with a single row and column representing the merged cluster. The new row (column) can be computed using the minimum of the values in the previously deleted pair of rows (columns) in M. This is because the distance of the other clusters to the merged cluster is the minimum of their distances to the individual clusters in the best-linkage scenario. For any other cluster $k \neq i, j$, this is equal to $\min\{M_{ik}, M_{jk}\}$ (for rows), and $\min\{M_{ki}, M_{kj}\}$ (for columns). The indices of the rows and columns are then updated to account for the deletion of the two clusters and their replacement with a new one. The best linkage approach is one of the instantiations of agglomerative methods that is very good at discovering clusters of arbitrary shape. This is because the data points in clusters of arbitrary shape can be successively merged with chains of data point pairs at small pairwise distances to each other. On the other hand, such chaining may also inappropriately merge distinct clusters when it results from noisy points.

2. *Worst (complete) linkage:* In this case, the distance between two groups of objects is equal to the maximum distance between all $m_i \cdot m_j$ pairs of objects in the two groups. This corresponds to the farthest pair in the two groups. Correspondingly, the matrix M is updated using the maximum values of the rows (columns) in this case. For any value of $k \neq i, j$, this is equal to $\max\{M_{ik}, M_{jk}\}$ (for rows), and $\max\{M_{ki}, M_{kj}\}$ (for columns). The worst-linkage criterion implicitly attempts to minimize the maximum

diameter of a cluster, as defined by the largest distance between any pair of points in the cluster. This method is also referred to as the *complete linkage* method.

3. *Group-average linkage:* In this case, the distance between two groups of objects is equal to the average distance between all $m_i \cdot m_j$ pairs of objects in the groups. To compute the row (column) for the merged cluster in M, a weighted average of the ith and jth rows (columns) in the matrix M is used. For any value of $k \neq i, j$, this is equal to $\frac{m_i \cdot M_{ik} + m_j \cdot M_{jk}}{m_i + m_j}$ (for rows), and $\frac{m_i \cdot M_{ki} + m_j \cdot M_{kj}}{m_i + m_j}$ (for columns).

4. *Closest centroid:* In this case, the closest centroids are merged in each iteration. This approach is not desirable, however, because the centroids lose information about the relative spreads of the different clusters. For example, such a method will not discriminate between merging pairs of clusters of varying sizes, as long as their centroid pairs are at the same distance. Typically, there is a bias towards merging pairs of larger clusters because centroids of larger clusters are statistically more likely to be closer to each other.

5. *Variance-based criterion:* This criterion minimizes the *change* in the objective function (such as cluster variance) as a result of the merging. Merging always results in a worsening of the clustering objective function value because of loss of granularity. In this case, the clusters are merged, so that the change in the objective function as a result of merging is as little as possible.

6. *Ward's method:* Instead of using the change in variance, one might also use the (unscaled) sum of squared error as the merging criterion. Surprisingly, this approach is a variant of the centroid method. The objective function for merging is obtained by multiplying the (squared) Euclidean distance between centroids with the harmonic mean of the number of points in each of the pair. Because larger clusters are penalized by this additional factor, the approach performs more effectively than the centroid method.

The various criteria have different advantages and disadvantages. For example, the single linkage method is able to successively merge chains of closely related points to discover clusters of arbitrary shape. However, this property can also (inappropriately) merge two unrelated clusters, when the chaining is caused by noisy points between two clusters. Examples of good and bad cases for single-linkage clustering are illustrated in Figures 9.11(a) and (b), respectively. Therefore, the behavior of single-linkage methods depends on the impact and relative presence of noisy data points.

The complete (worst-case) linkage method attempts to minimize the maximum distance between any pair of points in a cluster. This quantification can implicitly be viewed as an approximation of the diameter of a cluster. Because of its focus on minimizing the diameter, it will try to create clusters so that all of them have a similar diameter. However, if some of the natural clusters in the data are larger than others, then the approach will break up the larger clusters. It will also be biased towards creating clusters of spherical shape irrespective of the underlying data distribution. Another problem with the complete linkage method is that it gives too much importance to data points at the noisy fringes of a cluster because of its focus on the maximum distance between any pair of points in the cluster. The group-average, variance, and Ward's methods are more robust to noise due to the use of multiple linkages in the distance computation.

The agglomerative method requires the maintenance of a heap of sorted distances to efficiently determine the minimum distance value in the matrix. The initial distance matrix

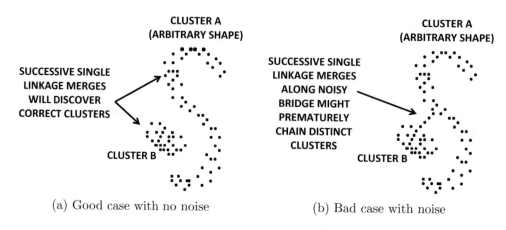

(a) Good case with no noise   (b) Bad case with noise

Figure 9.11: Good and bad cases for single-linkage clustering

computation requires $O(n^2 \cdot d)$ time, and the maintenance of a sorted heap data structure requires $O(n^2 \cdot \log(n))$ time over the course of the algorithm because there will be a total of $O(n^2)$ additions and deletions into the heap. Therefore, the overall running time is $O(n^2 \cdot d + n^2 \cdot \log(n))$. The required space for the distance matrix is $O(n^2)$. The space-requirement is particularly problematic for large data sets. In such cases, a similarity matrix M cannot be incrementally maintained, and the time complexity of many hierarchical methods will increase dramatically to $O(n^3 \cdot d)$. This increase occurs because the similarity computations between clusters need to be performed explicitly at the time of the merging.

Agglomerative hierarchical methods naturally lead to a binary tree of clusters. It is generally difficult to control the structure of the hierarchical tree with bottom-up methods as compared to top-down methods. Therefore, in cases where a taxonomy of a specific structure is desired, bottom-up methods are less desirable.

A problem with hierarchical methods is that they are sensitive to a small number of mistakes made during the merging process. For example, if an incorrect merging decision is made at some stage because of the presence of noise in the data set, then there is no way to undo it, and the mistake may further propagate in successive merges. In fact, some variants of hierarchical clustering, such as single-linkage methods, are notorious for successively merging neighboring clusters because of the presence of a small number of noisy points. Nevertheless, there are numerous ways to reduce these effects by treating noisy data points specially.

Agglomerative methods can become impractical from a *space- and time-efficiency* perspective for larger data sets. Therefore, these methods are often combined with sampling and other partitioning methods to efficiently provide solutions of high quality.

9.3.3   Top-down Divisive Methods

Although bottom-up agglomerative methods are typically distance-based methods, top-down hierarchical methods can be viewed as general-purpose meta-algorithms that can use almost any clustering algorithm as a subroutine. Because of the top-down approach, greater control is achieved on the global structure of the tree in terms of its degree and balance between different branches.

The overall approach for top-down clustering uses a general-purpose flat clustering algorithm $\mathcal{A}$ as a subroutine. The algorithm initializes the tree at the root-node containing

Algorithm *GenericTopDownClustering*(Data: $\mathcal{D}$, Flat Algorithm: $\mathcal{A}$)
begin
    Initialize tree $\mathcal{T}$ to root containing $\mathcal{D}$;
    repeat
        Select a leaf node L in $\mathcal{T}$ based on pre-defined criterion;
        Use algorithm $\mathcal{A}$ to split L into $L_1 \ldots L_k$;
        Add $L_1 \ldots L_k$ as children of L in $\mathcal{T}$;
    until termination criterion;
end

Figure 9.12: Generic top-down meta-algorithm for clustering

all the data points. In each iteration, the data set at a particular node of the current tree is split into multiple nodes (clusters). By changing the criterion for node selection, one can create trees balanced by height or trees balanced by the number of clusters. If the algorithm $\mathcal{A}$ is randomized, such as the k-means algorithm (with random seeds), it is possible to use multiple trials of the same algorithm at a particular node and select the best one. The generic pseudocode for a top-down divisive strategy is illustrated in Figure 9.12. The algorithm recursively splits nodes with a top-down approach until either a certain height of the tree is achieved or each node contains fewer than a predefined number of data objects. A wide variety of algorithms can be designed with different instantiations of the algorithm $\mathcal{A}$ and growth strategy. Note that the algorithm $\mathcal{A}$ can be any arbitrary clustering algorithm, and not just a distance-based algorithm.

9.3.3.1 Bisecting k-Means

The bisecting k-means algorithm is a top-down hierarchical clustering algorithm in which each node is split into exactly two children with a 2-means algorithm. To split a node into two children, several randomized trial runs of the split are used, and the split that has the best impact on the overall clustering objective is used. Several variants of this approach use different growth strategies for selecting the node to be split. For example, the heaviest node may be split first, or the node with the smallest distance from the root may be split first. These different choices lead to balancing either the cluster weights or the tree height.

9.3.4 Probabilistic Model-based Algorithms

Most clustering algorithms discussed in this book are *hard* clustering algorithms in which each data point is deterministically assigned to a particular cluster. Probabilistic model-based algorithms are *soft* algorithms in which each data point may have a non-zero assignment probability to many (typically all) clusters. A soft solution to a clustering problem may be converted to a hard solution by assigning a data point to a cluster with respect to which it has the largest assignment probability.

The broad principle of a mixture-based *generative* model is to assume that the data was generated from a mixture of k distributions with probability distributions $\mathcal{G}_1 \ldots \mathcal{G}_k$. Each distribution $\mathcal{G}_i$ represents a cluster and is also referred to as a *mixture component*. Each data point $\overline{X_i}$, where $i \in \{1 \ldots n\}$, is generated by this mixture model as follows:

1. Select a mixture component with prior probability $\alpha_i = P(\mathcal{G}_i)$, where $i \in \{1 \ldots k\}$. Assume that the rth one is selected.

2. Generate a data point from $\mathcal{G}_r$.

This generative model will be denoted by $\mathcal{M}$. The different prior probabilities α_i and the parameters of the different distributions $\mathcal{G}_r$ are not known in advance. Each distribution $\mathcal{G}_i$ is often assumed to be the Gaussian, although any arbitrary (and different) family of distributions may be assumed for each $\mathcal{G}_i$. The choice of distribution $\mathcal{G}_i$ is important because it reflects the user's *a priori* understanding about the distribution and shape of the individual clusters (mixture components). The parameters of the distribution of each mixture component, such as its mean and variance, need to be estimated from the data, so that the overall data has the maximum likelihood of being *generated* by the model. This is achieved with the *expectation-maximization (EM)* algorithm. The parameters of the different mixture components can be used to describe the clusters. For example, the estimation of the mean of each Gaussian component is analogous to determining the mean of each cluster center in a k-representatives algorithm. After the parameters of the mixture components have been estimated, the *posterior* generative (or assignment) probabilities of data points with respect to each mixture component (cluster) can be determined.

Assume that the probability density function of mixture component $\mathcal{G}_i$ is denoted by $f^i(\cdot)$. The probability (density function) of the data point $\overline{X}_j$ being generated by the model is given by the weighted sum of the probability densities over different mixture components, where the weight is the prior probability $\alpha_i = P(\mathcal{G}_i)$ of the mixture components:

$$f^{point}(\overline{X}_j|\mathcal{M}) = \sum_{i=1}^{k} \alpha_i \cdot f^i(\overline{X}_j) \tag{9.28}$$

Then, for a data set $\mathcal{D}$ containing n data points, denoted by $\overline{X_1} \dots \overline{X_n}$, the probability density of the data set being generated by the model $\mathcal{M}$ is the product of all the point-specific probability densities:

$$f^{data}(\mathcal{D}|\mathcal{M}) = \prod_{j=1}^{n} f^{point}(\overline{X}_j|\mathcal{M}) \tag{9.29}$$

The log-likelihood fit $\mathcal{L}(\mathcal{D}|\mathcal{M})$ of the data set $\mathcal{D}$ with respect to model $\mathcal{M}$ is the logarithm of the aforementioned expression and can be (more conveniently) represented as a sum of values over the different data points. The log-likelihood fit is preferred for computational reasons.

$$\mathcal{L}(\mathcal{D}|\mathcal{M}) = \log(\prod_{j=1}^{n} f^{point}(\overline{X}_j|\mathcal{M})) = \sum_{j=1}^{n} \log(\sum_{i=1}^{k} \alpha_i f^i(\overline{X}_j)) \tag{9.30}$$

This log-likelihood fit needs to maximized to determine the model parameters. A salient observation is that if the probabilities of data points being generated from different clusters were known, then it becomes relatively easy to determine the optimal model parameters separately for each component of the mixture. At the same time, the probabilities of data points being generated from different components are dependent on these optimal model parameters. This circularity is reminiscent of a similar circularity in optimizing the objective function of partitioning algorithms in Section 9.3.1. In that case, the knowledge of a *hard* assignment of data points to clusters provides the ability to determine optimal cluster representatives locally for each cluster. In this case, the knowledge of a *soft* assignment provides the ability to estimate the optimal (maximum likelihood) model parameters *locally* for each cluster. This naturally suggests an iterative EM algorithm, in which the model parameters and probabilistic assignments are iteratively estimated from one another.

Let Θ be a vector, representing the *entire set* of parameters describing all components of the mixture model. For example, in the case of the Gaussian mixture model, Θ contains all the component mixture means, variances, co-variances, and the *prior* generative probabilities $\alpha_1 \ldots \alpha_k$. Then, the EM algorithm starts with an initial set of values of Θ (possibly corresponding to random assignments of data points to mixture components), and proceeds as follows:

1. (E-step) Given the current value of the parameters in Θ, estimate the *posterior* probability $P(\mathcal{G}_i | \overline{X}_j, \Theta)$ of the component $\mathcal{G}_i$ having been selected in the generative process, given that we have observed data point $\overline{X}_j$. The quantity $P(\mathcal{G}_i | \overline{X}_j, \Theta)$ is also the soft cluster assignment probability that we are trying to estimate. This step is executed for each data point $\overline{X}_j$ and mixture component $\mathcal{G}_i$.

2. (M-step) Given the current probabilities of assignments of data points to clusters, use the maximum likelihood approach to determine the values of all the parameters in Θ that maximize the log-likelihood fit on the basis of current assignments.

The two steps are executed repeatedly in order to improve the maximum likelihood criterion. The algorithm is said to converge when the objective function does not improve significantly in a certain number of iterations. The details of the E-step and the M-step will now be explained.

The E-step uses the currently available model parameters to compute the probability density of the data point $\overline{X}_j$ being generated by each component of the mixture. This probability density is used to compute the Bayes probability that the data point $\overline{X}_j$ was generated by component $\mathcal{G}_i$ (with model parameters fixed to the current set of the parameters Θ):

$$P(\mathcal{G}_i | \overline{X}_j, \Theta) = \frac{P(\mathcal{G}_i) \cdot P(\overline{X}_j | \mathcal{G}_i, \Theta)}{\sum_{r=1}^{k} P(\mathcal{G}_r) \cdot P(\overline{X}_j | \mathcal{G}_r, \Theta)} = \frac{\alpha_i \cdot f^{i,\Theta}(\overline{X}_j)}{\sum_{r=1}^{k} \alpha_r \cdot f^{r,\Theta}(\overline{X}_j)} \quad (9.31)$$

A superscript Θ has been added to the probability density functions to denote the fact that they are evaluated for current model parameters Θ.

The M-step requires the optimization of the parameters for each probability distribution under the assumption that the E-step has provided the "correct" soft assignment. To optimize the fit, the partial derivative of the log-likelihood fit with respect to corresponding model parameters needs to be computed and set to zero. Without specifically describing the details of these algebraic steps, the values of the model parameters that are computed as a result of the optimization are described here.

The value of each α_i is estimated as the current weighted fraction of points assigned to cluster i, where a weight of $P(\mathcal{G}_i | \overline{X}_j, \Theta)$ is associated with data point $\overline{X}_j$. Therefore, we have:

$$\alpha_i = P(\mathcal{G}_i) = \frac{\sum_{j=1}^{n} P(\mathcal{G}_i | \overline{X}_j, \Theta)}{n} \quad (9.32)$$

In practice, in order to obtain more robust results for smaller data sets, the expected number of data points belonging to each cluster in the numerator is augmented by 1, and the total number of points in the denominator is $n + k$. Therefore, the estimated value is as follows:

$$\alpha_i = \frac{1 + \sum_{j=1}^{n} P(\mathcal{G}_i | \overline{X}_j, \Theta)}{k + n} \tag{9.33}$$

This approach is also referred to as *Laplacian smoothing*.

To determine the other parameters for component i, the value of $P(\mathcal{G}_i | \overline{X}_j, \Theta)$ is treated as a weight of that data point. Consider a Gaussian mixture model in d dimensions, in which the distribution of the ith component is defined as follows:

$$f^{i,\Theta}(\overline{X}_j) = \frac{1}{\sqrt{|\Sigma_i|}(2 \cdot \pi)^{(d/2)}} e^{-\frac{1}{2}(\overline{X}_j - \overline{\mu_i})\Sigma_i^{-1}(\overline{X}_j - \overline{\mu_i})} \tag{9.34}$$

Here, $\overline{\mu_i}$ is the d-dimensional mean vector of the ith Gaussian component, and Σ_i is the $d \times d$ covariance matrix of the generalized Gaussian distribution of the ith component. The notation $|\Sigma_i|$ denotes the determinant of the covariance matrix. It can be shown[5] that the maximum-likelihood estimation of $\overline{\mu_i}$ and Σ_i yields the (probabilistically weighted) means and covariance matrix of the data points in that component. These probabilistic weights were derived from the assignment probabilities in the E-step. Interestingly, this is exactly how the representatives and covariance matrices of the Mahalanobis k-means approach are derived in Section 9.3.1. The only difference was that the data points were not weighted because hard assignments were used by the deterministic k-means algorithm. Note that the term in the exponent of the Gaussian distribution is the square of the Mahalanobis distance.

The E-step and the M-step can be iteratively executed to convergence to determine the optimal parameter set Θ. At the end of the process, a probabilistic model is obtained that describes the entire data set in terms of a generative model. The model also provides soft assignment probabilities $P(\mathcal{G}_i | \overline{X}_j, \Theta)$ of the data points, on the basis of the final execution of the E-step.

In practice, to minimize the number of estimated parameters, the non-diagonal entries of Σ_i are often set to 0. In such cases, the determinant of Σ_i simplifies to the product of the variances along the individual dimensions. This is equivalent to using the square of the *Minkowski* distance in the exponent. If all diagonal entries are further constrained to have the same value, then it is equivalent to using the Euclidean distance, and all components of the mixture will have spherical clusters. Thus, different choices and complexities of mixture model distributions provide different levels of flexibility in representing the probability distribution of each component.

This two-phase iterative approach is similar to representative-based algorithms. The E-step can be viewed as a soft version of the *assign* step in distance-based partitioning algorithms. The M-step is reminiscent of the *optimize* step, in which optimal component-specific parameters are learned on the basis of the fixed assignment. The distance term in the exponent of the probability distribution provides the natural connection between probabilistic and distance-based algorithms.

The E-step is structurally similar to the *Assign* step, and the M-step is similar to the *Optimize* step in k-representative algorithms. Many mixture component distributions can be expressed in the form $K_1 \cdot e^{-K_2 \cdot Dist(\overline{X}_i, \overline{Y}_j)}$, where K_1 and K_2 are regulated by distribution parameters. The log-likelihood of such an exponentiated distribution directly maps to an addi-

---

[5]This is achieved by setting the partial derivative of $\mathcal{L}(\mathcal{D}|\mathcal{M})$ (see Equation 9.30) with respect to each parameter in $\overline{\mu_i}$ and Σ to 0.

tive distance term $Dist(\overline{X_i}, \overline{Y_j})$ in the M-step objective function, which is structurally identical to the corresponding additive optimization term in k-representative methods. For many EM models with mixture probability distributions of the form $K_1 \cdot e^{-K_2 \cdot Dist(\overline{X_i}, \overline{Y_j})}$, a corresponding k-representative algorithm can be defined with a distance function $Dist(\overline{X_i}, \overline{Y_j})$.

9.3.5 Kohonen Self-Organizing Map

The Kohonen self-organizing map [103] constructs a 1-dimensional or 2-dimensional embedding in which a 1-dimensional string-like or 2-dimensional lattice-like structure is imposed on the neurons. The dimensionality of the embedding is the same as the dimensionality of the lattice. It is also possible to create 3-dimensional embeddings with the appropriate choice of lattice structure. We will consider the case in which a 2-dimensional lattice-like structure is imposed on the neurons. As we will see, this type of lattice structure enables the mapping of all points to 2-dimensional space for visualization. An example of a 2-dimensional lattice structure of 25 neurons arranged in a 5×5 rectangular grid is shown in Figure 9.13(a). A hexagonal lattice containing the same number of neurons is shown in Figure 9.13(b). The shape of the lattice affects the shape of the 2-dimensional regions in which the clusters will be mapped.

The idea of using the lattice structure is that the values of $\overline{W}_i$ in adjacent lattice neurons tend to be similar. Here, it is important to define separate notations to distinguish between the distance $||\overline{W}_i - \overline{W}_j||$ and the distance on the lattice. The distance between adjacent pairs of neurons on the lattice is exactly one unit. For example, the distance between the neurons i and j based on the lattice structure in Figure 9.13(a) is 1 unit, and the distance between neurons i and k is $\sqrt{2^2 + 3^2} = \sqrt{13}$. The vector-distance in the original input space (e.g., $||\overline{X} - \overline{W}_i||$ or $||\overline{W}_i - \overline{W}_j||$) is denoted by a notation like $Dist(\overline{W}_i, \overline{W}_j)$. On the other hand, the distance between neurons i and j along the lattice structure is denoted by $LDist(i, j)$. Note that the value of $LDist(i, j)$ is dependent only on the indices (i, j), and is independent of the values of the vectors $\overline{W}_i$ and $\overline{W}_j$.

The learning process in the self-organizing map is regulated in such a way that the closeness of neurons i and j (based on lattice distance) will also bias their weight vectors to be more similar. In other words, *the lattice structure of the self-organizing maps acts as a regularizer in the learning process.* As we will see later, imposing this type of 2-dimensional structure on the learned weights is helpful for visualizing the original data points with a 2-dimensional embedding.

The overall self-organizing map training algorithm proceeds in a similar way to competitive learning by sampling $\overline{X}$ from the training data, and finding the winner neuron based on the Euclidean distance. The weights in the winner neuron are updated in a manner similar to the vanilla competitive learning algorithm. However, the main difference is that a damped version of this update is also applied to the lattice-neighbors of the winner neuron. In fact, in soft variations of this method, one can apply this update to all neurons, and the level of damping depends on the lattice distance of that neuron to the winning neuron. The damping function, which always lies in $[0, 1]$, is typically defined by a Gaussian kernel:

$$Damp(i, j) = \exp\left(-\frac{LDist(i, j)^2}{2\sigma^2}\right) \tag{9.35}$$

Here, σ is the bandwidth of the Gaussian kernel. Using extremely small values of σ reverts to pure winner-take-all learning, whereas using larger values of σ leads to greater regularization in which lattice-adjacent units have more similar weights. For small values of σ, the damping

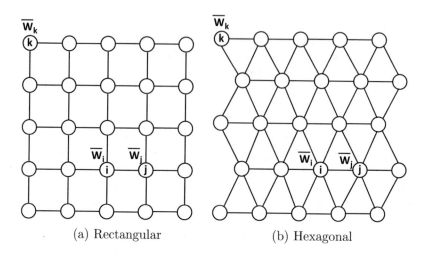

(a) Rectangular   (b) Hexagonal

Figure 9.13: An example of a 5×5 lattice structure for the self-organizing map. Since neurons i and j are close in the lattice, the learning process will bias the values of $\overline{W}_i$ and $\overline{W}_j$ to be more similar. The rectangular lattice will lead to rectangular clustered regions in the resulting 2-dimensional representation, whereas the hexagonal lattice will lead to hexagonal clustered regions in the resulting 2-dimensional representation

function will be 1 only for the winner neuron, and it will be 0 for all other neurons. Therefore, the value of σ is one of the parameters available to the user for tuning. Note that many other kernel functions are possible for controlling the regularization and damping. For example, instead of the smooth Gaussian damping function, one can use a thresholded step kernel, which takes on a value of 1 when $LDist(i,j) < \sigma$, and 0, otherwise.

The training algorithm repeatedly samples $\overline{X}$ from the training data, and computes the distances of $\overline{X}$ to each weight $\overline{W}_i$. The index p of the winning neuron is computed. Rather than applying the update only to $\overline{W}_p$ (as in winner-take-all), the following update is applied to each $\overline{W}_i$:

$$\overline{W}_i \Leftarrow \overline{W}_i + \alpha \cdot Damp(i,p) \cdot (\overline{X} - \overline{W}_i) \quad \forall i \tag{9.36}$$

Here, $\alpha > 0$ is the learning rate. It is common to allow the learning rate α to reduce with time. These iterations are continued until convergence is reached. Note that weights that are lattice-adjacent will receive similar updates, and will therefore tend to become more similar over time. *Therefore, the training process forces lattice-adjacent clusters to have similar points, which is useful for visualization.*

Using the Learned Map for 2D Embeddings

The self-organizing map can be used in order to induce a 2-dimensional embedding of the points. For a $k \times k$ grid, all 2-dimensional lattice coordinates will be located in a square in the positive quadrant with vertices $(0,0)$, $(0, k-1)$, $(k-1, 0)$, and $(k-1, k-1)$. Note that each grid point in the lattice is a vertex with integer coordinates. The simplest 2-dimensional embedding is simply by representing each point $\overline{X}$ with its closest grid point (i.e., winner neuron). However, such an approach will lead to overlapping representations of points. Furthermore, a 2-dimensional representation of the data can be constructed and each coordinate is one of $k \times k$ values from $\{0 \ldots k-1\} \times \{0 \ldots k-1\}$. This is the reason that the self-organizing map is also referred to as a *discretized* dimensionality reduction method. It is

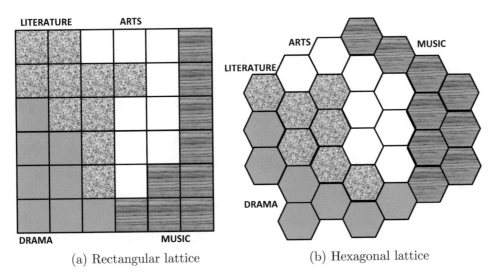

| (a) Rectangular lattice | (b) Hexagonal lattice |

Figure 9.14: Examples of 2-dimensional visualization of documents belonging to four topics

possible to use various heuristics to disambiguate these overlapping points. When applied to high-dimensional document data, a visual inspection often shows documents of a particular topic being mapped to a particular local regions. Furthermore, documents of related topics (e.g., politics and elections) tend to get mapped to adjacent regions. Illustrative examples of how a self-organizing map arranges documents of four topics with rectangular and hexagonal lattices are shown in Figures 9.14(a) and (b), respectively. The regions are colored differently, depending on the majority topic of the documents belonging to the corresponding region.

Self-organizing maps have a strong neurobiological basis in terms of their relationship with how the mammalian brain is structured. In the mammalian brain, various types of sensory inputs (e.g., touch) are mapped onto a number of folded planes of cells, which are referred to as *sheets* [59]. When parts of the body that are close together receive an input (e.g., tactile input), then groups of cells that are physically close together in the brain will also fire together. Therefore, proximity in (sensory) inputs is mapped to proximity in neurons, as in the case of the self-organizing map. This type of neurobiological inspiration is used in many neural architectures, such as *convolutional neural networks* for image data [107, 110].

9.3.6 Spectral Clustering

Spectral clustering [129, 164] combines nonlinear dimensionality reduction with k-means clustering, which allows one to learn clusters of arbitrary shapes. Spectral clustering combines row and column similarities by first using row-wise similarities to create a matrix and then using dimensionality reduction to create a new representation that incorporates column-wise similarities as well. Spectral clustering uses the following steps:

1. **(Breaking inter-cluster links):** Let $S = [s_{ij}]$ be a symmetric $n \times n$ similarity matrix defined over n data points, in which s_{ij} is the similarity between data points i and j. It is not necessary for the data points to be multidimensional. The similarity matrix might be created with the use of a domain-specific similarity function such as a *string subsequence kernel* [118] in the case of sequence or text data. The diagonal

entries of S are set to 0. All pairs (i, j) are identified such that data points i and j are *mutual* κ-nearest neighbors of each other according to the similarity matrix S. Such similarity values, s_{ij}, are retained in S. Otherwise, the value of s_{ij} is set to 0. This step sparsifies the similarity matrix, and intuitively tries to "break" the inter-cluster links, so that the resulting points are less likely to be close to one another in the engineered representation. The number of nearest neighbors, κ, regulates the sparsity of the similarity matrix.

2. **(Normalizing for dense and sparse regions):** For each row i, the sum of each row in the symmetric matrix S is computed as follows:

$$S_i = \sum_j s_{ij}$$

Intuitively, the value of S_i quantifies the "density" in the locality of data point i. Then, each similarity value is normalized using the following relation:

$$s_{ij} \Leftarrow \frac{s_{ij}}{\sqrt{S_i \cdot S_j}} = \frac{s_{ij}}{\text{GEOMETRIC-MEAN}(S_i, S_j)}$$

The basic idea is to normalize the similarities between data points with the geometric mean of the "densities" at their end points. Therefore, the similarity is *relative* to the *local* data distribution. For example, the similarity between two modestly similar data points in a local region belonging to a sparsely populated cluster becomes magnified, whereas the similarity between two data points in a dense region is de-emphasized. This type of adjustment makes the similarity function more adaptive to the statistics of data locality. For example, if a data point is in a very dense region, it facilitates the creation of a larger number of fine-grained clusters in that region. At the same time, it becomes possible to create fewer clusters with more widely separated points in sparse regions. An intuitive way of understanding this (in the context of a spatial application) is that population clusters in sparsely-populated Alaska would be geographically larger than those in densely-populated California.

3. **(Explicit feature engineering):** The resulting similarity matrix S is diagonalized to $S = Q \Delta Q^T$, where the columns of Q contain the eigenvectors, and Δ is a diagonal matrix containing the eigenvalues. Only the largest $r \ll n$ eigenvectors (columns) of Q need to be computed to create a smaller $n \times r$ matrix Q_0. Furthermore, each row of Q_0 is scaled to unit norm, so that all engineered points (i.e., rows of Q_0) lie on the unit sphere. At this point, the k-means algorithm is applied on the normalized and engineered points with the Euclidean distance.

The first two steps change the similarity matrix in a data-dependent way because aggregated statistics from multiple points are used to change the entries. The various adjustments to the engineered representation such as the dropping of lower-order eigenvectors help in improving the representation of the data for clustering. Spectral clustering is able to discover nonlinearly shaped, entangled clusters from the data that are not possible to discover using methods like k-means, which discover only spherically shaped clusters. For example, spectral clustering will be able to discover both clusters in Figure 9.11, since the portions of the similarity matrix corresponding to two clusters will have non-zero entries. Furthermore, the bridge of points shown in Figure 9.11(b) will often be disconnected in the sparsification step with the proper choice of similarity function.

9.4 Why Unsupervised Learning Is Important

Most of human and animal learning is unsupervised learning, which serves as a base for other forms of learning. For example, humans learn the nature of their environment all the time, as they take in sensory inputs and file away useful bits of information all the time. This information is then used for learning more specialized tasks. For example, it is much easier for a person to learn the laws of physics after having experienced how the world works via observation. A similar observation holds true in machine learning, wherein unsupervised learning often makes it easier to perform specialized tasks like classification. Unsupervised methods are used often for feature engineering, pretraining, and semisupervised learning. In the following, we will provide several examples of unsupervised models for supervised learning.

9.4.1 Feature Engineering for Machine Learning

The class of methods, referred to as *kernel methods* use unsupervised feature engineering in order to perform classification. It is noteworthy that feature engineering is also used to change the behavior of unsupervised methods like clustering, when it is desired for the method to show particular types of characteristics. A specific example of feature engineering for clustering is the use of spectral methods, where the diagonalization of a similarity matrix provides a new set of features on which k-means clustering is performed. By changing the nature of the similarity matrix, one can change the behavior of the underlying algorithm.

In kernel methods for classification, a similarity matrix is constructed from a data set with the use of a similarity function that is more sensitive to distances than the dot product. Recall from the discussion earlier in this chapter that if we have an $n \times d$ data set, then the eigenvectors of the $n \times n$ similarity matrix DD^T yield a rotated version of the data set which the same as the representation provided by SVD. The (i, j)th entry of $S = DD^T$ is the dot product similarity between the ith and jth rows of D. In other words, if $S = [s_{ij}]$ is the similarity matrix, and $\overline{X}_i$ is the ith row of D, then we have the following:

$$s_{ij} = \overline{X}_i \cdot \overline{X}_j$$

Singular value decomposition creates the embedding $U = Q\Sigma$ for diagonal matrix Σ, so that the following is satisfied:

$$S = DD^T = Q\Sigma^2 Q^T = \underbrace{(Q\Sigma)(Q\Sigma)^T}_{U} = UU^T$$

Here, Q, Σ, and U are all $n \times n$ matrices. An important property of U in vanilla SVD is that at most d columns are nonzero, and therefore the new embedding is also (at most) d-dimensional. This is not particularly surprising, since SVD simply rotates the data set.

Kernel methods use a nonlinear version of singular value decomposition in which the matrix S is constructed with a more sensitive similarity function (e.g., the Gaussian kernel) rather than with the use of dot products:

$$s_{ij} \propto \exp(-\|\overline{X}_i - \overline{X}_j\|^2/\sigma^2) \tag{9.37}$$

The use of a more sensitive similarity function results in a new embedding $S = UU^T$, in which it is possible for all n columns of U to be nonzero. Therefore, the new embedding might be of much higher dimensionality than vanilla SVD. The larger dimensionality of the data set makes it easier to separate the classes in the new representation with a linear

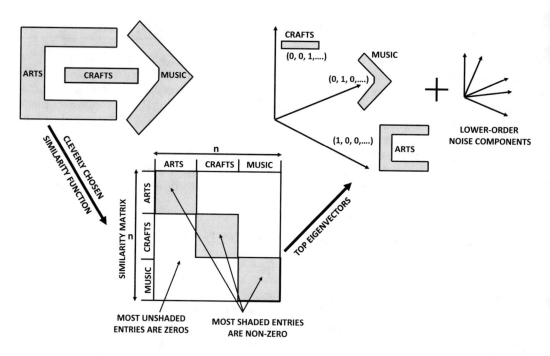

Figure 9.15: Explaining the rationale for nonlinear dimensionality reduction

hyperplane. As a result, one often transforms the data to this new space, before performing classification. In order to understand why nonlinearly separable clusters become linearly separable with the use of a more distance sensitive similarity function, we will use an example.

Consider a setting in which the data matrix D is an $n \times d$ matrix containing the frequencies of d words in each of n documents. The data set contains three classes of related topics corresponding to *Arts*, *Crafts*, and *Music*, and we wish to build a classifier separating these classes. The classes are naturally clustered in the underlying data, as shown in Figure 9.15. Unfortunately, the classes are not linearly separable, and a linear classifier will have difficulty in separating any particular class from the other classes.

Now imagine that we could somehow define a similarity matrix in which most of the similarities between documents of different topics are close to zero, whereas most of the similarities between documents of the same topic are nearly 1s. This can be achieved with the Gaussian kernel of Equation 9.37, provided that the bandwidth σ is chosen appropriately. The resulting similarity matrix S is shown in Figure 9.15 with a natural block structure. What type of embedding U will yield the factorization $S = UU^T$? First let us consider the absolutely perfect similarity function in which the entries in all the shaded blocks are 1s and all the entries outside shaded blocks are 0s. In such a case, it can be shown (after ignoring zero eigenvalues) that every document in *Arts* will receive an embedding of $(1, 0, 0)$, every document in *Music* will receive an embedding of $(0, 1, 0)$, and every document in *Crafts* will receive an embedding of $(0, 0, 1)$. Of course, in practice, we will never have a precise block structure of 1s and 0s, and there will be significant noise/finer trends within the block structure. These variations will be captured by the lower-order eigenvectors shown in Figure 9.15. Even with these additional noise dimensions, this new representation will typically be linearly separable for classification. The key idea here is that dot product similarities are sometimes not very good at capturing the *detailed* structure of the data, which other similarity functions with sharper locality-centric variations can sometimes capture. The purpose of using the Gaussian kernel is to precisely accentuate these locality-centric variations appropriately. The only supervision here is in choosing the bandwidth of the kernel based on out-of-sample performance. Note that it is not necessary to use all the columns of U; rather one can drop the smaller eigenvectors as noisy dimensions.

Armed with this basic idea, we provide an example of a kernel classification algorithm, based on an $n \times d$ data matrix D that contains *both* the training and the test rows:

Diagonalize $S = Q\Sigma^2 Q^T$;
Extract the n-dimensional embedding in rows of $U = Q\Sigma$;
Partition rows of U into U_{train} and U_{test};
Apply linear SVM on training rows of U_{train} and class labels to learn model $\mathcal{M}$;
Apply $\mathcal{M}$ on each row of U_{test} to yield predictions;

This model is almost the same as the kernel SVM, with the only difference being that a kernel SVM performs the feature engineering with only the training rows, and then fits the test rows into the training space. Generalizing nonlinear tranformations to out-of-sample rows in this way is referred to as the *Nystrom method*. The precise feature engineering analog to the kernel SVM is described in [8]. Furthermore, it is more common to use this type of feature engineering indirectly with the use of the *kernel trick* in traditional machine learning, rather than via feature engineering. Nevertheless, these kernel methods are roughly equivalent to the procedure described above.

9.4.2 Radial Basis Function Networks for Feature Engineering

A traditional feed-forward network contains many layers, and the nonlinearity is typically created by the repeated composition of activation functions. On the other hand, an RBF network typically uses only an input layer, a single hidden layer (with a special type of behavior defined by RBF functions), and an output layer. As in feed-forward networks, the input layer is not really a computational layer, and it only carries the inputs forward. The layers of the RBF network are designed as follows:

1. The input layer simply transmits from the input features to the hidden layers. Therefore, the number of input units is exactly equal to the dimensionality d of the data. As in the case of feed-forward networks, no computation is performed in the input layers. As in all feed-forward networks, the input units are fully connected to the hidden units and carry their input forward.

2. The computations in the hidden layers are based on comparisons with *prototype vectors*. Each hidden unit contains a d-dimensional prototype vector. Let the prototype vector of the ith hidden unit be denoted by $\overline{\mu}_i$. In addition, the ith hidden unit contains a bandwidth denoted by σ_i. Although the prototype vectors are always specific to particular units, the bandwidths of different units σ_i are often set to the same value σ. The prototype vectors and bandwidth(s) are usually learned either in an unsupervised way, or with the use of mild supervision.

   Then, for any input training point $\overline{X}$, the activation $\Phi_i(\overline{X})$ of the ith hidden unit is defined as follows:

$$h_i = \Phi_i(\overline{X}) = \exp\left(-\frac{||\overline{X} - \overline{\mu}_i||^2}{2 \cdot \sigma_i^2}\right) \quad \forall i \in \{1, \ldots, m\} \tag{9.38}$$

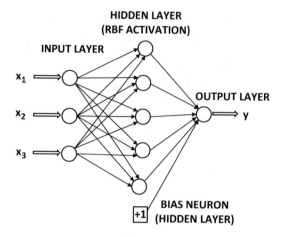

Figure 9.16: An RBF network: Note that the hidden layer is broader than the input layer, which is typical (but not mandatory)

The total number of hidden units is denoted by m. Each of these m units is designed to have a high level of influence on the particular cluster of points that is closest to its prototype vector. Therefore, one can view m as the number of clusters used for modeling, and it represents an important hyper-parameter available to the algorithm. For low-dimensional inputs, it is typical for the value of m to be larger than the input dimensionality d, but smaller than the number of training points n.

3. For any particular training point $\overline{X}$, let h_i be the output of the ith hidden unit, as defined by Equation 9.38. The weights of the connections from the hidden to the output nodes are set to w_i. Then, the prediction $\hat{y}$ of the RBF network in the output layer is defined as follows:

$$\hat{y} = \sum_{i=1}^{m} w_i h_i = \sum_{i=1}^{m} w_i \Phi_i(\overline{X}) = \sum_{i=1}^{m} w_i \exp\left(-\frac{||\overline{X} - \overline{\mu}_i||^2}{2 \cdot \sigma_i^2}\right)$$

The variable $\hat{y}$ has a circumflex on top to indicate the fact that it is a predicted value rather than observed value. If the observed target is real-valued, then one can set up a least-squares loss function, which is much like that in a feed-forward network. The values of the weights $w_1 \ldots w_m$ need to be learned in a supervised way.

An example of an RBF network is illustrated in Figure 9.16.

In the RBF network, there are two sets of computations corresponding to the hidden layer and the output layer. The parameters $\overline{\mu}_i$ of the hidden layer are learned in an unsupervised way, whereas those of the output layer are learned in a supervised way with gradient descent. The latter is similar to the case of the feed-forward network. The prototypes $\overline{\mu}_i$ may either be sampled from the data, or be set to be the m centroids of an m-way clustering algorithm. In other words, we can partition the training data into m clusters with an off-the-shelf clustering algorithm, and use the means of the m clusters as the m prototypes. The parameters σ_i are set to the same value of σ, which is often treated as a hyper-parameter. In other words, it is tuned on a portion of the data hat is held out in order to optimize classification accuracy.

An interesting special case is when the prototypes are set to the individual training points (and therefore the value m is the same as the number of training examples). In such cases, RBF networks can be shown to specialize to well-known *kernel methods* in machine learning. However, since RBF networks can choose different prototypes than the training points, it suggests that RBF networks have greater power and flexibility than do kernel methods.

9.4.3 Semisupervised Learning

The process by which most humans and animals learn is most closely reflected in semi-supervised learning. Not all experiences of humans are task focused; rather there is a level of background knowledge that one learns from day-to-day experiences. This background knowledge is often deployed in task-focused learning. The day-to-day experiences (which are not task-focused) can be viewed as unsupervised experiences. These experiences do, however, help in learning more focused tasks in most cases. An important observation is that humans train in an unsupervised way most of the time, whereas task-focused learning is only performed a small part of the time. A similar observation applies to machine learning tasks where one can improve the accuracy of supervised algorithms by using unsupervised data. In most cases, unsupervised data is copious, whereas the supervised data is very limited. The goal is to restrict the model to small parts of the feature space. This is particularly useful when a lot of unlabeled data is available, and only a small part of the data is labeled.

Many generic meta-algorithms, such as self-training, co-training, and pre-training, are often used for learning. The goal of generic meta-algorithms is to use existing classification algorithms to enhance the classification process with unlabeled data. The simplest method is *self-training*, in which the smoothness assumption is used to incrementally expand the labeled portions of the training data. The major drawback of this approach is that it might lead to overfitting. One way of avoiding overfitting is by using *co-training*. Co-training partitions the feature space and independently labels instances using classifiers trained on each of these feature spaces. The labeled instances from one classifier are used as feedback to the other, and vice versa.

9.4.3.1 Self-Training

The self-training procedure can use any existing classification algorithm $\mathcal{A}$ as input. The classifier $\mathcal{A}$ is used to incrementally assign labels to unlabeled examples for which it has the most confident prediction. As input, the self-training procedure uses the initial labeled set L, the unlabeled set U, and a user-defined parameter k that may sometimes be set to 1. The self-training procedure iteratively uses the following steps:

1. Use algorithm $\mathcal{A}$ on the current labeled set L to identify the k instances in the unlabeled data U for which the classifier $\mathcal{A}$ is the most confident.

2. Assign labels to the k most confidently predicted instances and add them to L. Remove these instances from U.

The major drawback of self-training is that the addition of predicted labels to the training data can lead to propagation of errors in the presence of noise. Another procedure, known as *co-training*, is able to avoid such overfitting more effectively.

9.4.3.2   Co-Training

In co-training, it is assumed that the feature set can be partitioned into two *disjoint* groups F_1 and F_2, such that each of them is sufficient to learn the target classification function. It is important select the two feature subsets so that they are as independent from one another as possible. Two classifiers are constructed, such that one classifier is constructed on each of these groups. These classifiers are not allowed to interact with one another directly for prediction of unlabeled examples though they are used to build up training sets for each other. This is the reason that the approach is referred to as *co-training*.

Let L be the labeled training data and U be the unlabeled data. Let L_1 and L_2 be the labeled sets for each of these classifiers. The sets L_1 and L_2 are initialized to the available labeled data L, except that they are represented in terms of disjoint feature sets F_1 and F_2, respectively. Over the course of the co-training process, as different examples from the initially unlabeled set U are added to L_1 and L_2, respectively, the training instances in L_1 and L_2 may vary from one another. Two classifier models $\mathcal{A}_1$ and $\mathcal{A}_2$ are constructed using the training sets L_1 and L_2, respectively. The following steps are then iteratively applied:

1. Train classifier $\mathcal{A}_1$ using labeled set L_1, and add k most confidently predicted instances from unlabeled set $U - L_2$ to training data set L_2 for classifier $\mathcal{A}_2$.

2. Train classifier $\mathcal{A}_2$ using labeled set L_2, and add k most confidently predicted instances from unlabeled set $U - L_1$ to training data set L_1 for classifier $\mathcal{A}_1$.

In many implementations of the method, the most confidently labeled examples *for each class* are added to the training sets of the other classifier. This procedure is repeated until all instances are labeled. The two classifiers are then retrained with the expanded training data sets. This approach can be used to label not only the unlabeled data set U, but also unseen test instances. At the end of the procedure, two classifiers are returned. For an unseen test instance, each classifier may be used to determine the class label scores. The score for a test instance is determined by combining the scores of the two classifiers. For example, if the Bayes method is used as the base classifier, then the product of the posterior probabilities returned by the two classifiers may be used.

The co-training approach is more robust to noise because of the disjoint feature sets used by the two algorithms. An important assumption is that of *conditional independence* of the features in the two sets with respect to a particular class. In other words, after the class label is fixed, the features in one subset are conditionally independent of the other. The intuition for this is that instances generated by one classifier appear to be randomly distributed to the other, and vice versa. As a result, the approach will generally be more robust to noise than the self-training method.

9.4.3.3   Unsupervised Pretraining in Multilayer Neural Networks

A neural network with multiple layers is referred to as a *deep network*. The early layers learn features that are used by later layers. Deep networks are inherently hard to train because the magnitudes of the gradients in various layers are often quite different. As a result, the different layers of the neural network do not get trained at the same rate. The multiple layers of the neural network cause distortions in the gradient, which make them hard to train.

Although the depth of the neural network causes challenges, the problems associated with depth are also heavily dependent on how the network is initialized. A good initialization point can often solve many of the problems associated with reaching good solutions. A

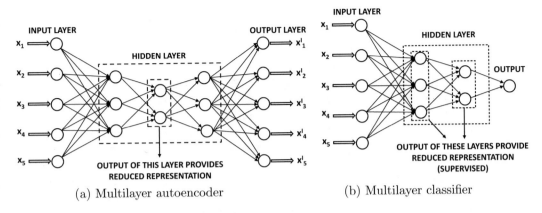

(a) Multilayer autoencoder   (b) Multilayer classifier

Figure 9.17: Both the multilayer classifier and the multilayer autoencoder use a similar pretraining procedure

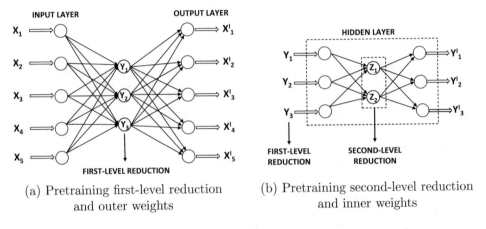

(a) Pretraining first-level reduction and outer weights   (b) Pretraining second-level reduction and inner weights

Figure 9.18: Pretraining a neural network

ground-breaking break-through in this context was the use of unsupervised pretraining in order to provide robust initializations [80]. This initialization is achieved by training the network greedily in layer-wise fashion. The approach was originally proposed in the context of deep belief networks, but it was later extended to other types of models such as autoencoders [146, 189]. In this chapter, we will study the autoencoder approach because of its simplicity. First, we will start with the dimensionality reduction application, because the application is unsupervised and it is easy to show how to use unsupervised pretraining in this case. However, unsupervised pretraining can also be used for supervised applications like classification with minor modifications.

In pretraining, a greedy approach is used to train the network one layer at a time by learning the weights of the outer hidden layers first and then learning the weights of the inner hidden layers. The resulting weights are used as starting points for a final phase of traditional neural network backpropagation in order to fine-tune them.

Consider the autoencoder and classifier architectures shown in Figure 9.17. Since these architectures have multiple layers, randomized initialization can sometimes cause challenges. However, it is possible to create a good initialization by setting the initial weights layer by layer in a greedy fashion. First, we describe the process in the context of the autoencoder shown in Figure 9.17(a), although an almost identical procedure is relevant to the classifier of Figure 9.17(b). We have intentionally chosen neural architectures in the two cases so that the hidden layers have similar numbers of nodes.

The pretraining process is shown in Figure 9.18. The basic idea is to assume that the two (symmetric) outer hidden layers contain a first-level reduced representation of larger dimensionality, and the inner hidden layer contains a second-level reduced representation of smaller dimensionality. Therefore, the first step is to learn the first-level reduced representation and the corresponding weights associated with the outer hidden layers using the simplified network of Figure 9.18(a). In this network, the middle hidden layer is missing and the two outer hidden layers are collapsed into a single hidden layer. The assumption is that the two outer hidden layers are related to one another in a symmetric way like a smaller autoencoder. In the second step, the reduced representation in the first step is used to learn the second-level reduced representation (and weights) of the inner hidden layers. Therefore, the inner portion of the neural network is treated as a smaller autoencoder in its own right. Since each of these pretrained subnetworks is much smaller, the weights can be learned more easily. This initial set of weights is then used to train the entire neural network with backpropagation. Note that this process can be performed in layerwise fashion for a deep neural network containing any number of hidden layers.

So far, we have only discussed how we can use unsupervised pretraining for unsupervised applications. A natural question arises as to how one can use pretraining for supervised applications. Consider a multilayer classification architecture with a single output layer and k hidden layers. During the pretraining stage, the output layer is removed, and the representation of the final hidden layer is learned in an unsupervised way. This is achieved by creating an autoencoder with $2 \cdot k - 1$ hidden layers, where the middle layer is the final hidden layer of the supervised setting. For example, the relevant autoencoder for Figure 9.17(b) is shown in Figure 9.17(a). Therefore, an additional $(k - 1)$ hidden layers are added, each of which has a symmetric counterpart in the original network. This network is trained in exactly the same layer-wise fashion as discussed above for the autoencoder architecture. The weights of only the encoder portion of this autoencoder are used for initialization of the weights entering into all hidden layers. The weights between the final hidden layer and the output layer can also be initialized by treating the final hidden layer and output nodes as a single-layer network. This single-layer network is fed with the reduced representations of the final hidden layer (based on the autoencoder learned in pretraining). After the weights of all the layers have been learned, the output nodes are re-attached to the final hidden layer. The backpropagation algorithm is applied to this initialized network in order to fine-tune the weights from the pretrained stage. Note that this approach learns all the initial hidden representations in an unsupervised way, and only the weights entering into the output layer are initialized using the labels. Therefore, the pretraining can still be considered to be largely unsupervised.

Unsupervised pretraining helps even in cases where the amount of training data is very large. It is likely that this behavior is caused by the fact that pretraining helps in issues beyond model generalization. One evidence of this fact is that in larger data sets, even the error on the training data seems to be high, when methods like pretraining are not used. In these cases, the weights of the early layers often do not change much from their initializations, and one is using only a small number of later layers on a random transformation of the data (defined by the random initialization of the early layers). As a result, the trained portion of the network is rather shallow, with some additional loss caused by the random transformation. In such cases, pretraining also helps a model realize the full benefits of depth, thereby facilitating the improvement of prediction accuracy on larger data sets.

Another way of understanding pretraining is that it provides insights into the repeated patterns in the data, which are the features learned from digits by putting together these frequent shapes. However, these shapes also have discriminative power with respect to recognizing digits. Expressing the data in terms of a few features then helps in recognizing how these features are related to the class labels. This is at the heart of the idea of unsupervised learning, which uses copiously available labeled data to recognize the frequent patterns. This principle is summarized by Geoff Hinton [79] in the context of image classification as follows: "*To recognize shapes, first learn to generate images.*" This type of regularization preconditions the training process in a semantically relevant region of the parameter space, where several important features have already been learned, and further training can fine-tune and combine them for prediction.

9.5 Summary

Unsupervised learning methods use a variety of techniques to develop compressed representations from data. This compressed representation could take the form of a data set with reduced dimensionality, or it could take the form of the representatives of the various clusters in the data set. Linear dimensionality reduction is a form of matrix factorization, in which a data matrix can be represented as a product of two matrices. Some forms of linear dimensionality reduction (such as nonnegative matrix factorization) are closely related to clustering.

Unsupervised methods are used for various types of feature engineering algorithms. The core idea of feature engineering is to create a new representation of the data on which existing supervised algorithms work effectively. Examples include kernel methods and radial basis function networks. One can also use unsupervised methods in order to improve the accuracy of supervised methods. Unsupervised methods are able to learn the manifolds on which the data points lie. The knowledge of the structure of the manifold reduces the amount of labeled data that is then required for classification.

9.6 Further Reading

Methods for dimensionality reduction and matrix factorization are discussed in detail in [8]. A detailed book on data clustering may be found in [10]. A discussion of the use of autoencoders for dimensionality reduction may be found in [6]. Discussions of feature engineering and semisupervised learning may be found in [8, 10]. The Kohonen self-organizing map is discussed in detail in [103]. Detailed discussions of pretraining methods for supervised and unsupervised learning may be found in [6].

9.7 Exercises

1. Use singular value decomposition to show the *push-through identity* for any $n \times d$ matrix D:
$$(\lambda I_d + D^T D)^{-1} D^T = D^T (\lambda I_n + D D^T)^{-1}$$

2. Let D be an $n \times d$ data matrix, and $\overline{y}$ be an n-dimensional column vector containing the dependent variables of linear regression. The regularized solution to linear regression predicts the dependent variables of a test instance $\overline{Z}$ using the following equation:
$$\text{Prediction}(\overline{Z}) = \overline{Z}\,\overline{W} = \overline{Z}(D^T D + \lambda I)^{-1} D^T \overline{y}$$

Here, the vectors $\overline{Z}$ and $\overline{W}$ are treated as $1 \times d$ and $d \times 1$ matrices, respectively. Show using the result of Exercise 1, how you can write the above prediction purely in terms of similarities between training points or between $\overline{Z}$ and training points.

3. Suppose that you are given a truncated SVD $D \approx Q\Sigma P^T$ of rank-k. Show how you can use this solution to derive an alternative rank-k decomposition $Q'\Sigma'P'^T$ in which the unit columns of Q (or/and P) might not be mutually orthogonal and the truncation error is the same.

4. **Recommender systems:** Let D be an $n \times d$ matrix in which only a small subset of the entries are specified. This is commonly the case with recommender systems. Show how you can adapt the algorithm for unconstrained matrix factorization to this case, so that only observed entries are used to create the factors. How would you change the matrix-based updates of Equation 9.6 to this case.

5. **Biased matrix factorization:** Consider the factorization of an incomplete $n \times d$ matrix D into an $n \times k$ matrix U and a $d \times k$ matrix V:

$$D \approx UV^T$$

Suppose you add the constraint that all entries of the penultimate column of U and the final column of V are fixed to 1. Discuss the similarity of this model to that of the addition of bias to classification models. How is gradient descent modified?

6. The text of the book discusses gradient descent updates (cf. Equation 9.6) for unconstrained matrix factorization $D \approx UV^T$. Suppose that the matrix D is symmetric, and we want to perform the symmetric matrix factorization $D \approx UU^T$. Formulate the objective function and gradient descent steps of symmetric matrix factorization in a manner similar to the asymmetric case.

7. Discuss why the following integer matrix factorization is equivalent to the objective function of the k-means algorithm for an $n \times d$ matrix D, in which the rows contain the data points:

$$\text{Minimize}_{U,V} \|D - UV^T\|_F^2$$
$$\text{subject to:}$$
$$\text{Columns of } U \text{ are mutually orthogonal}$$
$$u_{ij} \in \{0, 1\}$$

8. What is the maximum number of possible clusterings of a data set of n points into k groups? What does this imply about the convergence behavior of algorithms whose objective function is guaranteed not to worsen from one iteration to the next?

9. Suppose that you represent your data set as a graph in which each data point is a node, and the weight of the edge between a pair of nodes is equal to the Gaussian kernel similarity between them. Edges with weight less than a particular threshold are dropped. Interpret the single-linkage clustering algorithm in terms of this similarity graph.

10. The text of the chapter shows how one can transform any linear classifier into recognizing nonlinear decision boundaries by using a feature engineering phase in which the eigenvectors of an appropriately chosen similarity matrix are used to create new features. Discuss the impact of this type of preprocessing on the nature of the clusters found by the k-means algorithm.

Chapter 10

Reinforcement Learning

"Human beings, viewed as behaving systems, are quite simple. The apparent complexity of our behavior over time is largely a reflection of the complexity of the environment in which we find ourselves." – Herbert Simon's ant hypothesis

10.1 Introduction

Learning in humans is a continuous experience-driven process in which decisions are made, and the reward/punishment received from the environment are used to guide the learning process for future decisions. In other words, learning in intelligent beings is by reward-guided *trial and error*. Almost all of biological intelligence, as we know it, originates in one form or other through an interactive process of trial and error with the environment. Since the goal of artificial intelligence is to simulate biological intelligence, it is therefore natural to draw inspirations from the successes of biological trial-and-error in simplifying the design of highly complex learning algorithms. We have already seen one form of this trial and error in the chapter on adversarial search (cf. Chapter 3), where Monte Carlo trees are used to learn the best chess moves through trial and error.

A reward-driven trial-and-error process, in which a system learns to interact with a complex environment to achieve rewarding outcomes, is referred to in machine learning parlance as *reinforcement learning*. In reinforcement learning, the process of trial and error is driven by the need to maximize the expected rewards over time. These rewards are the same types of utility functions that wee discussed in the context of deductive reasoning systems in earlier chapters. However, in the case of reinforcement learning, the agent's actions need to be done in a data-driven manner rather than through the use of domain-specific heuristics. For example, a deductive system for chess leverages human-specified utility functions on the chess position in order to make choices (cf. Chapter 3). On the other hand, an inductive system experiments with various positions in order to *decide from its own experience*, which moves provide the most likelihood of winning.

© Springer Nature Switzerland AG 2021
C. C. Aggarwal, *Artificial Intelligence*, https://doi.org/10.1007/978-3-030-72357-6_10

Reinforcement learning has been used in recent years to create game-playing algorithms, chatbots, and even intelligent robots that interact with the environment. Some examples of reinforcement learning systems that have been developed in recent years are as follows:

1. Deep learners have been trained to play video games by using only the raw pixels of the video console as feedback. The reinforcement learning algorithm predicts the actions based on the display and inputs them into the video game console. Initially, the computer algorithm makes many mistakes, which are reflected in the virtual rewards received from the system. As the learner gains experience from its mistakes, it makes better decisions. This is exactly how humans learn to play video games. The performance of a recent algorithm on the Atari platform has been shown to surpass human-level performance for a large number of games [69, 126, 127]. Video games are excellent test beds for reinforcement learning algorithms, because they can be viewed as highly simplified representations of the choices one has to make in various decision-centric settings. Simply speaking, video games represent toy microcosms of real life.

2. DeepMind has trained a deep learning algorithm *AlphaGo* [166] to play chess, shogi, and *Go* by using computer self-play. *AlphaGo* has not only convincingly defeated top human players, but has contributed to innovations in the style of human play by using unconventional strategies in defeating these players. These innovations were a result of the reward-driven experience gained by *AlphaGo* by playing itself over time.

3. In recent years, deep reinforcement learning has been harnessed in self-driving cars by using the feedback from various sensors around the car to make decisions. Although it is more common to use supervised learning (or *imitation learning*) in self-driving cars, the option of using reinforcement learning has also been recognized [208].

4. The quest for creating self-learning robots is a task in reinforcement learning [114, 116, 159]. For example, robot locomotion turns out to be surprisingly difficult in nimble configurations. In the reinforcement learning paradigm, we only incentivize the robot to get from point A to point B as efficiently as possible using its available limbs and motors [159]. Through reward-guided trial and error, robots learn to roll, crawl, and eventually walk.

Reinforcement learning is appropriate for tasks *that are simple to evaluate but hard to reason explicitly with a deductive system*. For example, it is easy to evaluate a player's performance at the end of a complex game like chess, but it is hard to specify the precise action in every situation with the use of a reasoning methodology. Any human-designed evaluation function is likely to be riddled with approximations and inaccuracies, which eventually show up in the quality of game playing. This is at the heart of the weaknesses in many traditional chess-playing software like *Stockfish*. In many cases, intuition plays an important role in the choices, which is hard for a human to enunciate and encode in the knowledge base or domain-specific evaluation function of a board position. As in biological organisms, reinforcement learning provides a path to the *simplification of learning complex behaviors* by only defining the reward and letting the algorithm *learn* reward-maximizing behaviors (rather than specifying it through explicit reasoning). The resulting function that is learned may not be expressible easily in an understandable way.

Chapter Organization

This chapter is organized as follows. The next section introduces multi-armed bandits, which is the simplest setting in reinforcement learning. The notion of states is introduced in section 10.3. The simplest algorithm for reinforcement learning, which uses direct simulations, is discussed in section 10.4. The notion of bootstrapping is introduced in section 10.5. Policy gradient methods are discussed in section 10.6. The use of Monte Carlo tree search strategies is discussed in 10.7. A number of case studies are discussed in section 10.8. The safety issues associated with deep reinforcement learning methods are discussed in section 10.9. A summary is given in section 10.10.

10.2 Stateless Algorithms: Multi-Armed Bandits

The simplest example of a reinforcement learning setting is the *multi-armed bandit problem*, which addresses the problem of a gambler choosing one of many slot machines in order to maximize his or her payoff. The gambler suspects that the (expected) rewards from the various slot machines are not the same, and therefore it makes sense to play the machine with the largest expected reward. Since the expected payoffs of the slot machines are not known in advance, the gambler has to *explore* different slot machines by playing them and also *exploit* the learned knowledge to maximize the reward. Although exploration of a particular slot machine might gain some additional knowledge about its payoff, it incurs the risk of the (potentially fruitless) cost of playing it. Multi-armed bandit algorithms provide carefully crafted strategies to optimize the trade-off between exploration and exploitation.

The key trade-off between exploration and exploitation is as follows. Trying the slot machines randomly is wasteful but helps in gaining experience. Trying the slot machines for a very small number of times and then always picking the best machine might lead to solutions that are poor in the long-term. How should one navigate this trade-off between exploration and exploitation? Note that every trial provides the same probabilistically distributed reward as previous trials for a given action, and therefore *there is no notion of state* in such a system (as is the case in chess where the action depends on the board state). Needless to say, such settings cannot capture more complex situations that reinforcement learning is inherently designed for. In a computer video game, moving the cursor in a particular direction has a reward that heavily depends on the state of the video game, and a chess move depends on the state of the chess board.

There are a number of strategies that the gambler can use to regulate the trade-off between exploration and exploitation of the search space. In the following, we will briefly describe some of the common strategies used in multi-armed bandit systems. All these methods are instructive because they provide the basic ideas and framework, which are used in generalized settings of reinforcement learning. In fact, some of these stateless algorithms are also used to define state-specific policies in general forms of reinforcement learning. Therefore, it is important to explore this simplified setting.

10.2.1 Naïve Algorithm

In this approach, the gambler plays each machine for a fixed number of trials in the exploration phase. Subsequently, the machine with the highest payoff is used forever in the exploitation phase. Although this approach might seem reasonable at first sight, it has a number of drawbacks. The first problem is that it is hard to determine the number of trials at which one can confidently predict whether a particular slot machine is better than another machine. The process of estimation of payoffs might take a long time, especially in cases where the payoff events are rare compared to non-payoff events. Using many exploratory trials will waste a significant amount of effort on suboptimal strategies. Furthermore, if the wrong strategy is selected in the end, the gambler will use the wrong slot machine forever.

Therefore, the approach of fixing a particular strategy forever is unrealistic in real-world problems.

10.2.2 ϵ-Greedy Algorithm

The ϵ-greedy algorithm is designed to use the best strategy as soon as possible, without wasting a significant number of trials. The basic idea is to choose a random slot machine for a fraction ϵ of the trials. These exploratory trials are also chosen at random (with probability ϵ) from all trials, and are therefore fully interleaved with the exploitation trials. In the remaining $(1 - \epsilon)$ fraction of the trials, the slot machine with the best average payoff so far is used. An important advantage of this approach is that one is guaranteed to not be trapped in the wrong strategy forever. Furthermore, since the exploitation stage starts early, one is often likely to use the best strategy a large fraction of the time.

The value of ϵ is an algorithm parameter. For example, in practical settings, one might set $\epsilon = 0.1$, although the best choice of ϵ will vary with the application at hand. It is often difficult to know the best value of ϵ to use in a particular setting. Nevertheless, the value of ϵ needs to be reasonably small in order to gain significant advantages from the exploitation portion of the approach. However, at small values of ϵ it might take a long time to identify the correct slot machine. A common approach is to use *annealing*, in which large values of ϵ are initially used, with the values declining with time.

10.2.3 Upper Bounding Methods

Even though the ϵ-greedy strategy is better than the naïve strategy in dynamic settings, it is still quite inefficient at learning the payoffs of new slot machines. In upper bounding strategies, the gambler does not use the mean payoff of a slot machine. Rather, the gambler takes a more optimistic view of slot machines that have not been tried sufficiently, and therefore uses a slot machine with the best *statistical upper bound* on the payoff. Therefore, one can consider the upper bound U_i of testing a slot machine i as the sum of expected reward Q_i and one-sided confidence interval length C_i:

$$U_i = Q_i + C_i \qquad (10.1)$$

The value of C_i is like a bonus for increased uncertainty about that slot machine in the mind of the gambler. The value C_i is proportional to the standard deviation of the *mean* reward of the tries so far. According to the central limit theorem, this standard deviation is inversely proportional to the square-root of the number of times the slot machine i is tried (under the i.i.d. assumption). One can estimate the mean μ_i and standard deviation σ_i of the ith slot machine and then set C_i to be $K \cdot \sigma_i / \sqrt{n_i}$, where n_i is the number of times the ith slot machine has been tried. Here, K decides the level of confidence interval. Therefore, rarely tested slot machines will tend to have larger upper bounds (because of larger confidence intervals C_i) and will therefore be tried more frequently.

Unlike ϵ-greedy, the trials are no longer divided into two categories of exploration and exploitation; the process of selecting the slot machine with the largest upper bound has the dual effect of encoding both the exploration and exploitation aspects within each trial. One can regulate the trade-off between exploration and exploitation by using a specific level of statistical confidence. The choice of $K = 3$ leads to a 99.99% confidence interval for the upper bound under the Gaussian assumption. In general, increasing K will give large bonuses C_i for uncertainty, thereby causing exploration to comprise a larger proportion of the plays compared to an algorithm with smaller values of K.

10.3 Reinforcement Learning Framework

The bandit algorithms of the previous section are stateless. In other words, the decision made at each time stamp has an identical environment, and the actions in the past only affect the knowledge of the agent (not the environment itself). This is not the case in generic reinforcement learning settings like video games or self-driving cars, which have a notion of state (as introduced in Chapter 1). Therefore, the reinforcement learning systems work within the same broader framework as search-based (or deductive reasoning) systems.

While playing a video game, the reward of a particular action depends on the state of the video game screen. In a video game, the reward might often take on the form of points being given to the player for fulfilling a small condition in a particular action, and points would be continually accumulated over time. In a self-driving car, the reward for violently swerving a car in a normal state would be different from that of performing the same action in a state that indicates the danger of a collision. In other words, we need a way to quantify the reward of each action in a way that is specific to a particular system state. This is the main domain-specific knowledge that needs to be introduced (in inherently inductive systems) like reinforcement learning. In some applications like self-driving cars, the choice of rewards can be quite subjective, whereas in other settings like video games or chess, the rules for moves, wins, and losses (rewards) are concrete, and they do not depend on the subjective choices of a domain expert. Some examples of states and corresponding rewards are as follows:

1. *Game of tic-tac-toe, chess, or Go:* The state is the position of the board at any point, and the actions correspond to the moves made by the agent. The reward is +1, 0, or −1 (depending on win, draw, or loss), *which is received at the end of the game.* Note that rewards are often not received immediately after strategically astute actions.

2. *Robot locomotion:* The state corresponds to the current configuration of robot joints and its position. The actions correspond to the torques applied to robot joints. The reward at each time stamp is a function of whether the robot stays upright and the amount of forward movement from point A to point B.

3. *Self-driving car:* The states correspond to the sensor inputs from the car, and the actions correspond to the steering, acceleration, and braking choices. The reward is a hand-crafted function of car progress and safety.

Some effort usually needs to be invested in defining the state representations and corresponding rewards. However, once these choices have been made, reinforcement learning frameworks are end-to-end systems.

Consider a set up of video games, in which the player is the agent, and moving the joystick in a certain direction in a video game is an action. The environment is the entire set up of the video game itself. These actions change the state of the environment, which is shown by the display of the video game. The environment gives the agent rewards, such as the points awarded in a video game. The consequence of an action is sometimes long lasting. For example, the player might have cleverly positioned a cursor at a particularly convenient point a few moves back, and the current action may yield a high reward only because of that characteristic of the current state. Therefore, not only does the current state-action pair need to be credited, but the past state-action pair (e.g., positioning the cursor) needs to be appropriately credited when a reward (e.g., video game point) is received. Furthermore, the reward for an action might not be deterministic (as in card games), which adds to complexity. *One of the primary goals of reinforcement learning is to identify the inherent*

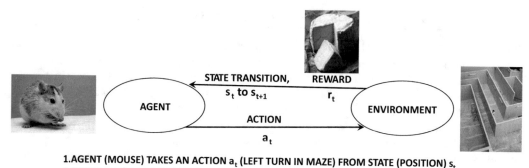

STATE TRANSITION,   REWARD
s_t to s_{t+1} r_t

AGENT   ENVIRONMENT

ACTION
a_t

1.AGENT (MOUSE) TAKES AN ACTION a_t (LEFT TURN IN MAZE) FROM STATE (POSITION) s_t
2.ENVIRONMENT GIVES MOUSE REWARD r_t (CHEESE/NO CHEESE)
3.THE STATE OF AGENT IS CHANGED TO s_{t+1}
4.MOUSE'S NEURONS UPDATE SYNAPTIC WEIGHTS BASED ON WHETHER ACTION EARNED CHEESE

OVERALL: AGENT LEARNS OVER TIME TO TAKE STATE-SENSITIVE ACTIONS THAT EARN REWARDS

Figure 10.1: The broad framework of reinforcement learning

values of actions in different states, irrespective of the timing and stochasticity of the reward.
The agent can then choose actions based on these values.

This general principle is drawn from how reinforcement learning works in biological organisms. Consider a mouse learning a path through a maze to earn a reward. The value of a particular action of the mouse (e.g., left turn) depends on where it is in the maze. When a reward is earned by reaching the goal, the synaptic weights in the mouse's brain adjust to reflect all past actions in various positions and not just the final step. This is exactly the approach used in deep reinforcement learning, where a neural network is used to predict values of actions from sensory inputs (e.g., pixels of video game), and the values of past actions in various states are indirectly updated depending on the received rewards (by updating the weights of the neural network). This relationship between the agent and the environment is shown in Figure 10.1.

The entire set of states and actions and rules for transitioning from one state to another is referred to as a *Markov decision process*. The main property of a Markov decision process is that the state at any particular time stamp encodes all the information needed by the environment to make state transitions and assign rewards based on agent actions. Finite Markov decision processes (e.g., tic-tac-toe) terminate in a finite number of steps, which is referred to as an *episode*. Infinite Markov decision processes (e.g., continuously working robots) do not have finite length episodes and are referred to as *non-episodic* or *continuous*. A Markov decision process can be represented as a sequence of actions, states, and rewards as follows:

$$s_0 a_0 r_0 s_1 a_1 r_1 \ldots s_t a_t r_t \ldots$$

Note that s_t is the state *before* performing action a_t, and performing the action a_t causes a reward of r_t and transition to state s_{t+1}. This is the time-stamp convention used throughout this chapter (and several other sources), although the convention in other sources outputs r_{t+1} in response to action a_t in state s_t (which slightly changes the subscripts in all the results).

The reward r_t only corresponds to the specific amount received at time t, and we need a notion of *cumulative future reward over the long term* for each state-action pair in order to estimate its inherent value. The cumulative expected reward $E[R_t|s_t, a_t]$ for state-action pair (s_t, a_t) is given by the discounted sum of all future expected rewards at discount factor

$\gamma \in (0, 1)$:

$$E[R_t|s_t, a_t] = E[r_t + \gamma \cdot r_{t+1} + \gamma^2 \cdot r_{t+2} + \gamma^3 \cdot r_{t+3} \ldots |s_t, a_t] = \sum_{i=0}^{\infty} \gamma^i E[r_{t+i}|s_t, a_t] \quad (10.2)$$

The discount factor $\gamma \in (0, 1)$ regulates how myopic we want to be in allocating rewards. The value of γ is an application-specific parameter less than 1, because future rewards are considered less important than immediate rewards. Choosing $\gamma = 0$ will result in myopically setting the full reward R_t to r_t and nothing else. Larger values of γ will have a better long-term perspective but will need more data for robust learning. *If the expected cumulative rewards of all state-action pairs can be learned, it provides a basis for a reinforcement learning algorithm of selecting the best action in each state.* There are, however, numerous challenges in learning these values for the following reasons:

- If we define a state-sensitive exploration-exploitation *policy* (like ϵ-greedy in multiarmed bandits) in order to create randomized action sequences for estimation of $E[R_t|s_t, a_t]$, the estimated value of $E[R_t|s_t, a_t]$ will be sensitive to the policy used. This is because actions have long-lasting consequences, which can interact with subsequent choices of the policy. For example, a heavily exploratory policy will not learn a large value for $E[R_t|s_t, a_t]$ for those actions a_t that bring a robot near the edge of a cliff, even if a_t is an optimal action in state s_t. It is customary to use the notation $E^p[R_t|s_t, a_t]$ to show that the expected values are specific to a policy p. Fortunately, $E^p[R_t|s_t, a_t]$ is still quite helpful in predicting high-quality actions for reasonable choices of policies.

- The reinforcement learning system might have a very large number of states (such as the number of positions in chess). Therefore, explicit tabulation of $E^p[R_t|s_t, a_t]$ is no longer possible, and it needs to learned as a parameterized *function* of (s_t, a_t), and can therefore be predicted even for unseen state-action pairs. This task of model generalization is the primary function of machine learning modules built into the system.

We will discuss the challenges of reinforcement learning in greater deal in subsequent sections.

10.4 Monte Carlo Sampling

The simplest method for reinforcement learning is to use Monte Carlo sampling in which sequences of actions are sampled using a policy p that exploits currently estimated values of $E^p[R_t|s_t, a_t]$, while improving these estimates as well. Each sampled sequence (episode) is referred to as a *Monte Carlo rollout*. This approach is a generalization of the multiarmed bandit algorithm, in which different arms are sampled repeatedly to learn the values of actions. In stateless multi-armed bandits, rewards only depend on actions. In general reinforcement learning, we also have a notion of state, and therefore, one must estimate $E^p[R_t|s_t, a_t]$ for state-action *pairs*, rather than simply actions. Therefore, the key idea is to apply a generic randomized sampling policy, such as the ϵ-greedy policy, in a state-sensitive way. In the following section, we will introduce a simple Monte Carlo algorithm with the ϵ-greedy policy, although one can design similar algorithms with other policies such as a randomized variant of the upper-bounding policy or biased sampling. *Note that the learned values of state-action pairs are sensitive to the policy used or even the value of ϵ.*

10.4.1 Monte Carlo Sampling Algorithm

The ϵ-greedy algorithm in the previous section is the simplest example of a Monte Carlo sampling algorithm, where one is simulating slot machines to decide which actions are rewarding. In this section, we will show how one can generalize the stateless ϵ-greedy algorithm in the previous section to a setting with states (using the game of tic-tac-toe as an example). Recall that the gambler (of multiarmed bandits) continuously tries different arms of the slot machine in order to learn more profitable actions in the longer term. However, the tic-tac-toe environment is no longer stateless, and the choice of move depends on the current state of the tic-tac-toe board. In this case, each board position is a state, and the action corresponds to placing 'X' or 'O' at a valid position. The number of valid states of the 3×3 board is bounded above by $3^9 = 19683$, which corresponds to three possibilities ('X', 'O', and blank) for each of 9 positions. Note that all these 19683 positions may not be valid, and therefore they may not be reached in an actual game.

Instead of estimating the value of each (stateless) action in multi-armed bandits, we now estimate the value of each state-action *pair* (s, a) based on the historical performance of action a in state s against a human opponent who is assumed to be the 'O' player (while the agent plays 'X'). Shorter wins are preferred at discount factor $\gamma < 1$, and therefore the *unnormalized* value of action a in state s is increased with γ^{r-1} in case of wins and $-\gamma^{r-1}$ in case of losses after r moves (including the current move). Draws are credited with 0. The normalized values, which represent estimates of $E^p[R_t|s_t, a_t]$, are obtained by dividing the unnormalized values with the number of times the state-action pair was updated (which is maintained separately). The table starts with small random values, and the action a in state s is chosen greedily to be the action with the highest normalized value with probability $1 - \epsilon$, and is chosen to be a random action otherwise. All moves in a game are credited after the termination of each game. In settings where rewards are given after each action, it is necessary to update all past actions in the rolled out episode with time-discounted values. Over time, the values of all state-action pairs for the 'X' player will be learned and the resulting moves will also adapt to the play of the human opponent.

The above description assumes that crowdsourced human players are available to train the system. This can be problem in settings where a large number of players are unavailable for training. Reinforcement learning traditionally requires large amounts of data, and it is often not realistic to have access to large numbers of players. One disadvantage of this method is also that the resulting moves will also adapt to the style of play of the specific human players that are used. For example, if the specific human players are not experts and repeatedly make suboptimal moves, there will be limitations to how well the system learns from its experience. This is similar to the human experience, where one learns best by playing against stronger or equally strong players at any moment in time. Furthermore, one can even use self-play to generate these tables optimally without human effort. When self-play is used, separate tables are maintained for the 'X' and 'O' players. The tables are updated from a value in $\{-\gamma^r, 0, \gamma^r\}$ depending on win/draw/loss *from the perspective of the player for whom moves are made*. Over many rollouts, the value of ϵ is often annealed to 0. At inference time, the move with the highest normalized value from either the 'X' table or the 'O' table is selected (while setting $\epsilon = 0$).

The overarching goal of the Monte Carlo sampling algorithm for tic-tac-toe is to learn the inherent *long-term* value of each state-action pair, since the rewards are received long after valuable actions are performed. By playing through each game, some state-action pairs are more likely to win than others, and this fact will eventually be reflected in the tabular statistics of state-action pairs. In the early phases of training, only states that are very close to win/loss/draws (i.e., one or two moves lead to the outcome), will have accurate long-term values, whereas the early positions on the board will not have accurate values. However, as these statistics improve, the Monte Carlo rollouts also become increasingly accurate in terms of making correct choices, and the values of early-stage positions also start becoming increasingly accurate. The goal of the training process is, therefore, to perform the *value discovery* task of identifying which actions are truly beneficial in the long-term at a particular state. For example, making a clever move in tic-tac-toe might set a trap, which eventually results in assured victory. Examples of two such scenarios are shown in Figure 10.2(a) (although the trap on the right is somewhat less obvious). Therefore, one needs to credit a *strategically* good move favorably in the table of state-action pairs and not just the final winning move. The trial-and-error technique based on the ϵ-greedy method of section 10.4.1 will indeed assign high values to clever traps. Examples of typical values from such a table are shown in Figure 10.2(b). Note that the less obvious trap of Figure 10.2(a) has a slightly lower value because moves assuring wins after longer periods are discounted by γ, and ϵ-greedy trial-and-error might have a harder time finding the win after setting the trap.

Although the above approach described reinforcement learning in an adversarial setting, it can also be leveraged for non-adversarial settings. For example, the case in which human opponents are used for training can be considered more similar to a non-adversarial setting, since self-play is not needed. One can use this type of approach for many applications like learning to play video games or for training robots to walk (by defining appropriate rewards).

10.4.2 Monte Carlo Rollouts with Function Approximators

The aforementioned algorithm for tic-tac-toe directly stores the statistics associated with individual states, which provides an unnecessarily high level of granularity for data collection. The earliest algorithms for reinforcement learning used this type of state-action statistics collection. The main problem with this approach is that the number of states in many reinforcement learning settings is too large to tabulate explicitly. For example, the number of possible states in a game of chess is so large that the set of all known positions by humanity is a minuscule fraction of the valid positions. In fact, the algorithm of section 10.4.1 is a refined form of *rote learning* in which Monte Carlo simulations are used to refine and remember the long-term values of *seen* states. One learns about the value of a trap in tic-tac-toe only because previous Monte Carlo simulations have experienced victory many times *from that exact board position*. In most challenging settings like chess, one must *generalize* knowledge learned from prior experiences to a state that the learner has not seen before. Therefore, this approach is not very useful in most settings where the player will usually encounter board positions (states) that *have never been seen before*. All forms of learning (including reinforcement learning) are most useful when they are used to generalize known experiences to unknown situations. In such cases, the table-centric forms of reinforcement learning are woefully inadequate.

Machine learning models serve the role of *function approximators*. Instead of learning and *tabulating* the values of all moves in all positions (using reward-driven trial and error), one learns the value of each move as a *function* of the input state, based on a *trained model* using the outcomes of prior positions. The idea is that the learner can discover *important patterns on the board* and integrate them into an evaluation of a particular position. In other words, *explicit tabulation of the explosive number of possibilities is replaced with a compressed machine learning model.* Without this approach, reinforcement learning cannot be used

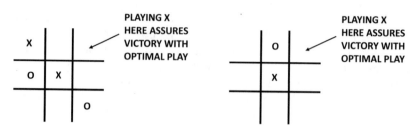

(a) Two examples from tic-tac-toe assuring victory down the road.

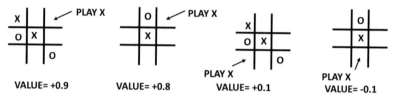

(b) Four entries from the table of state-action values in tic-tac-toe. Trial-and-error learns that moves assuring victory have high value.

(c) Positions from two different games between *Alpha Zero* (white) and *Stockfish* (black) [168]: On the left, white sacrifices a pawn and concedes a passed pawn in order to trap black's light-square bishop behind black's own pawns. This strategy eventually resulted in a victory for white after many more moves than the horizon of a conventional chess-playing program like *Stockfish*. In the second game on the right, white has sacrificed material to incrementally cramp black to a position where all moves worsen the position. Incrementally improving positional advantage is the hallmark of the very best human players rather than chess-playing software like *Stockfish*, whose hand-crafted evaluations sometimes fail to accurately capture subtle differences in positions. The neural network in reinforcement learning, which uses the board state as input, evaluates positions in an integrated way without any prior assumptions. The data generated by trial-and-error provides the only experience for training a very complex evaluation function that is indirectly encoded within the parameters of the neural network. The trained network can therefore *generalize* these learned experiences to new positions. This is similar to how humans learn from previous games to better evaluate board positions.

Figure 10.2: Deep learners are needed for large state spaces like (c)

beyond toy settings like tic-tac-toe. Without this approach, reinforcement learning cannot be used beyond toy settings like tic-tac-toe. For example, a possible algorithm for chess might use the same Monte Carlo sampling algorithm of section 10.4.1, but the value of state-action pair (s_t, a_t) is *estimated* by using the board state after action a_t as input to a convolutional neural network. The output is the estimate of $E^p[R_t|s_t, a_t]$. The ϵ-greedy sampling algorithm is simulated to termination by exploiting the estimated value of $E^p[R_t|s_t, a_t]$. The discounted ground-truth value of each chosen move from $\{\gamma^{r-1}, 0, -\gamma^{r-1}\}$ depends on game outcome and number of moves r to termination. Instead of updating a table of state-action pairs for each chosen move in the simulation, the parameters of the neural network are updated by treating it as a training point. The output of the neural network is compared to the ground-truth value of the move from $\{\gamma^{r-1}, 0, -\gamma^{r-1}\}$ to update the parameters. At inference time, the move a_t with the largest $E^p[R_t|s_t, a_t]$ predicted by the neural network can be chosen without the need for ϵ-greedy exploration.

Although the aforementioned approach sounds somewhat naive, it is surprisingly effective with some additional enhancements like Monte Carlo trees and minimax "lookahead" evaluations. A sophisticated system with Monte Carlo tree search, known as *Alpha Zero*, has recently been trained [168] to play chess. This algorithm is a variant of the broader approach of using Monte Carlo rollouts. Two examples of positions [168] from different games in the match between *Alpha Zero* and a conventional chess program, *Stockfish-8.0*, are provided in Figure 10.2(c). In the chess position on the left, the reinforcement learning system makes a *strategically* astute move of cramping the opponent's bishop at the expense of immediate material loss, which most hand-crafted computer evaluations would not prefer. In the position on the right, *Alpha Zero* has sacrificed two pawns and a piece exchange in order to incrementally constrict black to a point where all its pieces are completely paralyzed. Even though *AlphaZero* (probably) never encountered these specific positions during training, its deep learner has the ability to extract relevant features and patterns from previous trial-and-error experience in other board positions. In this particular case, the neural network seems to recognize the primacy of spatial patterns representing subtle positional factors over tangible material factors (much like a human's neural network). It is noteworthy that the *Alpha Zero* algorithm uses a variant of Monte Carlo rollouts, referred to as *Monte Carlo tree search*. Monte Carlo tree search is described in Chapter 3, and it is closely related to the Monte Carlo rollout method discussed in this section. These connections are discussed in the next section.

In real-world settings, states are often described using sensory inputs. The deep learner uses this input representation of the state to learn the values of specific actions (e.g., making a move in a game) in lieu of the table of state-action pairs. Even when the input representation of the state (e.g., pixels) is quite primitive, neural networks are masters at squeezing out the relevant insights. This is similar to the approach used by humans to process primitive sensory inputs to define the *state* of the world and make decisions about *actions* using our biological neural network. We do not have a table of pre-memorized state-action pairs for every possible real-life situation. The deep-learning paradigm converts the forbiddingly large table of state-action values into a parameterized model mapping state-action pairs to values, which can be trained easily with backpropagation.

10.4.3 Connections to Monte Carlo Tree Search

The simple tic-tac-toe algorithm for using Monte Carlo simulations (which are also referred to as rollouts) is closely related to Monte Carlo tree search, as discussed in Chapter 3. Performing the rollouts and collecting the statistics for state-action pairs is very similar to how statistics are constructed in Monte Carlo search trees. The number of leaf nodes in the Monte Carlo tree is equal to the number of rollouts. The Monte Carlo algorithm (discussed in the previous section for playing tic-tac-toe) may be considered a very simplified version of a Monte Carlo search tree. By performing repeated rollouts and collecting statistics on wins and losses, one is in effect learning the statistics for each possible branch in a state. The main difference is that the tic-tac-toe algorithm consolidates the statistics for the duplicate state-action pairs that occur in different parts of the search tree, rather than treating them separately (as in Monte Carlo tree search). Consolidating the statistics of duplicate nodes in the tree leads to more robust results. Therefore, a Monte-Carlo search tree is not explicitly built by the tic-tac-toe algorithm, although the statistics collected are very similar. The exploration-exploitation strategy is also similar in the two cases in terms of favoring more promising actions; however, the tic-tac-toe algorithm uses the ϵ-greedy method in order to perform the exploration portion of the search, whereas Monte Carlo search trees generally use the upper-bounding method. This choice is, however, a minor difference in detail. In general, Monte Carlo search trees are very much an implementation of reinforcement learning that use rollouts with the use of a tree structure in order to regulate the exploration and exploitation process. Monte Carlo search trees have historically been leveraged in the context of adversarial game-playing settings like chess and Go. However, this is not a hard constraint, as it is also possible to leverage Monte Carlo tree search for traditional reinforcement learning settings (like training robots). The main difference is that the tree only needs to be constructed for the choices of the single agent rather than the choices of two adversarial agents in alternate levels of the tree. This change, in fact, simplifies the Monte Carlo tree construction.

10.5   Bootstrapping and Temporal Difference Learning

The Monte Carlo sampling approach does not work for *non-episodic settings*. In episodic settings like tic-tac-toe, a fixed-length sequence of at most nine moves can be used to characterize the full and final reward. In non-episodic settings like robots, one is forced to assign credit to an infinitely long sequence in the past based on the reward. Creating a sample of the ground-truth reward by Monte Carlo sampling also has *high variance* because a particular outcome is a very noisy estimate of what might happen *in expectation*. The noise increases with the length of the episode, which increases the number of required rollouts. In other words, we need a methodology for reducing randomness by sampling only a small number of actions. Unfortunately, after a small number of actions, one will not reach a terminal state, which is required for final value estimation (i.e., drawing a sample of $E[R_t|s_t, a_t]$ in Equation 10.2). However, it is possible to estimate the value of states *approximately* with the methodology of *bootstrapping*, which makes the assumption that *states further along the decision process always have existing value estimates that are better than those states that are earlier along the decision process.* Therefore, one can use the current estimate of the value of the non-terminal state after a few actions in order to better estimate the value of the current state. The idea of bootstrapping may, therefore, be summarized as follows:

Intuition 10.5.1 (Bootstrapping) *Consider a Markov decision process in which we have running estimates of the values (e.g., long-term rewards) of states. We can use a partial simulation of the future to improve the value estimation of the state at the current time-stamp by adding the discounted rewards over these simulated actions with the discounted value of the state reached at the end of the simulation.*

This class of methods is also referred to as *temporal difference learning*. The earliest example of this type of bootstrapping was Samuel's checkers program [155], and it did not combine bootstrapping with Monte Carlo rollouts. Rather, it combined the principle of bootstrapping with minimax trees; it used the difference in evaluation at the current position and the minimax evaluation obtained by looking several moves ahead with the same function as a "prediction error" in order to update the evaluation function. The idea is that the minimax evaluation from looking ahead is stronger than the one without lookahead and can therefore be used as a "ground truth" to compute the error. In general, value function learning also combines a host of strategies in combination with bootstrapping, some of which use partial Monte Carlo rollouts, whereas others use optimization methods like minimax trees or dynamic programming. This section will discuss a variety of such methods, two of the most important of which are *Q-Learning* and *SARSA*.

10.5.1 Q-Learning

Consider a Markov decision process with the following sequence of states, actions, and rewards denoted by the repeating sequence $s_t a_t r_t$ at time stamp t. It is assumed that rewards are earned with discount factor γ based in Equation 10.2. The *Q-function* or *Q-value* for the state-action pair (s_t, a_t) is denoted by $Q(s_t, a_t)$, and is a measure of the *inherent* (i.e., long-term) value of performing the action a_t in state s_t (under the best possible choice of actions). This value can be viewed as an optimized version $E^*[R_t|s_t, a_t]$ of the expected reward in Equation 10.2 in which one is using the theoretically optimal policy, and the specific choice of policy used for data collection/exploration does not matter. One can choose the next action of the agent by maximizing this value over the set A of all possible actions:

$$a_t^* = \text{argmax}_{a_t \in A} Q(s_t, a_t) \tag{10.3}$$

This predicted action is a good choice for the next move, although it is often combined with an exploratory component (e.g., ϵ-greedy policy) to improve long-term training outcomes. This is exactly similar to how actions are chosen in Monte carlo rollouts. The main difference is in terms of how $Q(s_t, a_t)$ is computed and in terms of finding an *optimal* policy.

Instead of using explicit rollouts to termination for crediting the values of state-value pairs, one computes $Q(s_t, a_t)$ by using a single step to s_{t+1}, and then using the best possible Q-value estimate at state s_{t+1} in order to update $Q(s_t, a_t)$. This type of update is referred to as *Bellman's equation*, which is a form of dynamic programming:

$$Q(s_t, a_t) \Leftarrow r_t + \gamma \text{max}_a Q(s_{t+1}, a) \tag{10.4}$$

The correctness of this relationship follows from the fact that the Q-function is designed to maximize the discounted future payoff. We are essentially looking at all actions one step ahead in order to create an improved estimate of $Q(s_t, a_t)$. *The above update of Equation 10.4 replaces the updates of Monte Carlo sampling, in order to create the Q-learning algorithm.* While one is using the actual outcomes resulting from a rollout in Monte Carlo sampling, the Q-learning approach is approximating the best possible outcome with a combination of bootstrapping and dynamic programming. This best action is used to continue the simulation of the state-space sampling (like in Monte Carlo sampling). However, an important point is that the *best possible action is not used for making steps*. Rather, the best possible move is made with probability $(1 - \epsilon)$ and a random move is made with probability ϵ in order to move to the next iteration of the algorithm. This is again done in order to be able to navigate the exploration-exploitation trade-off properly. The simulation is continued in order to improve the tabular estimates of $Q(s, a)$ over time. Unlike Monte Carlo sampling, there is a dichotomy between how steps are made (with randomized exploration) and how updates to $Q(s, a)$ are made (in an optimal way with dynamic programming). This dichotomy means that no matter what policy (e.g., ϵ-greedy or biased sampling) is used for simulations, one will always compute the same value of $E^*[R_t|s_t, a_t] = Q(s_t, a_t)$, which is the optimal policy.

In the case of episodic sequences, the Monte Carlo sampling method of the previous section improves the estimates of the values of state-action pairs that are later in the sequence (and closer to termination) first. This is also the case in Q-learning, where the updates for a state that is one step from termination is exact. It is important to set $\hat{Q}(s_{t+1}, a)$ to 0 in case the process terminates after performing a_t for episodic sequences. Therefore, the accuracy of the estimation of the value of the state-action pair will be propagated over time from states closer to termination to earlier states. For example, in a tic-tac-toe game, the value of a winning action in a state will be accurately estimated in a single iteration, whereas the value of a first move in tic-tac-toe will require the propagation of values from later states to earlier states via the Bellman equation. This type of propagation will require a few iterations.

In practice, *learning rates* are used to provide stability to the tabular updates. In cases where the state-space is very small (like tic-tac-toe), one can learn $Q(s_t, a_t)$ explicitly by using the Bellman equations (cf. Equation 10.4) at each move to update an *array* containing the explicit value of $Q(s_t, a_t)$. However, Equation 10.4 directly is too aggressive. More generally, gentle updates are performed using the learning rate $\alpha < 1$:

$$Q(s_t, a_t) \Leftarrow Q(s_t, a_t)(1 - \alpha) + \alpha(r_t + \gamma \max_a Q(s_{t+1}, a)) \qquad (10.5)$$

Using $\alpha = 1$ will result in Equation 10.4. Updating the array continually will result in a table containing the correct *strategic* value of each move; see, for example, Figure 10.2(a) for an understanding of the notion of strategic value. Figure 10.2(b) contains examples of four entries from such a table. Note that this is a direct alternative to the tabular approach to Monte Carlo sampling and it will result in the same values as in the final table as Monte Carlo sampling, but it will work only for toy settings like tic-tac-toe.

10.5.2   Using Function Approximators

As in the case of Monte Carlo sampling methods, function approximators are particularly useful when the number of states is too large to tabulate explicitly. This can occur frequently in settings like chess or while playing video games. Therefore, the updates in Equation 10.5 will be useless most of the time, as most of the variables on the right hand side would never have been updated even once even after a long period of learning.

For ease in discussion, we will work with the Atari video game setting [126] in which a fixed window of the last few snapshots of pixels provides the state s_t. Assume that the feature representation of s_t is denoted by $\overline{X}_t$. The neural network uses $\overline{X}_t$ as the input and outputs $Q(s_t, a)$ for each possible legal action a from the universe of actions denoted by the set A of actions.

Assume that the neural network is parameterized by the vector of weights $\overline{W}$, and it has $|A|$ outputs containing the Q-values corresponding to the various actions in A. In other words, for each action $a \in A$, the neural network is able to compute the function $F(\overline{X}_t, \overline{W}, a)$, which is defined to be the *learned estimate* of $Q(s_t, a)$:

$$\hat{Q}(s_t, a) = F(\overline{X}_t, \overline{W}, a) \qquad (10.6)$$

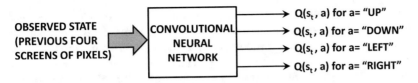

Figure 10.3: The Q-Network for the Atari video game setting

Note the circumflex on top of the Q-function in order to indicate that it is a predicted value using the learned parameters $\overline{W}$. Learning $\overline{W}$ is the key to using the model for deciding which action to use at a particular time-stamp. For example, consider a video game in which the possible moves are up, down, left, and right. In such a case, the neural network will have four outputs as shown in Figure 10.3. In the specific case of the Atari 2600 games, the input contains $m = 4$ spatial pixel maps in grayscale, representing the window of the last m moves [126, 127]. A convolutional neural network is used to convert pixels into Q-values. This network is referred to as a *Q-network*.

The weights $\overline{W}$ of the neural network need to be learned via training. Here, we encounter an interesting problem. We can learn the vector of weights only if we have *observed* values of the Q-function. With observed values of the Q-function, we could easily set up a loss in terms of $Q(s_t, a) - \hat{Q}(s_t, a)$ in order to perform the learning after each action. The problem is that the Q-function represents the maximum discounted reward over all *future* combinations of actions, and there is no way of observing it at the current time.

It is here that the bootstrapping trick is used for setting up the neural network loss function. According to Intuition 10.5.1, *we do not really need the observed Q-values in order to set up a loss function as long as we know an improved estimate of the Q-values by using partial knowledge from the future.* Then, we can use this improved estimate to create a surrogate "observed" value. This "observed" value is defined by the Bellman equation (cf. Equation 10.4) discussed earlier:

$$Q(s_t, a_t) = r_t + \gamma \max_a \hat{Q}(s_{t+1}, a) \tag{10.7}$$

One difference from Equation 10.4 is that the right-hand side uses the *predicted* value $\hat{Q}(s_{t+1}, a)$ for creating the "ground-truth" in learning rather than the *tabulated* value $Q(s_{t+1}, a)$. After all, tabulated values are no longer retained in settings that use function approximation. We can write this relationship in terms of our neural network predictions as well:

$$Q(s_t, a_t) = r_t + \gamma \max_a F(\overline{X}_{t+1}, \overline{W}, a) \tag{10.8}$$

Note that one must first wait to observe the state $\overline{X}_{t+1}$ and reward r_t by performing the action a_t, before we can compute the "observed" value at time-stamp t on the right-hand side of the above equation. This provides a natural way to express the loss L_t of the neural network at time stamp t by comparing the (surrogate) observed value $Q(s_t, a_t)$ to the predicted value $F(\overline{X}_t, \overline{W}, a_t)$ at time stamp t:

$$L_t = \left\{ Q(s_t, a_t) - F(\overline{X}_t, \overline{W}, a_t) \right\}^2$$

One can also write the loss function directly in terms of the neural network predictions:

$$L_t = \left\{ \underbrace{[r_t + \gamma \max_a F(\overline{X}_{t+1}, \overline{W}, a)]}_{\text{Ground-truth } Q(s_t, a_t)} - F(\overline{X}_t, \overline{W}, a_t) \right\}^2 \tag{10.9}$$

Therefore, we can now update the vector of weights $\overline{W}$ by computing the derivatives of this loss function with respect to the loss function. In the case of neural networks, this computation amounts to the use of the backpropagation algorithm. Here, it is important to note that the target values $\hat{Q}(s_t, a_t)$ at time t, which are estimated using bootstrapping on the predictions at time $(t+1)$ are treated as constant ground-truths by the backpropagation algorithm. Therefore, the derivative of the loss function will treat these estimated values as constants, even though they were obtained from the parameterized neural network with input $\overline{X}_{t+1}$. Not treating $F(\overline{X}_{t+1}, \overline{W}, a)$ as a constant will lead to poor results. This is because we are treating the prediction at $(t+1)$ to create an improved estimate $\hat{Q}(s_t, a_t)$ of the ground-truth at time t (based on the boot-strapping principle). Therefore, the weights need to be updated as follows:

$$\overline{W} \Leftarrow \overline{W} - \alpha \frac{\partial L_t}{\partial \overline{W}} \tag{10.10}$$

$$= \overline{W} + \alpha \left\{ \underbrace{[r_t + \gamma \max_a F(\overline{X}_{t+1}, \overline{W}, a)]}_{\text{Treat as constant ground-truth}} - F(\overline{X}_t, \overline{W}, a_t) \right\} \frac{\partial F(\overline{X}_t, \overline{W}, a_t)}{\partial \overline{W}} \tag{10.11}$$

Note that $F(\overline{X}_{t+1}, \overline{W}, a)$ is treated as a constant in the above derivative. In matrix-calculus notation, the partial derivative of a function $F()$ with respect to the vector $\overline{W}$ is essentially the gradient $\nabla_{\overline{W}} F$. At the beginning of the process, the Q-values estimated by the neural network are random because the vector of weights $\overline{W}$ is initialized randomly. However, the estimation gradually becomes more accurate with time, as the weights are constantly changed to reduce losses (thereby maximizing rewards).

We now provide a listing of the training steps used by the Q-Learning algorithm. At any given time-stamp t at which action a_t and reward r_t has been observed, the following training process is used for updating the weights $\overline{W}$:

1. Perform a forward pass through the network with input $\overline{X}_{t+1}$ to compute $\hat{Q}_{t+1} = \max_a F(\overline{X}_{t+1}, \overline{W}, a)$. The value is 0 in case of termination after performing a_t. *Treating the terminal state specially is important.* According to the Bellman equations, the Q-value at previous time-stamp t should be $r_t + \gamma \hat{Q}_{t+1}$ for observed action a_t at time t. Therefore, instead of using observed values of the target, we have created a *surrogate* for the target value at time t, and we pretend that this surrogate is an observed value given to us.

2. Perform a forward pass through the network with input $\overline{X}_t$ to compute $F(\overline{X}_t, \overline{W}, a_t)$.

3. Set up a loss function in $L_t = (r_t + \gamma \hat{Q}_{t+1} - F(\overline{X}_t, \overline{W}, a_t))^2$, and backpropagate in the network with input $\overline{X}_t$. Note that this loss is associated with neural network output node corresponding to action a_t, and the loss for all other actions is 0.

4. One can now use backpropagation on this loss function in order to update the weight vector $\overline{W}$. Even though the term $r_t + \gamma \hat{Q}_{t+1}$ in the loss function is also obtained as a prediction from input $\overline{X}_{t+1}$ to the neural network, it is treated as a (constant) observed value during gradient computation by the backpropagation algorithm.

Both the training and the prediction are performed simultaneously, as the values of actions are used to update the weights and select the next action. It is tempting to select the action with the largest Q-value as the relevant prediction. However, such an approach might

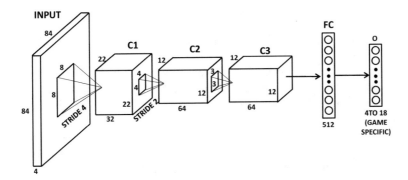

Figure 10.4: The convolutional neural network for the Atari setting

perform inadequate exploration of the search space. Therefore, one couples the optimality prediction with a policy such as the ϵ-greedy algorithm in order to select the next action. The action with the largest predicted payoff is selected with probability $(1 - \epsilon)$. Otherwise, a random action is selected. The value of ϵ can be annealed by starting with large values and reducing them over time. Therefore, the *target prediction value* for the neural network is computed using the best possible action in the Bellman equation (which might eventually be different from observed action a_{t+1} based on the ϵ-greedy policy). This is the reason that Q-learning is referred to as an *off-policy algorithm* in which the target prediction values for the neural network update are computed using actions that might be different from the actually observed actions in the future.

10.5.3 Example: Neural Network Specifics for Video Game Setting

For the convolutional neural network [126, 127], the screen sizes were set to 84×84 pixels, which also defined the spatial footprints of the first layer in the convolutional network. The input was in grayscale, and therefore each screen required only a single spatial feature map, although a depth of 4 was required in the input layer to represent the previous four windows of pixels. Three convolutional layers were used with filters of size 8×8, 4×4, and 3×3, respectively. A total of 32 filters were used in the first convolutional layer, and 64 filters were used in each of the other two, with the strides used for convolution being 4, 2, and 1, respectively. The convolutional layers were followed by two fully connected layers. The number of neurons in the penultimate layer was equal to 512, and that in the final layer was equal to the number of outputs (possible actions). The number of output layers varied between 4 and 18, and was game-specific. The overall architecture of the convolutional network is illustrated in Figure 10.4.

All hidden layers used the ReLU activation, and the output used linear activation in order to predict the real-valued Q-value. No pooling was used, and the strides in the convolution provided spatial compression. The Atari platform supports many games, and the same broader architecture was used across different games in order to showcase its generalizability. There was some variation in performance across different games, although human performance was exceeded in many cases. The algorithm faced the greatest challenges in games in which longer-term strategies were required. Nevertheless, the robust performance of a relatively homogeneous framework across many games was encouraging.

10.5.4   On-Policy Versus Off-Policy Methods: SARSA

The Q-Learning methodology belongs to the class of methods, referred to as *temporal difference learning*. In Q-learning, the actions are chosen according to an ϵ-greedy policy. However, the parameters of the neural network are updated based on the best possible action at each step with the Bellman equation. The best possible action at each step is not quite the same as the ϵ-greedy policy used to perform the simulation. Therefore, Q-learning is an *off-policy reinforcement learning method*. Choosing a different policy for executing actions from those for performing updates is a consequence of the fact that the Bellman updates are intended to find an *optimal policy* rather than *evaluating* a specific policy like ϵ-greedy (as is the case with Monte Carlo methods). In *on-policy methods* like Monte Carlo sampling, the actions are consistent with the updates, and therefore the updates can be viewed as policy *evaluation* rather than policy *optimization*. Therefore, changing the policy affects the predicted actions more significantly in Monte Carlo sampling than in Q-Learning. A bootstrapped approximation of Monte Carlo sampling can also be achieved with the use of the SARSA (State-Action-Reward-State-Action) algorithm, in which the reward in the next step is updated using the action a_{t+1} predicted by the ϵ-greedy policy rather than the optimal step from the Bellman equation. Let $Q^p(s, a)$ be the evaluation of policy p (which is ϵ-greedy in this case) for state-action pair (s, a). Then, after sampling action a_{t+1} using ϵ-greedy, the update is as follows:

$$Q^p(s_t, a_t) \Leftarrow Q^p(s_t, a_t)(1 - \alpha) + \alpha(r_t + \gamma Q(s_{t+1}, a_{t+1})) \qquad (10.12)$$

If action a_t at state s_t leads to termination (for episodic processes), then $Q^p(s_t, a_t)$ is simply set to r_t. Note that this update is different from the Q-Learning update of Equation 10.5, because action a_{t+1} includes the effect of exploration.

When using function approximators, the loss function for the next step is defined as follows:

$$L_t = \left\{ r_t + \gamma F(\overline{X}_{t+1}, \overline{W}, a_{t+1}) - F(\overline{X}_t, \overline{W}, a_t) \right\}^2 \qquad (10.13)$$

The function $F(\cdot, \cdot, \cdot)$ is defined in the same way as the previous section. The weight vector is updated based on this loss, and then the action a_{t+1} is executed:

$$\overline{W} \Leftarrow \overline{W} + \alpha \left\{ \underbrace{[r_t + \gamma F(\overline{X}_{t+1}, \overline{W}, a_{t+1})]}_{\text{Treat as constant ground-truth}} - F(\overline{X}_t, \overline{W}, a_t) \right\} \frac{\partial F(\overline{X}_t, \overline{W}, a_t)}{\partial \overline{W}} \qquad (10.14)$$

Here, it is instructive to compare this update with those used in Q-learning according to Equation 10.11. In Q-learning, one is using the *best possible* action at each state in order to update the parameters, even though the policy that is actually executed might be ϵ-greedy (which encourages exploration). In SARSA, we are using the action that was actually selected by the ϵ-greedy method in order to perform the update. Therefore, the approach is an *on-policy method*. Off-policy methods like Q-learning are able to decouple exploration from exploitation, whereas on-policy methods are not. Note that if we set the value of ϵ in the ϵ-greedy policy to 0 (i.e., vanilla greedy), then both Q-Learning and SARSA would specialize to the same algorithm. However, such an approach would not work very well because there is no exploration. SARSA is useful when learning cannot be done separately from prediction. Q-learning is useful when the learning can to be done offline, which is followed by exploitation of the learned policy with a vanilla-greedy method at $\epsilon = 0$ (and no need for further model updates). Using ϵ-greedy at inference time would be dangerous in Q-learning, because the policy never pays for its exploratory component (in the update) and therefore does not learn how to keep exploration safe. For example, a Q-learning based robot will take the shortest path to get from point A to point B even if it is along the edge of the cliff, whereas a SARSA-trained robot will not. This is because walking along the edge of a cliff will occasionally lead to falling from the cliff, when the exploratory component (with probability ϵ) is triggered. A SARSA-trained robot will learn the greater risks of the exploratory component of a path at the edge of the cliff, because of its on-policy (rather than greedy) updates, and will therefore it will be able to avoid such a path when learning is completed. Another point is that off-policy methods take longer to converge, as the updates are different from the policy used to choose the actions. For example, in the robot-on-edge-of-cliff example, the off-policy method will result in the robot falling off the edge of cliff often, but will be unable to learn from it in order to make learning more stable. In other words, the variance of the approach is higher. The trade-off is that off-policy methods tend to reach more optimal solutions in the longer term, provided that sufficient data is available. The appropriate choice of method depends on the application at hand. In some applications, such as physical robots, the costs of physical damage caused by exploratory accidents in off-policy methods is simply too large to allow the collection of sufficient data. SARSA can also be implemented with n-step lookaheads (rather than 1-step bootstrapping), which brings it even closer to the Monte Carlo sampling of section 10.4.

10.5.5 Modeling States Versus State-Action Pairs

A minor variation of the theme in the previous sections is to learn the value of a particular state (rather than state-action pair). One can implement all the methods discussed earlier by maintaining values of states rather than state-action pairs. For example, SARSA can be implemented by evaluating all the values of states resulting from each possible action and selecting a good one based on a pre-defined policy like ϵ-greedy. In fact, the earliest methods for temporal difference learning (or *TD-learning*) maintained values on states rather than state-action pairs. From an efficiency perspective, it is more convenient to output the values of all actions in one shot (rather than repeatedly evaluate each forward state) for value-based decision making. Working with state values rather that state-action pairs becomes useful only when the policy cannot be expressed neatly in terms of state-action pairs. For example, we might evaluate a forward-looking tree of promising moves in chess, and report some averaged value for bootstrapping. In such cases, it is desirable to evaluate states rather than state-action pairs. This section will therefore discuss a variation of temporal difference learning in which states are directly evaluated.

Let the value of the state s_t be denoted by $V(s_t)$. Now assume that you have a parameterized neural network that uses the observed attributes $\overline{X}_t$ (e.g., pixels of last four screens in Atari game) of state s_t to estimate $V(s_t)$. An example of this neural network is shown in Figure 10.5. Then, if the function computed by the neural network is $G(\overline{X}_t, \overline{W})$ with parameter vector $\overline{W}$, we have the following:

$$G(\overline{X}_t, \overline{W}) = \hat{V}(s_t) \tag{10.15}$$

Note that the policy being followed to decide the actions might use some arbitrary evaluation of forward-looking states to decide actions. For now, we will assume that we have some reasonable heuristic policy for choosing the actions that uses the forward-looking state values in some way. For example, if we evaluate each forward state resulting from an action and select one of them based on a pre-defined policy (e.g., ϵ-greedy), the approach discussed below is the same as SARSA.

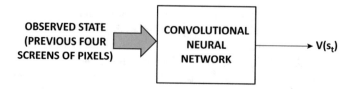

Figure 10.5: Estimating the value of a state with temporal difference learning

If the action a_t is performed with reward r_t, the resulting state is s_{t+1} with value $V(s_{t+1})$. Therefore, the bootstrapped ground-truth estimate for $V(s_t)$ can be obtained with the help of this lookahead:

$$V(s_t) = r_t + \gamma V(s_{t+1}) \tag{10.16}$$

This estimate can also be stated in terms of the neural network parameters:

$$G(\overline{X}_t, \overline{W}) = r_t + \gamma G(\overline{X}_{t+1}, \overline{W}) \tag{10.17}$$

During the training phase, one needs to shift the weights so as to push $G(\overline{X}_t, \overline{W})$ towards the improved "ground truth" value of $r_t + \gamma G(\overline{X}_{t+1}, \overline{W})$. As in the case of Q-learning, we work with the boot-strapping pretension that the value $r_t + \gamma G(\overline{X}_{t+1}, \overline{W})$ is an observed value given to us. Therefore, we want to minimize the *TD-error* defined by the following:

$$\delta_t = \underbrace{r_t + \gamma G(\overline{X}_{t+1}, \overline{W})}_{\text{"Observed" value}} - G(\overline{X}_t, \overline{W}) \tag{10.18}$$

Therefore, the loss function L_t is defined as follows:

$$L_t = \delta_t^2 = \left\{ \underbrace{r_t + \gamma G(\overline{X}_{t+1}, \overline{W})}_{\text{"Observed" value}} - G(\overline{X}_t, \overline{W}) \right\}^2 \tag{10.19}$$

As in Q-learning, one would first compute the "observed" value of the state at time stamp t using the input $\overline{X}_{t+1}$ into the neural network to compute $r_t + \gamma G(\overline{X}_{t+1}, \overline{W})$. Therefore, one would have to wait till the action a_t has been observed, and therefore the observed features $\overline{X}_{t+1}$ of state s_{t+1} are available. This "observed" value (defined by $r_t + \gamma G(\overline{X}_{t+1}, \overline{W})$) of state s_t is then used as the (constant) target to update the weights of the neural network, when the input $\overline{X}_t$ is used to predict the value of the state s_t. Therefore, one would need to move the weights of the neural network based on the gradient of the following loss function:

$$\overline{W} \Leftarrow \overline{W} - \alpha \frac{\partial L_t}{\partial \overline{W}}$$

$$= \overline{W} + \alpha \left\{ \underbrace{[r_t + \gamma G(\overline{X}_{t+1}, \overline{W})]}_{\text{"Observed" value}} - G(\overline{X}_t, \overline{W}) \right\} \frac{\partial G(\overline{X}_t, \overline{W})}{\partial \overline{W}}$$

$$= \overline{W} + \alpha \delta_t (\nabla G(\overline{X}_t, \overline{W}))$$

This algorithm is a special case of the $TD(\lambda)$ algorithm with λ set to 0. This special case only updates the neural network by creating a bootstrapped "ground-truth" for the current time-stamp based on the evaluations of the next time-stamp. This type of ground-truth is an inherently myopic *approximation*. For example, in a chess game, the reinforcement learning system might have inadvertently made some mistake many steps ago, and it is suddenly showing high errors in the bootstrapped predictions without having shown up earlier. The errors in the bootstrapped predictions are indicative of the fact that we have received new information about each past state $\overline{X}_k$, which we can use to alter its prediction. One possibility is to bootstrap by looking ahead for multiple steps (see Exercise 7). Another solution is the use of $TD(\lambda)$, which explores the continuum between perfect Monte Carlo ground truth and single-step approximation with smooth decay. The adjustments to older predictions are increasingly discounted at the rate $\lambda < 1$. In such a case, the update can be shown to be the following [181]:

$$\overline{W} \Leftarrow \overline{W} + \alpha \delta_t \sum_{k=0}^{t} \underbrace{(\lambda\gamma)^{t-k} (\nabla G(\overline{X}_k, \overline{W}))}_{\text{Alter prediction of } \overline{X}_k} \qquad (10.20)$$

At $\lambda = 1$, the approach can be shown to be equivalent to a method in which Monte-Carlo evaluations (i.e., rolling out an episodic process to the end) are used to compute the ground-truth [181]. This is because we are always using new information about errors to fully correct our past mistakes without discount at $\lambda = 1$, thereby creating an unbiased estimate. Note that λ is only used for discounting the steps, whereas γ is also used in computing the TD-error δ_t according to Equation 10.18. The parameter λ is *algorithm-specific*, whereas γ is *environment-specific*. Using $\lambda = 1$ or Monte Carlo sampling leads to lower bias and higher variance. For example, consider a chess game in which agents Alice and Bob each make three errors in a single game but Alice wins in the end. This single Monte Carlo rollout will not be able to distinguish the impact of each specific error and will assign the discounted credit for final game outcome to each board position. On the other hand, an n-step temporal difference method (i.e., n-ply board evaluation) might see a temporal difference error for each board position in which the agent made a mistake and was detected by the n-step lookahead. It is only with sufficient data (i.e., more games) that the Monte Carlo method will distinguish between different types of errors. However, choosing very small values of λ will have difficulty in learning openings (i.e., greater bias) because errors with long-term consequences will not be detected. Such problems with openings are well documented [17, 187].

Temporal difference learning was used in Samuel's celebrated checkers program [155], and also motivated the development of TD-Gammon for Backgammon by Tesauro [185, 186]. A neural network was used for state value estimation, and its parameters were updated using temporal-difference bootstrapping over successive moves. The final inference was performed with minimax evaluation of the improved evaluation function over a shallow depth such as 2 or 3. TD-Gammon was able to defeat several expert players. It also exhibited some unusual strategies of game play that were eventually adopted by top-level players.

10.6 Policy Gradient Methods

The value-based methods like Q-learning attempt to predict the value of an action with the neural network and couple it with a generic policy (like ϵ-greedy). On the other hand, policy gradient methods estimate the *probability* of each action at each step with the goal of maximizing the overall reward. Therefore, the policy is itself parameterized, rather than using the value estimation as an intermediate step for choosing actions. Furthermore, policy gradient methods cannot be designed as tabular methods like value-based methods. This is

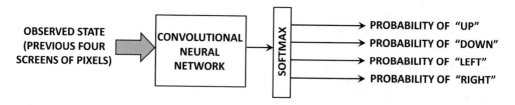

Figure 10.6: The policy network for the Atari video game setting. It is instructive to compare this configuration with the Q-network of Figure 10.3

because such methods focus of finding parameters directly by using their gradients; obviously a discrete, tabular method is not parameterized and cannot be used in conjunction with a policy gradient method. This makes policy gradient methods fundamentally different from temporal difference methods or Monte Carlo sampling methods in that function approximators are inherently tied into the model from the very beginning.

The neural network for estimating the policy is referred to as a *policy network* in which the input is the current state of the system, and the output is a set of probabilities associated with the various actions in the video game (e.g., moving up, down, left, or right). As in the case of the Q-network, the input can be an observed representation of the agent state. For example, in the Atari video game setting, the observed state can be the last four screens of pixels. An example of a policy network is shown in Figure 10.6, which is relevant for the Atari video game setting. It is instructive to compare this policy network with the Q-network of Figure 10.3. Given an output of probabilities for various actions, we throw a biased die with the faces associated with these probabilities, and select one of these actions. Therefore, for each action a, observed state representation $\overline{X}_t$, and current parameter $\overline{W}$, the neural network is able to compute the function $P(\overline{X}_t, \overline{W}, a)$, which is the probability that the action a should be performed. One of the actions is sampled, and a reward is observed for that action. If the policy is poor, the action will more likely to be a mistake and the reward will be poor as well. Based on the reward obtained from executing the action, the weight vector $\overline{W}$ is updated for the next iteration. The update of the weight vector is based on the notion of policy gradient with respect to the weight vector $\overline{W}$. One challenge in estimating the policy gradient is that the reward of an action is often not observed immediately, but is tightly integrated into the future sequence of rewards. Often *Monte Carlo policy roll-outs* must be used in which the neural network is used to follow a particular policy to estimate the discounted rewards over a longer horizon.

We want to update the weight vector of the neural network along the policy gradient, so that the modified policy results in increasing of the expected discounted reward over time. As in Q-Learning, the expected discounted rewards over a given horizon H is computed using a truncated version of Equation 10.2:

$$J = \sum_{i=0}^{H} \gamma^i E[r_{t+i}|s_t, a_t] \tag{10.21}$$

Therefore, the goal is to update the weight vector as follows:

$$\overline{W} \Leftarrow \overline{W} + \alpha \nabla J \tag{10.22}$$

The main problem in estimating the gradient ∇J is that the neural network only outputs probabilities. The observed rewards are only Monte Carlo samples of these outputs, whereas we want to compute the gradients of *expected* rewards (cf. Equation 10.21). Common policy gradients methods include *finite difference methods*, *likelihood ratio methods*, and *natural policy gradients*. In the following, we will only discuss policy gradients.

10.6.1 The Likelihood Ratio Principle

Likelihood-ratio methods were proposed by Williams [203] in the context of the REINFORCE algorithm. Consider the case in which we are following the policy with probability vector $\overline{p}$ and we want to maximize $E[Q^p(s,a)]$, which is the long-term expected value of state s and each sampled action a from the neural network. Consider the case in which the probability of action a is $p(a)$ (which is output by the neural network). In such a case, we want to find the gradient of $E[Q^p(s,a)]$ with respect to the weight vector $\overline{W}$ of the neural network for stochastic gradient ascent. Finding the gradient of an expectation from sampled events is non-obvious. However, the log-probability trick allows us to bring the expectation outside the gradient, which is additive over the samples of state-action pairs:

$$\nabla E[Q^p(s,a)] = E[Q^p(s,a)\nabla \log(p(a))] \tag{10.23}$$

We show the proof of the above result in terms of the partial derivative with respect to a single neural network weight w under the assumption that a is a discrete variable:

$$\frac{\partial E[Q^p(s,a)]}{\partial w} = \frac{\partial \left[\sum_a Q^p(s,a)p(a)\right]}{\partial w} = \sum_a Q^p(s,a)\frac{\partial p(a)}{\partial w} = \sum_a Q^p(s,a)\left[\frac{1}{p(a)}\frac{\partial p(a)}{\partial w}\right]p(a)$$

$$= \sum_a Q^p(s,a)\left[\frac{\partial \log(p(a))}{\partial w}\right]p(a) = E\left[Q^p(s,a)\frac{\partial \log(p(a))}{\partial w}\right]$$

The above result can also be shown for the case in which a is a continuous variable (cf. Exercise 1). Continuous actions occur frequently in robotics (e.g., distance to move arm).

It is easy to use this trick for neural network parameter estimation. Each action a sampled by the simulation is associated with the long-term reward $Q^p(s,a)$, which is obtained by Monte Carlo simulation. Based on the relationship above, the gradient of the expected advantage is obtained by multiplying the gradient of the log-probability $\log(p(a))$ of that action (computable from the neural network in Figure 10.6 using backpropagation) with the long-term reward $Q^p(s,a)$ (obtained by Monte Carlo simulation).

Consider a simple game of chess with a win/loss/draw at the end and discount factor γ In this case, the long-term reward of each move is simply obtained as a value from $\{+\gamma^{r-1}, 0, -\gamma^{r-1}\}$, when r moves remain to termination. The value of the reward depends on the final outcome of the game, and number of remaining moves (because of reward discount). Consider a game containing at most H moves. Since multiple roll-outs are used, we get a whole bunch of training samples for the various input states and corresponding outputs in the neural network. For example, if we ran the simulation for 100 roll-outs, we would get at most $100 \times H$ different samples. Each of these would have a long-term reward drawn from $\{+\gamma^{r-1}, 0, -\gamma^{r-1}\}$. For each of these samples, the reward serves as a weight during a gradient-ascent update of the log-probability of the sampled action.

$$\overline{W} \Leftarrow \overline{W} + Q^p(s,a)\nabla \log(p(a)) \tag{10.24}$$

Here, $p(a)$ is the neural network's output probability of the sampled action. The gradients are computed using backpropagation, and these updates are similar to those in Equation 10.22. This process of sampling and updating is carried through to convergence.

Note that the gradient of the log-probability of the ground-truth class is often used to update softmax classifiers with cross-entropy loss in order to increase the probability of the correct class (which is intuitively similar to the update here). The difference here is that we are weighting the update with the Q-values because we want to push the parameters more aggressively in the direction of highly rewarding actions. One could also use mini-batch gradient ascent over the actions in the sampled roll-outs. Randomly sampling from different roll-outs can be helpful in avoiding the local minima arising from correlations because the successive samples from each roll-out are closely related to one another.

Reducing Variance with Baselines: Although we have used the long-term reward $Q^p(s, a)$ as the quantity to be optimized, it is more common to subtract a baseline value from this quantity in order to obtain its *advantage* (i.e, differential impact of the action over expectation). The baseline is ideally state-specific, but can be a constant as well. In the original work of REINFORCE, a constant baseline was used (which is typically some measure of average long-term reward over all states). Even this type of simple measure can help in speeding up learning because it reduces the probabilities of less-than-average performers and increases the probabilities of more-than-average performers (rather than increasing both at differential rates). A constant choice of baseline does not affect the bias of the procedure, but it reduces the variance. A *state-specific* option for the baseline is the value $V^p(s)$ of the state s immediately *before* sampling action a. Such a choice results in the advantage $(Q^p(s, a) - V^p(s))$ becoming identical to the temporal difference error. This choice makes intuitive sense, because the temporal difference error contains *additional* information about the differential reward of an action beyond what we would know before choosing the action. Discussions on baseline choice may be found in [141, 160].

Consider an example of an Atari game-playing agent, in which a roll-out samples the move UP and output probability of UP was 0.2. Assume that the (constant) baseline is 0.17, and the long-term reward of the action is $+1$, since the game results in win (and there is no reward discount). Therefore, the score of every action in that roll-out is 0.83 (after subtracting the baseline). Then, the gain associated with all actions (output nodes of the neural network) other than UP at that time-step would be 0, and the gain associated with the output node corresponding to UP would be $0.83 \times \log(0.2)$. One can then backpropagate this gain in order to update the parameters of the neural network.

Adjustment with a state-specific baseline is easy to explain intuitively. Consider the example of a chess game between agents Alice and Bob. If we use a baseline of 0, then each move will only be credited with a reward corresponding to the final result, and the difference between good moves and bad moves will not be evident. In other words, we need to simulate a lot more games to differentiate positions. On the other hand, if we use the value of the state (before performing the action) as the baseline, then the (more refined) temporal difference error is used as the advantage of the action. In such a case, moves that have greater state-specific impact will be recognized with a higher advantage (within a single game). As a result, fewer games will be required for learning.

10.6.2   Combining Supervised Learning with Policy Gradients

Supervised learning is useful for initializing the weights of the policy network before applying reinforcement learning. For example, in a game of chess, one might have prior examples of expert moves that are already known to be good. In such a case, we simply perform gradient ascent with the same policy network, except that each expert move is assigned the fixed credit of 1 for evaluating the gradient according to Equation 10.23. This problem becomes identical to that of softmax classification, where the goal of the policy network is to predict the same move as the expert. One can sharpen the quality of the training data with some examples of bad moves with a negative credit obtained from computer evaluations. This approach would be considered supervised learning rather than reinforcement learning because we are simply using prior data, and not generating/simulating the data that we learn from (as is common in reinforcement learning). This general idea can be extended to any reinforcement learning setting, where some prior examples of actions and associated rewards are available. Supervised learning is extremely common in these settings for initialization because of the difficultly in obtaining high-quality data in the early stages of the process. Many published works also interleave supervised learning and reinforcement learning in order to achieve greater data efficiency [114].

10.6.3 Actor-Critic Methods

So far, we have discussed methods that are either dominated by *critics* or by *actors* in the following way:

1. The Q-learning and $TD(\lambda)$ methods work with the notion of a value function that is optimized. This value function is a critic, and the policy (e.g., ϵ-greedy) of the actor is directly derived from this critic. Therefore, the actor is subservient to the critic, and such methods are considered *critic-only* methods.

2. The policy-gradient methods do not use a value function at all, and they directly learn the probabilities of the policy actions. The values are often estimated using Monte Carlo sampling. Therefore, these methods are considered *actor-only* methods.

Note that the policy-gradient methods do need to evaluate the advantage of intermediate actions, and this estimation has so far been done with the use of Monte Carlo simulations. The main problem with Monte Carlo simulations is its high complexity and inability to use in an online setting.

However, it turns out that one can learn the advantage of intermediate actions using value function methods. As in the previous section, we use the notation $Q^p(s_t, a)$ to denote the value of action a, when the policy p followed by the policy network is used. Therefore, we would now have two coupled neural networks– a policy network and a Q-network. The policy network learns the probabilities of actions, and the Q-network learns the values $Q^p(s_t, a)$ of various actions in order to provide an estimation of the advantage to the policy network. Therefore, the policy network uses $Q^p(s_t, a)$ (with baseline adjustments) to weight its gradient ascent updates. The Q-network is updated using an on-policy update as in SARSA, where the policy is controlled by the policy network (rather than ϵ-greedy). The Q-network, however, does not directly decide the actions as in Q-learning, because the policy decisions are outside its control (beyond its role as a critic). Therefore, the policy network is the actor and the value network is the critic. To distinguish between the policy network and the Q-network, we will denote the parameter vector of the policy network by $\overline{\Theta}$, and that of the Q-network by $\overline{W}$.

We assume that the state at time stamp t is denoted by s_t, and the observable features of the state input to the neural network are denoted by $\overline{X}_t$. Therefore, we will use s_t and $\overline{X}_t$ interchangeably below. Consider a situation at the tth time-stamp, where the action a_t has been observed after state s_t with reward r_t. Then, the following sequence of steps is applied for the $(t + 1)$th step:

1. Sample the action a_{t+1} using the current state of the parameters in the policy network. Note that the current state is s_{t+1} because the action a_t is already observed.

2. Let $F(\overline{X}_t, \overline{W}, a_t) = \hat{Q}^p(s_t, a_t)$ represent the estimated value of $Q^p(s_t, a_t)$ by the Q-network using the observed representation $\overline{X}_t$ of the states and parameters $\overline{W}$. Estimate $Q^p(s_t, a_t)$ and $Q^p(s_{t+1}, a_{t+1})$ using the Q-network. Compute the TD-error δ_t as follows:

$$\delta_t = r_t + \gamma \hat{Q}^p(s_{t+1}, a_{t+1}) - \hat{Q}^p(s_t, a_t)$$
$$= r_t + \gamma F(\overline{X}_{t+1}, \overline{W}, a_{t+1}) - F(\overline{X}_t, \overline{W}, a_t)$$

3. **[Update policy network parameters]:** Let $P(\overline{X}_t, \overline{\Theta}, a_t)$ be the probability of the action a_t predicted by policy network. Update the parameters of the policy network as follows:

$$\overline{\Theta} \leftarrow \overline{\Theta} + \alpha \hat{Q}^p(s_t, a_t) \nabla_\Theta \log(P(\overline{X}_t, \overline{\Theta}, a_t))$$

Here, α is the learning rate for the policy network and the value of $\hat{Q}^p(s_t, a_t) = F(\overline{X}_t, \overline{W}, a_t)$ is obtained from the Q-network.

4. **[Update Q-Network parameters]:** Update the Q-network parameters as follows:

$$\overline{W} \Leftarrow \overline{W} + \beta \delta_t \nabla_W F(\overline{X}_t, \overline{W}, a_t)$$

Here, β is the learning rate for the Q-network. A caveat is that the learning rate of the Q-network is generally higher than that of the policy network.

The action a_{t+1} is then executed in order to observe state s_{t+2}, and the value of t is incremented. The next iteration of the approach is executed (by repeating the above steps) at this incremented value of t. The iterations are repeated, so that the approach is executed to convergence. The value of $\hat{Q}^p(s_t, a_t)$ is the same as the value $\hat{V}^p(s_{t+1})$.

If we use $\hat{V}^p(s_t)$ as the baseline, the advantage $\hat{A}^p(s_t, a_t)$ is defined by the following:

$$\hat{A}^p(s_t, a_t) = \hat{Q}^p(s_t, a_t) - \hat{V}^p(s_t)$$

This changes the updates as follows:

$$\overline{\Theta} \leftarrow \overline{\Theta} + \alpha \hat{A}^p(s_t, a_t) \nabla_\Theta \log(P(\overline{X}_t, \overline{\Theta}, a_t))$$

Note the replacement of $\hat{Q}(s_t, a_t)$ in the original algorithm description with $\hat{A}(s_t, a_t)$. In order to estimate the value $\hat{V}^p(s_t)$, one possibility is to maintain another set of parameters representing the value network (which is different from the Q-network). The TD-algorithm can be used to update the parameters of the value network. However, it turns out that a single value-network is enough. This is because we can use $r_t + \gamma \hat{V}^p(s_{t+1})$ in lieu of $\hat{Q}(s_t, a_t)$. This results in an advantage function, which is the same as the TD-error:

$$\hat{A}^p(s_t, a_t) = r_t + \gamma \hat{V}^p(s_{t+1}) - \hat{V}^p(s_t)$$

In other words, we need the single value-network (cf. Figure 10.5), which serves as the critic. The above approach can also be generalized to use the $TD(\lambda)$ algorithm at any value of λ.

10.6.4 Continuous Action Spaces

The methods discussed to this point were all associated with discrete action spaces. For example, in a video game, one might have a discrete set of choices such as whether to move the cursor up, down, left, and right. However, in a robotics application, one might have continuous action spaces, in which we wish to move the robot's arm a certain distance. One possibility is to discretize the action into a set of fine-grained intervals, and use the midpoint of the interval as the representative value. One can then treat the problem as one of discrete choice. However, this is not a particularly satisfying design choice. First, the ordering among the different choices will be lost by treating inherently ordered (numerical) values as categorical values. Second, it blows up the space of possible actions, especially if the action space is multidimensional (e.g., separate dimensions for distances moved by the robot's arm and leg). Such an approach can cause overfitting, and greatly increase the amount of data required for learning.

A commonly used approach is to allow the neural network to output the parameters of a continuous distribution (e.g., mean and standard deviation of Gaussian), and then sample from the parameters of that distribution in order to compute the value of the action in the next step. Therefore, the neural network will output the mean μ and standard deviation σ for the distance moved by the robotic arm, and the actual action a will be sampled from the Gaussian $\mathcal{N}(\mu, \sigma)$ with this parameter:

$$a \sim \mathcal{N}(\mu, \sigma) \qquad (10.25)$$

In this case, the action a represents the distance moved by the robot arm. The values of μ and σ can be learned using backpropagation. In some variations, σ is fixed up front as a hyper-parameter, with only the mean μ needing to be learned. The likelihood ratio trick also applies to this case, except that we use the logarithm of the density at a, rather than the discrete probability of the action a.

10.6.5 Advantages and Disadvantages of Policy Gradients

Policy gradient methods represent the most natural choice in applications like robotics that have continuous sequences of states and actions. For cases in which there are multidimensional and continuous action spaces, the number of possible combinations of actions can be very large. Since Q-learning methods require the computation of the maximum Q-value over all such actions, this step can turn out to be computationally intractable. Furthermore, policy gradient methods tend to be stable and have good convergence properties. However, policy gradient methods are susceptible to local minima. While Q-learning methods are less stable in terms of convergence behavior than are policy-gradient methods, and can sometimes oscillate around particular solutions, they have better capacity to reach near global optima.

10.7 Revisiting Monte Carlo Tree Search

Monte Carlo tree search has already been discussed in section 3 as a probabilistic alternative to the deterministic minimax trees that are used by conventional game-playing software (although the applicability is not restricted to games). In this section, we will revisit this method with a specific focus on the game of Go in order to provide a case study of how such methods are used. This will also provide a flavor of how rollout methods are often used in practical settings.

As discussed in Chapter 3, each node in the Monte Carlo tree corresponds to a state, and each branch corresponds to a possible action. The Monte Carlo tree search method is a variation on the Monte Carlo rollout method, where a tree is built explicitly in order to store the statistics of promising moves on tree branches, rather than with state-action pairs (as in Monte Carlo rollouts). The tree construction approach does not consolidate the statistics on duplicate state-action pairs within the tree (although it is possible to do so with some additional bookkeeping. The tree grows over time during the search as new states are encountered. The goal of the tree search is to select the best branch to recommend the predicted action of the agent. Each branch is associated with a value based on previous outcomes in tree search from that branch as well as an upper bound "bonus" that reduces with increased exploration. This value is used to set the priority of the branches during exploration. The learned goodness of a branch is adjusted after each exploration, so that branches leading to positive outcomes are favored in later explorations.

In the following, we will describe the Monte Carlo tree search used in *AlphaGo* as a case study for exposition for the game of Go. One can view this description as a more specific version of the approach described in Chapter 3. Assume that the probability $P(s, a)$ of each action (move) a at state (board position) s can be estimated using a policy network. At the same time, for each move we have a quantity $Q(s, a)$, which is the quality of the move a at state s. For example, the value of $Q(s, a)$ increases with increasing number of wins by following action a from state s in simulations. The *AlphaGo* system uses a more sophisticated algorithm that also incorporates some neural evaluations of the board position after a few moves (cf. section 10.8.1). Then, in each iteration, the "upper bound" $u(s, a)$ of the quality of the move a at state s is given by the following:

$$u(s, a) = Q(s, a) + K \cdot \frac{P(s, a) \sqrt{\sum_b N(s, b)}}{N(s, a) + 1}$$

(10.26)

Here, $N(s, a)$ is the number of times that the action a was followed from state s over the course of the Monte Carlo tree search. In other words, the upper bound is obtained by starting with the quality $Q(s, a)$, and adding a "bonus" to it that depends on $P(s, a)$ and the number of times that branch is followed. The idea of scaling $P(s, a)$ by the number of visits is to discourage frequently visited branches and encourage greater exploration. The Monte Carlo approach is based on the strategy of selecting the branch with the largest upper bound, as in multi-armed bandit methods (cf. section 10.2.3). Here, the second term on the right-hand side of Equation 10.26 plays the role of providing the confidence interval for computing the upper bound. As the branch is played more and more, the exploration "bonus" for that branch is reduced, because the width of its confidence interval drops. The hyperparameter K controls the degree of exploration. Large values of K increase the tendency of explore, whereas small values of K increase the tendency to exploit.

At any given state, the action a with the largest value of $u(s, a)$ is followed. This approach is applied recursively until following the optimal action does not lead to an existing node. This new state s' is now added to the tree as a leaf node with initialized values of each $N(s', a)$ and $Q(s', a)$ set to 0. Note that the simulation up to a leaf node is fully deterministic, and no randomization is involved because $P(s, a)$ and $Q(s, a)$ are deterministically computable. Monte Carlo simulations are used to estimate the value of the newly added leaf node s'. Specifically, Monte Carlo rollouts from the policy network (e.g., using $P(s, a)$ to sample actions) return either $+1$ or -1, depending on win or loss. After evaluating the leaf node, the values of $Q(s'', a'')$ and $N(s'', a'')$ on all edges (s'', a'') on the path from the current state s to the leaf s' are updated. The value of $Q(s'', a'')$ is maintained as the average value of all the evaluations at leaf nodes reached from that branch during the Monte Carlo tree search. After multiple searches have been performed from s, the most visited edge is selected as the relevant one, and is reported as the desired action.

Use in Bootstrapping

Traditionally, Monte Carlo tree search provides an improved estimate $Q(s, a)$ of the value of a state-action pair by performing repeated Monte Carlo rollouts. However, approaches that work with rollouts can often be implemented with bootstrapping instead of Monte Carlo rollouts (Intuition 10.5.1). Monte Carlo tree search provides an excellent alternative to n-step temporal-difference methods. One point about on-policy n-step temporal-difference methods is that they explore a single sequence of n-moves with the ϵ-greedy policy, and therefore tend to be too weak (with increased depth but not width of exploration). One way to strengthen them is to examine all possible n-sequences and use the optimal one with an off-policy technique (i.e., generalizing Bellman's 1-step approach). In fact, this was the approach used in Samuel's checkers program [155], which used the best option in the minimax tree for bootstrapping (and later referred to as *TD-Leaf* [17]). This results in increased complexity of exploring all possible n-sequences. Monte Carlo tree search can provide a robust alternative for bootstrapping, because it can explore multiple branches from a node to generate averaged target values. For example, the lookahead-based ground truth can use the averaged performance over all the explorations starting at a given node.

AlphaGo Zero [168] bootstraps policies rather than state values, which is extremely rare. *AlphaGo Zero* uses the relative visit probabilities of the branches at each node as *posterior* probabilities of the actions at that state. In other words, the visit counts of the various branches at a node are used to create visit probabilities. These posterior probabilities are improved over the probabilistic outputs of the policy network by virtue of the fact that the visit decisions use knowledge about the future (i.e., evaluations at deeper nodes of the Monte Carlo tree). The posterior probabilities are therefore bootstrapped as ground-truth values with respect to the policy network probabilities and used to update the weight parameters (cf. section 10.8.1.1).

10.8 Case Studies

In the following, we present case studies from real domains to showcase different reinforcement learning settings. We will present examples of reinforcement learning in *Go*, robotics, conversational systems, self-driving cars, and neural-network hyperparameter learning.

10.8.1 AlphaGo: Championship Level Play at Go

Go is a two-person board game like chess. The complexity of a two-person board game largely depends on the size of the board and the number of valid moves at each position. The simplest example of a board game is tic-tac-toe with a 3×3 board, and most humans can solve it optimally without the need for a computer. Chess is a significantly more complex game with an 8×8 board, although clever variations of the brute-force approach of *selectively* exploring the minimax tree of moves up to a certain depth can perform significantly better than the best human today. *Go* occurs at the extreme end of complexity because of its 19×19 board.

Players play with white or black *stones*, which are kept in bowls next to the *Go* board. An example of a *Go* board is shown in Figure 10.7. The game starts with an empty board,

Figure 10.7: Example of a *Go* board with stones

and it fills up as players put stones on the board. Black makes the first move and starts with 181 stones in her bowl, whereas white starts with 180 stones. The total number of junctions is equal to the total number of stones in the bowls of the two players. A player places a stone of her color in each move at a particular position (from the bowl), and does not move it once it is placed. A stone of the opponent can be captured by encircling it. The objective of the game is for the player to control a larger part of the board than her opponent by encircling it with her stones.

Whereas one can make about 35 possible moves (i.e., tree branch factor) in a particular position in chess, the average number of possible moves at a particular position in *Go* is 250, which is almost an order of magnitude larger. Furthermore, the average number of sequential moves (i.e., tree depth) of a game of *Go* is about 150, which is around twice as large as chess. All these aspects make *Go* a much harder candidate from the perspective of automated game-playing. The typical strategy of chess-playing software is to construct a minimax tree with all combinations of moves the players can make up to a certain depth, and then evaluate the final board positions with chess-specific heuristics (such as the amount of remaining material and the safety of various pieces). Suboptimal parts of the tree are pruned in a heuristic manner. This approach is simply a improved version of a brute-force strategy in which all possible positions are explored up to a given depth. The number of nodes in the minimax tree of *Go* is larger than the number of atoms in the observable universe, even at modest depths of analysis (20 moves for each player). As a result of the importance of spatial intuition in these settings, humans always perform better than brute force strategies at *Go*. The use of reinforcement learning in *Go* is much closer to what humans attempt to do. We rarely try to explore all possible combinations of moves; rather, we visually learn patterns on the board that are predictive of advantageous positions, and try to make moves in directions that are expected to improve our advantage.

The automated learning of spatial patterns that are predictive of good performance is achieved with a convolutional neural network. The state of the system is encoded in the board position at a particular point, although the board representation in *AlphaGo* includes some additional features about the status of junctions or the number of moves since a stone was played. Multiple such spatial maps are required in order to provide full knowledge of the state. For example, one feature map would represent the status of each intersection, another would encode the number of turns since a stone was played, and so on. Integer feature maps were encoded into multiple one-hot planes. Altogether, the game board could be represented using 48 binary planes of 19×19 pixels.

AlphaGo uses its win-loss experience with repeated game playing (both using the moves of expert players and with games played against itself) to learn good policies for moves in various positions with a policy network. Furthermore, the evaluation of each position on the *Go* board is achieved with a value network. Subsequently, Monte Carlo tree search is used for final inference. Therefore, *AlphaGo* is a multi-stage model, whose components are discussed in the following sections.

Policy Networks

The policy network takes as its input the aforementioned visual representation of the board, and outputs the probability of action a in state s. This output probability is denoted by $p(s, a)$. Note that the actions in the game of *Go* correspond to the probability of placing a stone at each legal position on the board. Therefore, the output layer uses the softmax activation. Two separate policy networks are trained using different approaches. The two networks were identical in structure, containing convolutional layers with ReLU nonlinearities. Each network contained 13 layers. Most of the convolutional layers convolve with 3×3 filters, except for the first and final convolutions. The first and final filters convolve with 5×5 and 1×1 filters, respectively. The convolutional layers were zero padded to maintain their size, and 192 filters were used. The ReLU nonlinearity was used, and no maxpooling was used in order to maintain the spatial footprint.

The networks were trained in the following two ways:

- *Supervised learning:* Randomly chosen samples from expert players were used as training data. The input was the state of the network, while the output was the action performed by the expert player. The score (advantage) of such a move was always +1, because the goal was to train the network to *imitate* expert moves, which is also referred to as *imitation learning*. Therefore, the neural network was backpropagated with the log-likelihood of the probability of the chosen move as its gain. This network is referred to as the SL-policy network. It is noteworthy that these supervised forms of imitation learning are often quite common in reinforcement learning for avoiding cold-start problems. However, subsequent work [167] showed that dispensing with this part of the learning was a better option.

- *Reinforcement learning:* In this case, reinforcement learning was used to train the network. One issue is that *Go* needs two opponents, and therefore the network was played against itself in order to generate the moves. The current network was always played against a randomly chosen network from a few iterations back, so that the reinforcement learning could have a pool of randomized opponents. The game was played until the very end, and then an advantage of +1 or −1 was associated with each move depending on win or loss. This data was then used to train the policy network. This network was referred to as the RL-policy network.

Note that these networks were already quite formidable *Go* players compared to state-of-the-art software, and they were combined with Monte Carlo tree search to strengthen them.

Value Networks

This network was also a convolutional neural network, which uses the state of the network as the input and the predicted score in $[−1, +1]$ as output, where +1 indicates a perfect probability of 1. The output is the predicted score of the next player, whether it is white or black, and therefore the input also encodes the "color" of the pieces in terms of "player" or "opponent" rather than white or black. The architecture of the value network was very similar to the policy network, except that there were some differences in terms of the input and output. The input contained an additional feature corresponding to whether the next player to play was white or black. The score was computed using a single tanh unit at the end, and therefore the value lies in the range $[-1, +1]$. The early convolutional layers of the value network are the same as those in the policy network, although an additional convolutional layer is added in layer 12. A fully connected layer with 256 units and ReLU activation follows the final convolutional layer. In order to train the network, one possibility is to use positions from a data set [223] of *Go* games. However, the preferred choice was to generate the data set using self-play with the SL-policy and RL-policy networks all the way to the end, so that the final outcomes were generated. The state-outcome pairs were used to train the convolutional neural network. Since the positions in a single game are correlated, using them sequentially in training causes overfitting. It was important to sample positions from different games in order to prevent overfitting caused by closely related training examples. Therefore, each training example was obtained from a distinct game of self-play.

Monte Carlo Tree Search

A simplified variant of Equation 10.26 was used for exploration, which is equivalent to setting $K = 1/\sqrt{\sum_b N(s, b)}$ at each node s. Section 10.7 described a version of the Monte Carlo tree search method in which only the RL-policy network is used for evaluating leaf nodes. In the case of *AlphaGo*, two approaches are combined. First, fast Monte Carlo rollouts were used from the leaf node to create evaluation e_1. While it is possible to use the policy network for rollout, *AlphaGo* trained a simplified softmax classifier with a database of human Go games and some hand-crafted features for faster speed of rollouts. Second, the value network created a separate evaluation e_2 of the leaf nodes. The final evaluation e is a convex combination of the two evaluations as $e = \beta e_1 + (1 - \beta)e_2$. The value of $\beta = 0.5$ provided the best performance, although using only the value network also provided closely matching performance (and a viable alternative). The most visited branch in Monte Carlo tree search was reported as the predicted move.

10.8.1.1 AlphaZero: Enhancements to Zero Human Knowledge

A later enhancement of the idea, referred to as *AlphaGo Zero* [167], removed the need for human expert moves (or an SL-network). Instead of separate policy and value networks, a single network outputs both the policy (i.e., action probabilities) $p(s, a)$ and the value $v(s)$ of the position. The cross-entropy loss on the output policy probabilities and the squared loss on the value output were added to create a single loss. Whereas the original version of *AlphaGo* used Monte Carlo tree search only for inference from trained networks, the zero-knowledge versions also use the visit counts in Monte Carlo tree search for training. One can view the visit count of each branch in tree search as a policy *improvement* operator over $p(s, a)$ by virtue of its lookahead-based exploration. This provides a basis for creating boot-strapped ground-truth values (Intuition 10.5.1) for neural network learning. While temporal difference learning bootstraps state values, this approach bootstraps visit counts for learning policies. The predicted probability of Monte Carlo tree search for action a in board state s is $\pi(s, a) \propto N(s, a)^{1/\tau}$, where τ is a temperature parameter. The value of $N(s, a)$ is computed using a similar Monte Carlo search algorithm as used for *AlphaGo*, where the *prior* probabilities $p(s, a)$ output by the neural network are used for computing

Equation 10.26. The value of $Q(s, a)$ in Equation 10.26 is set to the average value output $v(s')$ from the neural network of the newly created leaf nodes s' reached from state s.

AlphaGo Zero updates the neural network by bootstrapping $\pi(s, a)$ as a ground-truth, whereas ground-truth *state values* are generated with Monte Carlo simulations. At each state s, the probabilities $\pi(s, a)$, values $Q(s, a)$ and visit counts $N(s, a)$ are updated by running the Monte Carlo tree search procedure (repeatedly) starting at state s. The neural network from the previous iteration is used for selecting branches according to Equation 10.26 until a state is reached that does not exist in the tree or a terminal state is reached. For each non-existing state, a new leaf is added to the tree with its Q-values and visit values initialized to zero. The Q-values and visit counts of all edges on the path from s to the leaf node are updated based on leaf evaluation by the neural network (or by game rules for terminal states). After multiple searches starting from node s, the *posterior* probability $\pi(s, a)$ is used to sample an action for self-play and reach the next node s'. The entire procedure discussed in this paragraph is repeated at node s' to recursively obtain the next position s''. The game is recursively played to completion and the final value from $\{-1, +1\}$ is returned as the ground-truth value $z(s)$ of uniformly sampled states s on the game path. Note that $z(s)$ is defined from the perspective of the player at state s. The ground-truth values of the probabilities are already available in $\pi(s, a)$ for various values of a. Therefore, one can create a training instance for the neural network containing the input representation of state s, the bootstrapped ground-truth probabilities in $\pi(s, a)$, and the Monte Carlo ground-truth value $z(s)$. This training instance is used to update the neural network parameters. Therefore, if the probability and value outputs for the neural network are $p(s, a)$ and $v(s)$, respectively, the loss for a neural network with weight vector $\overline{W}$ is as follows:

$$L = [v(s) - z(s)]^2 - \sum_a \pi(s, a) \log[p(s, a)] + \lambda ||\overline{W}||^2 \qquad (10.27)$$

Here, $\lambda > 0$ is the regularization parameter.

Further advancements were proposed in the form of *Alpha Zero* [168], which could play multiple games such as *Go*, shogi, and chess. *AlphaZero* has handily defeated the best chess-playing software, *Stockfish*, and has also defeated the best shogi software (*Elmo*). The victory in chess was particularly unexpected by most top players, because it was always assumed that chess required too much domain knowledge for a reinforcement learning system to win over a system with hand-crafted evaluations.

Comments on Performance

AlphaGo has shown extraordinary performance against a variety of computer and human opponents. Against a variety of computer opponents, it won 494 out of 495 games [166]. Even when *AlphaGo* was handicapped by providing four free stones to the opponent, it won 77%, 86%, and 99% of the games played against (the software programs named) *Crazy Stone, Zen,* and *Pachi*, respectively. It also defeated notable human opponents, such as the European champion, the World champion, and the top-ranked player.

A more notable aspect of its performance was the way in which it achieved its victories. In several of its games, *AlphaGo* made many unconventional and brilliantly unorthodox moves, which would sometimes make sense only in hindsight after the victory of the program. There were cases in which the moves made by *AlphaGo* were contrary to conventional wisdom, but eventually revealed innovative insights acquired by *AlphaGo* during self-play. After this match, some top *Go* players reconsidered their approach to the entire game.

The performance of *Alpha Zero* in chess was similar, where it often made material sacrifices in order to incrementally improve its position and constrict its opponent. This type of behavior is a hallmark of human play and is very different from conventional chess software (which is already much better than humans). Unlike hand-crafted evaluations, it seemed to have no pre-conceived notions on the material values of pieces, or on when a king was safe in the center of the board. Furthermore, it discovered most well-known chess openings on its own using self-play, and seemed to have its own opinions on which ones were "better." In other words, it had the ability to discover knowledge on its own. A key difference of reinforcement learning from supervised learning is that *it has the ability to innovate beyond known knowledge through learning by reward-guided trial and error.* This behavior represents some promise in other applications.

10.8.2   Self-Learning Robots

Self-learning robots represent an important frontier in artificial intelligence, in which robots can be trained to perform various tasks such as locomotion, mechanical repairs, or object retrieval by using a reward-driven approach. For example, consider the case in which one has constructed a robot that is *physically* capable of locomotion (in terms of how it is constructed and the movement choices available to it), but it has to learn the precise *choice* of movements in order to keep itself balanced and move from point A to point B. As bipedal humans, we are able to walk and keep our balance naturally without even thinking about it, but this is not a simple matter for a bipedal robot in which an incorrect choice of joint movement could easily cause it to topple over. The problem becomes even more difficult when uncertain terrain and obstacles are placed in the way of a robot.

This type of problem is naturally suited to reinforcement learning, because it is easy to judge whether a robot is walking correctly, but it is hard to specify precise rules about what the robot should do in every possible situation. In the reward-driven approach of reinforcement learning, the robot is given (virtual) rewards every time it makes progress in locomotion from point A to point B. Otherwise, the robot is free to take any actions, and it is not pre-trained with knowledge about the specific choice of actions that would help keep it balanced and walk. In other words, the robot is not seeded with any knowledge of what walking looks like (beyond the fact that it will be rewarded for using its available actions for making progress from point A to point B). This is a classical example of reinforcement learning, because the robot now needs to learn the specific sequence of actions to take in order to earn the goal-driven rewards. Although we use locomotion as a specific example in this case, this general principle applies to any type of learning in robots. For example, a second problem is that of teaching a robot manipulation tasks such as grasping an object or screwing the cap on a bottle. In the following, we will provide a brief discussion of both cases.

10.8.2.1   Deep Learning of Locomotion Skills

In this case, locomotion skills were taught to virtual robots [160], in which the robot was simulated with the *MuJoCo* physics engine [212], which stands for *Multi-Joint Dynamics with Contact*. It is a physics engine aiming to facilitate research and development in robotics, biomechanics, graphics, and animation, where fast and accurate simulation is needed without having to construct an actual robot. Both a humanoid and a quadruped robot were used. An example of the biped model is shown in Figure 10.8. The advantage of this type of simulation is that it is inexpensive to work with a virtual simulation, and one avoids the natural safety and expense issues that arise with the physical damages in an experimentation framework that is likely to be marred by high levels of mistakes/accidents. On the flip

Figure 10.8: Example of the virtual humanoid robot. Original image is available at [212]

side, a physical model provides more realistic results. In general, a simulation can often be used for smaller scale testing before building a physical model.

The humanoid model has 33 state dimensions and 10 actuated degrees of freedom, while the quadruped model has 29 state dimensions and 8 actuated degrees of freedom. Models were rewarded for forward progress, although episodes were terminated when the center of mass of the robot fell below a certain point. The actions of the robot were controlled by joint torques. A number of features were available to the robot, such as sensors providing the positions of obstacles, the joint positions, angles, and so on. These features were fed into the neural networks. Two neural networks were used; one was used for value estimation, and the other was used for policy estimation. Therefore, a policy gradient method was used in which the value network played the role of estimating the advantage. Such an approach is an instantiation of an actor-critic method.

A feed-forward neural network was used with three hidden layers, with 100, 50, and 25 tanh units, respectively. The approach in [160] requires the estimation of both a policy function and a value function, and the same architecture was used in both cases for the hidden layers. However, the value estimator required only one output, whereas the policy estimator required as many outputs as the number of actions. Therefore, the main difference between the two architectures was in terms of how the output layer and the loss function was designed. The generalized advantage estimator (GAE) was used in combination with trust-based policy optimization (TRPO). The bibliographic notes contain pointers to specific details of these methods. On training the neural network for 1000 iterations with reinforcement learning, the robot learned to walk with a visually pleasing gait. A video of the final results of the robot walking is available at [211]. Similar results were also later released by Google DeepMind with more extensive abilities of avoiding obstacles or other challenges [75].

10.8.2.2 Deep Learning of Visuomotor Skills

A second and interesting case of reinforcement learning is provided in [114], in which a robot was trained for several household tasks such as placing a coat hanger on a rack, inserting a block into a shape-sorting cube, fitting the claw of a toy hammer under a nail with various grasps, and screwing a cap onto a bottle. Examples of these tasks are illustrated in

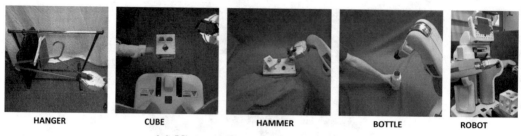

(a) Visuomotor tasks learned by robot

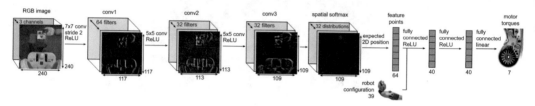

(b) Architecture of the convolutional neural network

Figure 10.9: Deep learning of visuomotor skills. These figures appear in [114]. (©2016 Sergey Levine, Chelsea Finn, Trevor Darrell, and Pieter Abbeel)

Figure 10.9(a) along with an image of the robot. The actions were 7-dimensional joint motor torque commands, and each action required a sequence of commands in order to optimally perform the task. In this case, an actual physical model of a robot was used for training. A camera image was used by the robot in order to locate the objects and manipulate them. This camera image can be considered the robot's eyes, and the convolutional neural network used by the robot works on the same conceptual principle as the visual cortex (based on Hubel and Wiesel's experiments). Even though this setting seems very different from that of the Atari video games at first sight, there are significant similarities in terms of how image frames can help in mapping to policy actions. For example, the Atari setting also works with a convolutional neural network on the raw pixels. However, there were some additional inputs here, corresponding to the robot and object positions. These tasks require a high level of learning in visual perception, coordination, and contact dynamics, all of which need to learned automatically.

A natural approach is to use a convolutional neural network for mapping image frames to actions. As in the case of Atari games, spatial features need to be learned in the layers of the convolutional neural network that are suitable for earning the relevant rewards in a task-sensitive manner. The convolutional neural network had 7 layers and 92,000 parameters. The first three layers were convolutional layers, the fourth layer was a spatial softmax, and the fifth layer was a fixed transformation from spatial feature maps to a concise set of two coordinates. The idea was to apply a softmax function to the responses across the spatial feature map. This provides a probability of each position in the feature map. The expected position using this probability distribution provides the 2-dimensional coordinate, which is referred to as a *feature point*. Note that each spatial feature map in the convolution layer creates a feature point. The feature point can be viewed as a kind of soft argmax over the spatial probability distribution. The fifth layer was quite different from what one normally sees in a convolutional neural network, and was designed to create a precise representation of the visual scene that was suitable for feedback control. The spatial feature points are concatenated with the robot's configuration, which is an additional input occurring only after the convolution layers. This concatenated feature set is fed into two fully connected layers, each with 40 rectified units, followed by linear connections to the torques. Note that only the observations corresponding to the camera were fed to the first layer of the convolutional neural network, and the observations corresponding to the robot state were fed to the first fully connected layer. This is because the convolutional layers cannot make much use of the robot states, and it makes sense to concatenate the state-centric inputs after the visual inputs have been processed by the convolutional layers. The entire network contained about 92,000 parameters, of which 86,000 were in the convolutional layers. The architecture of the convolutional neural network is shown in Figure 10.9(b). The observations consist of the RGB camera image, joint encoder readings, velocities, and end-effector pose.

The full robot states contained between 14 and 32 dimensions, such as the joint angles, end-effector pose, object positions, and their velocities. This provided a practical notion of a state. As in all policy-based methods, the outputs correspond to the various actions (motor torques). One interesting aspect of the approach discussed in [114] is that it transforms the reinforcement learning problem into supervised learning. A *guided policy search* method was used, which is not discussed in this chapter. This approach converts portions of the reinforcement learning problem into supervised learning. Interested readers are referred to [114], where a video of the performance of the robot (trained using this system) may also be found.

10.8.3 Self-Driving Cars

As in the case of the robot locomotion task, the car is rewarded for progressing from point A to point B without causing accidents or other undesirable road incidents. The car is equipped with various types of video, audio, proximity, and motion sensors in order to record observations. The objective of the reinforcement learning system is for the car to go from point A to point B safely irrespective of road conditions.

Driving is a task for which it is hard to specify the proper rules of action in every situation; on the other hand, it is relatively easy to judge when one is driving correctly. This is precisely the setting that is well suited to reinforcement learning. Although a fully self-driving car would have a vast array of components corresponding to inputs and sensors of various types, we focus on a simplified setting in which a single camera is used [25, 26]. This system is instructive because it shows that even a single front-facing camera is sufficient to accomplish quite a lot when paired with reinforcement learning. Interestingly, this work was inspired by the 1989 work of Pomerleau [142], who built the *Autonomous Land Vehicle in a Neural Network (ALVINN)* system, and the main difference from the work done over 25 years back was one of increased data and computational power. In addition, the work uses some advances in convolutional neural networks for modeling. Therefore, this work showcases the great importance of increased data and computational power in building reinforcement learning systems.

The training data was collected by driving in a wide variety of roads and conditions. The data was collected primarily from central New Jersey, although highway data was also collected from Illinois, Michigan, Pennsylvania, and New York. Although a single front-facing camera in the driver position was used as the primary data source for making decisions, the training phase used two additional cameras at other positions in the front to collect rotated and shifted images. These auxiliary cameras, which were not used for final decision making, were however useful for collecting additional data. The placement of the additional cameras ensured that their images were shifted and rotated, and therefore they could be

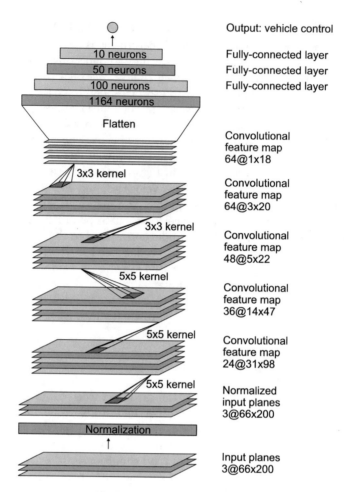

Figure 10.10: The neural network architecture of the control system in the self-driving car discussed in [25] (Courtesy NVIDIA)

used to train the network to recognize cases where the car position had been compromised. In short, these cameras were useful for data augmentation. The neural network was trained to minimize the error between the steering command output by the network and the command output by the human driver. Note that this approach tends to make the approach closer to supervised learning rather than reinforcement learning. These types of learning methods are also referred to as *imitation learning* [158]. Imitation learning is often used as a first step to buffer the cold-start inherent in reinforcement learning systems.

Scenarios involving imitation learning are often similar to those involving reinforcement learning. It is relatively easy to use reinforcement setting in this scenario by giving a reward when the car makes progress without human intervention. On the other hand, if the car either does not make progress or requires human intervention, it is penalized. However, this does not seem to be the way in which the self-driving system of [25, 26] is trained. One issue with settings like self-driving cars is that one always has to account for safety issues during training. Although published details on most of the available self-driving cars are limited, it seems that supervised learning has been the method of choice compared to reinforcement learning in this setting. Nevertheless, the differences between using supervised learning and reinforcement learning are not significant in terms of the broader architecture of the neural network that would be useful. A general discussion of reinforcement learning in the context of self-driving cars may be found in [216].

The convolutional neural network architecture is shown in Figure 10.10. The network consists of 9 layers, including a normalization layer, 5 convolutional layers, and 3 fully connected layers. The first convolutional layer used a 5×5 filter with a stride of 2. The next two convolutional layers each used non-strided convolution with a 3×3 filter. These convolutional layers were followed with three fully connected layers. The final output value was a control value, corresponding to the inverse turning radius. The network had 27 million connections and $250,000$ parameters. Specific details of how the deep neural network performs the steering are provided in [26].

The resulting car was tested both in simulation and in actual road conditions. A human driver was always present in the road tests to perform interventions when necessary. On this basis, a measure was computed on the percentage of time that human intervention was required. It was found that the vehicle was autonomous 98% of the time. A video demonstration of this type of autonomous driving is available in [215]. Some interesting observations were obtained by visualizing the activation maps of the trained convolutional neural network (based on the methodology discussed in Chapter 8). In particular, it was observed that the features were heavily biased towards learning aspects of the image that were important to driving. In the case of unpaved roads, the feature activation maps were able to detect the outlines of the roads. On the other hand, if the car was located in a forest, the feature activation maps were full of noise. Note that this does not happen in a convolutional neural network that is trained on a general-purpose image data set, because the feature activation maps would typically contain useful characteristics of trees, leaves, and so on. This difference in the two cases is because the convolutional network of the self-driving setting is trained in a goal-driven matter, and it learns to detect features that are relevant to driving. The specific characteristics of the trees in a forest are not relevant to driving.

10.9 Weaknesses of Reinforcement Learning

Simplifying the design of highly complex learning algorithms with reinforcement learning can sometimes have unexpected effects. By virtue of the fact that reinforcement learning systems have larger levels of freedom than other learning systems, it naturally leads to some safety related concerns. While biological greed is a powerful factor in human intelligence, it is also a source of many undesirable aspects of human behavior. The simplicity that is the greatest strength of reward-driven learning is also its greatest pitfall in biological systems. Simulating such systems therefore results in similar pitfalls from the perspective of artificial intelligence. For example, poorly designed rewards can lead to unforeseen consequences, because of the exploratory way in which the system learns its actions. Reinforcement learning systems can frequently learn unknown "cheats" and "hacks" in imperfectly designed video games, which tells us a cautionary tale of what might happen in a less-than-perfect real world. Robots learn that simply pretending to screw caps on bottles can earn faster rewards, as long as the human or automated evaluator is fooled by the action. In other words, the design of the reward function is sometimes not a simple matter.

Furthermore, a system might try to earn virtual rewards in an "unethical" way. For example, a cleaning robot might try to earn rewards by first creating messes and then cleaning them [13]. One can imagine even darker scenarios for robot nurses. Interestingly, these types of behaviors are sometimes also exhibited by humans. These undesirable similarities are a direct result of simplifying the learning process in machines by leveraging the simple greed-centric principles with which biological organisms learn. Striving for simplicity results in ceding more control to the machine, which can have unexpected effects. In some cases, there are ethical dilemmas in even designing the reward function. For example, if it becomes inevitable that an accident is going to occur, should a self-driving car save its driver or two pedestrians? Most humans would save themselves in this setting as a matter of reflexive biological instinct; however, it is an entirely different matter to incentivize a learning system to do so. At the same time, it would be hard to convince a human operator to trust a vehicle where her safety is not the first priority for the learning system. Another issue is that human operators have significantly higher thresholds on the safety requirements for systems that they cannot control as opposed to those that they can directly control (e.g., the process of manually driving a car). Therefore, a higher safety rating of a self-driving vehicle might still not be sufficient to convince a human operator to use the system (unless the difference is sufficiently large). Reinforcement learning systems are also susceptible to the ways in which their human operators interact with them and manipulate the effects of their underlying reward function; there have been occasions where a chatbot was taught to make offensive or racist remarks.

Learning systems have a harder time in generalizing their experiences to new situations. This problem is referred to as *distributional shift*. For example, a self-driving car trained in one country might perform poorly in another. Similarly, the exploratory actions in reinforcement learning can sometimes be dangerous. Imagine a robot trying to solder wires in an electronic device, where the wires are surrounded with fragile electronic components. Trying exploratory actions in this setting is fraught with perils. These issues tell us that we cannot build AI systems with no regard to safety. Indeed, some organizations like *OpenAI* [214] have taken the lead in these matters of ensuring safety. Some of these issues are also discussed in [13] with broader frameworks of possible solutions. In many cases, it seems that the human would have to be involved in the loop to some extent in order to ensure safety [157].

Finally, reinforcement learning algorithms require a *lot* of data, and work particularly well in *closed systems*, where it is easy to simulate and generate sufficient data. For example, one can generate unlimited amounts of data via simulation in games and virtual robots, but it is much harder to generate sufficient data with actual robots. As a result, while virtual simulators of robots do exceedingly well with reinforcement learning, it is harder to achieve similar results with real robots (where failed trials are expensive because of physical damage). The data-hungry nature of reinforcement learning continues to be a very serious impediment in its deployment in real-world applications.

10.10 Summary

This chapter studies the problem of reinforcement learning in which agents interact with the environment in a reward-driven manner in order to learn the optimal actions. There are several classes of reinforcement learning methods, of which the Q-learning methods and the policy-driven methods are the most common. Policy-driven methods have become increasingly popular in recent years. Many of these methods are end-to-end systems that integrate deep neural networks to take in sensory inputs and learn policies that optimize rewards. Reinforcement learning algorithms are used in many settings like playing video or other types of games, robotics, and self-driving cars. The ability of these algorithms to learn via experimentation often leads to innovative solutions that are not possible with other forms of learning. Reinforcement learning algorithms also pose unique challenges associated with safety because of the oversimplification of the learning process with reward functions.

10.11 Further Reading

An excellent overview on reinforcement learning may be found in the book by Sutton and Barto [182]. A number of surveys on reinforcement learning are available at [115]. David Silver's lectures on reinforcement learning are freely available on *YouTube* [213]. The method of temporal differences was proposed by Samuel in the context of a checkers program [155] and formalized by Sutton [181]. Q-learning was proposed by Watkins in [198]. The SARSA algorithm was introduced in [152]. The work in [185] developed TD-Gammon, which was a backgammon playing program.

In recent years, policy gradients have become more popular than Q-learning methods. Likelihood methods for policy gradients were pioneered by the REINFORCE algorithm [203]. A number of analytical results on this class of algorithms are provided in [183]. Policy gradients have been used in for learning in the game of *Go* [166], although the overall approach combines a number of different elements. Surveys are also available on specific types of reinforcement learning methods like actor-critic methods [68].

Monte Carlo tree search was proposed in [102]. Subsequently, it was used in the game of *Go* [166, 167]. A survey on these methods may be found in [31]. Later versions of *AlphaGo* dispensed with the supervised portions of learning, adapted to chess and shogi, and performed better with zero initial knowledge [167, 168]. Some TD-learning methods for chess, such as *NeuroChess* [187], *KnightCap* [17], and *Giraffe* [108] have been explored, but were not as successful as conventional engines. Several methods for training self-learning robots are presented in [114, 159, 160].

10.12 Exercises

1. The chapter gives a proof of the likelihood ratio trick (cf. Equation 10.23) for the case in which the action a is discrete. Generalize this result to continuous-valued actions.

2. Throughout this chapter, a neural network, referred to as the policy network, has been used in order to implement the policy gradient. Discuss the importance of the choice of network architecture in different settings.

3. You have two slot machines, each of which has an array of 100 lights. The probability distribution of the reward from playing each machine is an unknown (and possibly machine-specific) function of the pattern of lights that are currently lit up. Playing a slot machine changes its light pattern in some well-defined but unknown way. Discuss why this problem is more difficult than the multi-armed bandit problem. Design a deep learning solution to optimally choose machines in each trial that will maximize the average reward per trial at steady-state.

4. Consider the well-known game of rock-paper-scissors. Human players often try to use the history of previous moves to guess the next move. Would you use a Q-learning or a policy-based method to learn to play this game? Why? Now consider a situation in which a human player samples one of the three moves with a probability that is an unknown function of the history of 10 previous moves of each side. Propose a deep learning method that is designed to play with such an opponent. Would a well-designed

deep learning method have an advantage over this human player? What policy should a human player use to ensure probabilistic parity with a deep learning opponent?

5. Consider the game of tic-tac-toe in which a reward drawn from $\{-1, 0, +1\}$ is given at the end of the game. Suppose you learn the values of all states (assuming optimal play from both sides). Discuss why states in non-terminal positions will have non-zero values. What does this tell you about credit-assignment of intermediate moves to the reward value received at the end?

6. Write a Q-learning implementation that learns the value of each state-action pair for a game of tic-tac-toe by repeatedly playing against human opponents. No function approximators are used and therefore the entire table of state-action pairs is learned using Equation 10.4. Assume that you can initialize each Q-value to 0 in the table.

7. The two-step TD-error is defined as follows:

$$\delta_t^{(2)} = r_t + \gamma r_{t+1} + \gamma^2 V(s_{t+2}) - V(s_t)$$

(a) Propose a TD-learning algorithm for the 2-step case.

(b) Propose an on-policy n-step learning algorithm like SARSA. Show that the update is truncated variant of Equation 10.18 after setting $\lambda = 1$. What happens for the case when $n = \infty$?

(c) Propose an off-policy n-step learning algorithm like Q-learning and discuss its advantages/disadvantages with respect to (b).

Chapter 11

Probabilistic Graphical Models

"He who ignores the law of probabilities challenges an adversary that is seldom beaten." – Ambrose Bierce

11.1   Introduction

A probabilistic graphical model is a model in which the dependence between random variables is captured by a graph. The probabilistic graphical can be considered a special type of computational graph (cf. Chapter 7) in which the variables in the nodes correspond to random variables. Each variable in this probabilistic computational graph is generated in a conditional manner, based on the variables in the nodes that were generated in its incoming nodes. These conditional probabilities can be set either by domain experts, or they can be learned in a data-driven manner in the form parameters of probability distributions on the edges. These two different methods correspond to two primary schools of thought in artificial intelligence:

1. **Deductive school of thought:** In this case, the conditional probabilities on the edges are set by domain experts. This results in a probabilistic graph without a distinctive training phase, and which is used primarily for making inferences. An example of such a model is the *Bayesian network*.

2. **Inductive school of thought:** In this case, the edges typically correspond to *parameterized probability distributions*, in which the underlying parameters need to be learned in a data-driven manner. Each variable in the node is the outcome of sampling from the probability distribution defined on the incoming edges. The primary distinction from conventional neural networks is the probabilistic nature of the computation on the edge. Examples of such models include *Markov random fields, conditional random fields*, and *restricted Boltzmann machines*. Furthermore, many conventional models in machine learning like the expectation maximization algorithm, Bayes classifier, and logistic regression can also be considered special cases of such models. Since

© Springer Nature Switzerland AG 2021
C. C. Aggarwal, *Artificial Intelligence*, https://doi.org/10.1007/978-3-030-72357-6_11

most inductive models in machine learning can be reduced to a computational graph (cf. Chapter 7), the use of any kind of probabilistic model in the intermediate stages creates some kind of probabilistic graphical model.

The earliest types of probabilistic models were the Bayesian networks, in which probabilities were set in a domain-specific manner by domain experts. These networks were proposed by Judea Pearl [140]. Such networks are also referred to as *inference networks* or *causal networks*, and they were used as tools for probabilistic inference with the use of expert domain knowledge.

In later years, inductive forms of probabilistic networks became more popular. Most generative and discriminative models in traditional machine learning can also be considered forms of probabilistic graphical models, although they were not proposed as graphical models in the early years. However, such models are often depicted as graphical models in the form of *plate diagrams*, which can be considered as rudimentary forms of computational graphs.

Like general computational graphs in the deterministic settings, probabilistic graphical models can be either undirected or directed. In directed models, the inference only occurs in one direction, whereas in undirected models, the inference may occur in both directions. All forms of undirected computational graphs are always harder to train than directed models, because they implicitly contain cycles. Furthermore, many directed graphical models also contain cycles, which makes them harder to train. It is noteworthy that undirected computational graphs are extremely rare in the deterministic setting — all neural networks are directed computational graphs without cycles. However, in the probabilistic case, computational graphs with cycles are more common.

This chapter is organized as follows. The next section introduces basic forms of probabilistic graphical models, which correspond to Bayesian networks. The use of rudimentary probabilistic graphical models to explain conventional models in machine learning is discussed in Section 11.3. Boltzmann machines are introduced in Section 11.4. The specific case of restricted Boltzmann machines, which are special cases of Boltzmann machines, is discussed in Section 11.5. Applications of restricted Boltzmann machines are discussed in Section 11.6. A summary is given in Section 11.7.

11.2   Bayesian Networks

Bayesian networks are also referred to as causal networks or inference networks, in which the edges contain conditional probabilities and the variables in nodes are generated based on the values of the variables in incoming nodes. The values of these conditional probabilities, which associate the values of the variables at incoming nodes to those in outgoing nodes of edges are often set in a domain-specific manner by experts. It is noteworthy that the conditional probabilities may not necessarily be associated with individual edges. Bayesian networks are *directed acyclic graphs* (like neural networks), and the incoming nodes of a given node are referred to as its *parents*. The outgoing nodes of a given node are referred to as its *children*. All nodes from which a given node is reachable are referred to as its *ancestors*, and all nodes reachable from a given node are referred to as its *descendants*. An ancestor of a node that is not an immediate parent is referred to as an *indirect ancestor*. In the following, we will introduce Bayesian networks with Boolean variables in the nodes, and therefore each variable has a probability of taking on a *True* or *False* value. This probability is conditional on the combination of Boolean values at all its incoming nodes. It is noteworthy that the directed and acyclic nature of the graph is specific to conventional forms of Bayesian networks, and it does not apply to other types of graphical models. The directed acyclic nature of Bayesian

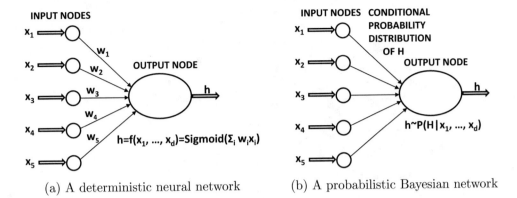

(a) A deterministic neural network   (b) A probabilistic Bayesian network

Figure 11.1: Comparing deterministic neural networks and Bayesian networks

networks is necessary for them to be used effectively and efficiently as causal models; the presence of cycles always makes it difficult to make inferences of node values from other node values because of the cyclic nature of the relationships. However, for many forms of probabilistic graphical models in inductive models, such as restricted Boltzmann machines, the graph is undirected, and therefore cycles exist within the graphical model.

The relationship between neural networks and Bayesian networks is quite noteworthy. Whereas neural networks use deterministic functions in the context of directed acyclic graphs, Bayesian networks use functions that effectively *sample* values from probability distributions. Consider a node with variable h in a neural network, in which the incoming values are $x_1, x_2, \ldots x_d$. In such a case, the neural network computes the following deterministic function $f(\cdot)$:

$$h = f(x_1, \ldots, x_d)$$

Examples of such functions include the linear operator, the sigmoid operator, or a composition of the two operators. On the other hand, in the case of a Bayesian network, the function is defined as a sampling operator, wherein the value h is sampled from a conditional probability distribution $P(H|x_1, \ldots, x_d)$ of the random variable H, given its inputs:

$$h \sim P(H|x_1, \ldots, x_d)$$

This natural way of defining Bayesian networks, and its relationship to neural networks is shown in Figure 11.1. It is noteworthy that this type of randomized network will produce different outputs for the same input, when one runs through the network multiple times. This is because each output is now an instantiation from a probability distribution rather than a deterministic function like $f(x_1, \ldots, x_d)$.

An important point is that conventional Bayesian networks assume that the probability distributions are fully specified in a domain-specific way. This assumption is obviously not practical, as the size of the Bayesian network increases. Joint probability distributions become increasingly cumbersome to specify in a domain-specific network, as the number of incoming nodes at a given node increases. As we will see later, much of machine learning moves beyond this original definition of Bayesian networks, and tries to learn the probabilities in a data-driven manner. Conventional Bayesian networks, in which probabilities are defined in a domain-specific manner, are essentially inference engines performing causal inference, and they correspond to the testing phase of traditional machine learning algorithms. When a learning component is included, such methods morph into the vast array

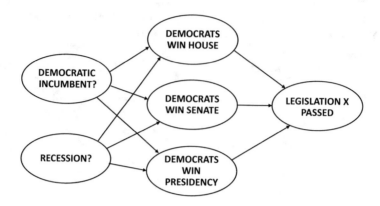

Figure 11.2: An example of a Bayesian network

of probabilistic graphical models used in machine learning. In fact, many of these probabilistic graphical models, such as restricted Boltzmann machines, are widely regarded as
probabilistic versions of neural networks.

The above way of defining the probabilistic functions computed at nodes implies that
Bayesian networks satisfy the *local Markov property*, which is also referred to simply as the
Markov property. According to this property, the value of the variable in a given node depends only on its immediate parents rather than its ancestors that are not directly connected
to it:

Definition 11.2.1 (Local Markov Property) *The probability of a variable is conditionally independent of its indirect ancestors, given the values of the variables in its immediate
parents.*

It is noteworthy that the Markov property is used in all types of sequence-centric models
in machine learning, such as *hidden Markov models*, and Bayesian networks represent the
natural generalization of these ideas, when the relationships among variables are not purely
sequential. For example, consider a Bayesian network, in which an edge exists from variable
h_1 to variable h_2, and also from variable h_2 to h_3, but no edge exists from variable h_1 to
h_3. In such a case, the local Markov property implies the following:

$$P(h_3|h_2, h_1) = P(h_3|h_2)$$

Note, however, that the above condition would not hold, if an edge exists from h_1 to h_3.
The local Markov property is true in the case of Bayesian networks, simply because the
value of a variable is defined as a sampling function of the variable values in its *directly*
incoming nodes; after all indirect ancestors do define the values of the parents, which are
the only inputs a node needs. Each node in the Bayesian network is associated with a
table of probabilities containing 2^k probability values, where k is the indegree of the node.
Therefore, a table of probability values is associated with each node, corresponding to a
joint probability distribution at the node, rather than a single conditional probability at
each edge.

In order to understand the local Markov property of a Bayesian network, we will start
with a simple example of a Bayesian network. Consider the network shown in Figure 11.2,
in which a network containing six nodes are shown. The Bayesian network in Figure 11.2
illustrates a particular scenario within the US political process, corresponding to the two

parties, which are the Democrats and the Republicans. The modeled scenario corresponds to the probability of a particular piece of legislation, denoted X, which is favored by Democrats but opposed by Republicans. It is assumed that this piece of legislation will be up for consideration after the election, which decides the House, Senate, and the Presidency. The possibility of the legislation being passed is inherently uncertain, since one does not know the outcome of the election. In particular, the probability of the legislation being passed heavily depends on whether Democrats or Republicans take control[1] of the House, Senate, or the Presidency after the election. It is assumed that a simple Bayesian model suggests that the probability of either party winning the house, senate, or presidency depends on two factors. The first factor is whether a Republican or Democrat incumbent president is *currently* in office (i.e., before the election), and the second is on whether or not the economy is doing well. In general, voters tend to blame the party belonging to the incumbent president in a poor economy, which has a high predictive power towards all electoral races in the United States. Therefore, one uses domain knowledge to model the probabilities of various outcomes of the presidential, house, and senate elections using a table containing $2^2 = 4$ entries, since each node has an indegree of 2. The corresponding table of probabilities is shown below:

Democratic incumbent	Recession?	Democrats win house	Democrats win senate	Democrats win presidency
True	True	0.1	0.05	0.25
True	False	0.95	0.55	0.8
False	True	0.9	0.65	0.9
False	False	0.25	0.1	0.3

Note that the two Boolean values are inputs to the model, based on the facts that are already known. The numerical values in table are probabilities that lie between 0 and 1. Of course, this is a rather crude model, but is illustrative of the point we are trying to make. If we assume that the legislation X is favored by Democrats, but opposed by Republicans, then it would have the highest chance of being passed when Republicans control a large number of these institutions after an election. The corresponding table is illustrated below and it has the $2^3 = 8$ entries, since the legislation node contains three incoming edges. Therefore, the next stage of the modeling takes as its input the outputs (probabilistic outcomes) of the previous stage, just as a neural network would do. However, the inputs are Boolean, since we work with the *outcomes* from the previous stage:

Democrats win house	Democrats win senate	Democrats win presidency	Probability that legislation passes
True	True	True	0.8
True	True	False	0.7
True	False	True	0.2
True	False	False	0.1
False	True	True	0.6
False	True	False	0.4
False	False	True	0.1
False	False	False	0.05

These probabilities can then be used to sample an outcome of the final output. This approach can be repeated multiple times to create multiple outputs. These multiple outputs can be converted into a probability that the legislation will be passed.

---

[1]The House and the Senate are the two institutions of the legislative branch of the US Government, containing elected representatives.

11.3 Rudimentary Probabilistic Models in Machine Learning

Many of the traditional machine learning models we have seen in earlier chapters are special cases of probabilistic graphical models. In these cases, the edges correspond to parameterized probability distributions and the training phase corresponds to the process of learning the parameters of these probability distributions. In this point of view, the expectation maximization algorithm, the Bayes classifier, and the logistic regression classifier can be cast as rudimentary forms of probabilistic graphical models. A key point is that the training of such models is simpler than the more general forms of probabilistic graphical models discussed in later sections.

In order to understand how rudimentary models in machine learning can be cast as probabilistic graphical models, we will introduce the notion of *plate diagrams*, which are frequently used in order to represent these different types of models. Plate diagrams show how the different variables in the models are generated, conditional on one another.

Although a plate diagram is not quite the same as a computational graph, it is conceptually related; in fact, plate diagrams can be used in many cases in order to quickly construct an equivalent computational graph. The basic way in which plate diagrams are constructed is shown in Figure 11.3(a). Each node in a plate diagram contains a variable (or vector of variables), just like a computational graph. The plate diagram shows dependencies between the different variables, when they generate one another. Note that this is exactly what a Bayesian network does. Plate diagrams clearly distinguish between hidden and visible variables by shading the visible nodes and not shading the hidden nodes. An edge between two nodes shows the probabilistic dependency between two nodes. When a node has many incoming edges, the variable in it is generated conditional on the variables in all the incoming nodes. Furthermore, when the same generative process is repeated multiple times, it is shown with a plate [see lower right illustration of Figure 11.3(a)] with the number of instantiations of the variable indicated within the plate.

Further examples of plate diagrams are shown in Figures 11.3(b), (c), and (d). The diagram of Figure 11.3(b) shows the clustering algorithm with expectation-maximization, and it requires two generative steps. First, the component of the mixture is selected, and then a data point is generated from a Gaussian distribution that is conditional on this chosen variable. The number of times that the generative process is repeated defines the number of data points. This number is included in the lower-right corner of the plate diagram of Figure 11.3(b). Note that this generative process is discussed in Section 9.3.4 of Chapter 9. The expectation-maximization algorithm is closely related to the Bayes classifier (cf. Section 6.7 of Chapter 6), where each class is analogous to a cluster. The generative process ie *exactly* the same as that of the expectation maximization algorithm, except that the class variable is visible in this case. This plate diagram is shown in Figure 11.3(c). Note that the main difference between Figures 11.3(b) and (c) is that the variable corresponding to the class/cluster identifier is not shaded in the case of clustering, but it is shaded in the case of classification. This is because the cluster identifier of data points is not known a priori in clustering, whereas it is known in classification (for training data points). Finally, the plate diagram for logistic regression is shown in Figure 11.3(d). Note that in this case, the difference from the Bayes classifier is that ordering of the generation of feature variables and class variable is reversed. This is because the class variable is generated conditional on the feature variables in the case of logistic regression. Furthermore, the nature of the input parameters is different; in the case of logistic regression, the weight parameter is input in

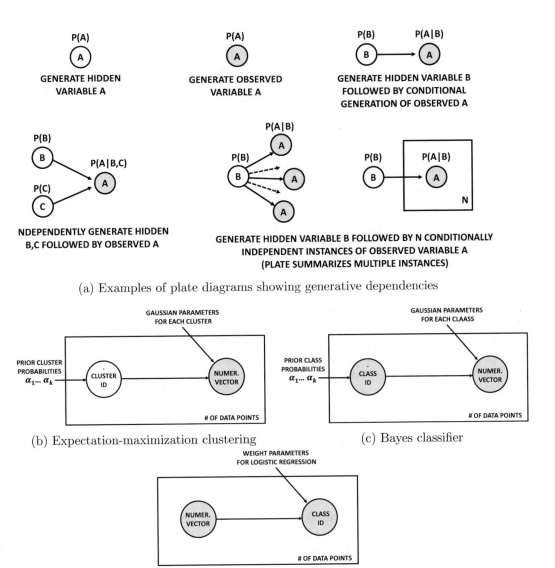

(a) Examples of plate diagrams showing generative dependencies

(b) Expectation-maximization clustering   (c) Bayes classifier

(d) Logistic Regression

Figure 11.3: The basic scheme for plate diagrams is shown in (a). Examples of plate diagrams for the expectation-maximization algorithm, Bayes classifier, and logistic regression, are shown in (b), (c), and (d)

order to generate the class variable. Both variables are shaded in this case, as they are visible during generation.

The discussion in this section shows that many rudimentary models of machine learning are special cases of probabilistic graphical models, although they are rarely viewed in this way. This is analogous to the fact that many machine learning models like support vector machines are special cases of deterministic neural networks. Therefore, expectation-maximization clustering, the Bayes classifier, and logistic regression can all be viewed as special cases of probabilistic graphical models. Many forms of matrix factorization, such as *probabilistic latent semantic analysis* and *latent Dirichlet allocation* can also be considered rudimentary special cases of probabilistic graphical models. Some models like logistic regression have the distinction of also being considered special cases of deterministic neural networks, when a sigmoid activation function is used in the output node. However, all these models are relatively rudimentary, since the structure of the graph is relatively simple and is of a directed nature. In the next section, we will introduce some undirected probabilistic graphical models, which are much more challenging.

11.4   The Boltzmann Machine

A Boltzmann machine is an undirected network, in which the q units (or neurons) are indexed by values drawn from $\{1 \ldots q\}$. Each connection is of the form (i, j), where each i and j is a neuron drawn from $\{1 \ldots d\}$. Each connection (i, j) is undirected, and is associated with a weight $w_{ij} = w_{ji}$. Although all pairs of nodes are assumed to have connections between them, setting w_{ij} to 0 has the effect of dropping the connection (i, j). The weight w_{ii} is set to 0, and therefore there are no self-loops. Each neuron i is associated with state s_i. An important assumption in the Boltzmann machine is that each s_i is a binary value drawn from $\{0, 1\}$, although one can use other conventions such as $\{-1, +1\}$. The ith node also has a bias b_i associated with it; large values of b_i encourage the ith state to be 1. The Boltzmann machine is an undirected model of symmetric relationships between attributes, and therefore the weights always satisfy $w_{ij} = w_{ji}$. The *visible* values of the states represent the binary attribute values from a training example, whereas the hidden states are probabilistic in nature, and are used in order to enable the construction of more complex models. The weights in the Boltzmann machine are its parameters; large positive weights between pairs of states are indicative of high degree of positive correlation in state values, whereas large negative weights are indicative of high negative correlation.

   Throughout this section, we assume that the Boltzmann machine contains a total of $q = (m + d)$ states, where d is the number of visible states and m is the number of hidden states. Therefore, it is assumed that the training data set contains d binary values corresponding to the relevant attributes. A particular state configuration is defined by the value of the state vector $\overline{s} = (s_1 \ldots s_q)$. If one explicitly wants to demarcate the visible and hidden states in $\overline{s}$, then the state vector $\overline{s}$ can be written as the pair $(\overline{v}, \overline{h})$, where $\overline{v}$ denotes the set of visible units and $\overline{h}$ denotes the set of hidden units. The states in $(\overline{v}, \overline{h})$ represent exactly the same set as $\overline{s} = \{s_1 \ldots s_q\}$, except that the visible and hidden units are explicitly demarcated in the former.

   The objective function of a Boltzmann machine is also referred to as its *energy function*, which is analogous to the loss function of a traditional feed-forward neural network. The energy function of a Boltzmann machine is set up in such a way that minimizing this function encourages nodes pairs connected with large positive weights to have similar states, and pairs connected with large negative weights to have different states. The training phase of a Boltzmann machine, therefore, learns the weights of edges in order to minimize the energy when the visible states in the Boltzmann machine are fixed to the binary attribute values in the individual training points. Therefore, learning the weights of the Boltzmann machine implicitly builds an unsupervised *model* of the training data set. The energy E of a particular combination of states $\overline{s} = (s_1, \ldots s_q)$ of the Boltzmann machine can be defined as follows:

$$E = -\sum_i b_i s_i - \sum_{i,j:i<j} w_{ij} s_i s_j \tag{11.1}$$

The term $-b_i s_i$ encourages units with large biases to be on. Similarly, the term $-w_{ij} s_i s_j$ encourages s_i and s_j to be similar when $w_{ij} > 0$. In other words, positive weights will cause state "attraction" and negative weights will cause state "repulsion."

The energy gap of the ith unit is defined as the difference in energy between its two configurations (with other states being fixed to pre-defined values):

$$\Delta E_i = E_{s_i=0} - E_{s_i=1} = b_i + \sum_{j:j\neq i} w_{ij} s_j \tag{11.2}$$

A Boltzmann machine assigns a *probability* to s_i depending on the energy gap. Positive energy gaps are assigned probabilities that are larger than 0.5. The probability of state s_i is defined by applying the sigmoid function to the energy gap:

$$P(s_i = 1 | s_1, \ldots s_{i-1}, s_{i+1}, s_q) = \frac{1}{1 + \exp(-\Delta E_i)} \tag{11.3}$$

Note that the state s_i is now a Bernoulli random variable and a zero energy gap leads to a probability of 0.5 for each binary outcome of the state.

For a particular set of parameters w_{ij} and b_i, the Boltzmann machine defines a probability distribution over various state configurations. The energy of a particular configuration $\bar{s} = (\bar{v}, \bar{h})$ is denoted by $E(\bar{s}) = E([\bar{v}, \bar{h}])$, and is defined as follows:

$$E(\bar{s}) = -\sum_i b_i s_i - \sum_{i,j:i<j} w_{ij} s_i s_j \tag{11.4}$$

However, these configurations are only probabilistically known in the case of the Boltzmann machine (according to Equation 11.3). The conditional distribution of Equation 11.3 follows from a more fundamental definition of the unconditional probability $P(\bar{s})$ of a particular configuration $\bar{s}$:

$$P(\bar{s}) \propto \exp(-E(\bar{s})) = \frac{1}{Z} \exp(-E(\bar{s})) \tag{11.5}$$

The normalization factor Z is defined so that the probabilities over all possible configurations sum to 1:

$$Z = \sum_{\bar{s}} \exp(-E(\bar{s})) \tag{11.6}$$

The normalization factor Z is also referred to as the *partition function*. In general, the explicit computation of the partition function is hard, because it contains an exponential number of terms corresponding to all possible configurations of states. Because of the intractability of the partition function, exact computation of $P(\bar{s}) = P(\bar{v}, \bar{h})$ is not possible. Nevertheless, the computation of many types of conditional probabilities (e.g., $P(\bar{v}|\bar{h})$) is possible, because such conditional probabilities are ratios and the intractable normalization factor gets canceled out from the computation. For example, the conditional probability of Equation 11.3 follows from the more fundamental definition of the probability of a configuration (cf. Equation 11.5) as follows:

$$P(s_i = 1 | s_1, \ldots s_{i-1}, s_{i+1}, s_q) = \frac{P(s_1, \ldots s_{i-1}, \overbrace{1}^{s_i}, s_{i+1}, s_q)}{P(s_1, \ldots s_{i-1}, \underbrace{1}_{s_i}, s_{i+1}, s_q) + P(s_1, \ldots s_{i-1}, \underbrace{0}_{s_i}, s_{i+1}, s_q)}$$

$$= \frac{\exp(-E_{s_i=1})}{\exp(-E_{s_i=1}) + \exp(-E_{s_i=0})} = \frac{1}{1 + \exp(E_{s_i=1} - E_{s_i=0})}$$

$$= \frac{1}{1 + \exp(-\Delta E_i)} = \text{Sigmoid}(\Delta E_i)$$

This is the same condition as Equation 11.5. One can also see that the logistic sigmoid function finds its roots in notions of energy from statistical physics.

One way of thinking about the benefit of setting these states probabilistically is that we can now sample from these states to create new data points that look like the original data. This makes Boltzmann machines probabilistic models rather than deterministic ones. Many generative models in machine learning (e.g., Gaussian mixture models for clustering) use a sequential process of first sampling the hidden state(s) from a prior, and then generating visible observations conditionally on the hidden state(s). This is not the case in the Boltzmann machine, in which the dependence between all pairs of states is *undirected*; the visible states depend as much on the hidden states as the hidden states depend on visible states. As a result, the generation of data with a Boltzmann machine can be more challenging than in many other generative models.

11.4.1   How a Boltzmann Machine Generates Data

In a Boltzmann machine, the dynamics of the data generation is complicated by the circular dependencies among the states based on Equation 11.3. Therefore, we need an iterative process to generate sample data points from the Boltzmann machine so that Equation 11.3 is satisfied for all states. A Boltzmann machine iteratively samples the states using a conditional distribution generated from the state values in the previous iteration until *thermal equilibrium* is reached. The notion of thermal equilibrium means that the observed frequencies of sampling various attribute values represent their long-term steady-state probability distributions. The process of reaching thermal equilibrium works as follows. We start at a random set of states, use Equation 11.3 to compute their conditional probabilities, and then sample the values of the states again using these probabilities. Note that we can iteratively generate s_i by using $P(s_i|s_1 \ldots s_{i-1}, s_{i+1}, \ldots s_q)$ in Equation 11.3. After running this process for a long time, the sampled values of the visible states provide us with random samples of generated data points. The time required to reach thermal equilibrium is referred to as the *burn-in time* of the procedure. This approach is referred to as *Gibbs sampling* or *Markov Chain Monte Carlo (MCMC) sampling*.

At thermal equilibrium, the generated points will represent the model captured by the Boltzmann machine. Note that the dimensions in the generated data points will be correlated with one another depending on the weights between various states. States with large weights between them will tend to be heavily correlated. For example, in a text-mining application in which the states correspond to the presence of words, there will be correlations among words belonging to a topic. Therefore, if a Boltzmann machine has been trained properly on a text data set, it will generate vectors containing these types of word correlations at thermal equilibrium, even when the states are randomly initialized. It is noticeable that even generating a set of data points with the Boltzmann machine is a more complicated process compared to many other probabilistic models. For example, generating data points from a Gaussian mixture model only requires to sample points directly from the probability distribution of a sampled mixture component. On the other hand, the undirected nature of the Boltzmann machine forces us to run the process to thermal equilibrium just to generate samples. It is, therefore, an even more difficult to task to learn the weights between states for a given training data set.

11.4.2   Learning the Weights of a Boltzmann Machine

In a Boltzmann machine, we want to learn the weights in such a way so as to maximize the log-likelihood of the specific training data set at hand. The log-likelihoods of individual states are computed by using the logarithm of the probabilities in Equation 11.5. Therefore, by taking the logarithm of Equation 11.5, we obtain the following:

$$\log[P(\overline{s})] = -E(\overline{s}) - \log(Z) \tag{11.7}$$

Therefore, computing $\frac{\partial \log[P(\overline{s})]}{\partial w_{ij}}$ requires the computation of the negative derivative of the energy, although we have an additional term involving the partition function. The energy function of Equation 11.4 is linear in the weight w_{ij} with coefficient of $-s_i s_j$. Therefore, the partial derivative of the energy with respect to the weight w_{ij} is $-s_i s_j$. As a result, one can show the following:

$$\frac{\partial \log[P(\overline{s})]}{\partial w_{ij}} = \langle s_i, s_j \rangle_{data} - \langle s_i, s_j \rangle_{model} \tag{11.8}$$

Here, $\langle s_i, s_j \rangle_{data}$ represents the averaged value of $s_i s_j$ obtained by running the generative process of Section 11.4.1, when the visible states are clamped to attribute values in a training point. The averaging is done over a mini-batch of training points. Similarly, $\langle s_i, s_j \rangle_{model}$ represents the averaged value of $s_i s_j$ at thermal equilibrium without fixing visible states to training points and simply running the generative process of Section 11.4.1. In this case, the averaging is done over multiple instances of running the process to thermal equilibrium. Intuitively, we want to strengthen the weights of edges between states that tend to be turned on together when the visible states are fixed to the training data points. This is precisely what is achieved by the update above, which uses the data- and model-centric difference in the value of $\langle s_i, s_j \rangle$. From the above discussion, it is clear that two types of samples need to be generated in order to perform the updates:

1. **Data-centric samples:** The first type of sample fixes the visible states to a randomly chosen vector from the training data set. The hidden states are initialized to random values drawn from Bernoulli distribution with probability 0.5. Then the probability of each hidden state is recomputed according to Equation 11.3. Samples of the hidden states are regenerated from these probabilities. This process is repeated for a while, so that thermal equilibrium is reached. The values of the hidden variables at this point provide the required samples. Note that the visible states are clamped to the corresponding attributes of the relevant training data vector, and therefore they do not need to be sampled.

2. **Model samples:** The second type of sample does not put any constraints on fixing states to training data points, and one simply wants samples from the unrestricted model. The approach is the same as discussed above, except that both the visible and hidden states are initialized to random values, and updates are continuously performed until thermal equilibrium is reached.

These samples help us create an update rule for the weights. From the first type of sample, one can compute $\langle s_i, s_j \rangle_{data}$, which represents the correlations between the states of nodes i and j, when the visible vectors are fixed to a vector in the training data $\mathcal{D}$ and the hidden states are allowed to vary. Since a mini-batch of training vectors is used, one obtains multiple samples of the state vectors. The value of $\langle s_i, s_j \rangle$ is computed as the average product over all

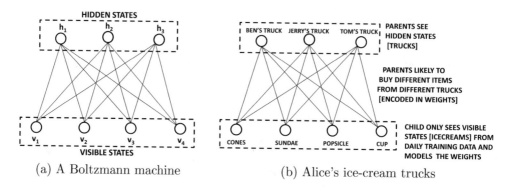

(a) A Boltzmann machine   (b) Alice's ice-cream trucks

Figure 11.4: A Restricted Boltzmann machine. Note the *restriction* of there being no inter-actions among either visible or hidden units

such state vectors that are obtained from Gibbs sampling. Similarly, one can estimate the value of $\langle s_i, s_j \rangle_{model}$ using the average product of s_i and s_j from the model-centric samples obtained from Gibbs sampling. Once these values have been computed, the following update is used:

$$w_{ij} \Leftarrow w_{ij} + \alpha \quad \underbrace{(\langle s_i, s_j \rangle_{data} - \langle s_i, s_j \rangle_{model})}_{\text{Partial derivative of log probability}} \quad (11.9)$$

The update rule for the bias is similar, except that the state s_j is set to 1. One can achieve this by using a dummy bias unit that is visible and is connected to all states:

$$b_i \Leftarrow b_i + \alpha \left(\langle s_i, 1 \rangle_{data} - \langle s_i, 1 \rangle_{model} \right) \quad (11.10)$$

Note that the value of $\langle s_i, 1 \rangle$ is simply the average of the sampled values of s_i for a mini-batch of training examples from either the data-centric samples or the model-centric samples.

The main problem with the aforementioned update rule is that it is slow in practice. This is because of the Monte Carlo sampling procedure, which requires a large number of samples in order to reach thermal equilibrium. There are faster approximations to this tedious process. In the next section, we will discuss this approach in the context of a simplified version of the Boltzmann machine, which is referred to as the *restricted* Boltzmann machine.

11.5 Restricted Boltzmann Machines

In the Boltzmann machine, the connections among hidden and visible units can be arbi-trary. For example, two hidden states might contain edges between them, and so might two visible states. This type of generalized assumption creates unnecessary complexity. A natural special case of the Boltzmann machine is the *restricted* Boltzmann machine (RBM), which is bipartite, and the connections are allowed only between hidden and visible units. An example of a restricted Boltzmann machine is shown in Figure 11.4(a). In this partic-ular example, there are three hidden nodes and four visible nodes. Each hidden state is connected to one or more visible states, although there are no connections between pairs of hidden states, and between pairs of visible states. The restricted Boltzmann machine is also referred to as a *harmonium* [170].

We assume that the hidden units are $h_1 \ldots h_m$ and the visible units are $v_1 \ldots v_d$. The bias associated with the visible node v_i be denoted by $b_i^{(v)}$, and the bias associated with hidden node h_j is denoted by $b_j^{(h)}$. Note the superscripts in order to distinguish between the biases of visible and hidden nodes. The weight of the edge between visible node v_i and hidden node h_j is denoted by w_{ij}. The notations for the weights are also slightly different for the restricted Boltzmann machine (compared to the Boltzmann machine) because the hidden and visible units are indexed separately. For example, we no longer have $w_{ij} = w_{ji}$ because the first index i always belongs to a visible node and the second index j belongs to a hidden node. It is important to keep these notational differences in mind while extrapolating the equations from the previous section.

In order to provide better interpretability, we will use a running example throughout this section, which we refer to as the example of "Alice's ice-cream trucks" based on the Boltzmann machine in Figure 11.4(b). Imagine a situation in which the training data corresponds to four bits representing the ice-creams received by Alice from her parents each day. These represent the visible states in our example. Therefore, Alice can collect 4-dimensional training points, as she receives (between 0 and 4) ice-creams of different types each day. However, the ice-creams are bought for Alice by her parents from one[2] or more of three trucks shown as the hidden states in the same figure. The identity of these trucks is hidden from Alice, although she knows that there are three trucks from which her parents procure the ice-creams (and more than one truck can be used to construct a single day's ice-cream set). Alice's parents are indecisive people, and their decision-making process is unusual because they change their mind about the selected ice-creams after selecting the trucks and vice versa. The likelihood of a particular ice-cream being picked depends on the trucks selected as well as the weights to these trucks. Similarly, the likelihood of a truck being selected depends on the ice-creams that one intends to buy and the same weights. Therefore, Alice's parents can keep changing their mind about selecting ice-creams after selecting trucks and about selecting trucks after selecting ice-creams (for a while) until they reach a final decision each day. As we will see, this *circular* relationship is the characteristic of undirected models, and process used by Alice's parents is similar to Gibbs sampling.

The use of the bipartite restriction greatly simplifies inference algorithms in RBMs, while retaining the application-centric power of the approach. If we know all the values of the visible units (as is common when a training data point is provided), the probabilities of the hidden units can be computed in one step without having to go through the laborious process of Gibbs sampling. For example, the probability of each hidden unit taking on the value of 1 can be written directly as a logistic function of the values of visible units. In other words, we can apply Equation 11.3 to the restricted Boltzmann machine to obtain the following:

$$P(h_j = 1|\overline{v}) = \frac{1}{1 + \exp(-b_j^{(h)} - \sum_{i=1}^{d} v_i w_{ij})} \tag{11.11}$$

This result follows directly from Equation 11.3, which relates the state probabilities to the energy gap ΔE_j between $h_j = 0$ and $h_j = 1$. The value of ΔE_j is $b_j + \sum_i v_i w_{ij}$ when the visible states are observed. This relationship is also useful in creating a reduced representation of each training vector, once the weights have been learned. Specifically, for a Boltzmann machine with m hidden units, one can set the value of the jth hidden value to the probability computed in Equation 11.11. Note that such an approach provides a real-

---

[2]This example is tricky in terms of semantic interpretability for the case in which no trucks are selected. Even in that case, the probabilities of various ice-creams turn out to be non-zero depending on the bias. One can explain such cases by adding a dummy truck that is always selected.

valued reduced representation of the binary data. One can also write the above equation using a sigmoid function:

$$P(h_j = 1|\overline{v}) = \text{Sigmoid}\left(b_j^{(h)} + \sum_{i=1}^{d} v_i w_{ij}\right) \tag{11.12}$$

One can also use a sample of the hidden states to generate the data points in one step. This is because the relationship between the visible units and the hidden units is similar in the undirected and bipartite architecture of the RBM. In other words, we can use Equation 11.3 to obtain the following:

$$P(v_i = 1|\overline{h}) = \frac{1}{1 + \exp(-b_i^{(v)} - \sum_{j=1}^{m} h_j w_{ij})} \tag{11.13}$$

One can also express this probability in terms of the sigmoid function:

$$P(v_i = 1|\overline{h}) = \text{Sigmoid}\left(b_i^{(v)} + \sum_{j=1}^{m} h_j w_{ij}\right) \tag{11.14}$$

One nice consequence of using the sigmoid is that it is often possible to create a closely related feed-forward network with sigmoid activation units in which the weights learned by the Boltzmann machine are leveraged in a directed computation with input-output mappings. The weights of this network are then fine-tuned with backpropagation. We will give examples of this approach in the application section.

Note that the weights encode the affinities between the visible and hidden states. A large positive weight implies that the two states are likely to be on together. For example, in Figure 11.4(b), it might be possible that the parents are more likely to buy cones and sundae from Ben's truck, whereas they are more likely to buy popsicles and cups from Tom's truck. These propensities are encoded in the weights, which regulate both visible state selection and hidden state selection in a circular way. The *circular* nature of the relationship creates challenges, because the relationship between ice-cream choice and truck choice runs both ways; it is the raison d'etre for Gibb's sampling. Although Alice might not know which trucks the ice-creams are coming from, she will notice the resulting correlations among the bits in the training data. In fact, if the weights of the RBM are known by Alice, she can use Gibb's sampling to generate 4-bit points representing "typical" examples of ice-creams she will receive on future days. Even the weights of the model can be learned by Alice from examples, which is the essence of an unsupervised generative model. Given the fact that there are 3 hidden states (trucks) and enough examples of 4-dimensional training data points, Alice can learn the relevant weights and biases between the visible ice-creams and hidden trucks. An algorithm for doing this is discussed in the next section.

11.5.1 Training the RBM

Computation of the weights of the RBM is achieved using a similar type of learning rule as that used for Boltzmann machines. In particular, it is possible to create an efficient algorithm based on mini-batches. The weights w_{ij} are initialized to small values. For the current set of weights w_{ij}, they are updated as follows:

- *Positive phase:* The algorithm uses a mini-batch of training instances, and computes the probability of the state of each hidden unit in exactly one step using Equation 11.11. Then a single sample of the state of each hidden unit is generated from this probability. This process is repeated for each element in a mini-batch of training instances. The correlation between these different training instances of v_i and generated instances of h_j is computed; it is denoted by $\langle v_i, h_j \rangle_{pos}$. This correlation is essentially the average product between each such pair of visible and hidden units.

- *Negative phase:*

- *Negative phase:* In the negative phase, the algorithm starts with randomly initialized states and uses Equations 11.11 and 11.13 repeatedly to thermal equilibrium to compute the probabilities of the visible and hidden units. These probabilities are used to draw samples if v_i and h_j, and the entire process is repeated multiple times. The multiple samples are used to compute the average product $\langle v_i, h_j \rangle_{neg}$ in the same way as the positive phase.

- One can then use the same type of update as is used in Boltzmann machines:

$$w_{ij} \Leftarrow w_{ij} + \alpha \left(\langle v_i, h_j \rangle_{pos} - \langle v_i, h_j \rangle_{neg} \right)$$
$$b_i^{(v)} \Leftarrow b_i^{(v)} + \alpha \left(\langle v_i, 1 \rangle_{pos} - \langle v_i, 1 \rangle_{neg} \right)$$
$$b_j^{(h)} \Leftarrow b_j^{(h)} + \alpha \left(\langle 1, h_j \rangle_{pos} - \langle 1, h_j \rangle_{neg} \right)$$

Here, $\alpha > 0$ denotes the learning rate. Each $\langle v_i, h_j \rangle$ is estimated by averaging the product of v_i and h_j over the mini-batch, although the values of v_i and h_j are computed in different ways in the positive and negative phases, respectively. Furthermore, $\langle v_i, 1 \rangle$ represents the average value of v_i in the mini-batch, and $\langle 1, h_j \rangle$ represents the average value of h_j in the mini-batch.

It is helpful to interpret the updates above in terms of Alice's trucks in Figure 11.4(b). When the weights of certain visible bits (e.g., cones and sundae) are highly correlated, the above updates will tend to push the weights in directions that these correlations can be explained by the weights between the trucks and the ice-creams. For example, if the cones and sundae are highly correlated but all other correlations are very weak, it can be explained by high weights between each of these two types of ice-creams and a single truck. In practice, the correlations will be far more complex, as will the patterns of the underlying weights.

11.5.2 Contrastive Divergence Algorithm

One issue with the above approach is the time required to reach thermal equilibrium and generate negative samples. However, it turns out that it is possible to run the Monte Carlo sampling for only a short time *starting by fixing the visible states to a training data point from the mini-batch* and still obtain a good approximation of the gradient. The fastest variant of the contrastive divergence approach uses a *single* additional iteration of Monte Carlo sampling (over what is done in the positive phase) in order to generate the samples of the hidden and visible states. First, the hidden states are generated by fixing the visible units to a training point (which is already accomplished in the positive phase), and then the visible units are generated again (exactly once) from these hidden states using Monte Carlo sampling. The values of the visible units are used as the sampled states in lieu of the ones obtained at thermal equilibrium. The hidden units are generated again using these visible units. Thus, the main difference between the positive and negative phase is only of the number of iterations that one runs the approach starting with the same initialization of visible states to training points. In the positive phase, we use only half an iteration of simply computing the hidden states. In the negative phase, we use at least one *additional* iteration (so that visible states are recomputed from hidden states and hidden states generated again). This difference in the number of iterations is what causes the contrastive divergence between the state distributions in the two cases. The intuition is that an increased number of iterations causes the distribution to move away (i.e., diverge) from the data-conditioned states to what is proposed by the current weight vector. Therefore, the value of $(\langle v_i, h_j \rangle_{pos} - \langle v_i, h_j \rangle_{neg})$ in the update quantifies the amount of contrastive divergence. This fastest variant of the contrastive divergence algorithm is referred to as CD_1 because it uses a single (additional) iteration in order to generate the negative samples. Of course, using such an approach is only an approximation to the true gradient. One can improve the accuracy of contrastive divergence by increasing the number of additional iterations to k, in which the data is reconstructed k times. This approach is referred to as CD_k. Increased values of k lead to better gradients at the expense of speed.

In the early iterations, using CD_1 is good enough, although it might not be helpful in later phases. Therefore, a natural approach is to progressively increase the value of k, while applying CD_k in training. One can summarize this process as follows:

1. In the early phase of gradient-descent, the weights are very inexact. In each iteration, only one additional step of contrastive divergence is used. One step is sufficient at this point because only a rough direction of descent is needed to improve the inexact weights. Therefore, even if CD_1 is executed, one will be able to obtain a good direction in most cases.

2. As the gradient descent nears a better solution, higher accuracy is needed. Therefore, two or three steps of contrastive divergence are used (i.e., CD_2 or CD_3). In general, one can double the number of Markov chain steps after a fixed number of gradient descent steps. Another approach advocated in [173] is to create the value of k in CD_k by 1 after every 10,000 steps. The maximum value of k used in [173] was 20.

The contrastive divergence algorithm can be extended to many other variations of the RBM. An excellent practical guide for training restricted Boltzmann machines may be found in [78]. This guide discusses several practical issues such as initialization, tuning, and updates. In the following, we provide a brief overview of some of these practical issues.

11.5.3 Practical Issues and Improvisations

There are several practical issues in training the RBM with contrastive divergence. Although we have always assumed that the Monte Carlo sampling procedure generates binary samples, this is not quite the case. Some of the iterations of the Monte Carlo sampling directly use *computed* probabilities (cf. Equations 11.11 and 11.13), rather than *sampled* binary values. This is done in order to reduce the noise in training, because probability values retain more information than binary samples. However, there are some differences between how hidden states and visible states are treated:

- *Improvisations in sampling hidden states:* The final iteration of CD_k computes hidden states as probability values according to Equation 11.11 for positive and negative samples. Therefore, the value of h_j used for computing $\langle v_i, h_j \rangle_{pos} - \langle v_i, h_j \rangle_{neg}$ would always be a real value for both positive and negative samples. This real value is a fraction because of the use of the sigmoid function in Equation 11.11.

- *Improvisations in sampling visible states:* Therefore, the improvisations for Monte Carlo sampling of visible states are always associated with the computation of $\langle v_i, h_j \rangle_{neg}$ rather than $\langle v_i, h_j \rangle_{pos}$ because visible states are always fixed to the training data. For the negative samples, the Monte Carlo procedure *always* computes probability values of visible states according to Equation 11.13 over *all* iterations rather than using 0-1 values. This is not the case for the hidden states, which are always binary until the very last iteration.

Using probability values iteratively rather than sampled binary values is technically incorrect, and does not reach correct thermal equilibrium. However, the contrastive divergence algorithm is an approximation anyway, and this type of approach reduces significant noise at the expense of some theoretical incorrectness. Noise reduction is a result of the fact that the probabilistic outputs are closer to expected values.

The weights can be initialized from a Gaussian distribution with zero mean and a standard deviation of 0.01. Large values of the initial weights can speed up the learning, but might lead to a model that is slightly worse in the end. The visible biases are initialized to $\log(p_i/(1 - p_i))$, where p_i is the fraction of data points in which the ith dimension takes on the value of 1. The values of the hidden biases are initialized to 0.

The size of the mini-batch should be somewhere between 10 and 100. The order of the examples should be randomized. For cases in which class labels are associated with examples, the mini-batch should be selected in such a way that the proportion of labels in the batch is approximately the same as the whole data.

11.6 Applications of Restricted Boltzmann Machines

In this section, we will study several applications of restricted Boltzmann machines. These methods have been very successful for a variety of unsupervised applications, although they are also used for supervised applications. When using an RBM in a real-world application, a mapping from input to output is often required, whereas a vanilla RBM is only designed to learn probability distributions. The input-to-output mapping is often achieved by constructing a feed-forward network with weights derived from the learned RBM. In other words, one can often derive a traditional neural network that is *associated* with the original RBM.

Here, we will like to discuss the differences between the notions of the *state* of a node in the RBM, and the *activation* of that node in the associated neural network. The state of a node is a binary value sampled from the Bernoulli probabilities defined by Equations 11.11 and 11.13. On the other hand, the activation of a node in the associated neural network is the probability value derived from the use of the sigmoid function in Equations 11.11 and 11.13. Many applications use the activations in the nodes of the associated neural network, rather than the states in the original RBM after the training. Note that the final step in the contrastive divergence algorithm also leverages the activations of the nodes rather than the states while updating the weights. In practical settings, the activations are more information-rich and are therefore useful. The use of activations is consistent with traditional neural network architectures, in which backpropagation can be used. The use of a final phase of backpropagation is crucial in being able to apply the approach to supervised applications. In most cases, the critical role of the RBM is to perform unsupervised feature learning. Therefore, the role of the RBM is often only one of pretraining in the case of supervised learning. In fact, pretraining is one of the important historical contributions of the RBM.

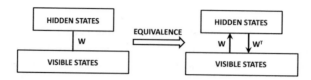

(a) Equivalence of directed and undirected relationships

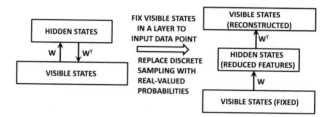

(b) Discrete graphical model to approximate real-valued neural network

Figure 11.5: Using trained RBM to approximate trained autoencoder

11.6.1   Dimensionality Reduction and Data Reconstruction

The most basic function of the RBM is that of dimensionality reduction and unsupervised feature engineering. The hidden units of an RBM contain a reduced representation of the data. However, we have not yet discussed how one can reconstruct the original representation of the data with the use of an RBM (much like an autoencoder). In order to understand the reconstruction process, we first need to understand the equivalence of the undirected RBM with directed graphical models [104], in which the computation occurs in a particular direction. Materializing a directed probabilistic graph is the first step towards materializing a traditional neural network (derived from the RBM) in which the discrete probabilistic sampling from the sigmoid can be replaced with real-valued sigmoid activations.

Although an RBM is an undirected graphical model, one can "unfold" an RBM in order to create a directed model in which the inference occurs in a particular direction. In general, an undirected RBM can be shown to be equivalent to a directed graphical model with an infinite number of layers. The unfolding is particularly useful when the visible units are fixed to specific values because the number of layers in the unfolding collapses to exactly twice the number of layers in the original RBM. Furthermore, by replacing the discrete probabilistic sampling with continuous sigmoid units, this directed model functions as a virtual autoencoder, which has both an encoder portion and a decoder portion. Although the weights of an RBM have been trained using discrete probabilistic sampling, they can also be used in this related neural network with some fine tuning. This is a heuristic approach to convert what has been learned from a Boltzmann machine (i.e., the weights) into the initialized weights of a traditional neural network with sigmoid units.

An RBM can be viewed as an undirected graphical model that uses the same weight matrix to learn $\overline{h}$ from $\overline{v}$ as it does from $\overline{v}$ to $\overline{h}$. If one carefully examines Equations 11.11 and 11.13, one can see that they are very similar. The main difference is that these equations uses different biases, and they use the transposes of each other's weight matrices. In other words, one can rewrite Equations 11.11 and 11.13 in the following form for some function $f(\cdot)$:

$$\overline{h} \sim f(\overline{v}, \overline{b}^{(h)}, W)$$
$$\overline{v} \sim f(\overline{h}, \overline{b}^{(v)}, W^T)$$

The function $f(\cdot)$ is typically defined by the sigmoid function in binary RBMs, which constitute the predominant variant of this class of models. Ignoring the biases, one can replace the undirected graph of the RBM with two directed links, as shown in Figure 11.5(a). Note that the weight matrices in the two directions are W and W^T, respectively. However, if we fix the visible states to the training points, we can perform just two iterations of these operations to reconstruct the visible states with *real-valued* approximations. In other words, we approximate this trained RBM with a traditional neural network by replacing discrete sampling with continuous-valued sigmoid activations (as a heuristic). This conversion is shown in Figure 11.5(b). In other words, instead of using the sampling operation of "$\sim$," we replace the samples with the probability values:

$$\overline{h} = f(\overline{v}, \overline{b}^{(h)}, W)$$
$$\overline{v}' = f(\overline{h}, \overline{b}^{(v)}, W^T)$$

Note that $\overline{v}'$ is the reconstructed version of $\overline{v}$ and it will contain real values (unlike the binary states in $\overline{v}$). In this case, we are working with real-valued activations rather than discrete samples. Because sampling is no longer used and all computations are performed in terms of expectations, we need to perform only one iteration of Equation 11.11 in order to learn the reduced representation. Furthermore, only one iteration of Equation 11.13 is required to learn the reconstructed data. The prediction phase works only in a single direction from the input point to the reconstructed data, and is shown on the right-hand side of Figure 11.5(b). We modify Equations 11.11 and 11.13 to define the states of this traditional neural network as real values:

$$\hat{h}_j = \frac{1}{1 + \exp(-b_j^{(h)} - \sum_{i=1}^{d} v_i w_{ij})} \tag{11.15}$$

For a setting with a total of $m \ll d$ hidden states, the real-valued reduced representation is given by $(\hat{h}_1 \ldots \hat{h}_m)$. This first step of creating the hidden states is equivalent to the encoder portion of an autoencoder, and these values are the expected values of the binary states. One can then apply Equation 11.13 to these *probabilistic values* (without creating Monte-Carlo instantiations) in order to reconstruct the visible states as follows:

$$\hat{v}_i = \frac{1}{1 + \exp(-b_i^{(v)} - \sum_j \hat{h}_j w_{ij})} \tag{11.16}$$

Although $\hat{h}_j$ does represent the expected value of the jth hidden unit, applying the sigmoid function again to this real-valued version of $\hat{h}_j$ only provides a rough approximation to the expected value of v_i. Nevertheless, the real-valued prediction $\hat{v}_i$ is an approximate reconstruction of v_i. Note that in order to perform this reconstruction we have used similar operations as traditional neural networks with sigmoid units rather than the troublesome discrete samples of probabilistic graphical models. Therefore, we can now use this related neural network as a good starting point for fine-tuning the weights with traditional backpropagation. This type of reconstruction is similar to the reconstruction used in the autoencoder architecture discussed in Chapter 9.

On first impression, it makes little sense to train an RBM when similar goals can be achieved with a traditional autoencoder. However, this broad approach of deriving a traditional neural network with a trained RBM is particularly useful when *stacking* multiple

RBMs together, much as one might create multiple layers of a neural network. The training of a stacked RBM does not face the same challenges as those associated with deep neural networks, especially the ones related with the vanishing and exploding gradient problems. Just as the simple RBM provides an excellent initialization point for the shallow autoencoder, the stacked RBM also provides an excellent starting point for a deep autoencoder [81]. This principle led to the development of the idea of pretraining with RBMs before conventional pretraining methods were developed without the use of RBMs. As discussed in this section, one can also use RBMs for other reduction-centric applications such as collaborative filtering and topic modeling.

11.6.2   RBMs for Collaborative Filtering

The previous section shows how restricted Boltzmann machines are used as alternatives to the autoencoder for unsupervised modeling and dimensionality reduction. However, dimensionality reduction methods are also used for a variety of related applications like collaborative filtering. In the following, we will provide an RBM-centric approach to recommender systems. This approach is based on the technique proposed in [154], and it was one of the ensemble components of the winning entry in the Netflix prize contest.

One of the challenges in working with ratings matrices is that they are incompletely specified. This tends to make the design of a neural architecture for collaborative filtering more difficult than traditional dimensionality reduction. The basic idea is to create a different training instance *and* a different RBM for each user, depending on which ratings are observed by that user. All these RBMs share weights. An additional problem is that the units are binary, whereas ratings can take on values from 1 to 5. Therefore, we need some way of working with the additional constraint.

In order to address this issue, the hidden units in the RBM are allowed to be 5-way softmax units in order to correspond to rating values from 1 to 5. In other words, the hidden units are defined in the form of a one-hot encoding of the rating. One-hot encodings are naturally modeled with softmax, which defines the probabilities of each possible position. The ith softmax unit corresponds to the ith movie and the probability of a particular rating being given to that movie is defined by the distribution of softmax probabilities. Therefore, if there are d movies, we have a total of d such one-hot encoded ratings. The values of the corresponding binary values of the one-hot encoded visible units are denoted by $v_i^{(1)}, \ldots v_i^{(5)}$. Note that only one of the values of $v_i^{(k)}$ can be 1 over fixed i and varying k. The hidden layer is assumed to contain m units. The weight matrix has a separate parameter for each of the multinomial outcomes of the softmax unit. Therefore, the weight between visible unit i and hidden unit j for the outcome k is denoted by $w_{ij}^{(k)}$. In addition, we have 5 biases for the visible unit i, which are denoted by $b_i^{(k)}$ for $k \in \{1, \ldots, 5\}$. The hidden units only have a single bias, and the bias of the jth hidden unit is denoted by b_j (without a superscript). The architecture of the RBM for collaborative filtering is illustrated in Figure 11.6. This example contains $d = 5$ movies and $m = 2$ hidden units. In this case, the RBM architectures of two users, Sayani and Bob, are shown in the figure. In the case of Sayani, she has specified ratings for only two movies. Therefore, a total of $2 \times 2 \times 5 = 20$ connections will be present in her case, even though we have shown only a subset of them to avoid clutter in the figure. In the case of Bob, he has four observed ratings, and therefore his network will contain a total of $4 \times 2 \times 5 = 40$ connections. Note that both Sayani and Bob have rated the movie *E.T.*, and therefore the connections from this movie to the hidden units will share weights between the corresponding RBMs.

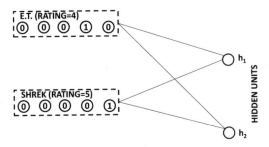

(a) RBM architecture for user Sayani (Observed Ratings: *E.T.* and *Shrek*)

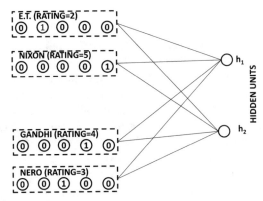

(b) RBM architecture for user Bob (Observed Ratings: *E.T.*, *Nixon*, *Gandhi*, and *Nero*)

Figure 11.6: The RBM architectures of two users are shown based on their observed ratings

The probabilities of the hidden states, which are binary, are defined with the use of the sigmoid function:

$$P(h_j = 1|\overline{v}^{(1)} \ldots \overline{v}^{(5)}) = \frac{1}{1 + \exp(-b_j - \sum_{i,k} v_i^{(k)} w_{ij}^k)} \qquad (11.17)$$

The main difference from Equation 11.11 is that the visible units also contain a superscript to correspond to the different rating outcomes. Otherwise, the condition is virtually identical. However, the probabilities of the visible units are defined differently from the traditional RBM model. In this case, the visible units are defined using the softmax function:

$$P(v_i^{(k)} = 1|\overline{h}) = \frac{\exp(b_i^{(k)} + \sum_j h_j w_{ij}^{(k)})}{\sum_{r=1}^5 \exp(b_i^{(r)} + \sum_j h_j w_{ij}^{(r)})} \qquad (11.18)$$

The training is done with Monte Carlo sampling. The main difference from the earlier techniques that the visible states are generated from a multinomial model. The corresponding updates for training the weights are as follows:

$$w_{ij}^{(k)} \Leftarrow w_{ij}^{(k)} + \alpha \left(\langle v_i^{(k)}, h_j \rangle_{pos} - \langle v_i^{(k)}, h_j \rangle_{neg} \right) \quad \forall k \qquad (11.19)$$

Note that only the weights of the *observed* visible units to all hidden units are updated for a single training example (i.e., user). In other words, the Boltzmann machine that is used is different for each user in the data, although the weights are shared across the different users. Examples of the Boltzmann machines for two different training examples are illustrated in Figure 11.6, and the architectures for Bob and Sayani are different. However, the weights for the units representing *E.T.* are shared. This type of model can also be achieved with a traditional neural architecture in which the neural network used for each training example is different [6]. The traditional neural architecture is equivalent to a matrix factorization technique. The Boltzmann machine tends to give somewhat different ratings predictions from matrix factorization techniques, although the accuracy is similar.

11.6.2.1 Making Predictions

Once the weights have been learned, they can be used for making predictions. However, the predictive phase works with real-valued activations rather than binary states, much like a traditional neural network with sigmoid and softmax units. First, one can use Equation 11.17 in order to learn the probabilities of the hidden units. Let the probability that the jth hidden unit is 1 be denoted by $\hat{p}_j$. Then, the probabilities of *unobserved* visible units are computed using Equation 11.18. The main problem in computing Equation 11.18 is that it is defined in terms of the values of the hidden units, which are only known in the form of probabilities according to Equation 11.17. However, one can simply replace each h_j with $\hat{p}_j$ in Equation 11.18 in order to compute the probabilities of the visible units. Note that these predictions provide the probabilities of each possible rating value of each item. These probabilities can also be used to compute the expected value of the rating if needed. Although this approach is approximate from a theoretical point of view, it works well in practice and is extremely fast. By using these real-valued computations, one is effectively converting the RBM into a traditional neural network architecture with logistic units for hidden layers and softmax units for the input and output layers. Although the original paper [154] does not mention it, it is even possible to tune the weights of this network with backpropagation (cf. Exercise 1).

The RBM approach works as well as the traditional matrix factorization approach, although it tends to give different types of predictions. This type of diversity is an advantage from the perspective of using an ensemble-centric approach. Therefore, the results can be combined with the matrix factorization approach in order to yield the improvements that are naturally associated with an ensemble method. Ensemble methods generally show better improvements when diverse methods of similar accuracy are combined.

11.6.3 Conditional Factoring: A Neat Regularization Trick

A neat regularization trick is buried inside the RBM-based collaborative filtering work of [154]. This trick is not specific to the collaborative filtering application, but can be used in any application of an RBM. This approach is not necessary in traditional neural networks, where it can be simulated by incorporating an additional hidden layer, but it is particularly useful for RBMs. Here, we describe this trick in a more general way, without its specific modifications for the collaborative filtering application. In some applications with a large number of hidden units and visible units, the size of the parameter matrix $W = [w_{ij}]$ might be large. For example, in a matrix with $d = 10^5$ visible units, and $m = 100$ hidden units, we will have ten million parameters. Therefore, more than ten million training points will be required to avoid overfitting. A natural approach is to assume a low-rank parameter structure of the weight matrix, which is a form of regularization. The idea is to assume that the matrix W can be expressed as the product of two low-rank factors U and V, which are of sizes $d \times k$ and $m \times k$, respectively. Therefore, we have the following:

$$W = UV^T \tag{11.20}$$

Here, k is the rank of the factorization, which is typically much less than both d and m. Then, instead of learning the parameters of the matrix W, one can learn the parameters of U and V, respectively. This type of trick is used often in various machine learning applications, where parameters are represented as a matrix. A specific example is that of factorization machines, which are also used for collaborative filtering [148]. This type of approach is not required in traditional neural networks, because one can simulate it by incorporating an additional linear layer with k units between two layers with a weight matrix of W between them. The weight matrices of the two layers will be U and V^T, respectively.

11.7 Summary

This chapter discusses probabilistic graphical models, which can be viewed as probabilistic variants of the types of computational graphs discussed in Chapter 7. These types of computational graphs work by generating data points from conditional distributions at nods of the graph, and they are closely related to the plate diagrams used in many classical models. Unlike deterministic computational graphs, probabilistic computational graphs are often undirected. All these characteristics make probabilistic computational graphs harder to train than deterministic graphs. Such models can be used for both unsupervised and supervised model, although their use in supervised learning is more common because of the inherently generative nature of these models.

11.8 Further Reading

A comprehensive treatment of probabilistic graphical models may be found in [104]. Overviews of Bayesian networks may be found in [35, 91, 140]. A treatment of probabilistic graphical models from the artificial intelligence point of view may be found in [153].

The earliest variant of the Boltzmann family of models was the Hopfield network [86]. The earliest algorithms for learning Boltzmann machines with the use of Monte Carlo sampling were proposed in [2, 82]. Discussions of Markov Chain Monte Carlo methods are provided in [63, 145], and many of these methods are useful for Boltzmann machines as well. A tutorial on energy-based models is provided in [112].

11.9 Exercises

1. This chapter discusses how Boltzmann machines can be used for collaborative filtering. Even though discrete sampling of the contrastive divergence algorithm is used for learning the model, the final phase of inference is done using real-valued sigmoid and softmax activations. Discuss how you can use this fact to your advantage in order to fine-tune the learned model with backpropagation.

2. Implement the contrastive divergence algorithm of a restricted Boltzmann machine. Also implement the inference algorithm for deriving the probability distribution of the hidden units for a given test example. Use Python or any other programming language of your choice.

3. Propose an approach for using RBMs for outlier detection.

Chapter 12

Knowledge Graphs

"Any fool can know. The point is to understand." – Albert Einstein

12.1 Introduction

Although knowledge graphs have existed since the early years of artificial intelligence in various forms, the term was popularized recently due to use of this idea in Google's search engine. The earliest form of a knowledge graph was referred to as a *semantic network*, which was designed to support deductive reasoning methods. These semantic networks are graphical representations of relationships among entities, in which nodes correspond to concepts, and the edges correspond to the relationships among them. There were several attempts during the early years to use these methods for natural language processing applications like word-sense disambiguation. Indeed, modern representations of vocabulary and the relationships among words are represented by knowledge graphs such as WordNet.

Many early forms of semantic networks were built in order to support propositional or first-order logic in traditional knowledge bases. Subsequently, variations of such semantic networks were proposed [90]. In semantic networks, each edge corresponds to one of a set of possible types of relationships. Therefore, all edges do not necessarily indicate the same type of relationship, and different edges correspond to different types of relationships. This representation was referred to as a "knowledge graph" in the early years. These early works formed the basis of the search methods that were eventually popularized by Google. It is important to understand that knowledge graphs were intended for artificial intelligence applications from the very beginning, and search represents an extended view of the ideas used in artificial intelligence applications. Search seems like a reasonable application for a knowledge graph, because many queries are associated with finding entities that are related to other known entities; after all, a knowledge graph is a rich source of information about entities and the relationships among them.

The term "knowledge graph" was subsequently popularized by the use of a graph-structured knowledge base by Google in order to enhance its search results. Google created

© Springer Nature Switzerland AG 2021
C. C. Aggarwal, *Artificial Intelligence*, https://doi.org/10.1007/978-3-030-72357-6_12

this graph-structured knowledge based by crawling various data sources on the Web, including semi-structured data on the Web (such as Wikipedia). This knowledge base is very useful to provide search results with a level of detail that is not possible with the use of pure content-based search. For example, a Google search for *"Abraham Lincoln"* yields an infobox containing several facts about Abraham Lincoln, such as his date of birth, the years that he served as president, and so on. This information is extracted from a knowledge base containing *entities* (e.g., people, places, and even dates), together with the *relations* (e.g., "born in") among them. Such a graph-structured repository of information, referred to more generally as a knowledge graph, is now used in many different contexts beyond Web search. For example, it includes various types of open-source, graph-structured databases, such as the *Semantic Web* and other linked entities within the Web. Some examples of knowledge graphs available openly on the Web include Wikidata, YAGO, and WordNet. Many of these open-source knowledge bases use inputs from one another during the construction process, whereas others are collaboratively edited and created.

Search providers like Google create their own internal knowledge bases that are tailored towards providing search results over different types of entities (e.g., people, places, and things). Such graphs are often constructed from semi-structured repositories crawled from the Web, and they are dominated by textual data, such as Wikipedia. In general, Google search results rely on open-source information available on the Web (which is often present in either unstructured or semi-structured form). Portions of this information are converted into knowledge bases in the form of knowledge graphs. In most cases, the knowledge graphs need to be constructed using a combination of linguistic and machine learning methods on semi-structured data crawled from the Web. An example of a knowledge graph is WordNet [124], which contains a semantic description of the relationships among words and their different forms. The corresponding knowledge graph contains nodes corresponding to similar classes of words, and the various types of relationships among them.

Knowledge graphs are closely related to the knowledge bases that arise in first-order logic. While the predicates in first-order logic define relations between any number of objects, knowledge graphs naturally encode relationships between pairs of objects in graphical form. Each edge between a pair of objects in a knowledge graph can be viewed as a predicate in first-order logic with two arguments. Stated more simply, the predicates in first-order knowledge can be n-ary relationships, whereas the edges in knowledge graphs are binary relationships. Therefore, the edges in knowledge graphs can be represented as *triplets*, which correspond to the two end-points of the edge together with the relationship type represented by the edge. While the inherently binary nature of the relations between objects might seem restrictive at first glance, the graphical nature of the representation it enables has numerous advantages. Much of the machinery belonging to data mining and machine learning can be more easily harnessed on such graph representations (rather than hyper-graph representations corresponding to n-ary relations). Therefore, knowledge graphs support inductive forms of learning more easily than do knowledge bases that are rooted in formal logic.

The notion of a knowledge graph has evolved slowly over the years through the efforts of diverse community of researchers from the fields of artificial intelligence, search, and the Semantic Web. This diversity is a direct consequence of the variety of applications that this concept supports. As a result, the concept is quite informal at present, and there has been some debate recently on how it should be appropriately defined [53, 139]. As elucidated by the discussions in [53, 139], there are different definitions of knowledge graphs floating around, which are suitable for different types of graph-structured databases, including the Web or the *Resource Descriptor Framework* used in the Semantic Web [56, 143]. Therefore, we try to use a definition that is as general as possible, and which does not rely on specific settings such as the Semantic Web. A useful definition is provided in Paulheim [139]:

"A knowledge graph (i) mainly describes real world entities and their interrelations, organized in a graph, (ii) defines possible classes and relations of entities in a schema, (iii) allows for potentially interrelating arbitrary entities with each other, and (iv) covers various topical domains."

Note that this definition only focuses on the organization of the knowledge, and it does not discuss how the knowledge graph is actually used. The classes corresponding to the entity are often arranged hierarchically as a tree structure, or more generally, as a directed acyclic graph. The classes are also referred to as *concepts*, and the underlying hierarchical structure is also referred to as an *ontology*. For example, in a movie database, a high-level concept for movies could be *fantasy*, and its subclass could be *science fiction*. However, the science fiction class could also be a subclass of the *science* category. Therefore, the arrangement of concepts is not always strictly hierarchical. A particular class may have restrictions on the values that are allowed inside it, and the ontology will contain this information as well.

An important part of the ontology is the schema. The concept of *schema* is similar to that used in traditional databases, which indicates how the different concepts are related to one another and eventually organized as database tables of relationships between entities. For example, a director concept can be related to a movie concept with the use of a *"directed in"* relation, but two movie concepts cannot be connected by such a relation. In other words, the schema is a "plan" of the database representation of the knowledge graph. The word "schema" is often used as a general term that subsumes the ontology, because it is a "blueprint" of the structure of the knowledge graph. The ontology of a knowledge graph is also referred to as the *TBox*. The actual instances of the entities/relations (e.g., Tom Hanks acting in *Saving Private Ryan*) is referred to as the *ABox*. A knowledge graph typically contains both an ABox and a TBox. Each node in the ABox is associated with a unique identification number for disambiguation purposes, since it is possible to have two distinct nodes with similar content. The ABox, which is the meat of the knowledge graph, is built on top of the TBox (which is the blueprint of the knowledge graph). The size of the ABox is typically orders of magnitude large than the TBox, and it is considered the more essential part of the knowledge graph for machine learning purposes. In some cases, the TBox (blueprint) is missing, when the knowledge graph contains only instances of objects.

Another definition in [53] is as follows:

"A knowledge graph acquires and integrates information into an ontology and applies a reasoner to derive new knowledge."

The second definition appears to be somewhat more restrictive, because this definition suggests that a *reasoner* must be used in order to derive new knowledge, whereas the definition in [139] does not indicate any constraints on how it must be used. The constraint on the use of a reasoner might possibly be a result of the fact that knowledge graphs were originally closely associated with traditional knowledge bases that are used in deductive reasoning. As we will see later, a knowledge graph can also be seen as a special case of a first-order knowledge base in which each edge of the knowledge graph can be considered a predicate has two inputs (nodes) and one output (relationship type). Like all knowledge bases, such knowledge graphs also contain assertions, which might be presented in if-then form. From this point of view, a knowledge graph might be viewed as a specialized form of a knowledge base, in which the graphical structure can be exploited by many types of reasoning and learning algorithms.

Many recent applications apply machine learning methods on knowledge graphs in order to infer and extract new knowledge. The use of machine learning on knowledge graphs has now become the dominant model of its use, as compared to the use of deductive reasoning methods. In fact, the entire area of *heterogeneous information network analysis* has focused on the development of such methods [178], even though information network analysis is considered a field in its own right. Information networks correspond to graphs of entities with various types of relationships among them, although the nodes are not always associated with a hierarchical taxonomy or a schema (ontology). In the most relaxed definition of a knowledge graph, it is not necessary to have ontologies associated with instances of classes. Any graph containing relationships between *instances* of these classes can also be considered a knowledge graph (albeit a somewhat incomplete one because it is missing the TBox). Most information networks belong to this category.

Similarly, a knowledge graph used by *Yahoo!* for making entity recommendations [22] can be considered a knowledge graph, although the amount of ontological information in it is limited. The definition in [53] rules out this structure as a knowledge graph, because the knowledge graph does not contain sufficient ontological information. Rather, it is simply a network of connection between entities. However, other works do consider such a structure a knowledge graph. Therefore, there are some differences in opinion across the broader literature as to what may be considered a knowledge graph. In this book, we choose to use the most relaxed definition possible in order to be inclusive of different classes of methods. In general, it is expected that a knowledge graph will contain both concepts from ontologies as well as their instances in an interconnected structure.

What are the applications of knowledge graphs? There is no consensus on the specific way in which knowledge graphs may be used. Although deductive reasoning methods were the dominant methods in artificial intelligence during the early years (and were used extensively with early versions of semantic networks), this is not the case today. Most of the great successes of deductive reasoning in the fields of chess and machine translation have now been superseded by inductive learning methods. This revolution is a direct result of the increased availability of data in many domains. It is envisioned that deductive reasoning methods will primarily play a helper role to inductive learning methods in the future of artificial intelligence; the primary goal of deductive reasoning will be to reduce data requirements with the use of background knowledge in those domains where sufficient data are not available. Indeed, knowledge graphs can and should be used with inductive learning methods; furthermore, they might allow the use of incomplete, inconsistent, and conflicting information to create *instances* of ontologies. This generalization can allow inductive learning methods to achieve the most powerful and non-obvious inferences from data that are inherently noisy but allow excellent inferences on an aggregate basis.

As discussed above, a knowledge graph is closely related to the notion of a heterogeneous information network, which contains a set of entities, together with the relations among them. The edges in a knowledge graph are of different types, corresponding to the various types of relationships among the different entities. Indeed, a heterogeneous information network framework is the most general representation of a knowledge graph, because it provides the most general graph structure that can be used in order to represent a knowledge graph. Therefore, we will use the most general definition of a knowledge graph as *any network structure of entities and different types of relations among them*. The nodes are often either instances of objects, or object types, which define the ontologies in the knowledge graph. Therefore, nodes are either *instance-nodes*, or they are *concept nodes*. The entity types of different objects in the graph (instance-type nodes) are usually specified with the use of hierarchical ontologies, although it is not necessary for such nodes to be present in the knowledge graph. Furthermore, in the definition used in this book, there is no constraint on how such a structure may be used, whether it is in conjunction with a deductive reasoning method or an inductive learning method.

This chapter is organized as follows. The next section introduces a knowledge graph, along with examples. Real-world examples of knowledge graphs are also provided in the next section. The process of constructing a knowledge graph is discussed in Section 12.3. Applications of knowledge graphs are discussed in Section 12.4. A summary is given in Section 12.5.

12.2 An Overview of Knowledge Graphs

We first motivate knowledge graphs with an example of entity-based search, which is where the use of knowledge graphs is most common. Search applications are particularly instructive, because many queries are associated with finding entities that are related to other known entities or for finding general information about specific entities. Consider the case where a user types in the search term "chicago bulls" in a purely content-centric search engine (without the use of knowledge graphs). This term corresponds to the well-known basketball team in Chicago, rather than a specific species of animal located in Chicago. If one did not use knowledge graphs, the search results would often not contain sufficiently relevant information to the basketball team. Even when relevant Web pages are returned, the context of the search results will fail to account for the fact that the results correspond to a specific type of entity. In other words, it is important to account for the specific context and relationships associated with entity-based search. On the other hand, if the reader tries this query on Google, it becomes evident that the proper entity is returned (together with useful information about entities related to the team), even when the search is not capitalized as a proper noun. Similarly, a search for "chinese restaurants near me" yields a list of Chinese restaurants near the GPS or network location of a user. This would often be hard to achieve with a content-centric search process. In all these cases, the search engine is able to identify entities with a specific name, or entities obeying specific relationships to other entities. In some cases, such as in the case of the search "Chicago Bulls," *infoboxes* are returned together with the results of the search. Infoboxes correspond to a tabular representation of the various properties of an entity. Infoboxes are returned with many types of named entities, such as people, places, and organizations. An example of an infobox for a search of the president *"John F. Kennedy"* shows the following results, which are extracted from the Wikipedia page[1] for John F. Kennedy:

Slot/Field	Value
Born	May 29, 1917
Political Party	Democratic
Spouse(s)	Jacqueline Bouvier
Parents	Joseph Kennedy Sr. Rose Kennedy
Alma mater	Harvard University
Positions	US House of Representatives US Senate US President
Military Service	Yes

Note that this type of information is quite rich, and it includes data about different types of family relations, affiliations, dates, and so on. This information is encoded in Google as

---

[1] https://en.wikipedia.org/wiki/John_F._Kennedy

a knowledge graph. The nature of the presented information depends on the type of search, and the returned infoboxes may also vary, depending on the type of search at hand. Google creates these knowledge graphs by harvesting open sources of data; this is an issue that will be revisited in a later section.

In the context of search, one often combines machine learning techniques with the structure of the graph in order to discover what other types of entities a user *might* be interested in. For example, if a user searches for the movie *Star Wars*, it often implies that they may be interested in particular types of actors starring in the movie (e.g., Ewan McGregor), or they may be interested in particular genres of movies (e.g., science fiction). In many cases, this type of additional information is returned together with the search results. The knowledge graph is, therefore, a rich representation that can be used in order to extract useful connections and relationships between the entities. This makes it particularly useful for answering search queries that are posed as relationship-centric questions, such as the following: "*Who played Obi Wan Kenobi in Star Wars?*" Then, by searching for the relations of the *Obi Wan Kenobi* and *Star Wars* entities, one can often infer that the entity *Ewan McGregor* is connected to these two entities with the appropriate types of edges. The key point in responding to such queries is to able to translate natural language queries into graph-structured queries. Therefore, a separate process is required to translate the unstructured language of free-form queries to the structured language of knowledge graphs. This is an issue that will be discussed later in this chapter. Such forms of search are also referred to as *question answering*, and the Watson artificial intelligence system is an example of such a system. Indeed, the original Watson system did use similar structured representations of knowledge bases for the search process (although current variations are quite diverse, depending on the application at hand).

A knowledge graph may be viewed as an advanced form of a knowledge base in which a network of relations among entities is represented, together with the hierarchical taxonomy of the entities. An ontology is built on top of basic objects, which serve as the leaf nodes of the taxonomy and belong to specific *classes*. The classes correspond to types of objects. From the perspective of a knowledge graph, each object or class may be represented as a node. The attributes corresponds to the types of properties that objects and classes might have. The upper levels of the taxonomy correspond to concept nodes, whereas the lower levels of the taxonomy correspond to instance (entity) nodes. The *relations* correspond to the ways in which the classes or instances of classes are related to one another. The relations are represented by the edges in the knowledge graph and they may correspond to edges between nodes at any level of the taxonomy. However, it is most common for the relationship edges to occur among instance nodes, and a few may exist among concept nodes. Like all knowledge bases, such knowledge graphs also contain assertions, which might be presented in if-then form. However, in many machine learning settings, such assertions may not be present at all.

In knowledge graphs, the relations associated with concept nodes of the ontology are often hierarchical in nature, and form a tree or (more commonly) a directed acyclic graph. This is because the concept nodes in a hierarchy are often organized from the general to the specific. For example, a movie is a type of object, and an action movie is a type of movie. On the other hand, the instance nodes of the knowledge graph (i.e., leaf nodes of the ontology) might be connected together with an arbitrary topology, because they correspond to the arbitrary way in which *instances* of objects are related to one another. The concept nodes (or classes) in a knowledge graph often naturally correspond to the classes in the object-oriented programming paradigm. Therefore, they may inherit various types of properties from their parent nodes. Furthermore, one can associate events with nodes in the knowledge graphs

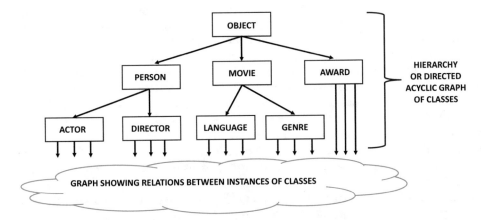

(a) The hierarchical class relations of a knowledge graph (concept nodes)

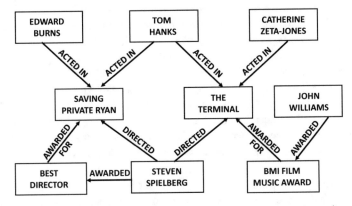

(b) A small snapshot of the knowledge graph between class instances (instance nodes)

Figure 12.1: A knowledge graph contains hierarchical class relations and the relationships among instances

with the use of class-related methods in object-oriented programming. In fact, the notion of object-oriented programming developed in the nineties, with a focus on developing such knowledge bases. An early proponent of object-oriented programming was Patrick Henry Winston, who is one of the fathers of the deductive school of artificial intelligence. In one of his early books on the C++ programming language, Winston describes the usefulness of being able to design hierarchical data types, together with the instances and methods around them. This type of programming paradigm is well suited to the construction of knowledge graphs, and is used in modern programming languages as well.

In order to understand the nature of knowledge graphs, we provide an example of a sample knowledge graph in Figure 12.1. The illustration of Figure 12.1(a) shows the hierarchy of the different object types in a movie database. Note that only a snapshot of the different object types is shown in Figure 12.1(a). Although Figure 12.1(a) shows a hierarchy of object types in the form of a tree structure, this hierarchy often exists as a directed acyclic graph (rather than a tree). Furthermore, different types of objects (e.g., movies and actors) would have different classifications with their own hierarchical structure, which suits their specific domain of discourse. The lowest level of leaf nodes in this hierarchy contains the instances of the objects. The instances of the objects are also connected by relations, depending on how one node is related to another for that specific pair of instances. For example, person instances may be connected to movie instances using the *"directed in"* relation. The instance-level graph may have a completely arbitrary structure as compared to the concept-level graph, which is a directed acyclic graph. The instance-level relations are shown by the cloud at the bottom of Figure 12.1(a). The nodes shown in Figure 12.1(a) correspond to the concept nodes, and they correspond to a tree structure in this particular example. However, it is more common for such nodes to be arranged in the form of a directed acyclic graph (where a node may not have a unique parent). An expansion of this cloud of Figure 12.1(a) is shown in Figure 12.1(b). In this case, we focus on a small snapshot of two movies, corresponding to *Saving Private Ryan* and *The Terminal*. Both movies were directed by Steven Spielberg, and the actor Tom Hanks stars in both movies. Steven Spielberg won the award for best director in *Saving Private Ryan*. Furthermore, John Williams received the BMI music award for *The Terminal*. These relationships are shown in Figure 12.1(b), and they correspond to edges of different types, such as *"acted in,"* *"directed,"* and so on. In a sense, each of these edges can be viewed as an assertion in a knowledge base, when one considers a knowledge graph as a specific example of a knowledge base. For example, the edge between a director and a movie could be interpreted as the assertion *"Steven Spielberg directed Saving Private Ryan."*

In many knowledge graphs, the class portion of the ontology is captured by a *type* attribute. Each object type has its own set of instances, methods, and relations. In fact, each object instance node of Figure 12.1(b) can be connected with an edge labeled "type" to a class node of Figure 12.1(b). Furthermore, one can associate additional attributes with entities depending on their type. This additional information creates a rich representation that can be useful in a wide variety of applications. One can view a knowledge graph as a generalized version of a concept referred to as an *information network* used in network science. This type of rich information is useful not only for answering direct queries (via a deductive reasoning methodology), but it can also be used to supplement data-driven examples in inductive learning. Many of the applications of knowledge graphs in recent years, including in the case of Google search, have focused on at least some level of machine learning on the knowledge graphs.

In many cases, knowledge graphs are represented formally with the use of the *Resource Description Framework*, also referred to as *RDF*. Edges are often represented as triplets,

Table 12.1: Examples of RDF triplets based on the knowledge graph of Figure 12.1

Subject	Predicate	Object
Edward Burns	Acted In	Saving Private Ryan
Tom Hanks	Acted In	Saving Private Ryan
Tom Hanks	Acted In	The Terminal
Catherine Zeta-Jones	Acted in	The Terminal
Steven Spielberg	Directed	Saving Private Ryan
Steven Spielberg	Directed	Saving Private Ryan
Steven Spielberg	Awarded	Best Director
Best Director	Awarded For	Saving Private Ryan
John Williams	Awarded	BMI Music Award
BMI Music Award	Awarded For	The Terminal

containing the source (entity) of the edge, the destination (entity) of the edge, and the type of relation between the two entities. This triplet is sometimes referred to as *subject*, *predicate*, and *object*. Examples of triplets based on Figure 12.1 are shown in Table 12.1. Note that the subject is the source and the object is the destination of the directed link in the knowledge graph. It is occasionally possible for the edges in the knowledge graph to be associated with additional attributes. For example, if a relationship type between a person and place corresponds to an event, it is possible for the edge to be associated with the date of the event. Therefore, the edges can be associated with additional attributes, which can be represented by expanding the RDF triplets associated with relationships among entities. In general, knowledge graphs are quite rich, and they may contain a lot of information beyond what is represented in the form of RDF triplets.

It is possible to view a knowledge graph from the relational database perspective of RDF triplets, and borrow many database concepts to apply to knowledge graphs. A key point is that there are many different types of triplets corresponding to different relationship types and entity types. When using a knowledge graph, it is sometimes difficult for the end user to know what types of entities are connected with particular types of relationships. Relational databases are often described using *schema*, which provide information on which attributes are related to one another in various tables and how the tables are linked to one another via shared columns. In a knowledge graph, "tables" correspond to relationships (RDF triplets) of a particular type between particular types of entities, and one column may be shared by multiple tables. These "shared columns" correspond to nodes of a particular type that are incident on other nodes using various types of relationships. The relationship type is also one of the columns in the resulting table of triplets. For example, only a person-type can be a director (relationship type) of a movie type in the aforementioned knowledge graph. This information is important for understanding the structure of the knowledge graph. Therefore, knowledge graphs are often provided together with schema, which describe which types of entities link to one another with particular types of relationships. Not all knowledge graphs are provided together with schema. A schema is a desirable but not essential component of a knowledge graph. The main challenge in incorporating schemas is that databases require schemas to be set in stone up front, whereas incremental and collaboratively edited knowledge bases require a greater degree of flexibility in modifying and appending to existing schemas. This challenge was addressed in one of the earliest knowledge bases, *Freebase*, with the use of a novel graph-centric database design, referred to as *Graphd*. The idea was to allow the community contributor to modify existing schema based on the data that was added. More details on this knowledge base and underlying graph database are discussed in Section 12.2.4.

In some knowledge graphs, location and time are added for richness of representation. This is particularly important when the facts in the knowledge graph can change over time. Some facts in a knowledge graph do not change with time, whereas others do change with time. For example, even though the director of a movie is decided up front and remains fixed after the movie's release, the heads of all countries change over time, albeit at different time scales. Similarly, the locations of annual events (e.g., scientific conferences) might change from year to year. As a result, it is critical to allow knowledge graphs to be updateable with time, as new data comes in and entities/relationships are updated. Most graph-centric database design methods support this type of functionality. The actual information about the place and time stamp can be stored along with the corresponding RDF triplet.

In general, there is no single representation that is used to consistently represent all types of knowledge graphs. In most cases, they have a number of common characteristics:

- They always contain nodes representing entities.

- They always contain edges corresponding to relationships. These edges are represented by RDF triplets, which are *instances* of the objects. This portion of the knowledge graph is also referred to as the *ABox*.

- In most cases, hierarchical taxonomies and ontologies are associated with nodes. The ontological portion is sometimes referred to as the *TBox*. In such cases, schema may also be associated with the knowledge graph that set up the plan for the database tables of the RDF triplets.

Several other characteristics of knowledge graphs, such as the presence of logical rules are optional, and are not included in many modern knowledge bases. Most knowledge graphs will contain both an ABox and a TBox, although some simplified knowledge graphs (like heterogeneous information networks) might contain only an ABox.

Knowledge graphs can cover either a broad variety of domains, or they can be domain-specific. The former types of knowledge graphs are referred to as open-domain knowledge graphs, and often cover a wide variety of entities searchable over the Web (and can therefore be used in Web search). Such knowledge graphs can also be used in open-domain question answering systems like Watson. Examples of open-domain knowledge graphs include Freebase, Wikidata, and YAGO. On the other hand, a domain-specific knowledge graph like WordNet or Gene Ontology will cover the entities specific to a particular domain (like English words or gene information). Domain-specific knowledge graphs have also been constructed in various commercial settings, such as the Netflix knowledge graph, the Amazon product graph, as well as various types of tourism-centric knowledge graphs. Such knowledge graphs tend to be useful in somewhat narrower applications like protein search or movie search. In each case, the entities are chosen based on the application domain at hand. For example, an Amazon product graph might contain entities corresponding to products, manufacturers, brand names, book authors, and so on. The relationships among these different aspects of products can be very useful for product search as well as for performing customer recommendations. In fact, a product graph can be viewed as an enriched representation of the content of products, and can be used for designing content-based algorithms (which are inherently inductive learning algorithms). Similarly, a Netflix product graph will contain entities corresponding to movies, actors, directors, and so on. A tourism-centric knowledge graph will contain entities corresponding to cities, historical sites, museums, and so on. In each case, the rich connections in the knowledge graph can be leveraged to perform domain-specific inferences. In the following, we will provide some examples of real-world knowledge graphs in various types of settings.

12.2.1 Example: WordNet

WordNet is a lexical database of English nouns, verbs, adjectives, and adverbs. These words are grouped into sets of cognitive synonyms, which are also referred to as *synsets*. Although WordNet serves some of the same functions as a thesaurus, it captures richer relationships in terms of the complexity of the relationships it encodes. WordNet can be expressed as a network of relationships between words, which go beyond straightforward notions of similarity. The primary relationship among the words in WordNet is that of synonymy, which is represented by a total of about 117, 000 synsets. Polysemous words (i.e., words with multiple meanings) occur in multiple synsets, with one occurrence corresponding to each of its possible meanings.

Synsets also have encoded relationships between them, such as between the general and the specific. These types of relationships are the super-subordinate relations (also referred to as *hypernomy* or *hyponomy*. For example, a specific form of *"furniture"* is *"bed."* WordNet distinguishes between types and instances. For example, a *"bunkbed"* is a type of bed, whereas *"Bill Clinton"* is an instance of a president. An instance is always a leaf node in the hierarchy of relationships.

Meronymy corresponds to part-whole relationships. For example, a *leg* is a part of a furniture such as a chair. However, not all types of furniture necessarily have legs. Note that if a chair has legs then all specific types of chairs will have legs, but generalizations of chairs might not have legs. Therefore, parts are inherited downwards but not upwards.

The specific types of relationships depend deeply on the parts of speech of the constituent words. For example, verbs can have relationships corresponding to intensity (e.g., *"like"* and *"love"*) whereas adjectives can have relationships corresponding to antonymy (e.g., *"good"* and *"bad"*). Verbs have hierarchies based on their level of specificity. The verbs at the bottom of the tree are referred to as tropnyms, and they tend to be more specific towards the bottom of trees (near the leaf nodes). An example of a path down the tree could be *communicate-talk-whisper*. In some cases, verbs describing events are linked when the corresponding events entail one another. Examples include *buy-pay* and *succeed-try*. Note that in order to buy, one must pay; similarly, in order to succeed one must try.

There are also a few relationships across different parts of speech, such as words arising from the same stem. For example, *"paint"* and *"painting"* arise from the same stem but are different parts of speech. These represent *morphosemantic* links that arise between semantically similar words sharing a stem with the same meaning.

In general, one can view WordNet as a knowledge graph in which groups of synonyms (synsets) are nodes and edges are relations. At the same time, the relationships between individual words are encoded as well in one form or another. The rich nature of the relationships in WordNet has made it invaluable as a source for various types of machine learning applications in natural language processing. The information in WordNet which provides useful information for machine learning applications. For example, WordNet is frequently used for pairing with text machine learning applications, by enriching the applications with deeper knowledge about the relationships between words.

WordNet also feeds into other open-domain knowledge graphs. It is common for some knowledge graphs to be extracted from one or more distinct knowledge graphs. In this context, WordNet is particularly useful, as it has been used to extract many knowledge bases. An example is YAGO [177], which is derived from Wikipedia, WordNet, and GeoNames.

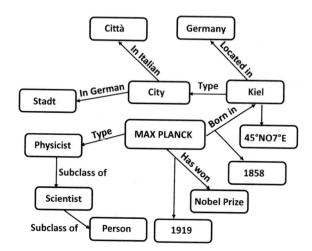

Figure 12.2: A snapshot of the YAGO knowledge graph adapted from the YAGO Website [Diagram has been redrawn for clarity]

12.2.2   Example: YAGO

The acronym *YAGO* stands for *Yet Another Great Ontology*, and this open-domain ontology was developed at the Max Planck Institute for Computer Sciences. This ontology has 10 million entities (as of 2020), and it contains about 120 million facts about these entities. YAGO contains information about person, places, and things, and it also integrates spatial and temporal information into the ontology. A tiny subset of the YAGO ontology (adapted from the Website [224] of the ontology) is shown in Figure 12.2. Note that some relationships are tagged with dates, which provides a temporal dimension to the ontology.

YAGO is linked to the DBpedia ontology, and both these ontologies were used by the Watson artificial intelligence system for question answering. Since YAGO has been constructed from open-source data (which could be erroneous), some of the information in it (e.g., relationships) might be spurious. The accuracy of its relationships has been manually evaluated, which was found to be about 95%. This type of validation of ontologies is extremely common in many open-domain settings, where a *trust* or *confidence* score is associated with each relationship. Each relationship in YAGO has been annotated with its confidence value, which is helpful in the context of various types of learning applications. YAGO has including extractions from thematic domains such as *music* and *science* from the WordNet hierarchy. The latest version of the knowledge graph, referred to as YAGO3 [119], uses information from Wikipedia in multiple languages in order to build the ontology. The YAGO ontology is freely downloadable, and has been widely used in various applications such as search and question-answering..

12.2.3   Example: DBpedia

Like YAGO, DBpedia [16] is extracted from open sources such as Wikipedia, and it contains a large number of entities, including persons, places, and things in various types of relationships. DBpedia contains very similar type of data as YAGO, and it is available in multiple languages. The English version of the DBpedia knowledge base is based on 4.58 million entities, out of which 4.22 million belong to an ontological hierarchy, including 1,445,000 persons, 735,000 places, 411,000 creative works (e.g., music albums, films, video games), 241,000 organizations (e.g., companies, educational institutions), 251,000 species and 6,000 diseases. When considering DBpedia over all languages, these versions together describe 38.3 million entities. The full DBpedia data set contains 38 million labels and abstracts in 125 different languages, 25.2 million links to images, 29.8 million links to external Web pages, 80.9 million links to Wikipedia categories, and 41.2 million links to YAGO categories. DBpedia is connected with various other knowledge graphs. The 2014 version of DBpedia consists of 3 billion relationships (RDF triplets), 20% of which were extracted from English language Wikipedia, and the remainder were extracted from the editions of Wikipedia in other languages. Like YAGO, DBpedia is freely downloadbale [225] for various applications like search and question answering. DBpedia is frequently used in enterprise search and information integration.

12.2.4 Example: Freebase

While Freebase is not currently available, it is an important knowledge graph in terms of historical significance. Freebase was a collaboratively created knowledge base, which was originally outlined by Danny Willis as far back as 2000. It was one of the earliest knowledge graphs that was created for large-scale use.

Freebase was constructed using some modifications to the original concepts in relational databases, and was constructed using a novel graph database, referred to as *Graphd*. This database overcame the challenges associated with representing graphs in a database format that was friendly to the natural steps required in creating knowledge graphs. The idea was to create a global interconnected "brain" of things and concepts. An important innovation introduced by Graphd was in terms of how database schemas are treated. With traditional databases, the schemas (i.e., how the tables are organized and related) are created up front, and the database tables are created and maintained on the basis of this plan. This was seen to be a liability, as one could not predict how future concepts would be related to the currently available concepts. Graphd avoided this type of up front schema creation, and allowed additions to the database (and corresponding adjustments to the schema) on an incremental basis. The adjustments to the schema could be executed by the user adding the new data to the knowledge base. This ability is critical for creating a community-sourced knowledge base where one cannot control what types of data will be added to the knowledge base at a future date. If a knowledge base is to be collaboratively edited and it also contains a schema, then effective and sophisticated schema-modification techniques are crucial. Graphd provided one of the first database frameworks for dynamic graph and schema editing in a collaborative fashion, and this was used to create a knowledge graph.

This effort grew into the knowledge graph Freebase [24], which was owned by the company MetaWeb. For collaborative growth, public read and write access to Freebase was allowed through an HTTP-based graph-query application programming interface using the *Metaweb Query Language*. Each entity in Freebase was associated with a unique identification number in order to enable disambiguation of entities with the same name (e.g., *Paris* in Texas versus *Paris* in France). Each such identification number was treated like a unique barcode. For example, George Clooney was given the identification number 014zcr. The original version of Graphd was restrictive in terms of how modifications could be performed; for example, rows could be appended to the table of relations, but such rows were treated as read-only objects, and could not be removed other than in exceptional cases. This company was eventually acquired by Google. As a result, Google's knowledge graph incorporated the knowledge available in Freebase. Eventually, Google shut down Freebase, and moved all of its data to Wikidata, which was a much larger open-source effort. Wikidata forms the basis of much of the open source semi-structured and unstructured data available on the Web today. Unlike publicly available knowledge graphs like Wikidata, Google's knowledge graph is a commercial system that is used by Google for resolving Web queries. There are several examples of such commercial knowledge graphs in various domains, such as those belonging to Facebook and Netflix.

12.2.5   Example: Wikidata

Wikidata is a collaboratively hosted knowledge base by the Wikimedia foundation, and its data is used by other Wikimedia projects such as Wikipedia and Wiki Commons. Wikipedia represents knowledge in unstructured format, whereas Wiki Commons is a repository of media objects, such as images, sounds, and video. Wikidata provides the facility to users to query the knowledge base with the use of the SPARQL query language (which it has popularized). Wikidata is collaboratively edited, and it provides users with the ability to add new facts to the knowledge base.

The SPARQL query language (pronounced as *sparkle*) is a database query language that is well suited to graph databases. SPARQL is a recursive acronym for S̲PARQL P̲rotocol a̲nd Query L̲anguage. The language has a structure and syntax that is very similar to the *structured query language* that is used in traditional multidimensional databases. However, it also has the functionality to support the querying and editing of graph databases. SPARQL has now become the de facto standard in working with knowledge graphs in general and RDF representations in particular.

Wikidata is one of the largest and most comprehensive sources of openly available knowledge bases today. Wikidata, Wikipedia, and Wiki Commons are closely related, and the data in one of these repositories is often used to augment data in the other repositories. Among these, only Wikidata can be considered a comprehensive knowledge graph containing all types of objects. The data from Wikidata is used in more than half of all Wikipedia's articles. Since many knowledge bases rely on data crawled from Wikipedia, it is clear that much of the data in a variety of knowledge bases is inherited by Wikidata. In this sense, the source of most of the existing knowledge bases comes in one form or another through collaboratively edited information, even though a knowledge base might be constructed through the use of semi-structured and unstructured data processing. This is not particularly surprising, given that the Web is itself a collaboratively edited endeavor at the most basic level. The construction of knowledge bases is an issue, which is discussed in detail in Section 12.3.

12.2.6   Example: Gene Ontology

The Gene Ontology (GO) knowledge base [15] is a knowledge graph containing information on the functions of genes. As with all knowledge graphs, the data is available in both human- and machine-readable format. The knowledge graph is fundamental to the computational analysis of large-scale molecular biology and genetics experiments in biomedical research. The knowledge graph also provides a lot of fundamental information on how various biochemical processes associated with genes interact with one another. The GO ontology mainly captures three aspects:

- *Molecular Function:* Gene products perform various types of activities, such as catalysis and transport. Molecular functions can be performed by individual gene products, such as protein, RNA, or molecular complexes.

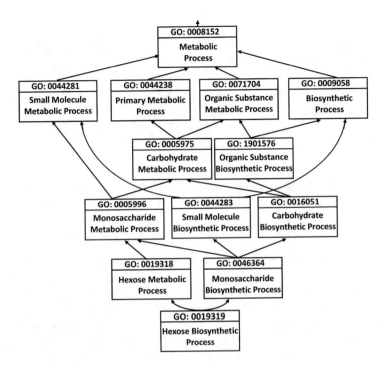

Figure 12.3: A small snapshot of the gene ontology adapted from the Gene Ontology Website [Diagram has been redrawn for clarity]

- *Cellular Component:* These represent the cellular locations in which a gene product performs a function, such as the mitochondrion or the ribosome.

- *Biological Processes:* These represent the larger processes accomplished by multiple molecular activities, such as DNA repair or signal transduction.

The gene ontology is an example of a knowledge graph in which the classes are hierarchical but not a tree structure. A node may have more than one parent, and therefore the ontology appears as a directed acyclic graph. A small snapshot of the gene ontology for biological processes is shown in Figure 12.3. This is obviously a gross simplification of the actual data, which contains much richer meta-information associated with nodes and links. The computational biology domain is particularly rich in various types of knowledge graphs, because of the richness of biological data and the various types of relationships among the underlying entities. Other examples of knowledge graphs in the biological domain include KEGG [226] and UniProt [14].

12.3 How to Construct a Knowledge Graph

A knowledge graph is a very rich and structured representation of the entities in the real world, and it needs to be explicitly curated or constructed from semi-structured to highly unstructured data. The way in which the knowledge graph is constructed depends on the sources from which the data is collected in order to construct the graph. The methodology of construction depends on the source from which the raw data is obtained. For open-source knowledge graphs like Wikidata, the effort is largely collaborative. On the other hand,

Table 12.2: Methodologies for knowledge graph construction, as presented in [130]

Construction method	Schema	Examples
Curated	Yes	Cyc/OpenCyc [113], WordNet [124], UMLS [23]
Collaborative	Yes	Wikidata [190], Freebase [24]
Auto. Semi-Structured	Yes	YAGO [177], DBPedia [16], Freebase [24]
Auto. Unstructured	Yes	Knowledge Vault [191], NELL [34], PATTY [192], PROSPERA [193], DeepDive/Elementary [132]
Auto. Unstructured	No	ReVerb [54], OLLIE [165], PRISMATIC [55]

knowledge graphs like WordNet are created via the process of curation by experts. A table of the different ways in which knowledge graphs are constructed is provided in [130]. We provide this information in Table 12.2.

Based on Table 12.2, we list the four main ways in which knowledge graphs are created:

1. In *curated methods*, the knowledge graph is created by a small group of experts. In other words, the group of people contributing is closed, and it is restricted to a small set of people. Presumably, the restriction to a small set of people, ensures that the recruited people are experts, and the resulting knowledge graph is of high quality. The main problem with this approach is that it does not scale particularly well to large knowledge bases. However, such an approach is particularly effective in specialized domains, where a standardized knowledge graph of high quality is required.

2. In *collaborative methods*, he methodology for knowledge graph construction is similar to that of curated methods, except that the people constructing the knowledge graph are an open group of volunteer. Even though the knowledge graph is open to contribution from the "wisdom of the crowds," there may be still some partial controls on who might contribute. This is done in order to avoid the effects of spam or other undesirable characteristics associated with open platforms. The collaborative approach does scale better than the curated approach, but the graph may sometimes contain errors or inconsistencies. As a result, confidence or trust values are often associated with the relationships in the constructed knowledge graphs.

3. In automated semi-structured approaches, the edges of the knowledge graph are extracted automatically from semi-structured text. This extraction can take on a variety of forms, such as domain-specific rules and machine learning methods. An example of this type of semi-structured data are the infoboxes on Wikipedia (which are themselves crowdsourced, albeit not in knowledge graph form).

4. In automated unstructured approaches, the edges are extracted automatically from unstructured text via machine learning and natural language processing techniques. Examples of such machine learning techniques include entity and relation extraction. This broader area is referred to as *information extraction* in natural language processing. Note that there are two separate entries in Table 12.2, depending on whether or not a database schema exists in conjunction with the knowledge graph.

The aforementioned list of methods is not exhaustive. In many cases, the data from diverse sources need to be combined in order to create the knowledge graph, or some of the above methods may be used in combination. In some cases, the knowledge graph may be constructed vi a combination of curation and collaborative effort. Similarly, even though a knowledge graph of movies can be constructed from the data corresponding to the movies, the data for various movies may come from diverse sources, such as relational data or unstructured data. This is because content from large producers may be available as relational data, whereas data about small home productions may need to be scraped from unstructured sources. In such cases, the knowledge graph needs to be meticulously constructed from the data obtained from diverse sources.

There are numerous special challenges when knowledge graphs need to be created from products for recommendation and related applications. This is because such graphs often cannot be created from open-source information, but one needs to rely on multiple retailers who might provide this information in a variety of different formats. Similarly, maintaining the freshness of the knowledge graph may be a challenge, as the information continually evolves with time. As a result, data integration and dynamic updates are critical in knowledge graphs. Therefore, most knowledge graphs are supported with graph databases that have the capabilities to perform these types of dynamic updates.

12.3.1 First-Order Logic to Knowledge Graphs

The above different ways of constructing knowledge graphs provide an understanding of the role that domain-specific rules may play in knowledge graph construction. These rules could have been extracted from a traditional knowledge base. An example of such a rule could be the following [174]:

$$\forall\, x, y\ [Married(x, y) \Rightarrow SameLocation(x, y)]$$

One could use this type of rule to rapidly populate edges in the knowledge base by repeatedly identifying pairs of nodes with *"married to"* relations between them, and then inserting the edge *"lives in same location"* between them. The reverse process of extracting rules with the use of machine learning methods on the graph is also possible. For example, if one identifies that a knowledge graph contains a *"lives in same location"* edge a vast majority of the times that a *"married to"* relation exists, the above rule can be extracted using association mining methods [4]. This can be achieved by creating a list of relations for each pair of entities, and then finding frequent patterns from these sets of entities. These patterns can be used to create rules using association mining methods discussed in [4]. Although the rules may not be absolute truths (as in the case of most deductive reasoning methods), a domain expert may often be used to decide which rules make sense from a semantic point of view. Subsequently, the extracted rules can be used to populate additional edges in the graph.

12.3.2 Extraction from Unstructured Data

Among the aforementioned methods, the extraction from unstructured data is the most interesting case, because the construction of the knowledge graph is itself a machine learning task. In the case of unstructured data extraction, even the entities may not be directly available, and they may need to be identified from unstructured data. These types of tasks lie within the ambit of the field of *natural language processing*. The specific area of natural language processing that is relevant to this problem is that of *information extraction*. There are two key steps that need to be performed in order to extract the nodes and edges of the knowledge graph:

1. *Named entity recognition:* In this case, important entities such as persons, places, and dates need to be identified from unstructured data (e.g., sentences in a piece of text).

These entities represent the nodes in the knowledge graph. For example, consider the following sentences:

> Bill Clinton lives in New York at a location that is a few miles away from an IBM building. Bill Clinton and his wife, Hillary Clinton, relocated to New York after his presidency.

For this text segment, it needs to be determined which tokens correspond to which type of entity. In this case, the system needs to recognize that *"New York"* is a location, *"Bill Clinton"* is a person, and *"IBM"* is an organization.

2. *Relation extraction:* Once the entities have been extracted, the relations among them need to be extracted in order to create the edges in the knowledge graph. These relations are used in order to create the edges in the knowledge graph. Examples of relationships may be as follows:

> **LocatedIn**(*Bill Clinton, New York*)
> **WifeOf**(*Bill Clinton, Hillary Clinton*)

The nature of the types of relations to be extracted will depend on the type of knowledge graph that one is trying to construct.

It is noteworthy that the hierarchical classes corresponding to the entities may also need to be extracted from a variety of sources. In many cases, the construction of the knowledge graph is an ad hoc effort by domain experts, which is as much of an art form, as it is a science. Nevertheless, since named entity recognition and relation extraction and important modules of this process in many cases, we provide a brief overview of these processes.

There are many different settings in which information extraction systems are used. An *open* information extraction task is unsupervised and has no idea about the types of entities to be mined up front. Furthermore, weakly supervised methods either expand a small set of initial relations, or they uses other knowledge bases from external sources in order to learn the relations in a corpus. Although such methods have recently been proposed in the literature, it is more common to use supervised methods. In this view, it is assumed that the types of entities and the relationships among them to be learned are *pre-defined*, and tagged training data (i.e., text segments) are available containing examples of such entities and/or relationships. Therefore, in named entity extraction, tagged examples of persons, locations, and organizations may be provided in the training data. In relationship extraction, examples of specific relationships to be mined may be provided along with the free text. Subsequently, the entities and relations are extracted from untagged text with the use of models learned on the training data. As a result, many of the important information extraction methods are supervised in nature, since they learn about specific types of entities and relationships from previous examples. A wide variety of machine learning methods, such as rule-based methods and *Hidden Markov Models* are used to extract entities and relationships from text data. A complete discussion of these methods is outside the scope of this book. We refer the reader to [7, 156] for a detailed description of these methods.

12.3.3   Handling Incompleteness

Because of the tedious and ad hoc nature of knowledge graph construction methods, there is an inherent incompleteness associated with knowledge graphs. For example, it is well known that key biographical characteristics (such as education and date of birth) are missing for many person entities in knowledge graphs. Furthermore, there are often errors in the knowledge graph, which could either be caused by manual errors (in the case of collaborative construction), by errors in the source text during automated extraction, or in the automated extraction process itself. Furthermore, automated methods to address missing links or values in the knowledge graphs can also be a source of error. In many cases, such automated methods are combined with curation and crowdsourcing in order to reduce the errors in the construction process.

There are many automated techniques that are used to handle incompleteness in knowledge graph construction. The basic principle is similar to that of incomplete data imputation in collaborative filtering applications. For example, in a recommender system, one can visualize a graph of users and items in which the edges connecting users and items are labeled with their ratings. One often uses matrix factorization (cf. page 308) in order to reconstruct this incomplete graph of users and items. Note that any graph is a matrix, which can be factorized, and the product of the factors reconstructs a corrected/completed graph. Similar to the case of collaborative filtering, one can also factorize the matrices associated with a knowledge graph. A knowledge graph is a *heterogeneous* information network, and therefore there are multiple $n \times n$ matrices $D_1 \ldots D_m$, one for each of the m edge types defined on the n entities. In other words, the matrix D_i contains only the weights for a link of a particular type (e.g., movie-actor) in the knowledge graph. Therefore, the number of possible matrices might be large, if the value of m is large. In such cases, one uses *shared* matrix factorization in order to create a latent representation of the knowledge graph:

$$D_i \approx U_i V^T \quad \forall i \in \{1 \ldots m\}$$

In this case, V is the $n \times k$ shared factor, and the matrices $U_1 \ldots U_m$ are the $n \times k$ factor matrices corresponding to the m different entity types. The model then minimizes the following objective function:

$$J = \|D_1 - U_1 V^T\|_F^2 + \sum_{i=2}^{m} \beta_i \|D_i - U_i V^T\|_F^2$$

One can optimize the parameters using gradient descent. Here, $\beta_2 \ldots \beta_m$ are the balancing factors that regulate the importance of different edge types. These are often learned in a data-driven manner by maximizing the accuracy of prediction of held-out edges of the knowledge graph. For gradient descent, the derivative of J with respect to the matrices is computed and used for the updates. Specifically, the gradient descent updates are as follows:

$$U_1 \Leftarrow U_1 - \alpha \frac{\partial J}{\partial U_1} = U_1 + \alpha(D_1 - U_1 V^T)V$$

$$U_i \Leftarrow U_i - \alpha \frac{\partial J}{\partial U_i} = U_i + \alpha \beta_i (D_i - U_i V^T)V \quad \forall i \geq 2$$

$$V \Leftarrow V - \alpha \frac{\partial J}{\partial V} = V + \alpha(D_1 - U_1 V^T)^T U_1 + \alpha \sum_{i=2}^{m} \beta_i (D_i - U_i V^T)^T U_i$$

Here, the nonnegative hyperparameter α denotes the learning rate. The updates are repeated to convergence. This general approach for computing gradients is similar to the matrix factorization techniques discussed in Chapter 9. In general, there are a wide variety of complex factorization techniques that can be used to reconstruct knowledge graphs. Detailed descriptions are provided in [117, 130].

12.4   Applications of Knowledge Graphs

The classical application with the use of knowledge graphs is that of *search*, since knowledge graphs can be used to respond to complex queries, such as finding specific relationships among entities. Many of the query responses in Google search use knowledge graphs in order to create search results that are sensitive to relationships among entities. In the early years, knowledge graphs were seen as variants of knowledge bases that could be used for different types of deductive reasoning methods. The algorithms in search are similar in many ways to deductive reasoning methods, although there also elements of inductive machine learning in many cases. The use of knowledge graphs has expanded to many other forms of machine learning techniques, such as clustering and classification. Indeed, the broad field of heterogeneous information networks explores the use of knowledge graphs for such applications. In the following, a brief overview of the different applications of knowledge graphs will be provided.

12.4.1   Knowledge Graphs in Search

Knowledge graphs are used in a wide variety of search applications. Indeed, the term "knowledge graph" was coined by Google in the context of search, although the broader idea was explored in many fields such as the Internet of Things, the Semantic Web, and artificial intelligence. Like the Google knowledge graph, Microsoft uses a knowledge base called Satori with its search engine. Furthermore, many domain-specific search applications use product graphs, which are specialized forms of knowledge graphs. In these cases, the search applications target other products or entities that are related to known products.

There are several ways in which search applications can be used in the context of knowledge graphs. For example, the search query *"Barack Obama's education"* yields a chronological list of the educational institutions attended by Obama, together with their images (cf. Figure 12.4). This type of response is hard to achieve with purely content-based search. Presumably, a knowledge graph was used to return the institutional entities to which Obama is connected via a link indicating affiliation.

The real issue in resolving such queries is in understanding the semantics of the query at hand. For example, in the case of the query on Obama's education, one needs to be able to infer that a portion of the query string, "Obama" refers to an entity, and the remaining portion refers to a relationship of the entity. This is often the most difficult part of resolving such queries. In many cases, complex natural language queries corresponding to complete sentences are used in order to query knowledge graphs. For example, consider the following search query: *"Find all actor-movie pairs have received an award for a movie directed by Steven Spielberg."* In such a case, the challenge is even greater in being able to provide responses to such queries. This problem can often be posed into a learning problem, where natural language queries are transformed into more structured queries that can be mechanically applied to the knowledge graph. One can view this problem to be somewhat similar to machine translation, in which a sentence in one language is transformed into a sentence of another. Note that machine translation methods are discussed in Section 8.8.2 of Chapter 8.

What type of structured query language is appropriate for a knowledge graph? The key language for querying RDF-based knowledge graphs is SPARQL, which was discussed briefly in an earlier section (cf. Section 12.2.5). The SPARQL language is similar in syntax to the SQL querying language, except that it is designed for RDF databases rather than relational databases. Like SQL, it contains commands like SELECT and WHERE in order to create a

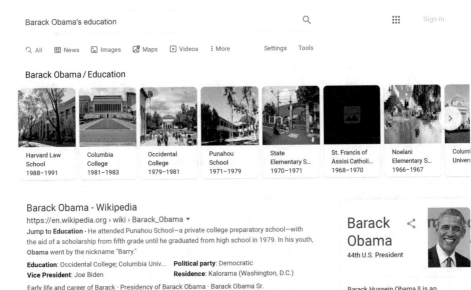

Figure 12.4: An example of the results yielded by the Google search query *"Barack Obama's education"*

clear syntax for what needs to be returned. Like any programming language, it can be easily understood by a parser and compiler in a non-ambiguous and unique way (unlike the case of natural language, which is always more challenging to understand). Many publicly available ontologies such as YAGO are tightly integrated with SPARQL-based query systems, and therefore it is relatively easy to build search functionalities into such knowledge bases.

In practice, one often wants to use natural language queries rather than SPARQL queries. Therefore, a natural step is to transform a natural language query into a SPARQL query. This can be performed by constructing a machine translation model between natural language queries and SPARQL queries. The training data for such a model can first be constructed manually (or via ad hoc translation methods) in order to handle the cold start problem in learning. For example, manually constructed rules by experts can be used to create candidate queries, which can be further curated (manually) by the human experts. The resulting pairs of natural language queries and SPARQL queries can be used to train a machine learning model, such as a sequence-to-sequence autoencoder [6]. Subsequently, implicit feedback from user clicks on search engine queries can be used to construct further training data from the outputs of this machine learning model. For example, for a natural language query (and correspondingly translated SPARQL query), if the user clicks on a particular search result, it is positive feedback for the SPARQL query generated to create that search result. This positive feedback can be used to generate further training data for the sequence-to-sequence learning algorithm. Some discussions on translating between natural language queries and SPARQL are provided in [47, 98, 194, 204]. Note that some of these techniques [98, 194] use traditional machine translation methods such as parse trees. However, if sufficient data are available, it makes more sense to construct a trained machine translation system that can provide more accurate results.

Training between pairs of query representations is, however, not the only approach used for responding to search queries in knowledge graphs. In many cases, one can directly train between pairs of questions and the subgraphs of the knowledge graph that represent answers to these questions. However, such an approach requires the learning system to have access to the knowledge graph in the first place, via a machine learning representation, such as a *memory network*. Examples of such methods are discussed in [27, 28].

12.4.2 Clustering Knowledge Graphs

Clustering knowledge graphs can be useful for creating concise summaries of the knowledge graph, and discovering related entities and relations. For example, in a knowledge graph of movies, one can discover similar types of movies, actors, or directors by clustering the knowledge graph. This problem is essentially identical to clustering of heterogeneous information networks. The problem of clustering homogeneous networks is an old one, and the classical methods for this problem include the Girvan-Newman algorithm [64], the Kernighan-Lin algorithm [100], and the METIS algorithm [97]. However, all these methods are designed for cases in which the graph contains edges of a single type. A higher quality clustering can be obtained when the type of link is explicitly incorporated into the clustering process. There are several ways of achieving this goal:

1. One can use shared matrix factorization in order to create representations of each entity. This shared matrix factorization approach is also discussed in an earlier section in order to create completed versions of incomplete knowledge graphs. We assume that we have a total of n link types and m entity types. For the $n \times n$ matrix D_i associated with the ith link type (out of a total of m link types), one performs the following factorization:

$$D_i \approx U_i V^T \quad \forall i \in \{1 \dots m\}$$

   In this case, V is the $n \times k$ shared factor, and the matrices $U_1 \dots U_m$ are the $n \times k$ factor matrices corresponding to the m different link types. By clustering the rows in V one can cluster the entities of different types in a single clustering that uses relationships of multiple types. Alternatively, the concatenation of the jth rows of $U_1 \dots U_m$ and the jth row of V provides a multidimensional representation of the jth entity. This expanded representation can also be used for clustering. The methodology for creating the embedding is discussed below.

2. In some cases, content attributes are associated with entity nodes. Therefore, text attributes have to be used as first-order citizens in the modeling process. These types of settings can also be addressed using shared matrix factorization methods, where separate matrices are set up for the content attributes. Examples of methods that use text attributes in conjunction with the network structure for clustering are discussed in [179, 195].

Irrespective of how the matrices are set up (with or without content), the following objective function is created:

$$J = \|D_1 - U_1 V^T\|_F^2 + \sum_{i=2}^{m} \beta_i \|D_i - U_i V^T\|_F^2$$

The concatenation of the jth rows of $U_1 \ldots U_m$ and the jth row of V provides a multidimensional representation of the jth entity. The parameters can be optimized with the use of gradient descent:

$$U_1 \Leftarrow U_1 - \alpha \frac{\partial J}{\partial U_1} = U_1 + \alpha(D_1 - U_1 V^T)V$$

$$U_i \Leftarrow U_i - \alpha \frac{\partial J}{\partial U_i} = U_i + \alpha\beta_i(D_i - U_i V^T)V \quad \forall i \geq 2$$

$$V \Leftarrow V - \alpha \frac{\partial J}{\partial V} = V + \alpha(D_1 - U_1 V^T)^T U_1 + \alpha \sum_{i=2}^{m} \beta_i(D_i - U_i V^T)^T U_i$$

The updates are repeated to convergence. Here, α is the learning rate. Note that this approach is almost identical to the methods used for handling incompleteness (cf. Section 12.3.3). In general, a wide variety of methods have been proposed for clustering heterogeneous information networks. An overview of these methods is provided in [178].

12.4.3 Entity Classification

The problem of entity classification is also referred to as that of collective classification in the field of information and social network analysis. Aside from its application-centric use, entity classification is used for completing the missing information in knowledge graphs. This application is often used to infer missing properties of nodes in information networks. For example, imagine a situation where we have a new person node being inserted into the information network, but we do not know whether it is an actor node or a director node. However, by analyzing its relationships with other nodes, it is possible to infer the type of the node with a high degree of certainty. For example, an actor node will be connected to movies by a different type of link (e.g., *acted in* link) than will a director node (e.g., *directed* link).

For homogeneous networks, the basic principle of collective classification relies on the principle of *homophily*. This principle suggests that nodes with similar labels are connected to one another. Therefore, nodes can be appropriately classified based on their (structural) distances to other labeled nodes. However, in the case of heterogeneous networks, such simplistic principles of homophily might not always work very well. The reason is that the properties of nodes in such heterogeneous networks depend on the specific pattern of the link types that emanate from a node, and the broader structural patterns associated with such link types.

The simplest approach for link prediction is to use an embedding method in order to transfer each node to a multidimensional representation. The approach used in the previous section (on clustering) can be used to create a multidimensional representation of each object. In other words, the matrix D_i for each link type is factorized as $D_i \approx U_i V^T$. Subsequently, the following objective function is created:

$$J = \|D_1 - U_1 V^T\|_F^2 + \sum_{i=2}^{m} \beta_i \|D_i - U_i V^T\|_F^2$$

This objective function is identical to that used for creating clustered embeddings (cf. Section 12.4.2). The matrix V contains a k-dimensional representation of the ith node in the ith row. Therefore, the problem can be transformed to that of multidimensional classification by using this k-dimensional representation in the rows of V in order to apply an off-the-shelf classifier. The labeled nodes correspond to the training data, whereas the unlabeled rows correspond to the test data. Any of the methods discussed in Chapter 6 can be used for the classification process. The parameters $\beta_2 \ldots \beta_m$ of the embedding can be chosen in out-of-sample fashion in order to maximize classification accuracy. Another interesting method for collective classification in such graphs is presented in [105]. It is also possible to incorporate some level of supervision into the embedding process.

12.4.4 Link Prediction and Relationship Classification

Finally, an important problem is link prediction and relationship classification, which are almost the same problem in the context of heterogeneous information networks. The link prediction problem predicts the different pairs of nodes between which links are most likely to appear for a particular link type. In homogeneous networks, one only has to predict the pairs of nodes between which links are most likely to appear in the future in link prediction. However, in heterogeneous information networks, the problem becomes much more complex, because one not only has to predict whether a link occurs, but one also has to predict the type of link between a pair of nodes. In relationship classification, one is given a pair of nodes with the information that a link does exist between them. Using this information, one has to predict the type of the link. The relationship classification problem is a simpler subproblem of the link prediction problem, wherein one already knows that a link appears between a pair of nodes, and one has to associate a specific relationship type with that link. Note that the link prediction problem automatically performs relationship classification because every predicted link has a relationship type associated with it.

As in the case of the previous problems discussed in this section, the embedding approach is a time-tested method to address most machine learning problems in knowledge graphs. One reason for this is that the inherent structure of knowledge graphs is rather ugly, with a multitude of different types of entities and links. However, the embedding approach transforms all nodes to the same multidimensional representation, and can therefore be leveraged in a unified way across most problems.

The link prediction problem is identical to the problem of handling incompleteness in knowledge graphs, as discussed in Section 12.3.3. The approach discussed in that section was based on matrix factorization. Furthermore, several applications discussed in this chapter, such as clustering and classification are based on a similar matrix factorization approach. As in the case of the previous applications discussed in this section, one can factorize the node-node link matrix for the ith link matrix as $D_i \approx U_i V^T$. Subsequently, one can examine the entries of each $U_i V^T$ in order to examine the propensity to form a link of the ith type. The entries with the largest values for $U_i V^T$ for each i can be reported[2] as the links of the ith type that are most likely to form in the future. Furthermore, in the case of the relationship classification problem, one can first start with matrix $D_1 \ldots D_m$ in which the entries have been normalized to the same mean value via scaling. Subsequently, after factorization, one can compare the values of the corresponding entries in $U_1 V^T \ldots U_m V^T$. If the (p, q)th entry of $U_j V^T$ has the largest value out of the set of matrices $U_1 V^T \ldots U_m V^T$, then the edge (p, q) can be classified to the link type j.

An alternative approach for relationship classification would be to create multidimensional representations of pairs of nodes by concatenating the representations of the individ-

---

[2]In most cases, one is interested in the incoming or outgoing links from a particular node. For outgoing links from the jth node, one can pull out the jth row $\overline{u}_{ij}$ of each U_i and multiply that row with V^T to create the row vector $\overline{u}_{ij} V^T$ of predictions. For links incoming into the jth node, once can pull out the jth row $\overline{v}_j$ of V, and create $U_i \overline{v}_j^T$ as a column vector of predictions.

ual nodes. Note that the multidimensional representation of the ith node can be obtained by using $d * m$ attributes for each node. For each if the m link types, there is one attribute for each node, which takes on a value of 1 if an edge exists from that node to the ith node. One can then create training data based on the labels of the corresponding node pairs. An off-the-shelf multidimensional classifier can be used to train on this data. For a given node pair for which the relationship is unknown, one can then use the trained classifier to predict the relationship type.

12.4.5 Recommender Systems

Knowledge graphs can be naturally used in conjunction with recommender systems, especially since recommender systems already use a variety of matrix factorization models. For example, consider a movie database in which one has a connected set of entities corresponding to actors, directors, movies, and so on. The $n \times n$ matrices corresponding to the m different link types are denoted by $D_1 \ldots D_m$. As before, we factorize each matrix $D_i = U_i V^T$, where U_i is of size $n \times k$, and V is of size $n \times k$.

In addition, one might have a ratings matrix R for the n entities. The ratings matrix is of size $u \times n$, since a total of u users have rated the n entities. A key point is that the ratings matrix is incompletely specified, which tends to make the learning process more challenging. The ratings matrix is then factorized as follows:

$$R \approx MV_1^T$$

Here, M is a $u \times k$ matrix containing the user factors. Therefore, the overall objective function of this recommendation problem is as follows:

$$J = \|R - MV^T\|_F^2 + \sum_{i=1}^{m} \beta_i \|D_i - U_i V^T\|_F^2$$

In the above optimization problem, there is some abuse of notation, because the Frobenius norm of $(R - UV^T)$ is aggregated only over the entries of R that are specified, and the missing entries are ignored. This optimization problem can be solved by using a similar gradient-descent technique to the techniques discussed in Section 12.3.3. However, care needs to be account for the fact that the ratings matrix is not fully specified, and therefore one can only use observed entries in order to make the updates. We leave the derivation of the gradient descent steps as an exercise (see Exercise 5). Note that we have m hyper-parameters $\beta_1 \ldots \beta_m$, which correspond to the weights of the different link types. These hyper-parameters can be set by estimating the accuracy of the recommendation model by holding out a subset of the ratings during gradient descent, and then setting the hyperparameters so that the accuracy of the prediction model on these held-out ratings is maximized. Once the matrices M, U_i, and V have been learned, the incomplete ratings matrix can be reconstructed as $R \approx MV^T$.

It is also possible to extend the approach to an entity of a single type (e.g., movies or directors). In such a case, the matrix R is of size $u \times n_1$, which contains the ratings of the u users for the n_1 items. Furthermore, we extract the n_1 rows of V corresponding to the rated entity type to create the smaller matrix V_1. In other words, V_1 is of size $n_1 \times k$, and it contains a subset of the rows of V. It is possible to modify the above objective function in this case, so that the first term in the objective function is the squared Frobenius norm of $(R - MV_1^T)$. The ratings matrix can then be reconstructed as $R \approx MV_1^T$.

12.5   Summary

Knowledge graphs have been used in diverse communities over the years for a variety of applications, such as the semantic Web, knowledge base representation, and heterogeneous information network analysis. Knowledge graphs are closely related to knowledge bases in first-order logic, except that they represent relationships (predicates) in the form of edges in a graph. Furthermore, knowledge graphs are greatly simplified compared to the restrictive jargon of first-order logic, which makes them much easier to use. Knowledge bases were popularized by Google in 2012 as a way of enhancing search and query processing. However, the broader principle of using heterogeneous information networks and/or ontologies precedes the literature on knowledge graphs. Knowledge graphs are constructed with the use of curation, rule-based methods, or via fully automated learning methods. In many cases, the graph may be constructed from semi-structured of unstructured data as a starting point. Once the knowledge graphs have been constructed, they can be used in a wide variety of applications, such as search, clustering, entity classification, relationship classification, and recommender systems. It is more common to use knowledge graphs for machine learning applications in recent years, as compared to applications based on reasoning. A unified theme in many of these machine learning methods is to be able to engineer multidimensional features from the underlying graph structure, which can then be used with off-the-shelf clustering and classification methods.

12.6   Further Reading

An overview of knowledge graphs in the context of machine learning may be found in [130]. An overview of heterogeneous information networks may be found in [178]. A discussion of automated methods for building knowledge graphs from unstructured text is provided in [7, 156]. An overview of knowledge graph methods for recommendations may be found in [135, 196, 197].

12.7   Exercises

1. Consider a repository of scientific articles containing articles published in various types of venues. You want to create a heterogeneous network containing three types of objects corresponding to articles, venues, and authors. Propose the various relationship types that you can construct from this heterogeneous information network. Suppose that you had additional information about authors and venues corresponding to their subject matter (which is hierarchically classified). Discuss how you can use this information to create an ontology to support the lower level instances in the knowledge graph.

2. Consider the DBLP publication database available at the URL https://dblp.uni-trier. de/xml/. Implement a program to create a heterogeneous information network discussed in Exercise 1. You may omit the step involving creation of the concept hierarchy.

3. Consider a repository of movies appearing in different countries. For each movie, you have a hierarchical classification corresponding to the genre. You want to create a heterogeneous network containing four types of objects corresponding to movies, country of origin, actors, and directors. Propose the various relationship types that you

can construct from this heterogeneous information network. Propose an ontology that is paired with this heterogeneous information network in order to create a knowledge graph.

4. Consider the IMDB movie database available at the URL https://www.imdb.com/interfaces/. Implement a program to create a heterogeneous information network discussed in Exercise 3. Include a concept hierarchy for movie objects based on the genres of the movies.

5. Compute the gradient-descent steps of the optimization model introduced in Section 12.4.5. Show that the gradient-descent steps are as follows:

$$M \Leftarrow M + \alpha EV$$

$$U_i \Leftarrow U_i + \alpha \beta_i (D_i - U_i V^T) V$$

$$V \Leftarrow V + \alpha E^T M + \alpha \sum_{i=1}^{m} \beta_i (D_i - U_i V^T)^T U_i$$

Here, α is the learning rate, and E is an error matrix $E = R - MV^T$, where the missing entries of R are set to values of 0 in E. In other words, if the (i, j)th entry of R is missing, then the (i, j)th entry of E is set to 0.

6. Show how you can perform the steps of the Exercise 5 with the use of stochastic gradient descent rather than gradient descent.

Chapter 13

Integrating Reasoning and Learning

"The temptation to form premature theories upon insufficient data is the bane of our profession."— The fictional character, Sherlock Holmes, in *The Valley of Fear*, authored by Arthur Conan Doyle

13.1 Introduction

In the previous chapters, we have discussed two primary schools of thought in artificial intelligence, which correspond to deductive reasoning and inductive learning, respectively. The class of deductive reasoning methods corresponds to techniques such as search, propositional logic, and first-order logic, whereas the class of inductive learning methods corresponds to techniques such as linear regression, support vector machines, and neural networks. The field of artificial intelligence was largely dominated by deductive reasoning methods and symbolic artificial intelligence during the early years. This emphasis was caused in part by the limited data availability and also the limited computational power, which served as impediments to learning-centric methods. However, as the availability of data and computational power increased, learning methods became increasingly popular. Furthermore, deductive reasoning methods failed to live up to their promise. Nevertheless, some deductive assumptions were always used to reduce data requirements in inductive learning methods, and therefore some elements of deductive reasoning were always involved in the overall process. These elements sometimes took on the form of *prior assumptions* that were used to reduce the data requirements of the learning process. An example of such a prior assumption in the optimization model of least-squares regression is as follows:

> When learning the parameter vector $\overline{W}$ in the linear regression prediction $y = \overline{W} \cdot \overline{X}$, one should select a parameter vector $\overline{W}$ that is as concise as possible (e.g., with small L_2-norm), when choosing between two values of $\overline{W}$ with almost similar predictive accuracy on the training data. The accuracy of the concise solution on the test data is often better, even when it is slightly worse on the training data.

© Springer Nature Switzerland AG 2021
C. C. Aggarwal, *Artificial Intelligence*, https://doi.org/10.1007/978-3-030-72357-6_13

This type of assumption is often baked into the objective function of learning algorithms as a form of *regularization*. Regularization adds a penalty to the objective function, which is proportional to the squared norm of the parameter vector $\overline{W}$. The key point is that it uses a *prior hypothesis* that assumes that parameter vectors with smaller norms are better. Therefore, regularization is a subtle form of deductive reasoning that is embedded into many learning algorithms (although it is rarely viewed from that perspective). One of the reasons for this is that using only the regularization term to perform the modeling results in $\overline{W} = \overline{0}$, which is obviously not a useful and informative result. There are also several informative ways of performing regularization, which often look more similar to deductive reasoning. For example, if constraints are imposed on the parameter space as a result of specific insights about the problem domain, it can often improve the accuracy of the underlying predictions by allowing a more accurate model to be constructed in the restricted space of solutions (with limited data). In such cases, one can often obtain somewhat more informative results (than norm-based regularization) even when no data is used in the modeling. In almost all such cases, the use of domain knowledge reduces the data requirements of learning algorithms at a given level of predictive accuracy.

In general, the core reason for incorporating some form of deductive reasoning within an inductive learning algorithm is to reduce the underlying data requirements. For example, purely deductive reasoning algorithms work *only* with domain knowledge and no data. However, such algorithms have a large amount of bias that is based on the prior assumptions of the domain expert. In recent years, learning algorithms have become increasingly popular because of increasing availability of data. However, in most cases, some type of assumption is always used to reduce data requirements.

Another reason for incorporating deductive reasoning into inductive learning is that the former class of methods is generally easily interpretable, whereas (purely) inductive learning methods are not quite as interpretable. Deductive reasoning methods are interpretable because the statements used in knowledge bases are often derived from interpretable facts about the domain of discourse. On the other hand, the hypotheses in machine learning models are often encoded as a cryptic function containing a myriad of parameters, which makes the overall model somewhat uninterpretable. This becomes particularly evident when one is working with complex models like neural networks. By combining inductive learning and deductive reasoning methods, one is often able to create methods that can learn from relatively few examples, and they also tend to be somewhat more interpretable.

The importance of data availability in making choices between inductive learning and deductive reasoning methods has been repeatedly confirmed by our experience with many game-playing programs. The earliest forms of chess-playing programs used minimax trees in combination with domain-specific evaluation functions that were hand crafted. These types of methods are discussed in Chapter 3, and they clearly belong to the deductive school of thought. Almost all the top chess programs were designed using this approach (until very recently). As time passed, evaluation functions for chess programs are designed using supervised learning, and then reinforcement learning methods were also designed. Recent methods for chess playing combine Monte Carlo search trees with statistical estimations of probabilities of success along each branch tree. In conventional chess programs, only human-designed evaluation functions are used in combination with such trees. Such a combination of search trees with data-driven analysis can be viewed as a hybrid method between the deductive and inductive schools of thought of artificial intelligence. Indeed, most learning methods use some assumption or the other in order to reduce data requirements by incorporating a *domain-specific bias* into the learning. Incorporating this type of bias can also cause errors because the domain-specific knowledge might not reflect many of the specific

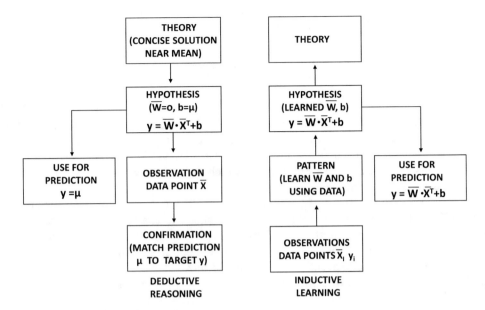

Figure 13.1: Revisiting the two schools of thought in artificial intelligence in the context of the regularization components and loss components of machine learning. Creating a predictor based on only the regularization (deductive) component will result in a predicted value of the domain mean μ for each test instance, which is obviously not accurate. At the same time, the inductive approach might give poor predictions for smaller data sets because of the specific nuances of the training data set (and a domain mean might turn out to be more accurate)

characteristics of the function that can be learned in a data-driven manner. While the domain knowledge is almost always useful in the presence of limited data, it is sometimes an impediment to the construction of more powerful models when a sufficient amount of data and computational power is present. Therefore, there is a natural trade-off between the *bias* caused by inappropriately strong domain-specific assumptions, and the *variance* (i.e., random error) caused by paucity of data. This trade-off is known in machine learning as the *bias-variance trade-off*. This trade-off will be discussed in the next section. It is noteworthy that this trade-off is also naturally present in human learning, where decisions are often achieved by a trade-off between prior beliefs and the additional knowledge obtained from observations in real life. Very strong prior biases about events can cause errors just as decisions based on few observations can also cause errors. The best decisions are often made using a combination of these two different ways of gaining knowledge.

In order to understand how combining the inductive and deductive schools of thought lead to a more robust model, we will use regularization as a test case. Consider a setting where one has a small amount of data for learning the parameters of a linear regression model. In this case, one wishes to learn the following model for n training pairs $(\overline{X}_i, y_i)$ and d-dimensional parameter vector $\overline{W}$ and bias b:

$$y_i \approx \overline{W} \cdot \overline{X}_i^T + b$$

One can then set up an optimization model as follows:

$$J = \sum_{i=1}^{n}(y_i - W \cdot \overline{X}_i^T - b)^2$$

One can learn $\overline{W}$ and b in a purely data-driven manner with gradient descent. This is classical inductive learning. Note that using a small amount of data to learn the linear model will lead to highly erroneous results on out-of-sample data, because different training data sets will have random nuances that will affect the predictions prominently for smaller data sets. This shows that there are cases in which inductive learning has a difficult time in making accurate predictions.

On the other hand, if the data is limited, one could set $\overline{W}$ and b using our knowledge of the domain at hand. For example, let μ be the mean of the target vector *based on the analyst's knowledge of the domain at hand* (rather than averaging[1] the target vectors). Then, one might consider making the *hypothesis* that unless one has additional knowledge about the effect of feature attributes, the vector $\overline{W}$ should have no effect (i.e., has small L_2-norm) and the bias b is the mean of the target attribute in the domain at hand. This results in setting $\overline{W} = \overline{0}, b = \mu$, which yields a prediction of $y_i = \mu$ for each point. Clearly, this solution is not too informative in terms of variability across various data points (although most deductive reasoning methods are more informative in general). However, *in the absence of sufficient data*, this is a reasonable starting point. In fact, if the number of training points is very small, this simplistic prediction might turn out to be more accurate than the purely inductive learning model obtained by minimizing $\sum_i \|y_i - \overline{W} \cdot \overline{X}_i^T - b\|^2$.

The relationships of the loss components and the regularization components to inductive learning and deductive reasoning are shown in Figures 13.1. In the case of deductive reasoning, it is common to have situations where the prediction cannot adjust to the true complexity of the underlying hypothesis (such as predicting every point to the domain mean). This occurs because deductive systems are often blunt instruments that are unable to adjust to the nuances to different parts of the feature space with the use of limited human knowledge — the knowledge that is implicit in the data is often far more complex than any human can interpret and encode into a knowledge base. The inflexibility of deductive reasoning systems in adapting to the complexity of different situations is a common problem, which is referred to as *bias*. On the other hand, purely inductive systems might have errors caused by too much variation across individual training data sets. In other words, if we change the training data set to a different one, the same test point might be predicted quite differently. Obviously, this type of instability is a problem as well, and it is referred to as *variance*, which increases when the data availability is small. A key point is that it is often possible to improve prediction accuracy (with limited data) by integrating inductive learning and deductive reasoning. The idea is to start with a reasonable hypothesis from the deductive reasoning system, and overrule it only when sufficient evidence is available from the inductive learning process. This is, in fact, exactly how humans continually gain knowledge/hypotheses, use them to make predictions, and update their knowledge/hypotheses with incremental learning when predictions do not match their prior hypotheses.

One way of achieving a happy medium between the two schools of thought is to create a combined objective function that uses both the data and the domain knowledge. Although we often view regularized machine learning models as purely inductive systems, they can also be viewed as a combination of inductive learning (by using the data) and deductive reasoning (by imposing the data-free conciseness/domain knowledge hypothesis). This leads

---

[1] In other words, we expect μ to be roughly equal to $\sum_{i=1}^{n} y_i/n$ when the value of n is large.

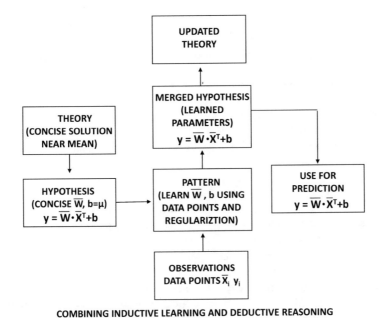

COMBINING INDUCTIVE LEARNING AND DEDUCTIVE REASONING

Figure 13.2: Combining the two schools of thought in artificial intelligence with regularized learning as an example

to the following objective function[2] of linear regression:

$$J = \underbrace{\frac{1}{2}\sum_{i=1}^{n}(y_i - \overline{W} \cdot \overline{X}_i^T - b)^2}_{\text{Learning}} + \underbrace{\frac{\lambda}{2}\|\overline{W}\|^2 + \frac{\lambda}{2}(b - \mu)^2}_{\text{Hypothesis}} \qquad (13.1)$$

Here, λ is the regularization parameter, which controls the influence of the deductive reasoning component. The choice of the regularization parameter λ controls how much weight one provides to each part of the hypothesis. This weight controls the trade-off between the errors caused by inappropriately strong domain-specific assumptions and the random errors caused by paucity of data. Therefore, this approach combines the known theories (e.g., concise solutions that predict near the domain mean are better) into inductive learning systems. The specific combination of inductive learning and deductive reasoning that is inherent in linear regression is shown in Figure 13.2.

The aforementioned exposition provides the most rudimentary example of combining inductive learning and deductive reasoning. However, in practice, far more complex forms of integration are possible. Some examples are as follows:

- One can use linguistic domain knowledge in an inductive learning system to reduce training data requirements in a sequence-to-sequence translation system between different languages. Modern translators between different languages are based on inductive learning with recurrent autoencoders, although there is significant scope in reducing data requirements with added linguistic domain knowledge.

---

[2]Normally, the domain mean is not used in linear regression, and the bias is not regularized. We have used the domain mean in order to make the use of domain knowledge more apparent. Even if the domain mean were not used, regularization would still be considered the use of a form of domain knowledge.

- One can use scientific laws to improve predictions, when inductive learning is used in the sciences like physics. When the attributes represent variables with known relationships among them, the underlying relationship can be used to improve predictions.

- One can use domain knowledge about images in order to reduce training data requirements for image classification. For example, a prior model for extraction of relevant features of the images can be used for a new and different task. This approach is the essence of an idea, referred to as *transfer learning*, where successive use of inductive learning systems to gain knowledge and update it is also seen as a form of integration of the two schools of thought.

In many cases, where there are known and "trustworthy" relationships between feature variables and the targets (e.g., the laws of physics), it would seem wasteful to not use such knowledge. Therefore, it is common to incorporate this type of knowledge in some form within the learning algorithm.

In the early years of artificial intelligence, the paucity of data was a more important factor, and even when sufficient data was available, computational power was limited. This situation contributed quite significantly to the preponderance of symbolic artificial intelligence and deductive reasoning systems over inductive learning systems like neural networks. However, inductive learning systems like deep neural networks have become increasingly popular in recent years, with increasing data availability and computational power. The integration of the two schools of thought further helps expand the scope of learning to increasingly complex scenarios. This chapter will focus on this process and the principles underlying it.

This chapter is organized as follows. The next section introduces the bias-variance trade-off in machine learning. The discussion of a generic meta-framework for combining reasoning and learning is provided in Section 13.3. This meta-framework can also be used to improve the accuracy of purely inductive learning methods. Transfer learning methods are given in Section 13.4. Lifelong learning methods are discussed in Section 13.5. Neuro-symbolic methods for artificial intelligence are discussed in Section 13.6. A summary is given in Section 13.7.

13.2   The Bias-Variance Trade-Off

The bias-variance trade-off provides the raison d'etre for why one might want to combine the inductive learning and deductive reasoning schools of thought. While learning methods can usually provide the most creative and accurate models *given sufficient data*, it would be a mistake to avoid the available background knowledge when it is indeed available. The regularization approach discussed in the previous section is a very primitive way of combining these ways of thinking and one can often obtain much more informative forms of background knowledge. If the amount of available data is limited, these types of background knowledge become particularly important in to order to make predictions for instances where similar examples are not available.

The bias-variance trade-off states that the squared error of a learning algorithm can be partitioned into three components:

1. *Bias:* The bias is the error caused by the simplifying assumptions in the model, which causes certain test instances to have consistent errors across different choices of training data sets. In purely deductive reasoning systems, no training data might be used,

and therefore a particular test instance will always be classified in exactly the same way every time it is presented to the system. For example, consider a case where one is trying to classify emails as *"spam"* or *"non-spam"* with the text of the email. One has the background knowledge available that many spam emails contain the phrase *"Free Money!."* A variety of other ad hoc rules might be used by a domain expert. In such cases, performing the classification will cause errors for test instances if the ad hoc rules are somewhat incorrect or if they are incomplete in terms of their ability to catch spam emails. Furthermore, even in inductive learning systems like linear regression, assumptions (e.g., linearity of relationship) are made about the nature of the relationship between feature variables and the dependent variables. Such assumptions are frequently erroneous and they lead to consistent errors on particular training instances, even if an infinite amount of data is available. In general, *any inappropriately strong assumption on the model will cause bias*. Most models in the real world do make such inappropriate assumptions. In the words of George Box, *"All models are wrong, but some are useful."* Almost all models have non-zero bias, which contributes to this principle.

2. *Variance:* Variance is caused by the inability to learn all the parameters of the model in a statistically robust way, especially when the data is limited and the model tends to have a larger number of parameters. The presence of higher variance is manifested by overfitting to the specific training data set at hand. For example, if one runs a linear regression model in d dimensions with fewer than d training data points, one will obtain wildly different predictions on the same test instance, when different training data sets are used. Obviously, all these different predictions cannot be correct – the reality is that almost all of them are likely to be highly erroneous on *out-of-sample test points*, because the model adjusts the vagaries and nuances of the specific training data set at hand. These vagaries are particularly noticeable for small training data sets, and the high level of variation in prediction of the same test instance over different training data sets is referred to as *variance*. It is noteworthy that a high-variance model will often provide a deceptively high accuracy on the training data, which does not generalize well to out-of-sample test data. It is even possible to conceive of settings in which one obtains 100% accuracy on the training data, whereas the performance on the test data is extremely poor. For example, training a linear regression model on d-dimensional data with less than d training instances will show this type of behavior.

3. *Noise:* The noise is caused by the inherent error in the data. For example, data collection mechanisms often incorporate all types of unintended errors in the data. For example, if one were to perform linear regression on a data set in which the target attribute is the temperature, then the errors in collecting the data because of the limitations of the hardware will cause the model predictions to vary from the observed values. This type of error is referred to as *noise*.

The above description provides a qualitative view of the bias-variance trade-off. In the following, we will provide a more formal and mathematical view.

13.2.1 Formal View

It is clear from the above discussion that bias is observed not only in models with a deductive reasoning component, but also in inductive systems where particular types of domain-specific assumptions are made. In general, incorporating any kind of premise (e.g., linear relationship between features and dependent variable) in an inductive learning system will lead to bias. Most inductive learning systems do have premises built into them either directly or indirectly. For example, choosing large levels of regularization can also cause bias in inductive learning systems. Reducing the weight of regularization will reduce the bias, but it will increase the variance. Choosing the correct level of regularization is important in regulating this trade-off appropriately.

In this section, we will examine the nature of this trade-off between bias and variance using at least a partially inductive setting in which some training data is available for influencing the classification (although some level of fixed domain knowledge or simplifying assumptions might add to the bias). It does not make sense to perform the analysis using a purely deductive system in which the same result will be output for a test instance, given fixed domain knowledge. We assume that the base distribution from which the training data set is generated is denoted by $\mathcal{B}$. One can generate a data set $\mathcal{D}$ from this base distribution:

$$\mathcal{D} \sim \mathcal{B} \tag{13.2}$$

One could draw the training data in many different ways, such as selecting only data sets of a particular size. For now, assume that we have some well-defined generative process according to which training data sets are drawn from $\mathcal{B}$. The analysis below does not rely on the specific mechanism with which training data sets are drawn from $\mathcal{B}$, nor does it rely on what the base distribution looks like.

Access to the base distribution $\mathcal{B}$ is equivalent to having access to an infinite resource of training data, because one can use the base distribution an unlimited number of times to generate training data sets. In practice, such base distributions (i.e., infinite resources of data) are not available. As a practical matter, an analyst uses some data collection mechanism to collect only *one finite instance* of $\mathcal{D}$. However, the conceptual existence of a base distribution from which other training data sets can be generated is useful in theoretically quantifying the sources of error in training on this finite data set.

Now imagine that the analyst had a set of t test instances in d dimensions, denoted by $\overline{Z}_1 \ldots \overline{Z}_t$. The dependent variables of these test instances are denoted by $y_1 \ldots y_t$. For clarity in discussion, let us assume that the test instances and their dependent variables were also generated from the same base distribution $\mathcal{B}$ by a third party, but the analyst was provided access only to the feature representations $\overline{Z}_1 \ldots \overline{Z}_t$, and no access to the dependent variables $y_1 \ldots y_t$. Therefore, the analyst is tasked with job of using the single finite instance of the training data set $\mathcal{D}$ in order to predict the dependent variables of $\overline{Z}_1 \ldots \overline{Z}_t$.

Now assume that the relationship between the dependent variable y_i and its feature representation $\overline{Z}_i$ is defined by the *unknown* function $f(\cdot)$ as follows:

$$y_i = f(\overline{Z}_i) + \epsilon_i \tag{13.3}$$

Here, the notation ϵ_i denotes the intrinsic noise, which is independent of the model being used. The value of ϵ_i might be positive or negative, although it is assumed that $E[\epsilon_i] = 0$. If the analyst knew what the function $f(\cdot)$ corresponding to this relationship was, then they could simply apply the function to each test point $\overline{Z}_i$ in order to approximate the dependent variable y_i, with the only remaining uncertainty being caused by the intrinsic noise.

The problem is that the analyst does not know what the function $f(\cdot)$ is in practice. Note that this function is used within the generative process of the base distribution $\mathcal{B}$, and the entire generating process is like an oracle that is unavailable to the analyst. The analyst only has examples of the input and output of this function. Clearly, the analyst would need to develop some type of *model* $g(\overline{Z}_i, \mathcal{D})$ using the training data in order to *approximate* this function in a data-driven way.

$$\hat{y}_i = g(\overline{Z}_i, \mathcal{D}) \tag{13.4}$$

Note the use of the circumflex (i.e., the symbol '^') on the variable $\hat{y}_i$ to indicate that it is a *predicted* value by a specific algorithm rather than the observed (true) value of y_i.

All prediction functions of learning models (including neural networks) are examples of the estimated function $g(\cdot, \cdot)$. Some algorithms (such as linear regression and the SVM) can be expressed in a concise and understandable way, although other learning algorithms might not be expressible in this way:

$$g(\overline{Z}_i, \mathcal{D}) = \underbrace{\overline{W} \cdot \overline{Z}_i^T}_{\text{Learn } \overline{W} \text{ with } \mathcal{D}} \quad [\text{Linear Regression}]$$

$$g(\overline{Z}_i, \mathcal{D}) = \underbrace{\text{sign}\{\overline{W} \cdot \overline{Z}_i^T\}}_{\text{Learn } \overline{W} \text{ with } \mathcal{D}} \quad [\text{SVM}]$$

Most neural networks are expressed algorithmically as compositions of multiple functions computed at different nodes. The choice of computational function includes the effect of its specific parameter setting, such as the coefficient vector $\overline{W}$ in linear regression or the SVM. Neural networks with a larger number of units will require more parameters to fully learn the function. This is where the variance in predictions arises on the same test instance; a model with a large parameter set $\overline{W}$ will learn very different values of these parameters, when a different choice of the training data set is used. Consequently, the prediction of the same test instance will also be very different for different training data sets. These inconsistencies add to the error. On the other hand, neural networks with few units will tend to construct inflexible models with high levels of bias. In some cases, simplifying assumptions (or domain-specific assumptions) can be added to inductive learning algorithms, which can increase bias, although these will usually reduce variance.

The goal of the bias-variance trade-off is to quantify the expected error of the learning algorithm in terms of its bias, variance, and the (data-specific) noise. For generality in discussion, we assume a numeric form of the target variable, so that the error can be intuitively quantified by the *mean-squared error* between the predicted values $\hat{y}_i$ and the observed values y_i. This is a natural form of error quantification in regression, although one can also use it in classification in terms of probabilistic predictions of test instances. The mean-squared error, MSE, of the learning algorithm $g(\cdot, \mathcal{D})$ is defined over the set of test instances $\overline{Z}_1 \ldots \overline{Z}_t$ as follows:

$$MSE = \frac{1}{t} \sum_{i=1}^{t} (\hat{y}_i - y_i)^2 = \frac{1}{t} \sum_{i=1}^{t} (g(\overline{Z}_i, \mathcal{D}) - f(\overline{Z}_i) - \epsilon_i)^2$$

The best way to estimate the error in a way that is independent of the specific choice of training data set is to compute the *expected* error over different choices of training data sets:

$$E[MSE] = \frac{1}{t} \sum_{i=1}^{t} E[(g(\overline{Z}_i, \mathcal{D}) - f(\overline{Z}_i) - \epsilon_i)^2]$$

$$= \frac{1}{t} \sum_{i=1}^{t} E[(g(\overline{Z}_i, \mathcal{D}) - f(\overline{Z}_i))]^2 + \frac{\sum_{i=1}^{t} E[\epsilon_i^2]}{t}$$

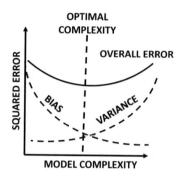

Figure 13.3: The trade-off between bias and variance usually causes a point of optimal model complexity

The second relationship is obtained by expanding the quadratic expression on the right-hand side of the first equation, and then using the fact that the average value of ϵ_i over a large number of test instances is 0.

The right-hand side of the above expression can be further decomposed by adding and subtracting $E[g(\overline{Z}_i, \mathcal{D})]$ within the squared term on the right-hand side:

$$E[MSE] = \frac{1}{t}\sum_{i=1}^{t} E[\{(f(\overline{Z}_i) - E[g(\overline{Z}_i, \mathcal{D})]) + (E[g(\overline{Z}_i, \mathcal{D})] - g(\overline{Z}_i, \mathcal{D}))\}^2] + \frac{\sum_{i=1}^{t} E[\epsilon_i^2]}{t}$$

One can expand the quadratic polynomial on the right-hand side to obtain the following:

$$E[MSE] = \frac{1}{t}\sum_{i=1}^{t} E[\{f(\overline{Z}_i) - E[g(\overline{Z}_i, \mathcal{D})]\}^2]$$

$$+ \frac{2}{t}\sum_{i=1}^{t} \{f(\overline{Z}_i) - E[g(\overline{Z}_i, \mathcal{D})]\}\{E[g(\overline{Z}_i, \mathcal{D})] - E[g(\overline{Z}_i, \mathcal{D})]\}$$

$$+ \frac{1}{t}\sum_{i=1}^{t} E[\{E[g(\overline{Z}_i, \mathcal{D})] - g(\overline{Z}_i, \mathcal{D})\}^2] + \frac{\sum_{i=1}^{t} E[\epsilon_i^2]}{t}$$

The second term on the right-hand side of the aforementioned expression evaluates to 0 because one of the multiplicative factors is $E[g(\overline{Z}_i, \mathcal{D})] - E[g(\overline{Z}_i, \mathcal{D})]$. On simplification, we obtain the following:

$$E[MSE] = \underbrace{\frac{1}{t}\sum_{i=1}^{t}\{f(\overline{Z}_i) - E[g(\overline{Z}_i, \mathcal{D})]\}^2}_{\text{Bias}^2} + \underbrace{\frac{1}{t}\sum_{i=1}^{t} E[\{g(\overline{Z}_i, \mathcal{D}) - E[g(\overline{Z}_i, \mathcal{D})]\}^2]}_{\text{Variance}} + \underbrace{\frac{\sum_{i=1}^{t} E[\epsilon_i^2]}{t}}_{\text{Noise}}$$

In other words, the squared error can be decomposed into the (squared) bias, variance, and noise. The variance is the key term that prevents neural networks from generalizing. In general, the variance will be higher for neural networks that have a large number of parameters. On the other hand, too few model parameters can cause bias because there are not sufficient degrees of freedom to model the complexities of the data distribution. This trade-off between bias and variance with increasing model complexity is illustrated in

Figure 13.3. Clearly, there is a point of optimal model complexity where the performance is optimized. Furthermore, paucity of training data will increase variance. However, careful choice of design can reduce overfitting. One way of reducing overfitting is to incorporate domain knowledge into the learning process.

13.3 A Generic Deductive-Inductive Ensemble

This section discusses a generic meta-framework for combining different models. Consider a situation, where we have two algorithms that compute the functions $f_1(\overline{X}_i)$ and $f_2(\overline{X}_i)$ as two different estimations of the output $\overline{y}_i$:

$$y_i \approx \hat{y}_i^1 = f_1(\overline{X}_i)$$
$$y_i \approx \hat{y}_i^2 = f_2(\overline{X}_i)$$

We assume that the function $f_1(\cdot)$ is computed using a deductive algorithm, whereas the function $f_2(\cdot)$ is computed using an inductive learning algorithm. In such a case, one can combine the two algorithms by using an *ensemble method*. In particular, the predictions of the two algorithms are combined to create a unified prediction with hyperparameter $\alpha \in (0, 1)$:

$$\hat{y}_i = \alpha \hat{y}_i^1 + (1 - \alpha)\hat{y}_i^2$$

Selecting $\alpha = 1$ provides a purely deductive algorithm, whereas selecting $\alpha = 0$ yields a purely inductive algorithm. Different values of α will provide different trade-offs between bias and variance.

The approach for combining the two algorithms works in the following way. A portion of the data is held out, and not used for training the inductive learning algorithm. This portion is usually quite small and may comprise a few hundred points. The remaining part of the data is used to learn the function $f_2(\cdot)$. At the same time, the domain knowledge is combined with deductive reasoning to define the function $f_1(\cdot)$. Once both $f_1(\cdot)$ and $f_2(\cdot)$ have been learned, the prediction of Equation 13.3 are applied to the held-out data for various values of α. For example, one might compute the predictions for $\alpha \in \{0, 0.2, 0.4, 0.5, 0.6, 0.8, 1.0\}$. For each of these different values of α the predicted value $\hat{y}_i$ is compared with the observed values y_i on the held out data, and an error metric (such as MSE for numerically dependent variables) is computed. The nature of the error metric depends on the nature of the dependent variable at hand. The value of α that provides the least error on the held out data is used to combine the inductive and deductive learning algorithms. It is noteworthy that regularized machine learning can be considered an example of a deductive-inductive ensemble.

13.3.1 Inductive Ensemble Methods

It is noteworthy that ensemble methods are not just used for combining inductive and deductive learning methods in machine learning. It is possible to construct the various model components using only inductive learning methods. It is not the goal of this chapter to discuss inductive ensemble methods in detail, since they do not represent a combination of induction and deduction methods. Therefore, we will discuss a few such methods very briefly, just to give an idea of how such methods work in practice. For detailed discussions of inductive ensemble methods, we refer the reader to [161]. Generally speaking ensemble methods can reduce either bias or variance, although the latter is more common. Therefore, we will first discuss variance reduction.

Variance Reduction

A particularly common ensemble for variance reduction is the *averaging* ensemble, wherein the predictions $f_1(\cdot) \ldots f_k(\cdot)$ of k algorithms of a similar nature are averaged:

$$\hat{y}_i = \frac{\sum_{j=1}^{k} f_j(\overline{X}_i)}{k}$$

The prediction $f_j(\cdot)$ can be a binary label, a real score for a binary label, or a real score for a regressand. There are many variations of this type of ensemble, depending on how each $f_j(\overline{X}_i)$ is constructed:

- When each $f_i(\cdot)$ is a prediction resulting from training on a random sample of the data, the ensemble is referred to as either *bagging* (sampling with replacement) or *subsampling* (sampling without replacement). Bagging was proposed in [30].

- When each $f_i(\overline{X}_i)$ is constructed using a random sample of the features in the training data, the resulting ensemble is referred to as *feature bagging*. Feature bagging methods are discussed in [83].

- When each $f_i(\cdot)$ is a decision tree with randomized splits, the resulting ensemble is referred to as a *random forest*. In this approach, the attribute to be used for a split at each node is selected as a best one from a randomly sampled bag of attributes. The randomly sampled bag of attributes may be different at each node of the decision tree. The score $f_i(\cdot)$ of an ensemble component is the fraction of instances belonging to the true class in the leaf node of the tree.

  Note that if the selected attributes are the same at each node of the tree, the resulting approach reduces to that of feature bagging. A detailed discussion of random forests is provided in Section 6.9.3 of Chapter 6.

In general, inductive ensemble methods are extremely popular, because they increase the robustness of predictions. Such inductive ensemble methods reduce the variance of prediction caused by random nuances in the specific data set. Therefore, they are particularly useful for smaller data sets, where they reduce the error of prediction caused by the specific vagaries of the data set at hand.

Bias Reduction

A well-known method for bias reduction is boosting. In boosting, a weight is associated with each training instance, and the different classifiers are trained with the use of these weights. The weights are modified iteratively based on classifier performance. In other words, the future models constructed are dependent on the results from previous models. Thus, each classifier in this model is constructed using a the same algorithm $\mathcal{A}$ on a weighted training data set. The basic idea is to focus on the incorrectly classified instances in future iterations by increasing the relative weight of these instances. The hypothesis is that the errors in these misclassified instances are caused by classifier bias. Therefore, increasing the instance weight of misclassified instances will result in a new classifier that corrects for the bias on these *particular* instances. By iteratively using this approach and creating a weighted combination of the various classifiers, it is possible to create a classifier with lower *overall* bias.

The most well-known approach to boosting is the *AdaBoost* algorithm. For simplicity, the following discussion will assume the binary class scenario. It is assumed that the class

Algorithm *AdaBoost*(Data Set: $\mathcal{D}$, Base Classifier: $\mathcal{A}$, Maximum Rounds: T)
begin
   $t = 0$;
   for each i initialize $W_1(i) = 1/n$;
   repeat
      $t = t + 1$;
      Determine weighted error rate ϵ_t on $\mathcal{D}$ when base algorithm $\mathcal{A}$
            is applied to weighted data set with weights $W_t(\cdot)$;
      $\alpha_t = \frac{1}{2}\log_e((1 - \epsilon_t)/\epsilon_t)$;
      for each misclassified $\overline{X_i} \in \mathcal{D}$ **do** $W_{t+1}(i) = W_t(i)e^{\alpha_t}$;
         else (correctly classified instance) **do** $W_{t+1}(i) = W_t(i)e^{-\alpha_t}$;
      for each instance $\overline{X_i}$ **do** normalize $W_{t+1}(i) = W_{t+1}(i)/[\sum_{j=1}^{n} W_{t+1}(j)]$;
   until $((t \geq T)$ OR $(\epsilon_t = 0)$ OR $(\epsilon_t \geq 0.5))$;
   Use ensemble components with weights α_t for test instance classification;
end

Figure 13.4: The *AdaBoost* algorithm

labels are drawn from $\{-1, +1\}$. This algorithm works by associating each training example with a *weight* that is updated in each iteration, depending on the results of the classification in the last iteration. The base classifiers therefore need to be able to work with weighted instances. Weights can be incorporated either by direct modification of training models, or by (biased) bootstrap sampling of the training data. The reader should revisit the section on rare class learning for a discussion on this topic. Instances that are misclassified are given higher weights in successive iterations. Note that this corresponds to intentionally biasing the classifier in later iterations with respect to the *global* training data, but reducing the bias in certain *local* regions that are deemed "difficult" to classify by the specific model $\mathcal{A}$.

In the tth round, the weight of the ith instance is $W_t(i)$. The algorithm starts with equal weight of $1/n$ for each of the n instances, and updates them in each iteration. In the event that the ith instance is misclassified, then its (relative) weight is increased to $W_{t+1}(i) = W_t(i)e^{\alpha_t}$, whereas in the case of a correct classification, the weight is decreased to $W_{t+1}(i) = W_t(i)e^{-\alpha_t}$. Here α_t is chosen as the function $\frac{1}{2}\log_e((1 - \epsilon_t)/\epsilon_t)$, where ϵ_t is the fraction of incorrectly predicted training instances (computed after weighting with $W_t(i)$) by the model in the tth iteration. The approach terminates when the classifier achieves 100% accuracy on the training data ($\epsilon_t = 0$), or it performs worse than a random (binary) classifier ($\epsilon_t \geq 0.5$). An additional termination criterion is that the number of boosting rounds is bounded above by a user-defined parameter T. The overall training portion of the algorithm is illustrated in Figure 13.4.

It remains to be explained how a particular test instance is classified with the ensemble learner. Each of the models induced in the different rounds of boosting is applied to the test instance. The prediction $p_t \in \{-1, +1\}$ of the test instance for the tth round is weighted with α_t and these weighted predictions are aggregated. The sign of this aggregation $\sum_t p_t \alpha_t$ provides the class label prediction of the test instance. Note that less accurate components are weighted less by this approach.

An error rate of $\epsilon_t \geq 0.5$ is as bad or worse than the expected error rate of a random (binary) classifier. This is the reason that this case is also used as a termination criterion. In some implementations of boosting, the weights $W_t(i)$ are reset to $1/n$ whenever $\epsilon_t \geq 0.5$, and the boosting process is continued with the reset weights. In other implementations, ϵ_t is allowed to increase beyond 0.5, and therefore some of the prediction results p_t for a test instance are effectively inverted with negative values of the weight $\alpha_t = \log_e((1 - \epsilon_t)/\epsilon_t)$.

Boosting primarily focuses on reducing the bias. The bias component of the error is reduced because of the greater focus on misclassified instances. The ensemble decision boundary is a complex combination of the simpler decision boundaries, which are each optimized to specific parts of the training data. For example, if the *AdaBoost* algorithm uses a linear SVM on a data set with a nonlinear decision boundary, it will be able to learn this boundary by using different stages of the boosting to learn the classification of different portions of the data. Because of its focus on reducing the bias of classifier models, such an approach is capable of combining many weak (high bias) learners to create a strong learner. Therefore, the approach should generally be used with simpler (high bias) learners with low variance in the individual ensemble components. In spite of its focus on bias, boosting can occasionally reduce the variance slightly when re-weighting is implemented with sampling. This reduction is because of the repeated construction of models on randomly sampled, albeit re-weighted, instances. The amount of variance reduction depends on the re-weighting scheme used. Modifying the weights less aggressively between rounds will lead to better variance reduction. For example, if the weights are not modified at all between boosting rounds, then the boosting approach defaults to bagging, which only reduces variance. Therefore, it is possible to leverage variants of boosting to explore the bias-variance trade-off in various ways. However, if one attempts to use the vanilla *AdaBoost* algorithm with a high-variance learner, severe overfitting is likely to occur.

Boosting is vulnerable to data sets with significant noise in them. This is because boosting assumes that misclassification is caused by the bias component of instances near the *incorrectly modeled decision boundary*, whereas it might simply be a result of the mislabeling of the *data*. This is the noise component that is intrinsic to the *data*, rather than the *model*. In such cases, boosting inappropriately overtrains the classifier to low-quality portions of the data. Indeed, there are many noisy real-world data sets where boosting does not perform well. Its accuracy is typically superior to bagging in scenarios where the data sets are not excessively noisy.

13.4   Transfer Learning

Transfer learning is the process by which previously learned inductive hypotheses are transferred to a new setting in which they are treated as predictive models. Although transfer learning is often seen as a purely inductive learning mechanism, it actually has at least some characteristics in common with the principles used in deductive reasoning. The key point of deductive reasoning is that it starts with a knowledge base that then becomes the basis for all further inferences. In transfer learning, this "knowledge base" is learned through another inductive mechanism on a different data set or data domain. This learned model then becomes "long-term" knowledge (i.e., hypothesis) that can be reused across a variety of settings. The key characteristic that is common between transfer learning and deductive reasoning is that the *hypotheses are provided by an external mechanism to the specific prediction scenario at hand*. A hypothesis can be seen as a long-term, reusable model, whereas the training process can be seen as a short-term, data-driven model. Note that all hypotheses used in deductive reasoning are also derived through scientific observations (i.e., are data-driven) in one way or the other. In this sense, transfer learning can be viewed to be closely related to settings in which inductive learning and deductive reasoning are combined. While they seem to be quite different in terms of core *algorithmic* mechanisms used for prediction, the principle of using pre-defined knowledge is very similar. It is not particularly surprising that one of the greatest benefits of transfer learning is the reduction in the data requirements that accrue from the use of external sources to learn models. However, unlike most deductive reasoning systems, transfer learning methods do not always lead to interpretable models. Furthermore, the predictive component (that uses the prior learned knowledge) is often not a reasoning system, even though it borrows the learned model from an external source.

Transfer learning is one of the most common forms of learning in biology, and is one of the keys to the great intelligence of living organisms. Some examples of transfer learning in biological settings are as follows:

1. Living organisms pass on their chromosomes from one generation to the next. Much of the neural structure of the brain is encoded directly or indirectly in this genetic material. This form of handover can be viewed as a form of transfer learning of intelligence that is learned in the biological cauldron of evolution. Recall that biological evolution is a form of inductive learning. In fact, all of biological intelligence is owed to the ability to continuously transfer knowledge from one generation to the next over time, while continuing the inductive process of evolution across generations (where experimental outcomes in biological survival lead to further improvements).

2. In various forms of science, scientists make hypotheses on the basis of observations in order to create theories. This is a form of inductive learning. However, once these hypotheses become theories they become "scientific knowledge" that is used to make predictions. Basic theories in physics, such as Newton's theory of gravitation and Einstein's theory of relativity fall in this category. For example, Einstein's theory of relativity was rooted in experimental observations on the constant speed of light. Creating a hypothesis and theory based on these types of empirical observations is an inductive process. These theories are then used to make predictions about real-world phenomena (e.g., the orbit of a satellite or of the planet Mercury). These predictions can be viewed as forms of deductive reasoning from available theories like Newtonian mechanics or relativity. However, the theory was itself learned via the process of inductive learning. This overall process can be viewed as a form of transfer learning in which inductively learned theories eventually become well-accepted components of deductive reasoning systems.

In general, transfer learning can be viewed as a form of long-term learning, which allows the reuse of knowledge from domains where data is copious. This data is turned into knowledge (transfer models), which are used repeatedly by different types of applications.

Transfer learning is particularly common in the context of feature engineering in some domains like images and text. The key idea is that these domains are often able to perform well with features that are learned in broadly generalizable settings. From a conceptual perspective, *a knowledge base can be viewed as a collection of knowledge that has been learned over the long term and will be used over the foreseeable future.* Although transfer learning methods are often viewed as inductive learning methods, they can also be viewed from the perspective that the learned knowledge often has a long-term use (just like knowledge bases). The learned knowledge is then combined with an inductive model on new data in order to perform predictions in a somewhat different setting. The resulting models are often applied to make predictions on new data. In some domains like image data, the learned knowledge from long-term use take the form of pre-trained neural network models (e.g., *AlexNet*) on large-scale data repositories like *ImageNet*. The data repository and its learned model can therefore be viewed in a similar way to a knowledge base, since it is repeatedly used in long-term fashion. In the following, we provide some examples of how transfer learning is used in different types of domains.

13.4.1 Image Data

A key point about image data is that the extracted features from a particular data set are highly reusable across data sources. For example, the way in which a cat is represented will not vary a lot if the same number of pixels and color channels are used in different data sources. In such cases, generic data sources, which are representative of a wide spectrum of images, are useful. For example, the *ImageNet* data set [217] contains more than a million images drawn from 1000 categories encountered in everyday life. The chosen 1000 categories and the large diversity of images in the data set are representative and exhaustive enough that one can use them to extract features of images for general-purpose settings. For example, the features extracted from the *ImageNet* data can be used to represent a completely different image data set by passing it through a pretrained convolutional neural network (like *AlexNet*) and extracting the multidimensional features from the fully connected layers. This new representation can be used for a completely different application like clustering or retrieval. This type of approach is so common, that *one rarely trains convolutional neural networks from scratch*. The extracted features from the penultimate layer are often referred to as FC7 features, which is an inheritance from the number of layers used in *AlexNet*. Even though the number of layers may be different in other models, the use of the term "FC7" has now become fairly standard.

This type of off-the-shelf feature extraction approach [147] can be viewed as a kind of transfer learning, because we are using a public resource like *ImageNet* to extract features, which can be viewed as storable "knowledge." This knowledge can be used to solve different problems in settings where enough training data is not available. Such an approach has become standard practice in many image recognition tasks, and many software frameworks like *Caffe* provide ready access to these features [227, 228]. In fact, *Caffe* provides a "zoo" of such pretrained models, which can be downloaded and used [228]. If some additional training data is available, one can use it to fine-tune only the deeper layers (i.e., layers closer to the output layer). The weights of the early layers (closer to the input) are fixed. The reason for training only the deeper layers, while keeping the early layers fixed, is that the earlier layers capture only primitive features like edges, whereas the deeper layers capture more complex features. The primitive features do not change too much with the application at hand, whereas the deeper features might be sensitive to the application at hand. For example, all types of images will require edges of different orientation to represent them (captured in early layers), but a feature corresponding to the wheel of a truck will be relevant to a data set containing images of trucks. In other words, early layers tend to capture highly generalizable features (across different computer vision data sets), whereas later layers tend to capture data-specific features.

The features available in FC7 can also be used for search applications. For example, if one wishes to search for an image that is similar to another image, all images can be mapped to the FC7 representation. The FC7 representation is multidimensional, and the distance-function similarity corresponds to semantic similarity. This is not the case for the original pixel representation of the image, where the distances between two pixels provide little idea of the semantics in most cases; for example, the orientation of a specific image might cause huge effects on the distance function. A discussion of the transferability of features derived from convolutional neural networks across data sets and tasks is provided in [134].

13.4.2 Text Data

Text is created out of words that have semantic significance, although a small collection may not contain this type of information about the relationships among words. For example, consider the following analog between words:

> King is to queen, as man is to woman.

In other words, the relationship between king and queen is similar to that between man and woman. This type of information is often hidden within the grammatical structure (i.e., distances of words in sentences). Furthermore, these types of semantic relationships are consistent and they do not change significantly from collection to collection. Is there a way in which one can create a numerical representation between words (from a large, standardized text corpus), so that the distances between these numerical representations reflect the underlying semantic distances? If one could create such a representation, it can be viewed as a "knowledge base" that can be used seamlessly across different settings and text collections.

An important property of languages is that the usage and semantic significance of words is roughly consistent across different collections with minor variations. As a result, the multidimensional representations of words that are obtained from one document collection can be used for another document collection. One can often use vast collections of documents available on the Web (such as online encyclopedias) in order to learn the multidimensional representations of words. These vast collections of documents, therefore, indirectly serve as knowledge bases, which are highly reusable for different types of applications. Since the structure of language is often embedded in the distances between words in sentences, it is natural to use the co-occurrence of words within specific windows in order to create embeddings. Alternatively, one can try to directly use a neural architecture in order to process complete sentences and extract word embeddings.

Consider the *weight matrix* W_{xh} from the input to hidden layer in the case of a recurrent neural network, as discussed in section of Chapter 8. The notation used here is based on Figure 8.10 of Chapter 8. The weight matrix W_{xh} is a $p \times d$ matrix for a lexicon of size d. This means that each of the d columns of this matrix contain a p-dimensional embedding for one of the words. The recurrent neural network can be applied to sentences extracted from a large online collection of sentences, such as an encyclopedia. The trained neural network can be used to extract the weight matrix, which in turn contains the embeddings.

The aforementioned neural embeddings provide only one of the alternatives to extraction of such representations. A different approach is to use *word2vec* embeddings, which can be obtained using a conventional neural architecture on windows of words. Although *word2vec* embeddings are not discussed in this book, we refer the reader to [6, 122, 123] for a detailed discussion of this embedding method. Therefore, one can learn these word embeddings on a large document collection like Wikipedia and then use the word embedding as a transferred representation for other tasks. In fact, the *word2vec* embedding is itself very useful, and is often available for download from various natural language toolkits as a pre-trained model.

13.4.3 Cross-Domain Transfer Learning

The previous examples of transfer learning focus on a single domain of data. A second type of transfer learning works across different domains. This type of approach is referred to as *translated learning*. In this approach, the key is to identify *correspondence data* is identified between the features of different domains using natural sources of co-occurrence of data involving the two domains. For example, an image co-occurring with a caption is illustrative of correspondence, and one can use this information to translate image features into text features and vice versa. This correspondence information is then used in order to make inferences in domains where the available data are limited. For example, consider a case where we have a labeled set of text documents, and we wish to associate images with these same labels even though no training data exists that contains labels of images. In such cases, the correspondence information between images and text is very useful, as it tells us how the images map to text documents. This information can be used in order to perform classification of the images. Correspondence between text and images can even be used for search. If both the text and images are embedded into a common representation, one can search for images using keywords, even though these images were not included in the correspondence information. One can view the translation between image features and text features as a kind of "knowledge base" using which one can perform the classification of images, even though no labels are available for images. In order to understand this point, consider the case where one has n pairs of corresponding images and text. The images have dimensionality m, whereas the text has dimensionality d. Therefore, the image matrix M can be represented in size $n \times m$, whereas the text matrix T is of size $n \times d$. The rows of the text matrix and the image matrix are sorted so that a one-to-one correspondence exists between the ith row of the text matrix and the ith row of the image matrix. Then, one can perform the following *shared* matrix factorization of rank $k \ll \min\{m, d\}$ (cf. Chapter 9):

$$M \approx UV^T$$
$$T \approx UW^T$$

The shared matrix U is of size $n \times k$, the matrix V is of size $m \times k$, and the matrix W is of size $d \times k$. Each row of U corresponds to a representation of one of the n pairs. One can then determine the matrices U, V, and W by optimizing the following objective function:

$$J = \|M - UV^T\|_F^2 + \beta\|T - UW^T\|_F^2$$

Here, β is a nonnegative balancing factor that decides the relative importance of the text and image matrices. One can use gradient descent in order to learn the matrices U, V, and W. Specifically, the following updates may be used:

$$U \Leftarrow U + \alpha(M - UV^T)V + \alpha\beta(T - UW^T)W$$
$$V \Leftarrow V + \alpha(M - UV^T)^T U$$
$$W \Leftarrow W + \alpha\beta(T - UW^T)^T U$$

Here, α is the learning rate. Note that $(M - UV^T)$ and $(T - UW^T)$ are the error matrices for the factorizations in the image and text domains, respectively. The above updates are repeatedly performed to convergence, which yields the three matrices U, V, and W. In practice, this type of learning is performed using stochastic gradient descent rather than gradient descent. In stochastic gradient descent, the updates are performed in entry-wise fashion in a manner similar to the matrix factorization techniques discussed in Chapter 9.

The matrix U is critical from the representational perspective, because it provides a shared space for both the text and the image representations. However, the representation in U is defined only for *in-sample* matrices for which the text and image data records are already available (and matched to one another) in the correspondence data. Note that the gradient-descent steps describe how to create the shared representations of image-text *pairs* in the correspondence data, rather than describing how to create the shared representation of a standalone image or a standalone text document (which does not have correspondence information from the other domain). In practice, one would be using this approach on *new data* in which the data items are standalone documents or images, and such correspondence information is not available. Therefore, a natural question arises how one would be able to create the shared representation of new data from *only* a single domain, once this model has been constructed. In order to understand this point, the key is to note that one only needs to store V and W in order to *approximately* reconstruct U. It is this approximate reconstruction that is used for out-of-sample data. The matrix V contains the mapping of the image features to the k-dimensional shared space, and is therefore referred to as the image feature representation matrix. Similarly, the matrix W is the text feature representation matrix W. These two matrices are very useful in approximately translating the original matrices (approximately) on a standalone basis.

Next, we describe how U can be approximately reconstructed from either the image matrix M and image feature matrix V, or from the text matrix T and text-feature representation matrix W. The reconstructions using individual domains will not be exactly the same (and will also not be as accurate as that obtained using both domains), but will provide a rough approximation to the true shared representation U. The first step is to construct the pseudo-inverses V^+ and W^+ of the matrices V and W, respectively:

$$V^+ = (V^T V)^{-1} V^T \tag{13.5}$$

$$W^+ = (W^T W)^{-1} W^T \tag{13.6}$$

It is noteworthy that $V^+ V = W^+ W = I$, but the (larger) matrices VV^+ and WW^+ are not equivalent to the identity matrix. In other words, V^+ and W^+ are left inverses of V and W respectively but they are not right inverses. The matrices VV^+ and WW^+ are referred to as *projection matrices*, multiplying with which might reduce the rank of the base matrix (and the identity matrix is a special case of the projection matrix).

One can extract the feature representations from M and T by right-multiplying these matrices with the transpose of the pseudo-inverses. By right-multiplying each of $M \approx UV^T$ and $T \approx UW^T$ with $[V^+]^T$ and $[W^+]^T$, respectively, we obtain the following:

$$M[V^+]^T \approx UV^T[V^+]^T = U[V^+V]^T = UI = U$$
$$T[W^+]^T \approx UW^T[W^+]^T = U[W^+W]^T = UI = U$$

In other words, once the matrices V and W have been learned, their pseudo-inverses can be used to extract the embedding. The first of the above extractions is obtained only from the image domain, whereas the second of the above extractions is obtained only from the text domain. Note that the two different extractions of U will be somewhat different, and will also not exactly match the value of U obtained from the above optimization model. Depending on the value of β, either the image or the text extraction might provide a more accurate approximation of the true shared representation. This is because the value of U is estimated using a weighted combination of the squared errors of the two matrix factorizations $M \approx UV^T$ and $T \approx UW^T$, where β defines the relative weights of the two approximations.

Although the inaccurate nature of such reconstructions might be undesirable for in-sample data, this type of approach might sometimes have practical uses when the number of data pairs is much greater than the dimensionality of either domain, and therefore one would rather prestore V and W over the longer term, rather than the matrix U (which is not reusable over new sets of objects). Note that the same embedding matrix U can be extracted either from the text matrix (by computing $T[W^+]^T$) or from the image matrix (by computing $M[V^+]^T$) — the matrices from both modalities are not needed during embedding extraction (since training has already created a shared space for both modalities with the use of pairwise correspondence information). One can apply this idea of multiplying an image or text matrix with the transpose of the pseudo-inverse to extract the representations of out-of-sample data in the shared space (even when the representation in the alternative modality is not available). Given a new image corresponding to row vector $\overline{X}$ or a text corresponding to row vector $\overline{Y}$, one can transform it to the k-dimensional row vectors $\overline{X}_1$ and $\overline{Y}_1$ as follows:

$$\overline{X}_1 = \overline{X}[V^+]^T$$
$$\overline{Y}_1 = \overline{Y}[W^+]^T$$

A key point is that $\overline{X}_1$ and $\overline{Y}_1$ are features in the same shared space, and their attribute values are directly comparable. Since a joint feature space is available, training data and models for the image domain also become models for the text domain, and vice versa. Therefore, consider the case where we have labeled text data T_l, unlabeled image data M_u, and correspondence data (M, T) between text and images. Note that the image matrix M for correspondence is typically different from the unlabeled image matrix M_u encountered in a specific application, and the two matrices might not even have the same number of rows (even though they have the same features). Similarly, the text matrix T used for correspondence may be different from the labeled text matrix T_l used in a specific application. In such a case, the steps for transfer learning are as follows:

1. Use the correspondence data (M, T) to extract the matrices V and W according to the above optimization model. The matrices V and W represent the key models learned up front, and they can be reused in a variety of applications. One can also store the pseudo-inverses V^+ and W^+, instead of V and W.

2. For the labeled text matrix T_l, create its translated representation $D_l = T_l[W^+]^T$.

3. For the unlabeled image matrix M_u, create its translated representation $D_u = M_u[V^+]^T$. Note that D_l and D_u lie in the same shared space with the same features, even though they were extracted from different data modalities.

4. Use any off-the-shelf classifier with D_l as the training data and D_u as the test data.

By using the matrices V and W, we are able to transform the image data and the text data into the same feature space. As a result, it now becomes possible to use an off-the-shelf classifier, which implicitly uses labeled text as training data and unlabeled images as test data. In a sense, the matrices V and W can be treated as "knowledge bases" derived from correspondence data, which can be used in any new learning or search situation, when training data from either domain are available. More details on translated learning methods may be found in [44, 136].

These methods are also used for cross-lingual transfer learning. In cross-lingual transfer learning, one creates a common attribute space in which documents of two different languages can be embedded. The general principle is similar to that of cross-domain transfer learning, except that the two different matrices correspond to the word frequencies of the documents in the two languages. One can then use the training data in one language in order to make predictions for the test data in a different language. The general principles of cross-lingual transfer learning in terms of first learning the matrices V and W, and then using them to translate to the same shared space is the same as in the case of cross-domain transfer learning. More details on cross-lingual transfer learning may be found in [136].

13.5 Lifelong Machine Learning

The discussion on transfer learning provides some insight into what artificial intelligence needs to be truly like in order to ensure that the knowledge gained from learning is not thrown away after each task but accumulated for future learning tasks. Lifelong learning, also referred to as *metalearning*, can be viewed as a massive generalization of the broader principles inherent in transfer learning, as it accumulates new knowledge (via learning) over time and then reuses this stored knowledge or hypotheses over time (as in deductive reasoning). However, life long learning is far more expansive than transfer learning, as it combines various types of unsupervised and supervised learning methods over a wide array of data sources in the learning process. A common feature of transfer learning is that it typically uses only two tasks, and leverages one task to learn from the other. On the other hand, lifelong learning is often associated with a large number of sequential tasks, and the first $(n - 1)$ tasks are used to learn the nth task, where the value of n continuously grows with time. It is also possible for these tasks to proceed simultaneously, with data being continuously received over time. Another important difference between transfer learning and lifelong learning is that the former often uses tasks that are very similar for the learning process. On the other hand, unsupervised learning might use both data sources and tasks that are significantly different from one another. For example, it is common to combine various types of unsupervised and supervised learning tasks in metalearning. This is similar to the diversity of experiences and learning processes that humans might experience over time.

Lifelong machine learning comes closest to how humans learn, in terms of continually processing sensory inputs abstracting them into the neural networks of the brain, and then using this gained knowledge over time. Most of the learning that goes on in humans is in the form of unsupervised learning, wherein data is continuously absorbed from the environment without any specific goal in mind. This type of learning results in an understanding of the relationships between different attributes of the data. These relationships can be useful for various types of goal-focused tasks. From this perspective, lifelong learning shares several aspects of semisupervised learning as well (cf. Section 9.4.3 of Chapter 9). It is noteworthy that even transfer learning with co-occurrence information can be viewed as a form of semi-supervised learning, where the co-occurrence data is a form of unsupervised input, which tells us about the shape of the manifold on which the data is distributed in the joint space of the two domains. In this section, we will first discuss a natural generalization of the two-task transfer learning phenomenon, and show how one can construct a mechanism for lifelong learning. In this section, we will describe a very generic mechanism that combines supervised learning, unsupervised learning, as well as co-occurrence data. We note that this framework is only intended to give a general understanding of how lifelong learning is supposed to work rather than a catch-all solution to all settings.

13.5.1 An Instructive Example of Lifelong Learning

This section provides an instructive example of lifelong learning that combines supervised learning, unsupervised learning, and co-occurrence learning in a single framework over a vast set of tasks and data sets. It is assumed that lifelong learning is defined by the following:

1. A set of data sets that reflect the diversity of the different scenarios in which a learning system is supposed to function. For example, humans function over a vast variety of "data sources" during their lifetime, such as vision, auditory, and various other sensory stimuli. For simplicity, we will assume that the data sources for machine learning applications can be represented as multidimensional matrices. Although this is a massive oversimplification of the vast complexity of machine learning data sets, it provides an understanding of the general principles of the approach. We will assume that we have a total of k data sets denoted by $D_1 \ldots D_k$. The size the matrix of the ith data set is given by $n_i \times d_i$.

2. For the ith data set, we have a total of t_i tasks. For simplicity, we assume here that we have one supervised learning task and one unsupervised learning task for each data set. The supervised learning task is assumed to be least-squares regression, and the unsupervised learning task is assumed to be dimensionality reduction. Therefore, for the ith data set, we have an n_i-dimensional numerical vector denoted by $\overline{y}_i$. Note that the target vector $\overline{y}_i$ need not have the same semantic interpretation across different domains (e.g., degree of like for an item), although the power of the approach[3] greatly increases when the target vectors have the same interpretation. The column vector $\overline{y}_i$ contains the n_i numerical targets for the n_i rows of D_i. In the case of unsupervised learning, we assume that all data sets need to be reduced to the same dimensionality s. In the case of humans, most of the learning occurs in the unsupervised mode, and the dimensionality reduction problem is a gross simplification of this abstract process.

3. Having different tasks or data sets is not useful for lifelong learning unless there is a way of relating the knowledge learned from these tasks. Humans do this naturally, as various stimuli are received simultaneously, and different pieces of knowledge are interrelated. Of course, not all pieces of data are always related, although there are usually enough interrelationships between different events so as to obtain a more integrated learning process. Here, we consider the simplified abstraction that we have co-occurrence matrices between a subset S of all possible $\binom{k}{2} + k$ pairs of data sets (including the k pairs of data sets in which i and j are the same):

$$S = \{(i,j) : 1 \leq i \leq j \leq k\}$$

For each pair $(i, j) \in S$, we have a co-occurrence matrix C_{ij}, which tells us about the similarity between the rows of data sets D_i and D_j. The matrix C_{ij} is of size $n_i \times n_j$, and the (p, q)th entry of C_{ij} tells us about the similarity between the pth row of D_i and the qth row of D_j. Note that it is assumed that self-similarity matrices (in which $i = j$) are always included in S. The ith self-similarity matrix is, therefore, denoted by C_{ii}. Therefore, the set S contains at least k (and typically many more) elements. The self-similarity matrices are symmetric and are key to the unsupervised learning associated with dimensionality reduction.

Given all these pieces of information, the goal of the learning model is to create an internal knowledge representation that can be used in order to create predictions for any of the supervised learning tasks. In other words, given a *new* data record $\overline{X}$ in n_r dimensions (which belongs to the same domain as D_r but the row vector $\overline{X}$ is not included in D_r),

---

[3]In many cases, it is possible to perform this type of learning even when the target vectors have different interpretations.

the goal is to predict the numerical target value for $\overline{X}$ based on the supervised model of the rth domain. Note that it is possible to perform the learning only from the supervised labels of the rth domain; however, as in transfer learning, this may not work very well when the available data labels in the rth domain are limited. It is noteworthy that this setup is essentially an expanded view of the transfer learning setting, and the main difference is that there might be a large number of data sets and tasks in the case of lifelong learning. Furthermore, lifelong learning is often incremental, although we will defer discussion of this issue to a later point. For now, we will simply discuss batch processing of the different tasks.

Multi-Task Learning Algorithm

In this section, we will introduce the multi-task learning algorithm needed to perform the learning. We assume that the data sets $D_1 \ldots D_k$ are represented by the reduced matrices $U_1 \ldots U_k$. The rth matrix U_r has size $n_r \times s$, where s is the reduced dimensionality of the data set. These learned matrices are engineered from the interrelationships between their features (unsupervised learning), the labels over the different data sets (supervised learning), and the co-occurrence data among different matrices. In the case of unsupervised learning, one uses the same type of symmetric matrix factorization that is used in standard dimensionality reduction (with the use of each self-similarity matrix C_{rr}):

$$C_{rr} \approx U_r U_r^T \quad \forall r \in \{1 \ldots k\}$$

Note that if C_{rr} is chosen to be the matrix containing the dot products between the rows of D_r, then the matrix U_r can yield an SVD of D_r with the use of eigendecomposition[4] of C_{rr} if no other constraints are imposed on the factorization. However, in this case, there are other constraints caused by the labels and by the co-occurrence data. First, note that the labels can be predicted using the engineered matrices $U_1 \ldots U_k$ with s-dimensional weight vectors $\overline{W}_1 \ldots \overline{W}_k$ (which are column vectors) as follows:

$$U_r \overline{W}_r \approx \overline{y}_r \quad \forall r \in \{1 \ldots k\}$$

Finally, the co-occurrence data is used in order to create matrices that satisfy the rules of similarity:

$$C_{ij} \approx U_i U_j^T \quad \forall (i,j) \in S$$

We note that the case $i = j$ is included in the above equation, although we have shown it separately (earlier) because it is a form of dimensionality reduction. It is also noteworthy that in many real-world applications all entries of C_{ij} may not be available (as in a ratings matrix). In such cases, the optimization model can be constructed only over the specified entries, as in recommender systems.

One can then combine all of the above predictions to create a single optimization-based objective function J as follows:

$$J = \sum_{i,j \in S} \|C_{ij} - U_i^T U_j\|_F^2 + \beta \sum_{r=1}^{k} \|U_r \overline{W}_r - \overline{y}_r\|^2$$

Note that both the unsupervised (dimensionality reduction) terms and the co-occurrence constraints are included in the first term of the above objective function. Both these terms

---

[4]The dot product matrix is positive semi-definite and can be diagonalized as $C_{rr} \approx Q\Sigma^2 Q^T$, where the $n_r \times s$ matrix Q has s orthogonal columns and Σ is an $s \times s$ diagonal matrix. In such a case, $U_r = Q\Sigma$ is the s-dimensional dimensionality reduction of D_r.

can be seen as forms of unsupervised learning. The second term corresponds to supervised learning. The hyper-parameter β regulates the level of trade-off between unsupervised and supervised learning. This optimization model can be solved with stochastic gradient descent, as in the case of transfer learning.

Once the feature matrices $U_1 \ldots U_k$ have been learned, they can be used to construct training models. Therefore, the training model is constructed in s-dimensional space rather than the original representation. Therefore, the models $\mathcal{M}_1 \ldots \mathcal{M}_k$ are constructed on these reduced representations. These models are essentially represented by the s-dimensional vectors $\overline{W}_1 \ldots \overline{W}_k$, which are leveraged for linear regression. In the event that a set of domains contain labels with the same semantic significance, the models for these domains can be consolidated into a single one. This can be done as long as the similarity values across objects in different domains are on the same scale, and they have the same semantic significance. As a practical matter, this is sometimes hard to achieve unless the similarity value has a particular semantic interpretation (e.g., degree of like for an object on the same rating scale).

As in the case of transfer learning, one can map the new row vector $\overline{X}$ from the rth domain into the reduced s-dimensional space (corresponding to row vector $\overline{u}_r$) by using of similarity relationships between the new vector and the vectors in U_r. Let $\overline{s}_r$ be a column vector of n_r similarities between each row of U_r and the row vector $\overline{u}_r$:

$$U_r \overline{u}_r^T = \overline{s}_r$$

Pre-multiplying both sides with the pseudo-inverse of U_r, one obtains the following;

$$\underbrace{[U_r^+ U_r]}_{I_r} \overline{u}_r^T = \overline{U}_r^+ \overline{s}_r$$

One can also write the above in row vector form as follows:

$$\overline{u}_r = \overline{s}_r^T [U_r^+]^T$$

Note that this transformation is similar to the mapping used in transfer learning (see page 456). Once the reduced representation has been constructed, the model $\mathcal{M}_r$ from the rth domain is used to classify it. In the case of least-squares regression, the numerical prediction is given by $\overline{W}_r \cdot \overline{u}_r^T$. Therefore, the overall approach can be summarized as follows:

1. Use similarity matrices and task-specific data to learn the k domain-specific embedding matrices $U_1, U_2 \ldots U_k$. This embedding is learned using the optimization model discussed above. If the labels and similarity matrices have the same semantic significance across some subsets of domains, the embeddings across different domains will result in comparable features in those sets of domains.

2. Create supervised models across each subset of domains in which similarity matrices and labels have the same semantic significance.

3. For any new test instance in particular domain, perform the supervised prediction using the model constructed on the appropriate (subset of) domain(s).

The above description is a simplification of the general process of lifelong learning. There are several complexities that arise in many settings. In the following, we discuss some of these complexities.

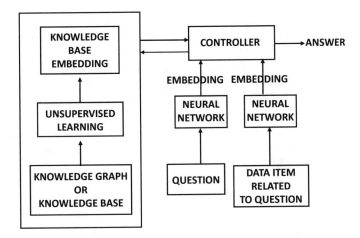

Figure 13.5: An example of an architecture of a question-answering system

Incremental Knowledge Acquisition

The approach discussed in this section assumes that all the data is available at one time in batch mode. In practice, lifelong learning is a sequential process, wherein new data from different domains arrive incrementally over time. Furthermore, the similarity values may also be highly incomplete (as in recommendation matrices). However, stochastic gradient descent methods can handle such situations relatively easily, since the updates work over individual entries of the similarity matrix rather than over rows of the matrix. Therefore, as new entries are received, they can be used in order to initiate stochastic gradient descent updates. The older entries can also be used to perform updates to ensure that their influence is not forgotten over time (unless this is specifically desired).

13.6 Neuro-Symbolic Artificial Intelligence

The previous section introduces a generic deductive-inductive ensemble, which is rather primitive in terms of the goals it achieves. In neuro-symbolic methods, symbolic methods (such as first-order logic-based methods or search methods) are tightly integrated with neural networks. At the same time, these methods tend to be somewhat more interpretable than purely machine learning methods because of the integration of symbolic methods into the process. Unlike the generic form of the inductive-deductive ensemble, neuro-symbolic methods tend to integrate symbolic methods with learning methods more tightly. This makes these methods more powerful than the generic form of the inductive-deductive ensemble. These methods have a program executor that needs to make discrete decisions in order to implement symbolic methods. These discrete decisions can be implemented in one of two ways. The first method uses a sequence of pre-defined steps based on optimization methods like search. The second uses reinforcement learning in order to learn the best sequence of decisions. The first of these methods is closer to a deductive reasoning method, whereas the second is an inductive learning method.

The general principle of neuro-symbolic learning is to create representations of knowledge bases that can be read by neural networks. The representations of the knowledge base can be constructed using conventional neural networks like convolutional neural networks, recurrent neural networks, or graph neural networks, although they would almost always be unsupervised methods like autoencoders. The precise choice of the neural network depends on the type of knowledge base or repository that needs to be converted to an internal representation. For example, a convolutional neural network would be appropriate in the case of image data, whereas a graph neural network would be appropriate in the case of a knowledge graph. This representation is then read by a controller that is often trained either using reinforcement learning or through a combination of reinforcement learning and other reasoning algorithms like search. In some applications, a knowledge base may not be used at all, when the question-answering system does not require a base of knowledge. For example, an image recognition-based question-answering system may not require a base of knowledge (beyond the training itself), whereas an open-domain question-answering system (like IBM's Watson) would require such a base of knowledge. This is because open-domain question-answering systems often require a large base of knowledge across a wide variety of domains — the main challenge is that this falls well beyond what can be reasonably encoded into the parameters of the system using only question-answer pairs.

The controller receives input from the representation of the knowledge base (which serves as a long-term memory) and the specific training of the controller is done using input-output examples. The input-output examples could be pairs of questions and answers in a specific application, along with data items related to the question. For example, the data item could be an image related to the question, although such an additional data item may not be required in many settings. The questions as well as the data items may also need to be converted to a neural embedding to enable the controller to use this information effectively. The controller may need to make discrete choices of how it needs to query the knowledge base in order to craft responses (answers) to inputs (questions) in a structured way. These choices can be made using a combination of reasoning and learning methods. The controller can also be trained in an end-to-end fashion with the use of question-answer pairs (together with data items). A number of options exist in this context, such as the use of reinforcement learning networks or the use of sequence-to-sequence recurrent neural networks. An overview of the overall approach is shown in Figure 13.5.

A key point is that the knowledge base is converted to a multidimensional embedding with the use of unsupervised learning. For example, a knowledge graph might create an embedding for each node of the graph. One can then store the knowledge graph as a matrix in which each row of the graph corresponds to the node, and its multidimensional attributes correspond to the embedding of the node. This type of embedding often encodes a lot of structural information, such as the position of a specific node in the graph, and the other nodes that it is related to. Such a knowledge base is very large in size, and most of its entries may not be relevant to a particular query. Therefore, It is important for the controller to be able to access the relevant parts of the knowledge graph in order to provide responses to queries. This can be achieved with the use of *attention mechanisms*, where the controller queries for the embeddings of specific nodes in the knowledge graph. For example, a question about the identity of the director of the movie *"Saving Private Ryan"* would require access to the portion of the knowledge graph corresponding to the nodes representing this movie and its immediate neighborhood. In other words, the controller somehow needs to access the embedding of nodes corresponding to this portion of the knowledge graph. In traditional search, one uses indexes to access portions of the knowledge base with reasoning algorithms. Attention mechanisms use the controller to issue commands related to the portions of the knowledge base that are relevant to the query at hand. These portions of the knowledge base can then be used by the controller to make further queries for other parts of the knowledge base. This type of iterative process can be naturally implemented by modeling the controller as a recurrent neural network. The recurrent neural network repeatedly outputs a location in the knowledge base from which node embeddings may need to be extracted. The optimal location can be output by a reinforcement learning algorithm like REINFORCE by the controller (see Chapter 10), and this location is used in order to access these parts of the knowledge base. However, the controller need not be trained using only reinforcement learning, which is an inductive learning mechanism. It is possible to train the controller to access relevant parts (nodes) of the embedded knowledge graph by using keyword matching between the question and the relevant labels of the nodes. In such cases, a search may be initiated in the locality of matching nodes in order to retrieve the relevant embeddings. These can be used as input to the neural network inside the controller to provide responses to the query. This type of approach has the advantage of requiring less training data — reinforcement learning algorithms are extremely data hungry. Since a search method is tightly integrated with a neural network embedding in this case, this approach can be viewed as a neuro-symbolic technique.

13.6.1 Question Answering on Images

In order to explain the broad concept of neuro-symbolic artificial intelligence, we will use an example of question answering on images. A formal approach to neuro-symbolic concept learning with the use of images and text has been proposed recently [120]. This approach is designed for learning question-answering on images, although the broader principle may be generalizable to other settings. This question-answering system does not require a knowledge base (like an open-domain question-answering system). The broad idea is that humans learn concepts by jointly associating visual perceptions with discussions about these visual concepts. The approach learns visual concepts, words, and semantic parsing of sentences by looking at images and reading paired questions and answers — while supervision is obtained from this type of higher-level feedback, no explicit supervision is performed in terms of actually tagging the concepts in an image in order to identify them. Examples of possible questions could include showing the program an image with a cube, and asking questions such as *"What is the color of the object?"* and *"Is there any cube?."* Like human concept learning, it starts by learning on simpler images with a smaller number of objects, and then it moves to more complex images with a larger number of concepts. This approach is similar to the idea of *curriculum learning* in which more complex models are incrementally constructed from simpler examples. This is exactly how a human learns a curriculum of study (from the simple to the complex). The neuro-symbolic concept learner also learns relational concepts by interpreting object referrals on the basis of these object based concepts. An example of a relevant question would be to ask whether there is an object to the right of the cylinder. More and more complex scenes are analyzed with the use of these types of question-answer pairs. One advantage of this approach is that it can learn visual concepts and their associations with symbolic representations of language. The learned concepts can be explicitly interpreted and deployed in other vision-language applications (such as image captioning).

The first step is to use a *visual perception module* to construct an object-based representation for a scene. Given an input, the image generates *object proposals* for the image. An object proposal is a list of possible objects in the image at hand. It is possible to train specific types of convolutional neural networks, such as the Mask R-CNN [74] to generate these types of proposals for objects within an image. For example, if an image contains a cube or a person, this approach can generate a bounding box that encloses those parts of the image. These proposals are then used in order to generate a latent representation of the image (which we refer to as the scene representation). The input question is then transformed into a domain-specific language. A quasi-symbolic program executor is used to infer the answer based on the scene representation. The program executor executes the program upon the derived scene representation and answers the question. The program executor is symbolic, deterministic, and it has a fully differentiable design with respect to the visual representations (from images) and concept representations (from text), so that gradient-based optimization can be supported during training.

The visual perception module extracts the bounding boxes from the images, and extracts their features with the use of a convolutional neural network. The features of the original image are extracted as well. These different representations are concatenated to represent the full object. In addition, visual features corresponding to the object's shape and color are also extracted. These types of features are implemented using different types of neural networks, which are referred to as "neural operators" in [120].

Another important component of this approach is the *semantic parsing module*, which translates a natural language question into a structured form, which is referred to as the *domain-specific language*. This domain-specific language depends on the specific type of domain on which the questions are posed; the question is typically composed with the help of a hierarchy of primitive operations, as in any querying language used in the field of databases. Note that any query language in computer science is almost always parsed as a tree structure by internal compilers. For example, each database SQL query can always be mapped to a tree structure, which is used to respond to queries in databases. Therefore, a key point is to be able to map a natural language query into a tree structure. In other words, we need a neural network that can translate sequences to trees. To achieve this goal, an advanced variant of a recurrent neural network, referred to as a *Gated Recurrent Unit (GRU)*, is used. The input to the GRU is the natural language question, and the output is the structured form of the query. One can view the embedding of this structured query as a "latent program" which can be input to the controller. The fundamental operations of the domain-specific language have the several capabilities for reasoning, such as filtering out objects with particular types of concepts or querying the attributes of an object such as its color. Therefore, when a natural language query is placed that filters an object based on its color and queries the shape of an object within the image, the translated domain-specific language will contain these concepts among its fundamental operations.

The parsed question in the domain-specific language is used by the controller. The controller needs to learn how to take the "latent program" as input and define a sequence of operations in order to respond to the query. In the field of databases, this is a simple matter, since the database representation is well-defined (as is the SQL query), and therefore one can created a purely reasoning algorithm in order to respond to queries. However, in this case, one is working with latent representations of knowledge bases, which may not be easily interpretable. Therefore, one needs to teach the controller to learn how to extract the relevant information available in the latent representation of the objects in the image. To achieve this goal, the intermediate results of the query are stored in a probabilistic manner. For example, when an object is filtered from the image based on its shape, a probability will be associated with each object in the image corresponding to whether it is the filtered object.

For the answers, the approach in [120] uses a multiple choice setting in which one of the answers from a fix set of answers is returned. For example, consider the case in which one has an image containing a man with a pen, and the question is as follows:

   Does this man have any pens on him?

The semantic parser converts the question into the following form using the domain-specific language discussed in [120]:

$$\text{Exist}\{\text{Filter}(\text{Man}, \text{Relate}[\text{Have}, \text{Filter}(\text{Pen})])\}$$

The terms "Exist," "Filter," and "Relate" correspond to the primitive operators used in the domain-specific language, just as he SQL query language uses the primitive operators "Select" and "Project" in the field of databases. The answer to this multiple-choice question is "Yes" in the case when the man is carrying a pen. In the setting used in [120], one out of a set of 18 candidate answers are returned; some of these answers are completely irrelevant, whereas the two relevant ones are "Yes" or "No." Each of these 18 possibilities is one out of 18 possible actions that is output by a reinforcement learning program. The controller, therefore, learns how to output discrete actions in response to the latent representation of a given question. This is achieved with the use of reinforcement learning, where the structured queries are mapped to actions. The REINFORCE algorithm (see Chapter 10) was used in [120] in order to provide responses to the queries.

13.7 Summary

This chapter discusses the potential integration of inductive learning and deductive reasoning methods with the use of ensembles. The motivation of such methods can be derived from the bias-variance trade-off. Deductive reasoning methods tend to have high bias, whereas the inductive learning methods tend of have high variance. The combination of the two methods can often have more accurate results, particularly for smaller data sets. For smaller data sets, one also combines multiple inductive methods for better robustness. Another important point is that deductive reasoning methods are interpretable, whereas inductive learning methods are not interpretable. A combination of the two methods often turns out to be more interpretable, while retaining the predictive power of inductive learning methods.

In the notion of ensembles is also useful for improving the predictive power of purely inductive methods, although these methods are not the focus of this chapter. Various ensemble methods such as bagging, subsampling, feature bagging, and random forests construct and combine predictors in different ways for more robust results. In recent years, a number of techniques such as neuro-symbolic learning and transfer learning have been developed in order to reduce the data requirements in artificial intelligence. In neuro-symbolic methods, knowledge bases are converted to internal representations by feature engineering with neural networks. These engineered features are then leveraged by a controller, which is separately trained via reinforcement learning to access the knowledge base so that it can be trained with the use of input-output pairs. This type of approach is particularly useful for question-answering systems.

13.8 Further Reading

This chapter introduces the bias-variance trade-off, which forms the theoretical basis for combining multiple models in machine learning. An overview of the bias-variance trade-off is provided in [161]. A detailed discussion of different types of ensemble methods is provided in [71]. Bagging methods and random forests were both introduced by Brieman [29, 30]. Feature bagging was introduced in [83]. The neuro-symbolic concept learner is proposed in [120]. A survey of transfer learning methods may be found in [136]. A method for integrating traditional machine learning with first-order logic is discussed in [60].

13.9 Exercises

1. Does the bias of a κ-nearest neighbor classifier increase or decrease with increasing value of κ? what happens to the variance? What does the classifier do, when one sets the value of κ to the data size n?

2. Does the bias of a decision tree increase or decrease by reducing the height of the tree via pruning? How about the variance?

3. Suppose that a model provides extremely poor (but similar) accuracies on both the training data and on the test data. What are the most likely sources of the error (among bias, variance, and noise)?

4. What effect does the use of Laplacian smoothing in the Bayes classifier have on the bias and variance?

5. Suppose that you modify an inductive rule-based classifier into a two-stage classifier. In the first stage, domain-specific rules are used to decide if the test instance matches these conditions. If it does, the classification is performed with the domain-specific rules. Otherwise, the second stage of the inductive rule-based classifier is used to classify the test instance. How does this modification affect the bias and the variance of the inductive rule-based classifier?

6. Suppose that the split at the top level of the decision tree is chosen using a domain-specific condition by a human expert. The splits at other levels are chosen in a data-driven manner. How does the bias and variance of this decision tree compare to that of a purely inductive decision tree?

7. Suppose that your linear regression model shows similar accuracy on the training and test data. How should you modify the regularization parameter?

8. How does the bias and variance of a bagged classifier compare to that of an individual classifier from the bagged set of classifiers?

Correction to: Artificial Intelligence

Correction to:
C. C. Aggarwal, *Artificial Intelligence,*
https://doi.org/10.1007/978-3-030-72357-6

This book was inadvertently published without extra supplementary material information. It has now been corrected to reflect the required information in the front matter of the book.

The updated original version for this book can be found at
https://doi.org/10.1007/978-3-030-72357-6

Bibliography

[1] B. Abramson. Expected-outcome: a general model of static evaluation. *IEEE transactions on PAMI* 12, pp. 182–193, 1990.

[2] D. Ackley, G. Hinton, and T. Sejnowski. A learning algorithm for Boltzmann machines. *Cognitive Science*, 9(1), pp. 147–169, 1985.

[3] C. Aggarwal. Data classification: Algorithms and Applications, *CRC Press*, 2014.

[4] C. Aggarwal. Data mining: The textbook, *Springer*, 2015.

[5] C. Aggarwal. Recommender systems: The textbook. *Springer*, 2016.

[6] C. Aggarwal. Neural networks and deep learning: A textbook. *Springer*, 2018.

[7] C. Aggarwal. Machine learning for text. *Springer*, 2018.

[8] C. Aggarwal. Linear algebra and optimization for machine learning: A textbook, *Springer*, 2020.

[9] C. Aggarwal, J. B. Orlin, and R. P. Tai. Optimized crossover for the independent set problem. *Operations Research*, 45(2), pp. 226–234, 1997.

[10] C. Aggarwal and C. Reddy. Data clustering: Algorithms and applications, *CRC Press*, 2013.

[11] A. Aho and J. Ullman. Foundations of computer science. *Computer Science Press*, 1992.

[12] R. Ahuja, T. Magnanti, and J. Orlin. Network flows: Theory, algorithms, and applications, *Prentice Hall*, 1993.

[13] D. Amodei *at al.* Concrete problems in AI safety. *arXiv:1606.06565*, 2016.
https://arxiv.org/abs/1606.06565

[14] R. Apweiler, *et al.* UniProt: the universal protein knowledgebase. *Nucleic acids research*, 32(1), D115–119, 2004. https://www.uniprot.org

[15] Ashburner *et al.* Gene ontology: tool for the unification of biology. *Nature Genetics*, 25(1), pp. 25–29, 2000.

[16] S. Auer, C. Bizer, G. Kobilarov, J. Lehmann, R. Cyganiak, and Z. Ives. DBpedia: A Nucleus for a Web of Open Data. *The Semantic Web*, Springer, Vol. 4825, pp. 722–735, 2007.

© Springer Nature Switzerland AG 2021

C. C. Aggarwal, *Artificial Intelligence*, https://doi.org/10.1007/978-3-030-72357-6

[17] J. Baxter, A. Tridgell, and L. Weaver. Knightcap: a chess program that learns by combining td (lambda) with game-tree search. *arXiv cs/9901002*, 1999.

[18] D. Bertsekas. Nonlinear programming. *Athena Scientific*, 1999.

[19] D. Bertsimas and J. Tsitsiklis. Simulated annealing. *Statistical Science*, 8(1), pp. 10–15, 1993.

[20] C. M. Bishop. Pattern recognition and machine learning. *Springer*, 2007.

[21] C. M. Bishop. Neural networks for pattern recognition. *Oxford University Press*, 1995.

[22] R. Blanco, B. B. Cambazoglu, P. Mika, and N. Torzec. Entity Recommendations in Web Search. *International Semantic Web Conference*, 2013.

[23] O. Bodenreider. The Unified Medical Language System (UMLS): Integrating biomedical terminology. *Nucleic Acids Research*, 32, pp. D267–270, 2004.

[24] K. Bollacker, C. Evans, P. Paritosh, T. Sturge, and J. Taylor. Freebase: a collaboratively created graph database for structuring human knowledge. *ACM SIGMOD Conference*, pp. 1247–1250, 2008.

[25] M. Bojarski *et al.* End to end learning for self-driving cars. *arXiv:1604.07316*, 2016.
https://arxiv.org/abs/1604.07316

[26] M. Bojarski *et al.* Explaining How a Deep Neural Network Trained with End-to-End Learning Steers a Car. *arXiv:1704.07911*, 2017.
https://arxiv.org/abs/1704.07911

[27] A. Bordes, S. Chopra, and J. Weston. Question answering with subgraph embeddings. *arXiv preprint arXiv:1406.3676*, 2014.

[28] A. Bordes, N. Usunier, S. Chopra, and J. Weston. Large-scale simple question answering with memory networks. *arXiv preprint arXiv:1506.02075*, 2015.

[29] L. Breiman. Random forests. *Journal Machine Learning archive*, 45(1), pp. 5–32, 2001.

[30] L. Breiman. Bagging predictors. *Machine Learning*, 24(2), pp. 123–140, 1996.

[31] C. Browne *et al.* A survey of monte carlo tree search methods. *IEEE Transactions on Computational Intelligence and AI in Games*, 4(1), pp. 1–43, 2012.

[32] A. Bryson. A gradient method for optimizing multi-stage allocation processes. *Harvard University Symposium on Digital Computers and their Applications*, 1961.

[33] M. Campbell, A. J. Hoane Jr., and F. H. Hsu. Deep blue. *Artificial Intelligence*, 134(1–2), pp. 57–83, 2002.

[34] A. Carlson, J. Betteridge, B. Kisiel, B. Settles, E. R. H. Jr, and T. M. Mitchell.Toward an Architecture for Never-Ending Language Learning. *Conference on Artificial Intelligence*, pp. 1306–1313, 2010.

[35] E. Charniak. Bayesian networks without tears. *AI magazine*, 12(4), 50, 1991.

[36] G. Chaslot *et al.* Progressive strategies for Monte-Carlo tree search. *New Mathematics and Natural Computation*, 4(03), pp. 343–357, 2008.

[37] W. Clocksin and C. Mellish. Programming in Prolog: Using the ISO standard. *Springer*, 2012.

[38] W. Cohen. Fast effective rule induction. *ICML Conference*, pp. 115–123, 1995.

[39] W. Cohen. Learning rules that classify e-mail. *AAAI Spring Symposium on Machine Learning in Information Access*, 1996.

[40] T. Cormen, C. Leiserson, R. Rivest, and C. Stein. Introduction to algorithms. *MIT Press*, 2009.

[41] C. Cortes and V. Vapnik. Support-vector networks. *Machine Learning*, 20(3), pp. 273–297, 1995.

[42] R. Coulom. Efficient selectivity and backup operators in Monte-Carlo tree search. *International Conference on Computers and Games*, pp. 72–83, 2006.

[43] T. Cover and P. Hart. Nearest neighbor pattern classification. *IEEE Transactions on Information Theory*, 13(1), pp. 1–27, 1967.

[44] W. Dai, Y. Chen, G. Xue, Q. Yang, and Y. Yu. Translated learning: Transfer learning across different feature spaces. *NIPS Conference*, pp. 353–360, 2008.

[45] M. Deisenroth, A. Faisal, and C. Ong. Mathematics for Machine Learning, *Cambridge University Press*, 2019.

[46] K. A. De Jong. Doctoral Dissertation: An analysis of the behavior of a class of genetic adaptive systems. *University of Michigan Ann Arbor*, MI, 1975.

[47] M. Dubey *et al.* Asknow: A framework for natural language query formalization in sparql. *European Semantic Web Conference*, 2016.

[48] R. Duda, P. Hart, W. Stork. *Pattern Classification*, Wiley Interscience, 2000.

[49] V. Dumoulin and F. Visin. A guide to convolution arithmetic for deep learning. *arXiv:1603.07285*, 2016.
https://arxiv.org/abs/1603.07285

[50] C. Eckart and G. Young. The approximation of one matrix by another of lower rank. *Psychometrika*, 1(3), pp. 211–218, 1936.

[51] S. Edelkamp, S. Jabbar, and A. Lluch-Lafuente. Cost-algebraic heur istic search. *AAAI* pp. 1362–1367, 2005.

[52] T. Fawcett. ROC Graphs: Notes and Practical Considerations for Researchers. *Technical Report HPL-2003-4*, Palo Alto, CA, HP Laboratories, 2003.

[53] L. Ehrlinger and W. Wob. Towards a Definition of Knowledge Graphs. *SEMANTiCS*, 48, 2016.

[54] A. Fader, S. Soderland, and O. Etzioni. Identifying relations for open information extraction. *Conference on Empirical Methods in Natural Language Processing*, pp. 1535–1545, 2011.

[55] J. Fan, D. Ferrucci, D. Gondek, and A. Kalyanpur. Prismatic: Inducing knowledge from a large scale lexicalized relation resource. *NAACL HLT 2010 First International Workshop on Formalisms and Methodology for Learning by Reading*, pp. 122-127 2010.

[56] M. Farber and A. Rettinger. A Statistical Comparison of Current Knowledge Bases. *CEUR Workshop Proceedings*, 2015.

[57] M. Fitting. First-order logic and automated theorem proving. *Springer*, 2012.

[58] K. Fukushima. Neocognitron: A self-organizing neural network model for a mechanism of pattern recognition unaffected by shift in position. *Biological Cybernetics*, 36(4), pp. 193–202, 1980.

[59] S. Gallant. Neural network learning and expert systems. *MIT Press*, 1993.

[60] A. Garcez, M. Gori, L. Lamb, L. Serafini, M. Spranger, and S. Tran. Neural-symbolic computing: An effective methodology for principled integration of machine learning and reasoning. *arXiv preprint arXiv:1905.06088*, 2019.

[61] M. Garey and D. Johnson. Computers and Intractability, *Freeman*, 2002.

[62] I. Gent, C. Jefferson, and P. Nightingale. Complexity of n-queens completion. *Journal of Artificial Intelligence Research*, 59, 815–848, 2017.

[63] W. Gilks, S. Richardson, and D. Spiegelhalter. Markov chain Monte Carlo in practice.*CRC Press*, 1995.

[64] M. Girvan and M. Newman. Community structure in social and biological networks. *Proceedings of the National Academy of Sciences*, 99(12), pp. 7821–7826, 2002.

[65] F. Glover. Tabu Search: A Tutorial. *Interfaces*, 1990.

[66] D. Goldberg. Genetic algorithms for search, optimization, and machine learning, *Addison Wesley*, 1989.

[67] I. Goodfellow, Y. Bengio, and A. Courville. Deep learning. *MIT Press*, 2016.

[68] I. Grondman, L. Busoniu, G. A. Lopes, and R. Babuska. A survey of actor-critic reinforcement learning: Standard and natural policy gradients. *IEEE Transactions on Systems, Man, and Cybernetics*, 42(6), pp. 1291–1307, 2012.

[69] X. Guo, S. Singh, H. Lee, R. Lewis, and X. Wang. Deep learning for real-time Atari game play using offline Monte-Carlo tree search planning. *Advances in NIPS Conference*, pp. 3338–3346, 2014.

[70] D. Hassabis, D. Kumaran, C. Summerfield, and M. Botvinick. Neuroscience-inspired artificial intelligence. *Neuron*, 95(2), pp. 245–258, 2017.

[71] T. Hastie, R. Tibshirani, and J. Friedman. The elements of statistical learning. *Springer*, 2009.

[72] K. He, X. Zhang, S. Ren, and J. Sun. Delving deep into rectifiers: Surpassing human-level performance on imagenet classification. *IEEE International Conference on Computer Vision*, pp. 1026–1034, 2015.

[73] K. He, X. Zhang, S. Ren, and J. Sun. Deep residual learning for image recognition. *IEEE Conference on Computer Vision and Pattern Recognition*, pp. 770–778, 2016.

[74] K. He, G. Gkioxari, P. Dollar, and R. Girshick. Mask R-CNN. *ICCV*, 2017.

[75] N. Heess *et al.* Emergence of Locomotion Behaviours in Rich Environments. *arXiv:1707.02286*, 2017.
https://arxiv.org/abs/1707.02286
Video 1 at: https://www.youtube.com/watch?v=hx_bgoTF7bs
Video 2 at: https://www.youtube.com/watch?v=gn4nRCC9TwQ&feature=youtu.be

[76] R. High. The era of cognitive systems: An inside look at IBM Watson and how it works. *IBM Corporation, Redbooks*, 2012.

[77] G. Hinton. Connectionist learning procedures. *Artificial Intelligence*, 40(1–3), pp. 185–234, 1989.

[78] G. Hinton. A practical guide to training restricted Boltzmann machines. *Momentum*, 9(1), 926, 2010.

[79] G. Hinton. To recognize shapes, first learn to generate images. *Progress in Brain Research*, 165, pp. 535–547, 2007.

[80] G. Hinton, S. Osindero, and Y. Teh. A fast learning algorithm for deep belief nets. *Neural Computation*, 18(7), pp. 1527–1554, 2006.

[81] G. Hinton and R. Salakhutdinov. Reducing the dimensionality of data with neural networks. *Science*, 313, (5766), pp. 504–507, 2006.

[82] G. Hinton and T. Sejnowski. Learning and relearning in Boltzmann machines. *Parallel Distributed Processing: Explorations in the Microstructure of Cognition*, MIT Press, 1986.

[83] T. K. Ho. The random subspace method for constructing decision forests. *IEEE Transactions on Pattern Analysis and Machine Intelligence*, 20(8), pp. 832–844, 1998.

[84] S. Hochreiter, Y. Bengio, P. Frasconi, and J. Schmidhuber. Gradient flow in recurrent nets: the difficulty of learning long-term dependencies, *A Field Guide to Dynamical Recurrent Neural Networks*, IEEE Press, 2001.

[85] J. Holland. Adaptation in natural and artificial systems: an introductory analysis with applications to biology, control, and artificial intelligence. *MIT Press*, 1992.

[86] J. J. Hopfield. Neural networks and physical systems with emergent collective computational abilities. *National Academy of Sciences of the USA*, 79(8), pp. 2554–2558, 1982.

[87] D. Hubel and T. Wiesel. Receptive fields of single neurones in the cat's striate cortex. *The Journal of Physiology*, 124(3), pp. 574–591, 1959.

[88] P. Hurley and L. Watson. A Concise Introduction to Logic. *Wadsworth*, 2007.

[89] H. Jaeger and H. Haas. Harnessing nonlinearity: Predicting chaotic systems and saving energy in wireless communication. *Science*, 304, pp. 78–80, 2004.

[90] P. James. Knowledge graphs. *Linguistic Instruments in Knowledge Engineering: proceedings of the 1991 Workshop on Linguistic Instruments in Knowledge Engineering*, Tilburg, The Netherlands, pp. 97–117, Elsevier, 1992.

[91] F. Jensen. An introduction to Bayesian networks. *UCL press*, 1996.

[92] C. Johnson. Logistic matrix factorization for implicit feedback data. *NIPS Conference*, 2014.

[93] R. Jozefowicz, W. Zaremba, and I. Sutskever. An empirical exploration of recurrent network architectures. *ICML Confererence*, pp. 2342–2350, 2015.

[94] A. Karpathy, J. Johnson, and L. Fei-Fei. Visualizing and understanding recurrent networks. *arXiv:1506.02078*, 2015.
https://arxiv.org/abs/1506.02078

[95] A. Karpathy. The unreasonable effectiveness of recurrent neural networks, *Blog post*, 2015.
http://karpathy.github.io/2015/05/21/rnn-effectiveness/

[96] A. Karpathy, J. Johnson, and L. Fei-Fei. Stanford University Class CS321n: Convolutional neural networks for visual recognition, 2016.
http://cs231n.github.io/

[97] G. Karypis, and V. Kumar. A fast and high quality multilevel scheme for partitioning irregular graphs. *SIAM Journal on scientific Computing*, 20(1), pp. 359–392, 1998.

[98] E. Kaufmann, A. Bernstein, and R. Zumstein. Querix: A natural language interface to query ontologies based on clarification dialogs. *International Semantic Web Conference*, pp. 980–981, 2006.

[99] H. J. Kelley. Gradient theory of optimal flight paths. *Ars Journal*, 30(10), pp. 947–954, 1960.

[100] B. Kernighan and S. Lin. An efficient heuristic procedure for partitioning graphs. *Bell System Technical Journal*, 1970.

[101] S. Kirkpatrick, C. Gelatt, and M. Vecchi. Optimization by simulated annealing. *Science*, 229(4598), pp. 671–680, 1983.

[102] L. Kocsis and C. Szepesvari. Bandit based monte-carlo planning. *European Conference on Machine Learning*, 2006.

[103] T. Kohonen. The self-organizing map. Neurocomputing, 21(1), pp. 1–6, 1998.

[104] D. Koller and N. Friedman. Probabilistic graphical models: principles and techniques. *MIT Press*, 2009.

[105] X. Kong *et al.* Meta path-based collective classification in heterogeneous information networks. *ACM CIKM Conference*, 2012.

[106] J. Koza. Genetic programming. *MIT Press*, 1994.

[107] A. Krizhevsky, I. Sutskever, and G. Hinton. Imagenet classification with deep convolutional neural networks. *NIPS Conference*, pp. 1097–1105. 2012.

[108] M. Lai. Giraffe: Using deep reinforcement learning to play chess. *arXiv:1509.01549*, 2015.

[109] A. Langville, C. Meyer, R. Albright, J. Cox, and D. Duling. Initializations for the nonnegative matrix factorization. *ACM KDD Conference*, pp. 23–26, 2006.

[110] Y. LeCun, L. Bottou, Y. Bengio, and P. Haffner. Gradient-based learning applied to document recognition. *Proceedings of the IEEE*, 86(11), pp. 2278–2324, 1998.

[111] Y. LeCun, K. Kavukcuoglu, and C. Farabet. Convolutional networks and applications in vision. *IEEE International Symposium on Circuits and Systems*, pp. 253–256, 2010.

[112] Y. LeCun, S. Chopra, R. M. Hadsell, M. A. Ranzato, and F.-J. Huang. A tutorial on energy-based learning. *Predicting Structured Data*, MIT Press, pp. 191–246,, 2006.

[113] D. Lenat. CYC: A large-scale investment in knowledge infrastructure. *Communications of the ACM*, 38(11), pp. 33–38, 1995.

[114] S. Levine, C. Finn, T. Darrell, and P. Abbeel. End-to-end training of deep visuomotor policies. *Journal of Machine Learning Research*, 17(39), pp. 1–40, 2016. **Video at:** https://sites.google.com/site/visuomotorpolicy/

[115] Y. Li. Deep reinforcement learning: An overview. *arXiv:1701.07274*, 2017. https://arxiv.org/abs/1701.07274

[116] L.-J. Lin. Reinforcement learning for robots using neural networks. *Technical Report*, DTIC Document, 1993.

[117] Y. Lin, Z. Liu, M. Sun, Y. Liu, and X. Zhu. Learning entity and relation embeddings for knowledge graph completion. *AAAI Conference*, 2015.

[118] H. Lodhi, C. Saunders, J. Shawe-Taylor, N. Cristianini, and C. Watkins. Text classification using string kernels. *Journal of Machine Learning Research*, 2, pp. 419–444, 2002.

[119] F. Mahdisoltani, J. Biega, and F. Suchanek. YAGO3: A Knowledge Base from Multilingual Wikipedias. *Conference on Innovative Data Systems Research*, 2015.

[120] J. Mao, C. Gan, P. Kohli, J. Tenenbaum, and J. Wu. The neuro-symbolic concept learner: Interpreting scenes, words, and sentences from natural supervision. *arXiv preprint arXiv:1904.12584*, 2019.

[121] P. McCullagh and J. Nelder. Generalized linear models *CRC Press*, 1989.

[122] T. Mikolov, K. Chen, G. Corrado, and J. Dean. Efficient estimation of word representations in vector space. *arXiv:1301.3781*, 2013.
https://arxiv.org/abs/1301.3781

[123] T. Mikolov, I. Sutskever, K. Chen, G. Corrado, and J. Dean. Distributed representations of words and phrases and their compositionality. *NIPS Conference*, pp. 3111–3119, 2013.

[124] G. Miller. WordNet: A Lexical Database for English. *Communocations of the ACM*, 38(11), pp. 39–41 1995.
https://wordnet.princeton.edu/

[125] M. Minsky and S. Papert. Perceptrons. An Introduction to Computational Geometry, *MIT Press*, 1969.

[126] V. Mnih *et al.* Human-level control through deep reinforcement learning. *Nature*, 518 (7540), pp. 529–533, 2015.

[127] V. Mnih, K. Kavukcuoglu, D. Silver, A. Graves, I. Antonoglou, D. Wierstra, and M. Riedmiller. Playing atari with deep reinforcement learning. *arXiv:1312.5602.*, 2013.
https://arxiv.org/abs/1312.5602

[128] A. Newell, J. Shaw, and H. Simon. Report on a general problem solving program. *IFIP Congress*, 256, pp. 64, 1959.

[129] A. Ng, M. Jordan, and Y. Weiss. On spectral clustering: Analysis and an algorithm. *NIPS Conference*, pp. 849–856, 2002.

[130] M. Nickel, K. Murphy, V. Tresp, and E. Gabrilovich. A review of relational machine learning for knowledge graphs. *Proceedings of the IEEE*, 104(1), pp. 11–33, 2015.

[131] N. J. Nilsson. Logic and artificial intelligence. *Artificial intelligence*, 47(1–3), pp. 31–56, 1991.

[132] F. Niu, C. Zhang, C. Re, and J. Shavlik. Elementary: Large-scale knowledge-base construction via machine learning and statistical inference. *International Journal on Semantic Web and Information Systems (IJSWIS)*, 8(3), pp. 42–73, 2012.

[133] P. Norvig. Paradigms in Artificial Intelligence Programming: Case Studies in Common LISP, *Morgan Kaufmann*, 1881.

[134] M. Oquab, L. Bottou, I. Laptev, and J. Sivic. Learning and transferring mid-level image representations using convolutional neural networks. *IEEE Conference on Computer Vision and Pattern Recognition*, pp. 1717–1724, 2014.

[135] E. Palumbo, G. Rizzo, and R. Troncy. Entity2rec: Learning user-item relatedness from knowledge graphs for top-n item recommendation. *ACM Conference on Recommender Systems*, pp. 32–36, 2017.

[136] S. Pan and Q. Yang. A survey on transfer learning. *IEEE Transactions on Knowledge and Data Engineering*, 22(10), pp. 1345–1359, 2009.

[137] R. Pascanu, T. Mikolov, and Y. Bengio. On the difficulty of training recurrent neural networks. *ICML Conference*, 28, pp. 1310–1318, 2013.

[138] R. Pascanu, T. Mikolov, and Y. Bengio. Understanding the exploding gradient problem. *CoRR, abs/1211.5063*, 2012.

[139] H. Paulheim. Knowledge Graph Re?nement: A Survey of Approaches and Evaluation Methods. *Semantic Web Journal*, 1–20, 2016.

[140] J. Pearl. Causality: models, reasoning and inference. *MIT press*, 1991.

[141] J. Peters and S. Schaal. Reinforcement learning of motor skills with policy gradients. *Neural Networks*, 21(4), pp. 682–697, 2008.

[142] D. Pomerleau. ALVINN: An autonomous land vehicle in a neural network. *Technical Report*, Carnegie Mellon University, 1989.

[143] J. Pujara, H. Miao, L. Getoor, and W. Cohen. Knowledge Graph Identification. *International Semantic Web Conference*, pp. 542–557, 2013.

[144] J. Quinlan. C4.5: programs for machine learning. *Morgan-Kaufmann Publishers*, 1993.

[145] R. M. Neal. Probabilistic inference using Markov chain Monte Carlo methods. *Technical Report CRG-TR-93-1*, 1993.

[146] M.' A. Ranzato, Y-L. Boureau, and Y. LeCun. Sparse feature learning for deep belief networks. *NIPS Conference*, pp. 1185–1192, 2008.

[147] A. Razavian, H. Azizpour, J. Sullivan, and S. Carlsson. CNN features off-the-shelf: an astounding baseline for recognition. *IEEE Conference on Computer Vision and Pattern Recognition Workshops*, pp. 806–813, 2014.

[148] S. Rendle. Factorization machines. *IEEE ICDM Conference*, pp. 995–100, 2010.

[149] F. Rosenblatt. The perceptron: A probabilistic model for information storage and organization in the brain. *Psychological Review*, 65(6), 386, 1958.

[150] D. Rumelhart, G. Hinton, and R. Williams. Learning internal representations by back-propagating errors. In *Parallel Distributed Processing: Explorations in the Microstructure of Cognition*, pp. 318–362, 1986.

[151] D. Rumelhart, G. Hinton, and R. Williams. Learning internal representations by back-propagating errors. In *Parallel Distributed Processing: Explorations in the Microstructure of Cognition*, pp. 318–362, 1986.

[152] G. Rummery and M. Niranjan. Online Q-learning using connectionist systems (Vol. 37). *University of Cambridge, Department of Engineering*, 1994.

[153] S. Russell, and P. Norvig. Artificial intelligence: a modern approach. *Pearson Education Limited*, 2011.

[154] R. Salakhutdinov, A. Mnih, and G. Hinton. Restricted Boltzmann machines for collaborative filtering. *ICML Confererence*, pp. 791–798, 2007.

[155] A. Samuel. Some studies in machine learning using the game of checkers. *IBM Journal of Research and Development*, 3, pp. 210–229, 1959.

[156] S. Sarawagi. Information extraction. *Foundations and Trends in Satabases*, 1(3), pp. 261–377, 2008.

[157] W. Saunders, G. Sastry, A. Stuhlmueller, and O. Evans. Trial without Error: Towards Safe Reinforcement Learning via Human Intervention. *arXiv:1707.05173*, 2017.
https://arxiv.org/abs/1707.05173

[158] S. Schaal. Is imitation learning the route to humanoid robots? *Trends in Cognitive Sciences*, 3(6), pp. 233–242, 1999.

[159] J. Schulman, S. Levine, P. Abbeel, M. Jordan, and P. Moritz. Trust region policy optimization. *ICML Conference*, 2015.

[160] J. Schulman, P. Moritz, S. Levine, M. Jordan, and P. Abbeel. High-dimensional continuous control using generalized advantage estimation. *ICLR Conference*, 2016.

[161] G. Seni and J. Elder. Ensemble methods in data mining: improving accuracy through combining predictions. *Synthesis lectures on data mining and knowledge discovery*, 2(1), pp. 1–126, 2010.

[162] E. Shortliffe. Computer-based medical consultations: MYCIN. *Elsevier*, 2002.

[163] P. Seibel. Practical common LISP. *Apress*, 2006.

[164] J. Shi and J. Malik. Normalized cuts and image segmentation. *IEEE Transactions on Pattern Analysis and Machine Intelligence*, 22(8), pp. 888–905, 2000.

[165] M. Schmitz *et al.* Open language learning for information extraction. *Joint Conference on Empirical Methods in Natural Language Processing and Computational Natural Language Learning*, pp. 523–534, 2012.

[166] D. Silver *et al.* Mastering the game of Go with deep neural networks and tree search. *Nature*, 529.7587, pp. 484–489, 2016.

[167] D. Silver *et al.* Mastering the game of go without human knowledge. *Nature*, 550.7676, pp. 354–359, 2017.

[168] D. Silver *et al.* Mastering chess and shogi by self-play with a general reinforcement learning algorithm. *arXiv*, 2017.
https://arxiv.org/abs/1712.01815

[169] K. Simonyan and A. Zisserman. Very deep convolutional networks for large-scale image recognition. *arXiv:1409.1556*, 2014.
https://arxiv.org/abs/1409.1556

[170] P. Smolensky. Information processing in dynamical systems: Foundations of harmony theory. *Parallel Distributed Processing: Explorations in the Microstructure of Cognition*, Volume 1: Foundations. pp. 194–281, 1986.

[171] R. Smullyan. First-order logic. *Courier Corporation*, 1995.

[172] J. Springenberg, A. Dosovitskiy, T. Brox, and M. Riedmiller. Striving for simplicity: The all convolutional net. *arXiv:1412.6806*, 2014.
https://arxiv.org/abs/1412.6806 h

[173] N. Srivastava, R. Salakhutdinov, and G. Hinton. Modeling documents with deep Boltzmann machines. *Uncertainty in Artificial Intelligence*, 2013.

[174] D. Stepanova, M. H. Gad-Elrab, and V. T. Ho. Rule induction and reasoning over knowledge graphs. *Reasoning Web International Summer School*, pp. 142–172, 2018.

[175] G. Strang. An introduction to linear algebra, Fifth Edition. *Wellseley-Cambridge Press*, 2016.

[176] G. Strang. Linear algebra and learning from data. *Wellesley-Cambridge Press*, 2019.

[177] F. M. Suchanek, G. Kasneci, and G. Weikum. Yago: A Core of Semantic Knowledge. *WWW Conference*, pp. 697–706, 2007.

[178] Y. Sun and J. Han. Mining heterogeneous information networks: principles and methodologies. *Synthesis Lectures on Data Mining and Knowledge Discovery*, 3(2), pp. 1–159, 2012.

[179] Y. Sun, C. Aggarwal, and J. Han. Relation strength-aware clustering of heterogeneous information networks with incomplete attributes. *Proceedings of the VLDB Endowment*, 5(5), pp. 394–405, 2012.

[180] I. Sutskever, J. Martens, G. Dahl, and G. Hinton. On the importance of initialization and momentum in deep learning. *ICML Confererence*, pp. 1139–1147, 2013.

[181] R. Sutton. Learning to Predict by the Method of Temporal Differences, *Machine Learning*, 3, pp. 9–44, 1988.

[182] R. Sutton and A. Barto. Reinforcement Learning: An Introduction. *MIT Press*, 1998.

[183] R. Sutton, D. McAllester, S. Singh, and Y. Mansour. Policy gradient methods for reinforcement learning with function approximation. *NIPS Conference*, pp. 1057–1063, 2000.

[184] C. Szegedy, W. Liu, Y. Jia, P. Sermanet, S. Reed, D. Anguelov, D. Erhan, V. Vanhoucke, and A. Rabinovich. Going deeper with convolutions. *IEEE Conference on Computer Vision and Pattern Recognition*, pp. 1–9, 2015.

[185] G. Tesauro. Practical issues in temporal difference learning. *Advances in NIPS Conference*, pp. 259–266, 1992.

[186] G. Tesauro. Td-gammon: A self-teaching backgammon program. *Applications of Neural Networks*, Springer, pp. 267–285, 1992.

[187] S. Thrun. Learning to play the game of chess *NIPS Conference*, pp. 1069–1076, 1995.

[188] A. Veit, M. Wilber, and S. Belongie. Residual networks behave like ensembles of relatively shallow networks. *NIPS Conference*, pp. 550–558, 2016.

[189] P. Vincent, H. Larochelle, Y. Bengio, and P. Manzagol. Extracting and composing robust features with denoising autoencoders. ICML Confererence, pp. 1096–1103, 2008.

[190] D. Vrandecic and M. Krotzsch. Wikidata: a free collaborative knowledgebase. *Communications of the ACM*, 57(1), pp. 78–85, 2014.

[191] X. Dong, E. Gabrilovich, G. Heitz, W. Horn, N. Lao, K. Murphy, T. Strohmann, S. Sun, and W. Zhang. Knowledge Vault: A Web-scale Approach to Probabilistic Knowledge Fusion. *ACM KDD Conference*, pp. 601–610, 2014.

[192] N. Nakashole, G. Weikum, and F. Suchanek. PATTY: A Taxonomy of Relational Patterns with Semantic Types. *Joint Conference on Empirical Methods in Natural Language Processing and Computational Natural Language Learning*, pp. 1135–1145, 2012.

[193] N. Nakashole, M. Theobald, and G. Weikum. Scalable knowledge harvesting with high precision and high recall. *WSDM Conference*, pp. 227–236, 2011.

[194] C. Wang *et al.* Panto: A portable natural language interface to ontologies. *European Semantic Web Conference*, 2007.

[195] C. Wang *et al.* Incorporating world knowledge to document clustering via heterogeneous information networks. *ACM KDD Conference*, 2015.

[196] H. Wang, F. Zhang, X. Xie, and M. Guo. DKN: Deep knowledge-aware network for news recommendation. *WWW Conference*, pp. 835–1844, 2018.

[197] X. Wang, D. Wang, C. Xu, X. He, Y. Cao, and T. S. Chua. Explainable reasoning over knowledge graphs for recommendation. *AAAI Conference*, 2019.

[198] C. J. H. Watkins. Learning from delayed rewards. *PhD Thesis*, King's College, Cambridge, 1989.

[199] K. Weinberger, B. Packer, and L. Saul. Nonlinear Dimensionality Reduction by Semidefinite Programming and Kernel Matrix Factorization. *AISTATS*, 2005.

[200] P. Werbos. The roots of backpropagation: from ordered derivatives to neural networks and political forecasting (Vol. 1). *John Wiley and Sons*, 1994.

[201] P. Werbos. Backpropagation through time: what it does and how to do it. *Proceedings of the IEEE*, 78(10), pp. 1550–1560, 1990.

[202] B. Widrow and M. Hoff. Adaptive switching circuits. *IRE WESCON Convention Record*, 4(1), pp. 96–104, 1960.

[203] R. J. Williams. Simple statistical gradient-following algorithms for connectionist reinforcement learning. *Machine Learning*, 8(3–4), pp. 229–256, 1992.

[204] M. Yahya *et al.* Natural language questions for the web of data. *Joint Conference on Empirical Methods in Natural Language Processing and Computational Natural Language Learning*, pp. 379–390, 2012.

[205] S. Zagoruyko and N. Komodakis. Wide residual networks. *arXiv:1605.07146*, 2016. https://arxiv.org/abs/1605.07146

[206] M. Zeiler and R. Fergus. Visualizing and understanding convolutional networks. *European Conference on Computer Vision*, Springer, pp. 818–833, 2013.

[207] D. Zhang, Z.-H. Zhou, and S. Chen. Non-negative matrix factorization on kernels. *Trends in Artificial Intelligence*, pp. 404–412, 2006.

[208] http://selfdrivingcars.mit.edu/

[209] http://www.bbc.com/news/technology-35785875

[210] https://deepmind.com/blog/exploring-mysteries-alphago/

[211] https://sites.google.com/site/gaepapersupp/home

[212] http://www.mujoco.org/

[213] https://www.youtube.com/watch?v=2pWv7GOvuf0

[214] https://openai.com/

[215] https://drive.google.com/file/d/0B9raQzOpizn1TkRIa241ZnBEcjQ/view

[216] https://www.youtube.com/watch?v=1L0TKZQcUtA&list=PLrAXtmErZgOeiKm4sg
NOknGvNjby9efdf

[217] http://www.image-net.org/

[218] http://www.image-net.org/challenges/LSVRC/

[219] http://code.google.com/p/cuda-convnet/

[220] https://www.cs.toronto.edu/~kriz/cifar.html

[221] https://arxiv.org/abs/1609.08144

[222] https://github.com/karpathy/char-rnn

[223] https://github.com/hughperkins/kgsgo-dataset-preprocessor

[224] https://www.mpi-inf.mpg.de/departments/databases-and-information-systems/research/
yago-naga/yago/

[225] https://wiki.dbpedia.org/

[226] https://www.genome.jp/kegg/

[227] http://caffe.berkeleyvision.org/gathered/examples/feature_extraction.html

[228] https://github.com/caffe2/caffe2/wiki/Model-Zoo

Index

Symbols

A^*-Search 52
ϵ-Greedy Algorithm 346

A

ABox 418
Abox 411
Action 13
Activation 216
Actuators 13
AdaBoost 448
Admissibility Condition for A^* 53
Agglomerative Clustering 320
AlexNet 255, 271
Alpha-Beta Pruning 88
AlphaGo 344, 371
AlphaGo Zero 374
AlphaZero 375
ALVINN Self-Driving System 379
And-Elimination 123
AND-OR Search Tree 73
Annihilation Law 113
Antecedent 109
Arity in First-Order Logic 138
Atomic Formula 146
Atomic Operands 107
Atomic Proposition 107
Autoencoders 313
Automated Theorem Proving 113, 122
Average-Pooling 265

B

Backpropagation 227
Backpropagation through Time 286
Backtracking Search 68
Backward Chaining 131, 163
Bagging 448
Bayesian Network 385
Best-First Search 51
Bisecting K-Means 325
Bound Variable in First-Order Logic 141
BPTT 286, 287
Breadth-First Search 41

C

Causal Networks 386
Chain Rule for Vectored Derivatives 235
Chain Rule of Calculus 235
CIFAR-10 256
Clause 116
Closed Formulas in First-Order Logic 141
CNF 117
Co-Training 338
Coefficient of Determination 204
Competitive Learning Algorithm 247
Complementarity Law 113
Completeness of Resolution 128
Conjunct 116
Conjunction 108
Conjunctive Normal Form 117, 159
Consequent 109

Printed in the United States
by Baker & Taylor Publisher Services